STATION 7

2003

Lymphocyte Development

Shiv Pillai

Lymphocyte Development

Cell Selection Events and Signals During Immune Ontogeny

Foreword by David Baltimore

With 156 Figures

Birkhäuser
Boston • Basel • Berlin

Shiv Pillai
MGH Cancer Center
Harvard Medical School
Building 149, 13th Street
Charlestown Navy Yard
Boston, MA 02129
USA
pillai@helix.mgh.harvard.edu

Library of Congress Cataloging-in-Publication Data
Pillai, Shiv.
Lymphocyte development : cell selection events and signals during immune ontogeny / Shiv Pillai.
p. ; cm.
Includes bibliographical references and index.
ISBN 0-8176-3853-9 (hardcover : alk. paper)
1. Lymphocyte transformation. 2. B cells—Differentiation. 3. T cells—Differentiation. I. Title.
[DNLM: 1. Lymphocytes—cytology. 2. Cell Differentiation. 3. Lymphocytes—immunology. 4. Signal Transduction. WH 200 P641L 2000]
QR 185.8.L9 P556 2000
616.07′9—dc21 99-052574

Printed on acid-free paper.

Birkhäuser

ISBN 0-8176-3853-9
ISBN 3-7643-3853-9 SPIN 10533712

Typeset by Best-set Typesetter Ltd., Hong Kong.
Printed and bound by Sheridan Books, Inc., Ann Arbor, MI.
Printed in the United States of America.

9 8 7 6 5 4 3 2 1

To
Honorine and Sohini

Foreword

Many years ago I decided to spend some of my time working on that great and global problem of biology, development. Many were at that time flocking to study early development decisions, often using model systems like *Drosophila* or *C. Elegans*. I thought that the other end of development—terminal differentiation—was equally interesting and it could best be studied in mammals. My choice of problem was lymphocyte development because it was such a rich system. The developing lymphocyte is involved in so many decisions from the earliest stage of determination to the final stages of becoming an effector cell that I expected many general statements about development to emerge from such studies.

Some years later, I received a letter from an Indian student, Shiv Pillai, who said that he wanted to work on B lymphocyte development with me. His understanding of the opportunities in this area was already very deep and although I rarely did this, I accepted him sight unseen. He went from my laboratory to the Massachusetts General Hospital where he remains. Now he has written a book bringing together the advances in what has become a very broad area of science. He continues to have a global perspective that can tie together the disparate elements of the problem. Today development involves the most complex areas of molecular biology—receptors, intracellular signaling, transcriptional responses, and cellular reprogramming. Shiv is able to integrate the advances in these areas into comprehensive stories.

For all that we know about lymphocyte development, it remains an unfinished story. Only when we can understand all of the decisions involved in allowing B and T cells to respond to foreign antigens will we have the understanding we need to deal with the unmet medical needs in this area. We still cannot make vaccines against many serious pathogens like tuberculosis and HIV. We still cannot understand and control autoimmune diseases like rheumatoid arthritis and multiple sclerosis. A full understanding of lymphocyte development should help us deal with these problems. It is therefore very good news that Shiv plans to keep the information up to date on the web.

Writing a book of this magnitude is a daunting undertaking. The scientific community is in the debt of people like Shiv who are willing to devote the time and energy needed to produce such a tome.

DAVID BALTIMORE
California Institute of Technology
February 2000

Preface

An understanding of lymphocyte development is necessary to obtain a clear picture of how the immune system is organized and disposes of pathogens. This book is intended for graduate students and advanced undergraduates with some basic knowledge of genetics and molecular biology. This book may prove of most value to the student or investigator who has perhaps read one of the excellent broader immunology textbooks (such as *Cellular and Molecular Immunology*, by Abbas, Lichtman, and Pober, or *Immunobiology*, by Janeway, Travers, Walport and Capra) but who seeks to go a step beyond.

There is no limit to the depth and detail in which many topics might have been examined. Clearly, entire books could well be devoted to the subject matter of each of the 11 chapters that follow the broad introduction in Chapter 1. I have drawn the line at the level of information that I consider necessary for a functional and up-to-date understanding of various issues. The amount of detail provided reflects the level of depth I expect from a serious student of the subject (such as a graduate student focusing on the immune system).

It is impossible to give due credit to all the primary work in all the fields covered in this book; the reference lists are neither exhaustive nor complete. I have cited recent reviews and selected articles published in the last few years wherever possible. I apologize at the outset to the innumerable investigators whose very important contributions I have failed to mention. A number of tables list the phenotypes of knockout mice relevant to lymphocyte development. These tables do include references; the reference lists at the end of each chapter are therefore somewhat biased toward papers that describe the consequences of gene inactivation. These tables are also on the web and will be regularly updated at the following website. http:www.mgh.harvard.edu/depts/cancercenter/tables.html

Acknowledgments

I have been fortunate in having received comments and suggestions from many close friends and associates. I am particularly grateful to Ivan Stamenkovic, Ranjan Sen, Ashok Venkitaraman, Bobby Cherayil, Andrew Luster, Katia Georgopoulos, Ananda Roy, Catherine Sabatos, Iain Macpherson, Aamir Aslam, Brenda Schulman, Chuenlei Parng, Annaiah Cariappa, and Luojing Chen. I am particularly indebted to the innumerable students over the years whose questions have helped me think through this subject. I would like to thank Michael Koy and Robin Smith at Birkhäuser and Springer-Verlag for their unfailing help during the production process. And finally, thanks to David Baltimore for getting me into lymphocyte development and for writing the foreword.

References

Abbas, A. K., Lichtman, A. H., and Pober, J. S. Cellular and Molecular Immunology. 3rd edition, 1997. W.B. Saunders, Philadelphia, PA.

Janeway, C. A., Travers, P., Walport, M., and Capra, J. D. Immunobiology: the immune system in health and disease. 4th Edition, 1999 Garland Publishing Levittown, PA.

SHIV PILLAI

Contents

1

Overview of Cell Fate Decisions During Lymphoid Ontogeny

The questions that we find particularly exciting when we think of development in general are also issues of obvious interest in the study of lymphoid differentiation. How is a stem cell committed to a particular differentiated fate? What are the mechanisms by which cellular asymmetry is set up during development? How are signals integrated to result selectively in proliferation, differentiation, or programmed cell death? Is development completely preprogrammed? These central, almost religious, issues are also topics of great relevance to students of lymphocyte development. Indeed, as we shall see in the course of this book, some of these questions can be very elegantly approached by manipulating genes in lymphocytes and their progenitors.

What sets lymphocyte development apart from the differentiation programs of other lineages? Unlike other cells, lymphocytes are subject to a number of specific cell-selection processes during development. The two central paradigms of the immune system are the *generation of diversity* and *self- nonself-discrimination*. Diversity of immune receptors is largely generated by the rearrangement of antigen-receptor gene segments. This rearrangement process is designed to involve the addition or loss of nucleotides at the junctions between recombined segments. This "inaccuracy" contributes to the tremendous versatility of the immune repertoire but demands that cells with out-of-frame rearrangements be allowed to drop out of differentiation and die, while cells that make productive rearrangements are "positively" selected and allowed to mature. In the process of generating extremely diverse immune repertoires, it is critical that we not permit the development of lymphocytes that can attack our own tissues. A major way in which self nonself-discrimination is achieved is by the elimination of cells that react with host antigens during a narrow developmental window in the course of lymphoid ontogeny. After lymphocytes mature and respond to foreign antigens, they are capable of a process of memory (remembering the antigen that turned them on in the first place) and may also (in the case of B lymphocytes) be further selected based on their affinity for a specific antigen.

Memory lymphocytes are generated after the initial exposure of naive lymphocytes to foreign antigen, may survive as long as the life span of the animal, and can be triggered more readily by foreign antigen than precursor naive cells. The activation of naive lymphocytes therefore leads to the generation of an "altered state" in which a lymphocyte is reprogrammed to express survival factors and to have certain genes in a partially "on" state so that they can be very easily triggered via antigen-receptor signaling. Memory cells are maintained in the absence of antigen, but may receive survival signals from low levels of circulating cytokines. The maintenance of naive lymphocytes in the periphery may depend on their antigen receptors being tickled by self-structures. "Tickling" is now accepted immunological jargon for the gentle stroking of antigen receptors by low-avidity ligands—a process that appears designed to keep lymphocytes in relative good humor. Full-blown "triggering" of antigen receptors has more dramatic and sometimes disturbing consequences. It may therefore be useful for lymphocyte repertoires to be positively selected by self-antigens that have low but discernible affinities for antigen receptors on lymphocytes.

Numerous other unique cellular processes are, of course, integral to lymphoid development, and these will be considered in some depth in later chapters. The process of regulated receptor-gene rearrangement occurs only in lymphocytes. Activated B cells turn on a distinct recombinational process that allows cells to switch from making one class of antibody to another. Activated B cells also exhibit a singular mutational mechanism that targets the regions of rearranged immunoglobulin (Ig) genes that contribute to antigen-binding activity, thus generating further immune diversity during an immune response.

Commitment

Pluripotent hematopoietic stem cells respond to extracellular signals and are committed to differentiate along specific cellular pathways. These stem cells give rise to the erythroid, myeloid, megakaryocytic, and lymphoid lineages. A progenitor cell that, in a poorly understood stochastic fashion, receives the appropriate combination of signals, may become committed to the lymphoid lineage. Commitment presumably occurs because these signals invoke the synchronized expression of a critical subset of transcriptional regulators that *irrevocably* commit this stem cell to a lymphoid fate. Issues that relate to the mechanisms involved in commitment will be considered in Chapter 2. As discussed there, note that lymphoid cells include B and T lymphocytes as well as natural killer (NK) cells. Lymphoid dendritic (LD) cells develop from the same oligopotent progenitor as these lymphocytes but represent nonlymphoid cells that develop from a "lymphoid" committed progenitor.

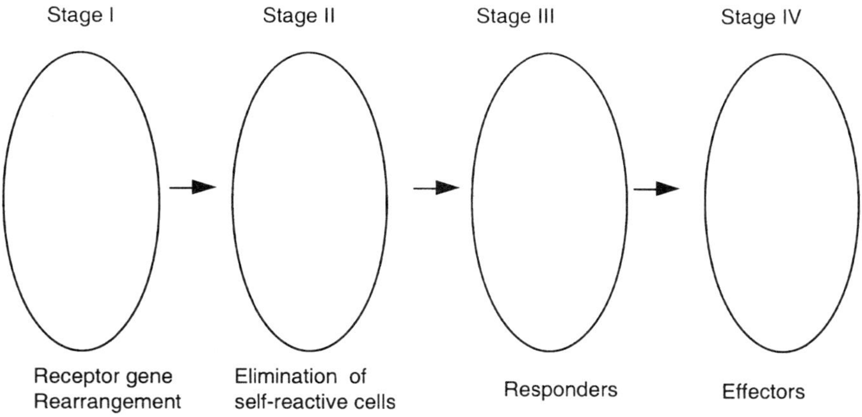

FIGURE 1.1. A simplified view of B and T lymphocyte development. Four broad functional stages are depicted.

Once a cell is commited to a B or a T lymphoid fate, it initiates rearrangement at either the Ig heavy chain locus (in the B lineage) or at T cell-receptor (TCR) β, γ, or δ chain loci (in T cells). Lymphoid differentiation, in a very simplified scheme (Figure 1.1), may be considered to be made up of four successive *functional* stages. Far more complex schemes involving many well-defined stages will begin to be considered in the next chapter.

During the first functional stage depicted in Figure 1.1, antigen-receptor gene rearrangement is initiated and completed. This corresponds to the pro-B and pre-B stages in the B lineage, and the pro-T and pre-T stages in the T lineage. Productive rearrangement of the Ig heavy chain or of the TCRβ-chain gene in a given developing lymphocyte leads to the formation of a membrane-bound complex known as a preantigen receptor, which provides cells at this stage with a signal. This signal not only permits survival, proliferation, and further differentiation of cells that have made in-frame rearrangements but also shuts off the process of rearrangement, ensuring that only one productive rearrangement (of the two receptor chain alleles) occurs in a given lymphocyte. This shut off of rearrangement is part of a process known as *allelic exclusion* that is geared toward ensuring that a given B cell or a T cell expresses a single functional receptor and is specific for a single antigen.

Central and Peripheral Lymphoid Organs and the Elimination of Self-Reactive Lymphocytes

The early stages of lymphocyte development occur in sites referred to as central lymphoid organs. B cells develop in an organ known as the bursa of Fabricius in birds and in the bone marrow in most other vertebrates. T lymphocytes mature in the thymus. Once cells have completed the process of rearrangement in the first functional stage, as considered above (roughly one in three cells may make in-frame rearrangements), the signal from the preantigen receptor permits movement to the next stage. Cells subsequently pass through a brief developmental window, during which time their signaling machinery is uniquely wired. During stage II (a circumscribed immature/transitional stage in the B lineage or the $CD4^+CD8^+$ double-positive T cell in the thymus), "complete" or "strong" signals delivered through the antigen receptor may result in the initiation of programmed cell death or apoptosis. Every lymphocyte passes through this window, where it is basically "tested" by self-antigens. If it possesses an antigen receptor that is cognate for a self-antigen, the signal delivered could result in its elimination. This is particularly true for T cells. It seems likely that B cells may also receive death signals during such a temporal "window" but that a more liberal outlook prevails, and these cells are first given an opportunity to reform themselves by the process of receptor editing and thus evade death. Lymphocytes that pass this "trial by death" progress into the responder lymphocyte stage (stage III in Figure 1.1) and make up the immune repertoire directed against the universe of nonself-antigens. These mature lymphocytes (mature IgM- and IgD-expressing B cells and either $CD4^+$ or $CD8^+$ T cells) are found in peripheral lymphoid organs, such as lymph nodes and the spleen. Here lymphocytes lie in wait of cognate antigen during brief interludes that punctuate their peripatetic wanderings from one lymph node to the next. The concept of a window for the *negative selection* of self-reactive lymphocytes is reiterated in Figure 1.2.

Activation of "responder" lymphocytes by antigen occurs in peripheral lymphoid organs. Antigen in infected tissues is transported either directly or in association with specialized antigen-presenting cells via minute lymphatic vessels to lymph nodes and other peripheral lymphoid sites. It is in these "peripheral" or "secondary" lymphoid organs that antigen and lymphocytes meet. Naive B and T lymphocytes are constantly routed through these crowded immune discotheques, so that the right partners may have an opportunity to get together and initiate an immune response. Lymphocyte activation leads to the generation of activated "memory" cells as well as more differentiated effector "end" cells, which are programmed to dispose of pathogens (recognized as nonself-antigens). Activated lymphocytes leave peripheral lymphoid tissues and are drawn to the nonlymphoid tissue sites that harbor the source of the cognate antigen.

T Cell Development Depends on Recognition of Peptides Bound to MHC Molecules

The appearance of T lymphocytes during evolution was apparently coincident with the appearance of B cells. What can T cells do that B lymphocytes cannot? The most satisfactory answer lies not just in understanding

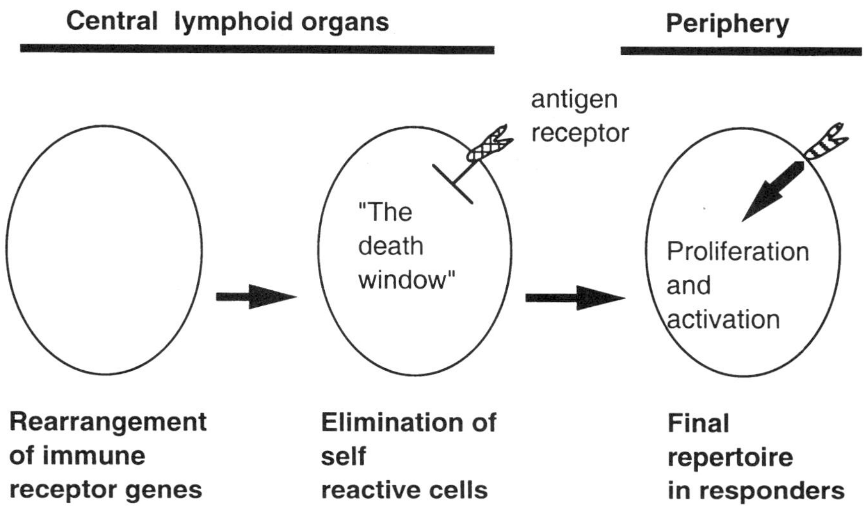

FIGURE 1.2. A developmental "death window" during lymphoid ontogeny facilitates the elimination or "negative selection" of self-reactive lymphocytes. As this process of self-tolerance occurs in central lymphoid organs, it is sometimes referred to as "central tolerance." In the B lineage, self-reactive cells may be afforded an opportunity to choose between death and reform, by opting to undergo "receptor editing" (discussed in Chapters 3 and 7).

what T cells are designed to see as opposed to what B cells recognize but also in appreciating *what T cells are designed to specifically ignore.*

Antibodies recognize antigens that may be free-floating or on the exterior of cells. They can recognize sugars, lipids, proteins, or any synthetic shape one can conceive of. When they recognize proteins, they may recognize "nonlinear" epitopes on native (properly folded) proteins or linear stretches of peptide on denatured or unfolded proteins. However, they lack two critical attributes. They are not designed to ignore any foreign shape. While they can recognize foreign structures that decorate a host cell, they cannot "look" into the interior of a live host cell to determine if a foreign invader lurks within.

The antigen receptor on T lymphocytes (the TCR) is designed only to see antigens on cell surfaces and to ignore structures that are soluble or free-floating (such as virus particles outside an infected cell). T cell receptors see small peptide fragments on cell surfaces, bound to major histocompatibility complex (MHC) molecules. A subset of T cells known as cytotoxic T cells can recognize foreign peptides bound to MHC class I molecules. These T cells actually recognize a "surface" that includes portions of specific MHC class I molecules and residues of the bound peptide that are exposed on the surface (Figure 1.3). The peptide bound to MHC class I molecules is generally derived from a protein synthesized by the host cell. Cytosolic

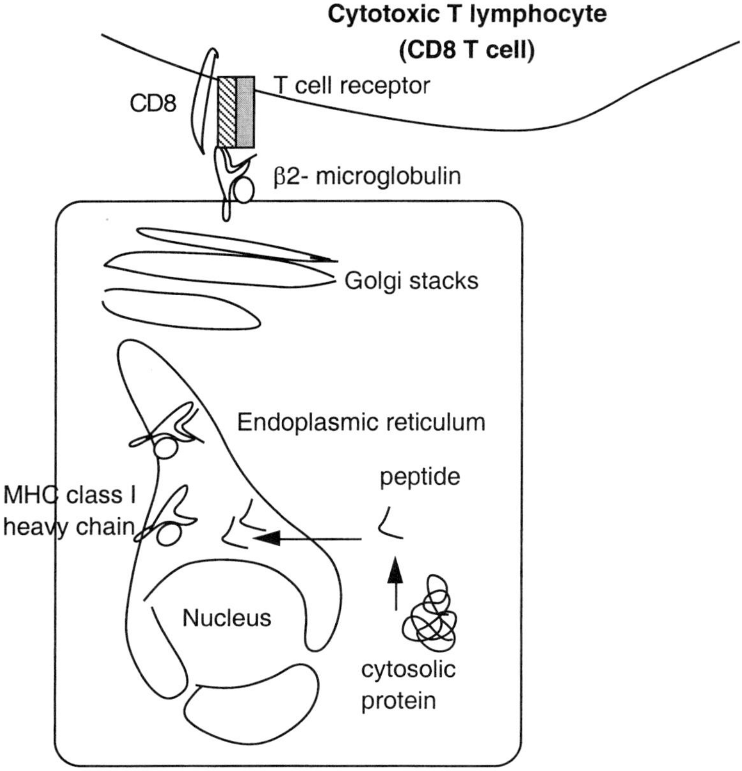

FIGURE 1.3. T cells see MHC–peptide complexes. T cell receptors on cytotoxic (CD8) T cells recognize peptides derived from the cytosol of the host cell, which are presented by MHC class I molecules. MHC class II molecules, in contrast, generally present peptides from extracellularly derived proteins that have been degraded in a postendocytic acidic compartment. Peptides presented by MHC class II molecules are recognized by helper (or CD4) T cells.

proteins are degraded in proteasomes, and 8-mer or 9-mer peptides are translocated into the endoplasmic reticulum (ER) by the transporter of antigen processing (TAP) peptide pump located in the ER membrane. These endogenously derived peptides are transported to the cell surface bound to MHC class I molecules. Proteins internalized by phagocytois or endocytosis are degraded in a specialized prelysosomal compartment and are presented by MHC class II molecules to helper T cells. T cells can therefore recognize both intracellularly located and extracellularly derived antigens.

An exclusively antibody-based immune system would likely be ineffectual when the host faces infection by viruses or by nonviral pathogens that reside intracellularly. During viral infections antibodies may be thoroughly distracted in an ocean of virus particles and are unlikely to be able to reach infected cells. T cells, however, can recognize infected cells particularly well, since their receptors are designed to see small protein fragments buried in specialized grooves that exist in proteins that are designed to constantly transport these intracellular peptides to the cell surface. T cell receptors ignore free antigen. It is therefore perhaps not surprising that T and B lymphocytes (and MHC molecules) apparently evolved together. Protective responses mediated by antibodies are considered to constitute humoral immunity as opposed to T lymphocyte-mediated cell-mediated immunity.

Positive and Negative Selection Events During B and T Cell Ontogeny

T cell clones recognize specific peptide–MHC complexes. However, no T cells that actually see free peptides or any structure other than an MHC–peptide complex are generated or permitted to mature during T lymphocyte development. This ability to ignore antigens that are not cell bound and to focus only on cells that are actually infected is probably why T cells are such an important component of the immune sytem.

Since T cell receptors in the mature T cell repertoire only see complexes of MHC molecules bound to nonself peptides, such a repertoire must be generated during T cell ontogeny. During Stage I of lymphoid development (the pro-T and pre-T stages in the scheme depicted in Figures 1.1 and 1.4), T cell-receptor genes are rearranged, and cells that have made productive rearrangements are positively selected by a signal generated from the pre-antigen receptor and mature into CD4/CD8 double-positive T cells (positive selection I). At this stage (Stage II in the scheme above), those double-positive T cells that have receptors that bind *avidly* to MHC–self-peptide complexes receive a death signal and are eliminated. In addition to this negative selection event, a second selection process occurs at this stage. It is important that the peripheral T cell repertoire be made up of cells with receptors that can recognize MHC-like shapes and not just any shape. This ensures that T cells only recognize cell-bound antigen and not free antigen. During the double-positive stage, cells bearing receptors that are not cognate for an MHC-like shape (if such receptors are ever generated) wither away because they receive no trophic signals. As pointed out above, cells bearing receptors that bind tightly to MHC–self-peptide complexes in the thymus receive a death signal. However, cells that bear receptors that bind weakly to MHC–self-peptide complexes in the thymus do receive qualitatively distinct signals, and they are positively selected (positive selection II) and differentiate into either single-positive CD4-bearing or CD8-bearing responder T cells. This positive selection process ensures that the peripheral T cell repertoire is "culled" so that it is finally made up of cells that bear receptors that can see self-MHC–peptide complexes (but not too well—the T cells that do so are eliminated). Individual T cells that emerge to form the peripheral repertoire bear receptors that will bind with high affinity to MHC molecules complexed with one or more nonself-peptides.

The antigen-receptor–mediated signal that positively selects T cells bearing receptors that can see MHC-like shapes also simultaneously drives differentiation to Stage III (the responder stage in the scheme in Figure 1.4). These selected cells leave the thymus and migrate to peripheral sites. In the B lineage, it would appear that no "culling" of the repertoire (beyond the elimination of self-reactive cells) should be required. Nonetheless, "immature" (stage II) B cells (which do not bear receptors for multivalent self-antigens and are therefore not eliminated) also receive an antigen-receptor–mediated signal that apparently selects cells that express complete receptor molecules. This putative positive selection event (positive selection II) leads to the emigration of B cells into the periphery and to the maintenance of mature B cells. As we will discuss in Chapters 7 and 9, positive selection may ensure that all naive lymphocytes have receptors that can receive trophic signals from self-antigens. Using self-structures to mould the repertoire of emerging lymphocytes (positive selection in a nutshell) may therefore be of strategic value to both B and T cells. This issue will be revisited in depth in Chapters 7 and 9.

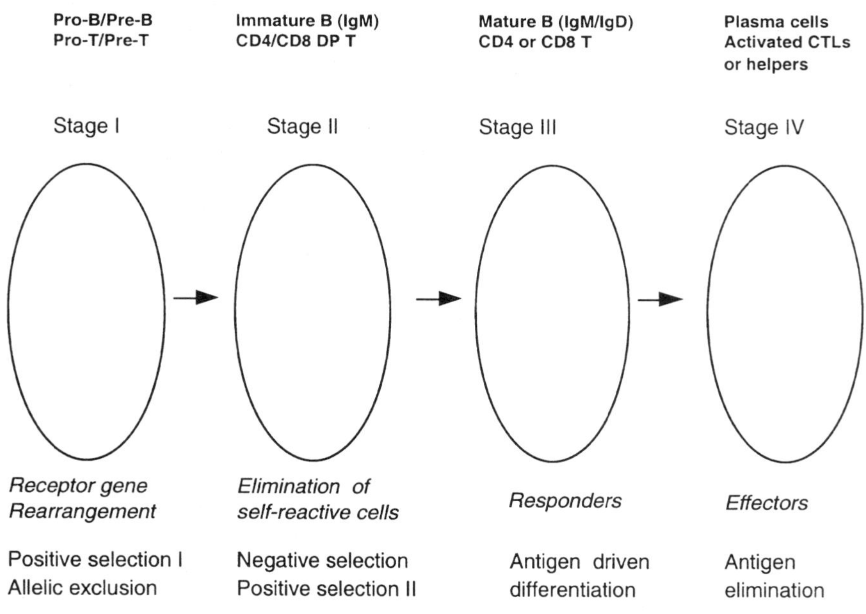

FIGURE 1.4. A bird's-eye view of lymphocyte development. The pro- and pre-T stages are also collectively referred to as the double-negative (lacking CD4 and CD8 on the surface) stage. CTL = cytotoxic T lymphocyte.

Activation of a responder B or T cell leads to the asymmetric generation of long-lived memory cells (which amplify future responses to the same antigenic challenge) and also of terminally differentiated effector cells. In the B lineage, these terminally differentiated cells are antibody-producing factories known as plasma cells. In the T lineage, the effector cells are either activated cytotoxic T cells or helper T cells.

A Bird's-Eye View of Lymphocyte Development

During the course of this book, the very simplified and unified model for B and T cell development that has been outlined in this chapter will be fleshed out. Indeed, in the case of the B lineage, this four-stage process represents only part of the picture; the entire generational process in the bone marrow that we have touched upon is re-created, in broad outline, in the lives of B cells that enter germinal centers. These issues will be dealt with in later chapters. We will address, in mechanistic detail, the nature of the signals that mediate lymphocyte proliferation, differentiation, and death. We will also visit phenomena that are unique to lymphocytes such as V(D)J recombination, isotype switching, and somatic mutation. The framework within which we will approach all these issues is the simplified model in Figure 1.1, which has been slightly expanded upon in Figure 1.4.

This book primarily examines the development of B and T lymphocytes and of NK cells. While the focus in this very brief introduction has been on presenting a simplified overview of the development of conventional B (the B-2 lineage) and T (the $\alpha\beta$ lineage) cells, in subsequent chapters, we will examine developmental branch points and the development of B-1 cells, $\gamma\delta$ T cells, and other less well-appreciated lymphoid lineages in some detail. When considering the issue of commitment in Chapter 2, it will become apparent that NK cells, as well as some specialized antigen-presenting cells, also develop from what is presumed to be a common lymphoid precursor. The development of NK cells will be considered in Chapter 11.

An Evolutionary Take on Lymphocyte Development: Focus on Mice and Men

Immunoglobulin genes, TCR α, β, γ, and δ genes, the *Rag* genes (which mediate *Ig* and TCR rearrangement), and *MHC* genes have not been identified in invertebrates, protochordates, and living jawless fish (agnathans) such as hagfish and lamprey. These genes are presumed to have evolved together in cartilaginous fishes, ancestors of modern-day skates and sharks. Lymphocyte development, as we picture it today, may therefore go back about 450 million years.

Cartilaginous fish rearrange their *TCR* genes but contain noncombinatorial *Ig* systems. All bony fish and tetrapods, however, rearrange their *Ig* and *TCR* genes. There are fairly significant differences in the ways different species generate antigen-receptor diversity during lymphocyte development. Considerable information is available on human lymphocyte development. For fairly obvious reasons, accidents of nature that compromise lymphocyte development have been most thoroughly investigated in humans. However, the day is near when every possible mouse gene that we know of will be inactivated by knockout, or more subtle and informative knock-in, approaches that depend on homologous recombination in embryonal stem cells. Even genes that may be required for embryonic development will, in the years to come, be selectively knocked out in lymphocytes, or perhaps even at selected stages of B and T cell development. The focus in this book is primarily on lymphocyte development in mice. When human disease reveals valuable insights, or when lymphocyte development in mice and in humans differs significantly, human lymphoid ontogeny will also be discussed. In Chapter 12, where naturally occurring defects in lymphocyte development will be discussed, the focus will largely be on human lymphocyte development.

2

Commitment to a Lymphoid Fate

Developing cells are preprogrammed to have specific fates. How the program that drives differentiation of any lineage is set up remains a mystery, although, clearly, all the information required to do so must be encoded in the germline.

The question of "commitment" is a central issue in the study of differentiation. The term refers to the point at which a cell has received a certain complement of signals that irreversibly dictates its fate. Although external signals are clearly of importance in this process, the critical event is presumably the stable expression of a certain subset of transcription factors that direct a cell to a specific phenotypic destiny. While it is attractive to consider the existence of master regulators or switches that in a singular fashion direct a lymphoid versus a myeloid fate, or a B cell versus a T cell fate, for many cells, such omnipotent regulators have yet to be discovered. Clearly, in some lineages, a more polytheistic view may be appropriate, and specific combinations of regulators may need to act in concert to switch a precursor into a more differentiated state. It appears that, in a temporal sense, commitment may not be as sharp an event as it is often portrayed or conceived of being. Commitment may actually be a multistep and relatively gradual courtship process during which opportunities exist for a cell to break its tryst with one set of commitment factors and to seek another destiny. A time does come, of course, when there is no going back, and commitment is complete.

A stem cell is usually operationally defined as a cell with a capacity for self-renewal that has the potential to differentiate into other cell types. A "multipotent" hematopoietic stem cell (HSC) may be preprogrammed to respond to a discrete set of signals that can commit the progeny of this cell to take on one of many blood cell fates. Such a stem cell is already irreversibly "committed" to a hematopoietic fate and cannot contribute to the development of muscle cells or of anything other than blood cells. These pluripotent cells respond to external signals that, by changing the pattern of gene expression in a given daughter cell, commit it irrevocably either into a less pluripotent progenitor of a smaller range of hematopoietic lineages (in a hierarchical model of hematopoiesis) or directly to a distinct hematopoietic lineage.

The prior expression of a certain subset of gene regulators presumably allows a given cell to become competent to respond to specific external signals, which may cause these cells either to proliferate or to change their phenotype. These latter differentiated cells may, in turn, express a distinct set of receptor molecules that allow them to respond to other signals and so on. A descriptive view of a cell's developmental program would depend on identifying sequential cycles of ligands, receptors, signaling pathways, and transcriptional regulators.

The entire process of lymphocyte development can be viewed as a succession of commitment steps. Obviously, there are many branch points during this process. In this chapter, we will take a general look at the properties of HSCs, which give rise to lymphoid-committed progenitors. We will then examine selected branch points during lymphocyte development and bring together some information about transcriptional regulation in general and lymphocyte-specific gene expression in particular. We will discuss current views of the commitment of cells to the lymphoid lineage as well as the commitment of established lymphoid precursors to a B cell or a T cell fate. In subsequent chapters, the further development of individual lineages will be described in detail.

Properties of Stem Cells

As noted above, a stem cell is operationally defined as a cell with a capacity for self-renewal that has the potential to differentiate. Three other properties are often ascribed to stem cells: (1) It is frequently assumed that stem cells must divide asymmetrically, with one daughter cell retaining stem cell properties and the other cell assuming a more differentiated phenotype. (2) Stem cells are also assumed to exist in a mitotically quiescent or dormant form and (3) are sometimes described as cells that can clonally regenerate all the cell types in a given tissue.

Self-renewal basically refers to the ability of cells to yield more identical cells. The capacity for self-renewal is often limited and does not necessarily cover the life span of the organism. As discussed below, many stem cells may actually lie dormant, and randomly selected cells from such a quiescent pool may be activated as the animal ages. Self-renewal capacity for the lifetime of an organism is not an absolute criterion for defining a stem cell.

Not all stem cells divide asymmetrically. Although evidence for asymmetric division exists in the case of stem cells examined in *Drosophila melanogaster* and *Caenorhabditis elegans*, in mammalian cells, it appears that stem cells may also divide symmetrically. Divisions may be either self-renewing or differentiative. Differentiative divisions driven by distinct

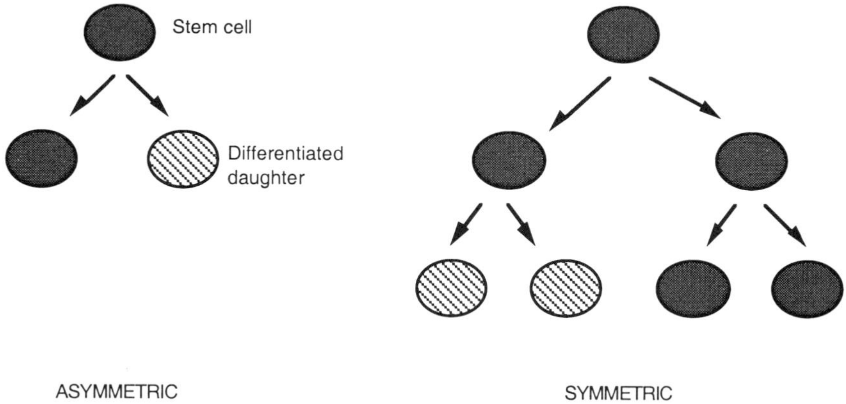

FIGURE 2.1. Stem cells may divide asymmetrically or symmetrically.

combinations of external factors may commit progeny to different fates (Figure 2.1).

HSCs probably exist in both dormant and cycling forms. It is probably inaccurate to characterize these cells as being intrinsically "dormant," especially since they must divide in order to be self-renewing. The view that the majority of HSCs may be mitotically quiescent arose from studies in which stem cells were retrovirally marked and transferred to new irradiated hosts. In these studies, a single or a few stem cells contributed to hematopoiesis in the recipients, a phenomenon described as "clonal dominance." This phenomenon implies that the majority of stem cells must be quiescent. More recent studies, however, suggest that multiple HSCs may contribute to hematopoiesis and that many HSCs divide regularly.

Not all stem cells can give rise to all the cell types in a given tissue. In terms of lymphocyte development, most B-1 cells (a type of B lymphocyte) and some subsets of $\gamma\delta$ T cell develop from fetal liver-derived progenitors and not from adult bone marrow stem cells. Absolute pluripotency is therefore not a criterion that can be readily applied to the stem cells involved in lymphocyte development.

HSCs and Primitive and Definitive Hematopoiesis

HSCs develop from the ventral mesoderm. In vertebrates, distinct patterns of hematopoiesis are observed in the early embryonic/larval stages, later during fetal life, and in adults. In early embryonic life in mammals, the first blood cells seen are very primitive nucleated erythrocytes that contain fetal hemoglobin. These cells are found in the blood islands of the yolk sac. This early generation of blood cells, mainly nucleated erythrocytes, before the formation of the fetal liver, is called "primitive" hematopoiesis. During later fetal life, the major site of hematopoiesis is the fetal liver. During this period, both primitive, nucleated, as well as fetal, non-nucleated, erythrocytes are produced, and the generation of most myeloid and lymphoid lineages is also initiated. The progenitors that develop in fetal liver are capable of "definitive" hematopoiesis, which involves the development of all blood lineages including lymphoid precursors. In adult vertebrates, the major site of hematopoiesis is the bone marrow. In some animals, the spleen can also be a site of hematopoiesis. All blood lineages are generated from HSCs in the bone marrow, and primitive erythrocytes are no longer made during adulthood.

In vertebrates, HSCs may develop initially in two anatomic locations, either in the embryo itself or extraembryonically in the mammalian yolk sac. In birds and amphibians, the ventral blood islands (VBIs) are equivalent in function to the mammalian yolk sac. HSCs that develop early in the yolk sac and VBIs, as well as in intraembryonic sites, are involved primarily, but not exclusively, in primitive hematopoiesis. Although for many years there were a number of different views as to where hematopoietic progenitors originated in different vertebrate species, some unifying ideas have now emerged. It is now clear that in all vertebrate studied, at least two types of HSCs exist: One type of precursor is capably only of primitive hematopoiesis, while the other can give rise to all hematopoietic lineages ("definitive" HSCs). In many vertebrates, "definitive" HSCs that emerge during fetal and adult life may have distinct properties that lead to different patterns of lymphocyte development.

Early in murine fetal life, HSCs develop in the body of the embryo, initially in the paraaortic splanchnopleura (from around day 8.5 postcoitus in the mouse), and soon after in the aorta, gonad, mesonephros (AGM) region. The structures in the AGM region all develop from the paraaortic splanchnopleura. In birds, the intraembryonic sites in which HSCs may develop include the dorsal mesentery, paraaortic sites, the anterior region of the mesonephros in association with the postcardinal veins, and the dorsal aorta. In *Xenopus*, the equivalent intraembryonic site is the dorsal lateral plate or (DLP). Studies in *Xenopus* have led to the conclusion that bipotential precursors of HSCs may develop in the primitive mesoderm, and, based on the inductive influences of the environment in which they develop, these may become dither primitive-type or definitive-type HSCs (Figure 2.2). In the AGM/DLP, the major inductive influence is for the development of definitive HSCs. In the yolk sac/VBI, the major inductive influences drive cells toward becoming primitive HSCs. Recent studies on mouse embryos suggest that on day 9 the yolk sac also contains definitive HSCs; the inductive event driving developing stem cells toward a primitive HSC fate in the yolk sac does not influence every HSC at this site. After fetal circulation is established, the yolk sac also harbors pluripotent HSCs, which might be derived from the AGM. These progenitor cells migrate to

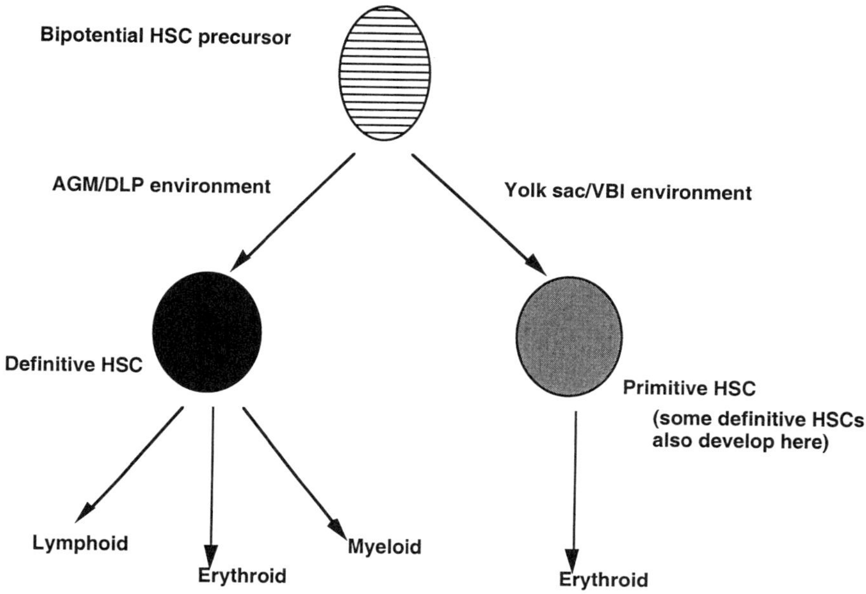

FIGURE 2.2. In early embryonic development, bipotential progenitors give rise to distinct HSCs, which mediate primitive and definitive hematopoiesis. The intraembryonic AGM/DLP environment favors the development of HSCs capable of definitive hematopoiesis, while the yolk sac/VBI milieu preferentially induces HSCs committed for primitive hematopoiesis.

the fetal liver, where most of fetal lymphopoiesis takes place. After birth, hematopoiesis no longer occurs in the liver, and the major site of blood cell formation in adults is the bone marrow. As discussed below, HSCs capable of definitive hematopoiesis that develop in the milieu of the fetal liver are functionally distinct from definitive-type HSCs that develop in the adult bone marrow.

Long-Term and Short-Term HSCs

HSCs originate as cells with a capacity for long-term (as long as the life span of the recipient animal) self-renewal capable of giving rise to all hematopoietic lineages. These cells then differentiate into a category of HSCs that have a short-term self-renewal capacity but that can still give rise to all hematopoietic lineages. Short-term HSCs subsequently differentiate into lymphomyeloid progenitors which are incapable of self-renewal but can give rise to myeloid and lymphoid cells. The lymphomyeloid progenitor then gives rise to a common lymphoid progenitor, which serves as the common precursor for B, T, natural killer (NK), and lymphoid dendritic (LD) cells.

Long-term and short-term HSCs can be identified by a characteristic surface marker profile. In mice, these cells are Lin^-, $Thy1.1^{lo}$, $c\text{-}kit^+$, $Sca\text{-}1^+$, $CD4^{lo\ to-}$. Lin^- refers to the fact that these cells do not express the characteristic markers for differentiated cells of different blood lineages or that they express low levels of markers such as Mac-1 or CD4. Human HSCs express CD34, a sialomucin of unknown function. While CD34 is a key marker of stem cells in humans, most murine stem cells capable of long-term reconstitution are found in the $CD34^{lo\ to-}$ population. CD34 has been shown to be present on some murine HSCs but not on others. In some studies, CD34 has been shown to be present on definitive murine HSCs but not on primitive HSCs. Mice in which the CD34 gene has been knocked out do not have a major defect in hematopoiesis. c-Kit, also known as CD117, is a receptor tyrosine kinase whose ligand is a secreted protein known as stem cell factor (SCF), or "Steel." SCF is a major factor in the proliferation and development of stem cells.

Genes Regulating HSC Development

Very little is known about the signals necessary for the development of HSCs. Members of the transforming growing factor β family, particular bone morphogenetic protein-4 (BMP-4) play an important role in the induction of ventral mesoderm from which HSCs are derived. Signaling components downstream of BMP-4 may also be involved in blood generation. In *Xenopus*, a paired-class homeobox gene, *Mix-1*, is probably downstream of BMP-4 and plays a role in the transcriptional induction or patterning of ventral mesoderm and in blood formation. It is likely that the stem cell leukemia (SCL/Tal-1) and GATA-motit binding-2 (GATA-2) transcription factors discussed below are downstream of the BMP-4/Mix-1 pathway.

The ability of HSCs to give rise to more differentiated blood cells also depends on the ability of these cells to seed organs such as the fetal liver, the spleen, the thymus, and the bone marrow. Integrins are heterodimeric cell surface molecules that contribute to cell–cell and cell–substrate interactions and influence the ability of cells to migrate. The role of these molecules in lymphocyte development will be considered in more depth in Chapter 6. In the absence of all β1 integrins, HSCs can still be generated and can differentiate *in vitro* into erythrocytes, myeloid lineage cells, and B lymphocytes. However, *in vivo* no blood cells develop, because HSCs cannot migrate into the fetal liver in the absence of β1 integrin expression. As discussed in the next section, a specific member of the β1 integrin family, the α4β1 molecule, is not required for fetal lymphopoiesis but is essential for the development of T and B lymphocytes after birth.

More information is presently available on transcription factors that are critical for the development of lymphoid progenitors and lymphocytes than the extracellular signals that induce or regulate these genes and their products. There is an apparent hierarchy to the transcriptional regulators of lymphocyte development (see Figure 2.3, later). Certain transcription factors, most notably SCL/Tal-1, acute myeloid leukemia-1/core binding factor A2 (AML1/CBFA2), core binding factor β (CBFβ), and GATA-2, are known to be critical for the development of hematopoietic precursors. AML1/CBFA2, CBFβ, and SCL/Tal-1 may be essential for the development of HSCs from ventral mesoderm, while GATA-2 is assumed to be crucial for the proliferation of hematopoietic progenitors derived from HSCs. As mentioned above, it is believed that the induction of both SCL/Tal-1 and GATA-2 occurs downstream of the TGF-β-type signaling pathway initiated by BMP-4, which may involve Mix-1.

The *Scl/Tal-1* gene was originally discovered in T-acute lymphocytic leukemia cells, where it is activated by translocations or by upstream interstitial deletions. The SCL/Tal-1 protein is a member of the basic helix–loop–helix (bHLH) family of DNA-binding proteins, which includes proteins critical for commitment to the B lineage, as discussed later in this chapter. bHLH proteins contain a conserved helix–loop–helix domain that mediates homo- or heterodimerization with bHLH polypeptides; the basic domain mediates sequence-specific DNA binding. Gene-targeting studies have revealed that SCL/Tal-1 is absolutely required for primitive erythropoiesis in yolk sac blood islands. It is also essential for definitive hematopoiesis of all blood lineages. It is likely that this transcriptional regulator is required very early in hematopoiesis—either at the time that cells in the ventral mesoderm acquire the ability to become blood precursors or subsequently as a maintenance or proliferation factor for HSCs themselves.

AML-1/CBFA2 was initially discovered in patients with acute myeloid leukemia as a chromosome 21-derived fusion partner of the putative transcription factors Eto and Evil in t(8:21) and t(3;21) translocations, respectively. In a significant fraction of pro-B acute lymphocytic leukemias, the Ets family gene *Tel* is also fused to the *Aml1* gene as a result of a t(12:21) translocation. The AML1 protein corresponds to the DNA-binding subunit of the heterodimeric polyoma enhancer core-binding factor. It is therefore also known as PEBP2aB1, or CBFA2. AML1/CBFA2 is the heterodimeric partner of CBFβ, which does not bind DNA. The DNA-binding domain of AML1 is homologous to the *Drosophila* pair-rule *runt* gene, and this domain is now known as a runt homology domain. A similar phenotype results from the absence of either CBFA2 or CBFβ, both of which are required for definitive hematopoiesis.

Hematopoietic cells include erythroid, myeloid, megakaryocytic, and lymphoid cells, and differentiation from an HSC into these lineages may involve the generation of precursors with a considerable degree of plasticity. The pathway of differentiation may not be as linear as suggested in Figure 2.3. The existence of bipotential B/myeloid progenitor cells, for

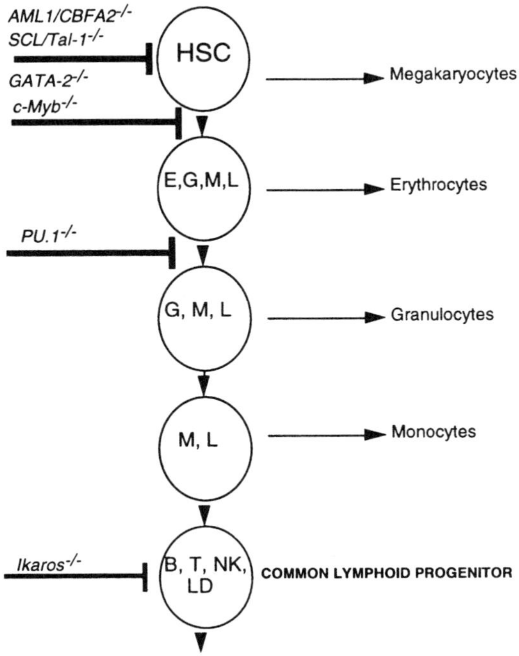

FIGURE 2.3. A linear, hierarchical overview of lymphocyte development from hematopoietic stem cells. HSC = hematopoietic stem cell, E = erythrocytes, G = granulocytes, M = monocytes, L = lymphocytes, LD = lymphoid dendritic cells, NK = natural killer cells. Transcription factors that are limiting for development before lymphoid commitment are included in this figure and are described in the text.

instance, is not accounted for in this kind of model, and these progenitors and their potential significance will be discussed in a later section. However, data from a number of gene-inactivation experiments suggest that increasingly restricted committed progenitors are likely to be present along the pathways of hematopoietic differentiation. Loss of c-Myb results in the absence of all hematopoietic cell types other than platelets (derived from megakaryocytes). c-Myb, like GATA-2, appears to be required for the proliferation of hematopoietic precursors. Inactivation of *PU.1* does not affect the development of erythroid cells and megakaryocytes but results in mice that lack granulocytes, monocytes, and lymphoid cells. Some of the properties of transcription factors that play a role in hematopoietic commitment and differentiation before lymphoid commitment are summarized in Table 2.1.

Lymphocyte development from HSCs occurs in waves, and there are differences in the pattern of development in lymphocytes from fetal progenitors and from adult progenitors. Before considering the roles played by specific transcription factors in lymphoid commitment, the differences in

TABLE 2.1. Transcription factors implicated in commitment and proliferation early in hematopoiesis

Factor	DNA-Binding Domain	DNA binding site	Knockout Phenotype	Reference
AML1/CBFA2/ PEBP2aB1	Runt	TGC/TGGT	No definitive hematopoiesis	Okuda et al., 1996; Wang et al., 1996b
CBFβ		non-DNA binding	No definitive hematopoiesis	Wang et al., 1996c
SCL/Tal-1	bHLH	CANNTG	No primitive or definitive hematopoiesis	Porcher et al., 1996
GATA-2	Zinc finger	A/TGATAA/G	No definitive hematopoiesis Death E10–11	Tsai et al., 1994
c-Myb	Myb	C/TAACG/TG	Reduction in definitive hematopoiesis Death E15	Mucenski et al., 1991

the contribution of progenitors from fetal liver and the postnatal bone marrow are presented.

Fetal Liver and Bone Marrow–Derived HSCs

There appear to be a number of differences in the potential of stem cells involved in hematopoiesis that arise in different sites and at distinct stages of development. For instance, early in embryonal development, most yolk sac-derived progenitors can contribute to primitive hematopoiesis but cannot give rise to nonerythroid lineages. The murine day-9 yolk sac also contains progenitors that give rise to all hematopoietic lineages. HSCs may develop from the paraaortic splanchnopleura and AGM regions and seed the yolk sac and also migrate into the fetal liver. These progenitor cells in fetal liver contribute to definitive hematopoiesis; they can generate all blood cells.

Lymphocyte development in the fetal liver occurs in waves. Waves of stem cells home to the thymus and give rise to conventional αβ and certain subsets of γδ T cells. A subset of B cells known as B-1a cells develop primarily from fetal liver-derived stem cells. During fetal development, lymphoid precursors lack an enzyme known as terminal deoxynucleotidyl transferase (TdT), which contributes to the diversity of immunoglobulins and T cell receptors (see Chapter 3). B and T cells generated during this ontological window are of restricted diversity, and this may be a significant factor (as we shall see in later chapters) in determining the types of lymphocyte populations that arise from fetal as opposed to adult progenitors.

A specific progenitor cell that may be responsible for the induction of peripheral lymphoid organs has been described. This $CD4^+CD3^-IL\text{-}7R\alpha^+$ LTβ-expressing cell is derived exclusively from HSCs that originate in fetal liver and is expressed in embryonic secondary lymphoid organs and in the gut. This cell is believed to be required for the generation of lymph nodes and Peyer's patches and is also a major precursor of NK cells. This $CD4^+CD3^-IL\text{-}7R\alpha^+$ cell is not present in the fetal thymus and is not required for the development of B and T lymphocytes. This precursor cell fails to develop in the absence of the bHLH inhibitor Id2, and as a result, lymph nodes and Peyer's patches are not generated and NK cells do not develop in $Id2^{-/-}$ mice. While this $CD4^+CD3^-IL\text{-}7R\alpha^+$ cell is apparently a critical source of LTβ during fetal life for the biogenesis of lymph nodes and Peyer's patches, some other source of LTβ is apparently required for the generation of splenic microarchitecture. B cells are likely to be a major source of LTβ during postnatal life and during germinal center formation in all secondary lymphoid organs. The development of peripheral lymphoid organs will be discussed in Chapter 6.

Specific molecules have been identified that are required for fetal liver-derived lymphopoiesis but are not essential for adult, bone marrow-dependent lymphopoiesis. Conversely, molecules have also been identified that are not essential for fetal lymphopoiesis but that are absolutely required for the later bone marrow-dependent phases of lymphocyte generation. An example of a molecule that is not absolutely required for the generation of fetal liver-derived B and T cells but is essential for the development of B and T cells after birth is the α4β1 integrin. The Ikaros transcription factor is absolutely required for lymphopoiesis from fetal liver HSCs, but after birth, a small fraction of αβ T cells and selected γδ T cells can develop in the absence of this protein.

Lymphocyte-Specific Gene Expression

Our current understanding of the transcriptional regulation of B and T lymphocyte development originated in studies that sought to ask how lymphoid-specific genes, such as the immunoglobulin (Ig) and T cell-receptor (TCR) genes among others, were expressed in a tissue-specific manner. This led to the identification of a large number of transcription factors, some of which were expressed in lymphocytes but not in all tissues, as well as others that were more widely expressed. Regulators that are themselves expressed only in lymphocytes as well as ubiquitously expressed transcription factors were discovered to be important for the lymphoid-specific expression of target genes. The inactivation of specific transcription factors involved in lymphoid-specific gene expression by gene targeting led to the discovery that some of these factors were essential for lymphocyte development. In this section, we will first examine some of the issues and mechanisms relevant to lymphoid-specific gene expression. We will then discuss the potential role of specific transcription factors in the processes of commitment to the lymphoid lineage as well as to the decisions that influence the development of B versus T cells. Regulators of gene expression that are required later in development—including nuclear proteins involved in lymphocyte proliferation, in differentiation after exposure to antigen, and in phenomena such as isotype switching—will be considered in subsequent chapters.

Regulatory Elements: Promoters, Enhancers, Locus Control Regions, Matrix Attachment Regions, and Silencers

Lymphoid-specific gene expression is regulated by the binding of transcription factors to cis-acting regulatory sequences found either in the promoter regions or in enhancers of genes expressed specifically in lymphocytes. Promoters refer to regulatory sequences at or near sites where transcription is initiated. Promoters may be composites of core promoter elements (responsible for the basal transcriptional activity in the vicinity of the transcriptional start site) and regulatory promoter elements that are located immediately upstream. For most genes, including a large fraction of lymphoid-specific genes, the transcriptional start site is approximately 30 basepairs downstream of a TATA box (usually a TAATA motif). A second category of RNA polymerase II–dependent genes was originally described in early B and T cells; these genes lack a TATA box but contain a distinct core promoter element known as an initiator sequence. Promoters of a sizable number of genes that are turned on early in lymphocyte differentiation—such as the $\lambda 5$ gene, the *TdT* gene, and a number of T cell-receptor *Vβ* genes—contain initiator motifs just upstream of the sites of transcriptional initiation.

Initiator motifs consist of the sequence YANA/TYY (where Y stands for a pyrimidine). Transcriptional initiation of TATA box–containing genes involves the formation of an initiation complex at the TATA box by a number of proteins of the basal transcription machinery including TFIID; this complex allows for the entry of RNA polymerase II. Initiator-dependent genes require a distinct initiation complex including a protein known as TFII-I. However, for lymphoid-specific genes, efficient transcription cannot be mediated efficiently by either type of initiation complex in the absence of additional transcription factors that bind cis-acting regulatory DNA sequences. These regulatory sequences may be immediately upstream of the initiation site when they may still be considered to be a part of the promoter but represent tissue-specific regulatory elements. RNA polymerase entry may also be regulated by distant cis-acting regulatory sequences, which may be upstream or downstream of the promoter. These distant regulatory elements are usually bunched together (often to facilitate cooperative interactions) and are called enhancers. Identical tissue-specific regulatory elements may be found in promoters and enhancers. Proteins that bind these motifs, whether they are operationally defined as being part of an enhancer or a promoter, might work in a similar fashion. Transcriptional activators, in addition to a DNA-binding domain, often contain motifs that mediate homo- or hetero-dimerization and also contain a separate domain that usually has a substantial proportion of either acidic residues or of prolines and that is known as an activation domain. This domain is essential for enhancing transcriptional activity and does so by directly or indirectly communicating with the basal transcriptional machinery.

Exactly how enhancers work from great distances is unclear. They most likely allow the formation of large loops of DNA bringing the enhancer-bound proteins close to the promoter. This presumably occurs as a result of protein–protein interactions between polypeptides directly or indirectly associated with enhancer and promoter elements, respectively. In addition to promoters and enhancers, locus control regions (LCRs) have been identified in some genes. LCRs probably exist for all genes, and these cis-regulatory elements may contribute to gene activation by altering the structure of chromatin in the vicinity of the gene, thus permitting transcriptional regulators to access cis-regulatory sites. LCRs were first decribed in the human *β-globin* locus and may operate over very large distances. They are operationally defined as regions that permit the expression in a cis-acting fashion of integrated transgenes and that, unlike enhancers, can upregulate transcription in a dose-dependent fashion. Enhancers may serve as components of LCRs, cooperating with other cis-acting sites. Apart from the *β-globin* gene, LCRs have been described for a macrophage-specific *lysozyme* locus, in the *Ig* heavy chain locus, in the *TCRα/δ* locus, and in the *CD2* locus. How LCRs actually mediate changes in chromatin structure and gene expression is completely obscure.

Another type of cis-acting regulatory region that can modulate transcription by influencing chromatin structure over long distances is known as a silencer. Silencers may recruit DNA-binding proteins, which in turn recruit, by protein–protein interactions, complexes with core histone-deacetylating activity and thus inhibit transcription. Silencers may influence chromatin structure in other ways as well, eventually contributing to chromatin condensation and heterochromatin generation. Silencers have been described for a few lymphoid loci including the TCRα/δ locus and the CD4 genes. The role of these silencers in T cell development and lineage commitment will be discussed in Chapter 9.

Matrix attachment regions (MARs) are AT-rich elements that are defined by their ability to physically tie chromosomal regions to the nuclear matrix. The nuclear matrix is operationally defined. It refers to structures in the nucleus that sediment after high salt and detergent washes. MARs may have diverse functions—potentially influencing transcription as components of LCRs or as silencers. They have been hypothesized to demarcate regions of chromatin, but their precise functions are unclear. They often contain motifs that facilitate the binding of DNA topoisomerase II. MARs have been described on either side of the Ig heavy chain intronic enhancer. These MARs collaborate with the intronic enhancer to function as an LCR and to thus open up the heavy chain locus at the chromatin level and enhance transcription at V gene promoters.

The regulation of gene expression may involve changes in chromatin structure. Specific protein complexes are involved in the remodeling of chromatin, which follows the "marking" of a region for local nucleosomal removal because of the binding to DNA of a specific DNA-binding protein or complex, which may be a transcription factor or, if it is not involved in transcription, may be considered an "accessibility" factor. Many transcriptional activators recruit enzymes that mediate the acetylation of histones and thus contribute to local changes in chromatin structure. Some transcriptional repressors are linked to complexes that contain histone-deacetylating activity. Another potentially "global" means of regulating gene expression is provided by methylation. CpG dinucleotides on inactive genes often contain methylated cytosines, while expressed genes are generally unmethylated. It is possible that demethylation is not merely a consequence of transcriptional activation but is independently regulated by cis-acting signals distinct from those that mediate transcriptional activation. However, it is generally considered likely that methylation may be a consequence of gene inactivation rather than a cause. Methylation remains a poorly understood regulatory mechanism for controlling gene expression or accessibility over relatively large distances.

Mechanisms of Tissue-Specific Gene Expression

It has been assumed that some lymphoid-specific genes are expressed in a tissue-specific manner because they are regulated by transcription factors that are themselves expressed in a parallel tissue-specific fashion. Unfortunately, no good examples directly support this very simple view of tissue-specific transcription. From a mechanistic standpoint, it is still unclear how transcriptional regulators specifically induce or repress gene expression in target organs or cell types. To better explain how transcription factors may drive lymphocyte development, we will briefly review existing models for the tissue-specific regulation of gene expression.

Tissue specificity of gene expression is determined in most cases by a combination of proteins, none of which is lymphoid specific, but all of which are *only* found together in lymphoid cells. These proteins may bind to individual cis-acting sites either individually or in concert (in a cooperative fashion); however, only the complete occupancy of all sites together permits tissue-specific expression (Figure 2.4a). In the case of the Ig heavy chain enhancer (considered below), concomitant occupancy of both the μE5 site (by an E2A bHLH protein) and the μE3 site (by the TFE3 bHLH factor) may be required for lymphoid-specific expression.

A very simple model would involve a truly tissue-specific transcription factor binding to a specific target sequence. In nonlymphoid cells, this site may either not be occupied or it may be occupied by a cognate protein that fails to activate transcription (Figure 2.4b). Unfortunately, no really clearcut example exists for this kind of model.

A third possibility is that a transcription factor might need to be post-translationally modified in a tissue-specific manner for it to be functional (Figure 2.4c). In the case of the E47 transcription factor, discussed later in this chapter, dephosphorylation of homodimers in lymphoid cells might contribute to lymphoid specificity.

Yet another way in which a ubiquitous DNA-binding protein might confer lymphoid specificity is by the expression, only in lymphoid cells, of an adaptor or co-activator protein that binds to this ubiquitous transcription factor (Figure 2.4d). An example of such a co-activator is the OBF-1/OCA-B/Bob-1 protein, which binds to the Oct-1 and Oct-2 transcription factors (discussed in the next section). Tissue- or cell-type specificity could also be attained by a circumscribed post-translational modification of a co-activator. The OBF-1 protein is phosphorylated and activated constitutively in B cells, but this event must be induced in T cells. A tissue-specific heterodimeric partner could confer specificity to a ubiquitous protein. Alternatively, the selective ability to form only homodimers in a specific cell type because of the specific absence of heterodimeric partners could also contribute to tissue-specific transcription.

Once the tissue-specific machinery is in place, how does it work? One view that is gaining ground is that tissue-specific transcription complexes directly recruit enzymes, such as the CBP/p300 proteins, that possess intrinsic histone acetyl transferase activity and that, by acetylating histones, contribute to the local opening of chromatin. Obviously, other models also need to be considered, but at present no data provide a convincing explanation for how tissue-specific transcription complexes contribute to more efficient transcriptional initiation or processivity.

Immunoglobulin Gene Expression

One of the earliest markers of lymphoid commitment is the generation of germline transcripts at the Ig heavy chain locus and at multiple TCR loci. Germline transcripts at the Ig heavy chain locus represent one of the earliest markers of commitment to the B lineage. A thorough understanding of the mechanisms by which Ig gene expression is regulated contributes to

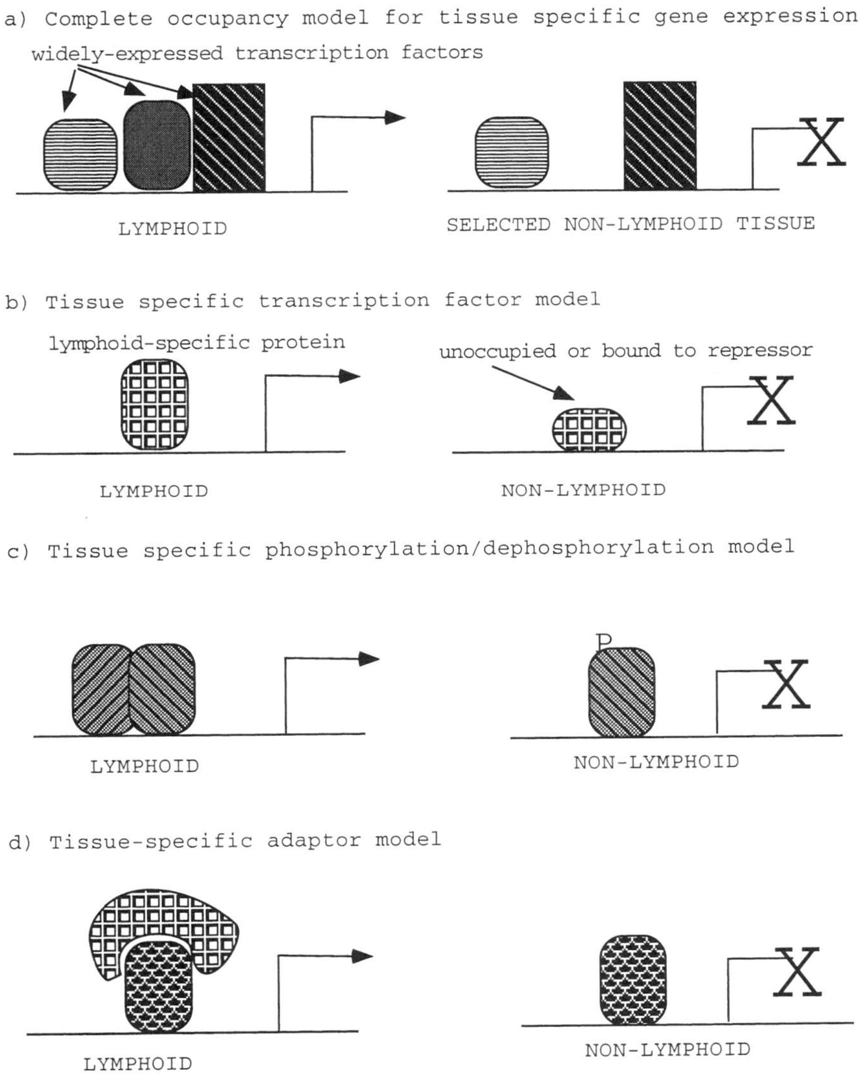

FIGURE 2.4. Models for tissue-specific gene expression. In model a, the right combination of widely expressed factors may only be found in a lymphoid cell. In model b, a lymphoid-specific factor determines specificity. In model c, a lymphoid-specific dephosphorylation event determines specificity; in this case the phosphorylated form of the transcription factor fails to dimerize. In model d, a lymphoid-specific adaptor or co-activator molecule is responsible for specificity. This co-activator may also be post-translationally modified in a cell-type specific manner.

knowledge of the processes by which hematopoietic progenitors are committed to develop into B cells. The IgV segment promoters and the Ig heavy and light chain intronic enhancers were among the earliest tissue-specific regulatory regions studied; the regulation of Ig gene expression remains one of the best paradigms for understanding tissue-specific gene expression in any organism.

Transcriptional activity at the Ig heavy chain and light chain loci correlates with accessibility for V(D)J recombination, as will be discussed in some depth in the next chapter. Tissue specificity of Ig gene expression is mediated by the positive regulation of specific elements in Ig promoters and enhancers in lymphoid cells, as well as by negative regulatory mechanisms in nonlymphoid cells. As shown below, multiple, presumably redundant, mechanisms contribute to lymphoid-specific expression of Ig genes. Immunoglobulin heavy chain enhancers are required not just for the expression of rearranged Ig heavy chain genes, but they also participate in V(D)J recombination, class switching, and somatic mutation. These latter functions of IgH enhancers will be considered in later chapters. The focus in this chapter is on commitment to the lymphoid lineage in general and to the B and T lineages in particular. Accordingly, we will primarily discuss the transcriptional regulation of the Ig heavy chain gene. The κ and λ light chain loci are rearranged and expressed later in B cell development, and their regulation will be discussed in Chapter 7. The mechanisms involved in enhancing the levels of expression of Ig genes in plasma cells will be considered in Chapter 8.

Germline Transcripts at the Heavy Chain Locus

The murine Ig heavy chain locus on chromosome 12 contains approximately 1000 V_H genes (the exact number remains to be determined), spread out over a megabase of DNA (Figure 2.5). More recent studies on the human V_H locus suggest that humans have only about 50 functional V_H genes and about 40 pseudogenes. The V_H genes in the mouse are distributed in at least 14 families; genes are considered to be in the same family if they are 80% identical at the nucleotide level. The most downstream V_H gene family (closest to the constant regions) is the V_H7183 gene family. This gene family has about 25 members, and the member that is located at the extreme 3′ end of the V_H locus is called V_{H81X}. This particular V_H gene is very frequently rearranged in fetal liver-derived pre-B cells. The most distal V gene family, located about a megabase away from the proximal V_{H7183} group, is the V_{HJ558} family of genes, which probably has a few hundred members (estimates using different approaches have ranged from 60 to 1000 members).

D segments are clustered downstream of the V_H locus. In the mouse, D segments may be subdivided into three families. The D_{SP2} family contains at least nine members, the D_{FL16} family has at least two members, and the

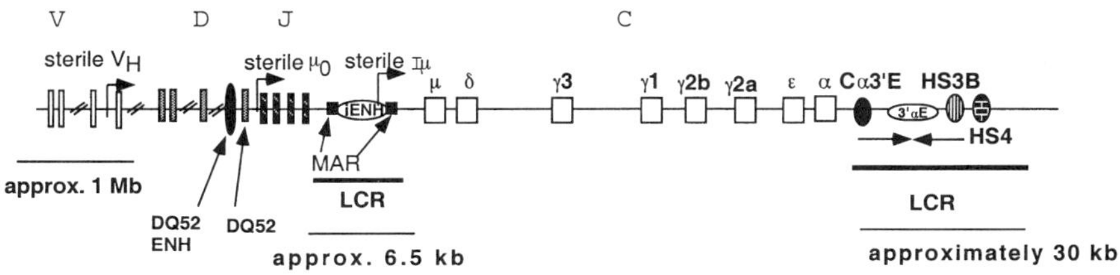

FIGURE 2.5. A schematic view of an unrearranged murine Ig heavy chain locus. iENH refers to the intronic Ig heavy chain enhancer, described in more detail in Figure 2.7. 3′αE refers to the 3′IgH enhancer, and this is described in more detail in Figure 2.8. Cα3′E (or HS3A for hypersensitivity site 3A) is a weak enhancer upstream of the 3′IgH enhancer (3′αE). An enhancer (HS3B), which is structurally similar to the Cα3′E, and another downstream enhancer segment (HS4) also contribute to a potential locus-controlling region (LCR). The opposing arrows in the 3′ region represent a large palindromic region. MAR refers to matrix-attachment region. The intronic enhancer and the flanking MARs also potentially constitute an LCR. D_{Q52} ENH refers to the D_{Q52} enhancer, which will be discussed in Chapter 3. The indicated sterile transcripts may play a role in locus accessibility for V(D)J recombination.

D_{Q52} family contains a single member. The D_{SP2} and D_{FL16} families are clustered in a 70-kb stretch. Approximately 19 kb downstream is the single D_{Q52} gene, which is only 696 bp 5′ of the first J_H segment. In the mouse, the J_H cluster contains four J_H segments.

Two prominent germline transcripts are generated from the heavy chain locus very early in development. Transcripts initiating from upstream of the unrearranged J segments are known as μ_0 transcripts, and those that originate from within the intronic heavy chain enhancer itself are known as Iμ transcripts. Both these transcripts are seen before the initiation of D-J rearrangement at the H-chain locus, and both probably require intronic enhancer activity. Since these germline transcripts do not appear to get translated, they are known as "sterile" transcripts. μ_0 Transcripts are seen in the earliest known B cell precursors, cells of Fraction A1 (AA4.1$^+$ CD43$^+$CD4^{lo} HSA$^-$B220$^+$) in the classification of Hardy (discussed in depth later). Transcripts that are similar to Iμ, and that originate from introns upstream of other isotypes (Iγ3, Iγ1 etc.), are of importance in class switching and will be discussed in Chapter 8. The Iμ transcript may depend on an octamer motif in the intronic enhancer (see below) for promoter activity; and the mature transcript—in addition to being encoded by downstream exons from the constant region of the μ heavy chain gene—also depends on a 5′ Iμ exon (derived from a portion of the J_H-C_H intron). Interestingly, the Iμ exon contains multiple-stop codons in all three reading frames, perhaps to ensure that the Iμ transcript does not produce a protein that could prematurely provide signals in a manner similar to those provided later by the pro-B or pre-B receptors.

Interestingly, V_H transcripts from proximal and distal *V* gene families may also be initiated in a temporally distinct manner, correlating with the accessibility of this locus for the V_H to DJ_H rearrangement event. Proximal V_H genes are transcribed earlier than distal V_H genes. It has become clear that germline transcripts from the distal V_{HJ558} family are influenced by signals from the interleukin-7 (IL-7) receptor, an issue that will be discussed in more detail in Chapter 3.

Transcripts have been noted to initiate from upstream of the D_{Q52} segment both before and after D to J_H rearrangement. Following this recombinational event (but not before), promoters upstream of the more distal D segments are activated, presumably because distal D segment promoters have been brought in proximity to the intronic heavy chain enhancer. These transcripts might reflect the generation of locus accessibility for V_H to DJ_H rearrangement, but some of them also probably subserve the function of encoding Dμ proteins (discussed in Chapter 7). An enhancer in the vicinity of the D_{Q52} segment has been described, and this might be involved in regulating accessibility of D_H segments during V(D)J recombination, an issue that will be considered in Chapter 3.

Transcriptional Regulation at V_H Promoters

Immunoglobulin promoters cooperate with Ig enhancers in mediating tissue-specific expression. Very similar promoter motifs are found upstream of each individual V gene segment (see Figure 2.6), but only the "chosen" V_H segment that is rearranged in a given B cell is in sufficiently close proximity to an enhancer to be capable of high-level transcriptional activation. All V_H promoters contain a conserved octanucleotide motif (ATTTGCAT), called the octamer, which is also found in the IgH intronic enhancer as well as in V_L gene promoters. A number of octamer-binding factors bind to these sites. The most prominent of these are Oct-1 (also known as OTF-1), which is ubiquitously expressed, and Oct-2 (sometimes referred to as OTF-2), which is largely expressed in lymphoid cells. Oct-1 and Oct-2 belong to the POU-domain family of DNA-binding proteins, so-named because the first three members of this family were a pituitary-specific transcription factor, Pit-1, the octamer binding proteins, and a regulator of *C. elegans* development, known as Unc 86. All these proteins share structurally similar DNA-binding domains known as POU domains. This domain is approximately 160 amino acids in length and consists of a POU-specific segment (found only in POU domain family members) and a homeobox homology segment (found in a variety of homeobox-containing developmental regulators first discovered in *Drosophila* but also found in all other metazoans).

While it was presumed that Oct-2 might somehow be responsible for lymphoid-specific gene expression of octamer motif-containing target genes, it has now become clear that tissue specificity is achieved by the recruitment of a B cell-specific protein known as Oca-B (or OBF-1, or Bob-1; see

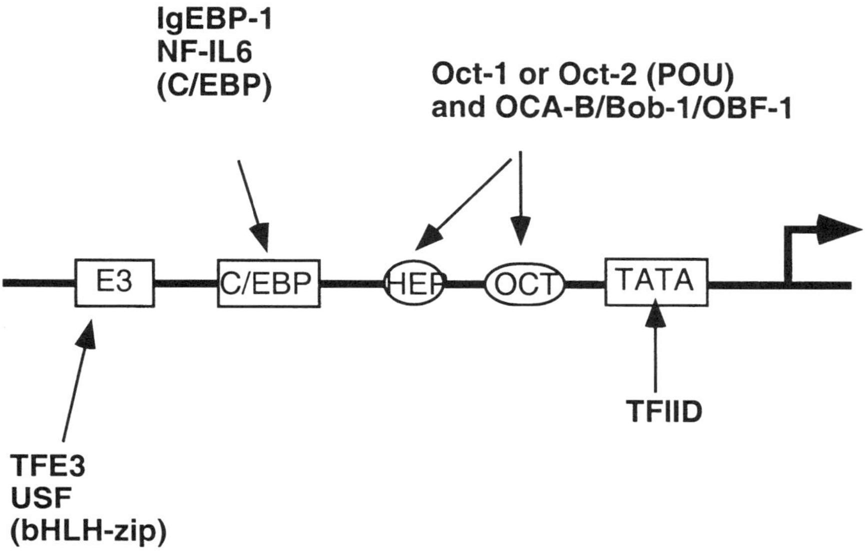

FIGURE 2.6. A schematic view of a V_H promoter region. The Oct-1 and Oct-2 proteins that bind to the octamer motif can also interact with Oca-B/Bob-1/OBF-1, which is a B-cell–specific protein. The heptamer (HEP) site upstream of the octamer (OCT) is present in V_H but not V_L promoters. Further details are provided in the text.

Figures 2.4d and 2.6). Oca-B functions as a transcriptional co-activator, allowing efficient RNA polymerase II entry. Oca-B binds to the POU domain of Oct-2 (and Oct-1) and alters the conformation of these transcription factors, allowing them to bind to a subset of octamer motifs. An additional level of tissue specificity is conferred by the phosphorylation of Oca-B on a specific serine residue (serine-184), which enhances the transactivation function of this protein. This phosphorylation event occurs constitutively in B cells but must be induced in T cells. The octamer is found both in Ig heavy and light chain promoters.

How is specificity for the heavy chain locus achieved when V_H genes are transcribed early in B cell development? V_H promoters, but not V_L promoters, contain a specific CTAATGA heptamer sequence upstream of the octamer site. Although this sequence does not bear any resemblance to the canonical octamer motif, nevertheless, the heptamer placed next to the octamer binds Oct-2 in a cooperative manner. Although the actual contribution of either the octamer or the heptamer sequence to the activation of germline genes has not been studied, it is possible that the cooperative interactions of octamer-binding proteins with these two adjacent regulatory sites might contribute to specific transcriptional activation. The *Oca-B* and *Oct-2* genes have been knocked out, and these mice both have defects in B cell activation, although early B cell development is unaffected. We will therefore return to the functions of Oct-2 and Oca-B in Chapter 8.

A pyrimidine-rich tract, C/EBP sites, and μE3 sites (which bind bHLH-Zip proteins such as TFE3) have been described in some V_H promoters. While the pyrimidine-rich element appears to be conserved in all V_H promoters examined, it is unclear whether the C/EBP site and the μE3 sites are conserved. bHLH-Zip proteins are a subclass of bHLH proteins (discussed below), which in addition to the basic domain and the helix-loop-helix domain, also contain a leucine zipper. These proteins contain two protein–protein interaction motifs (the hydrophobic portion of the HLH domain and the leucine zipper) and can form tetramers. Such tetramers have the potential ability to bring together distant DNA elements, and it is thought that proteins such as TFE3 might facilitate cooperative interactions between IgH promoters and the intronic IgH enhancer. The functional relevance of promoter elements other than the octamer for lymphoid-specific transcription remains to be established.

Germline transcripts from the V_H promoter do not require the intronic enhancer, although rearranged V_H promoters function in an enhancer-dependent manner. It has been speculated that germline V_H transcripts may depend on chromatin changes generated by the occupancy of sites in the IgH 3′LCR, but no evidence supports such a view. Other, yet to be discovered, regulatory elements may contribute to V_H accessibility.

The Intronic Heavy Chain Enhancer and Adjoining MARs

The heavy chain intronic enhancer binds both ubiquitous and tissue-specific proteins (Figure 2.7). This enhancer contributes to transcriptional activation of the IgH locus in B lineage cells and may function as a transcriptional silencer in nonlymphoid cells (though the evidence for a silencing role is somewhat limited). The enhancer is an amalgam of a number of cis-acting regulatory motifs. These include five E boxes, a C/EBP binding site, an octamer sequence, and two sites for Ets domain containing transcription factors. E boxes are characterized by a CANNTG motif and were originally found in the intronic IgH enhancer but are also found in regulatory sites for ubiquitously expressed genes. They bind members of the bHLH family of transcription factors either as homodimers or heterodimers. The two central nucleotides in the CANNTG motif and flanking nucleotides in specific sites determine whether a given site will bind to homodimers or heterodimeric forms of bHLH factors.

These E box-binding transcription factors contain a bHLH domain that includes a stretch of basic amino acids, followed by two amphipathic α-helices separated by an intervening spacer or loop. The basic amino acids mediate DNA binding, while the hydrophobic surfaces of the helices are responsible for dimerization.

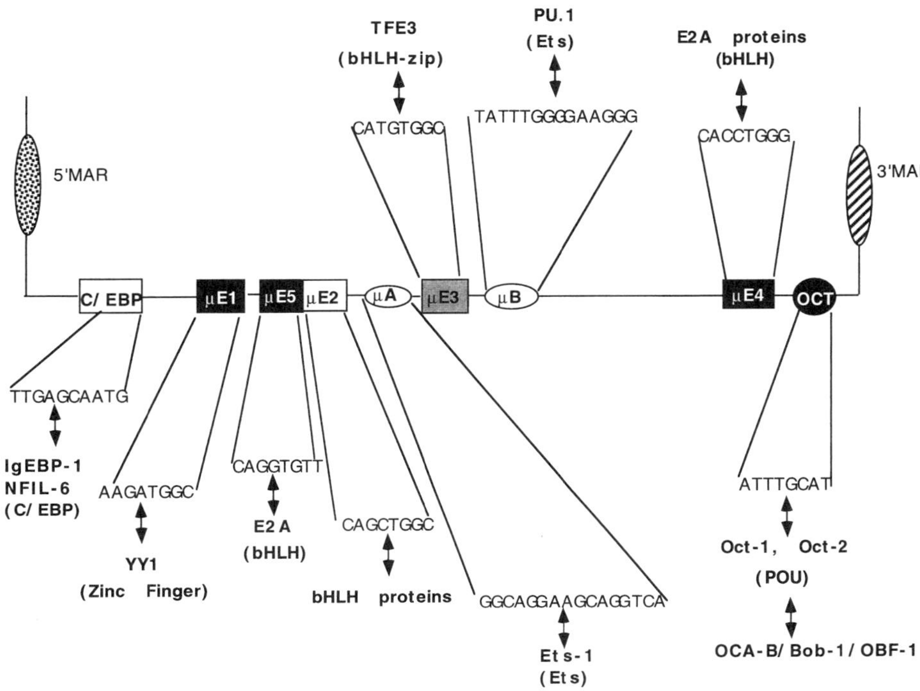

FIGURE 2.7. An outline of the murine intronic heavy chain enhancer. Transacting factors that bind to each of the sites delineated above are indicated. MARs are usually considered separately but functionally cooperate with the enhancer.

There are two broad subfamilies of bHLH proteins. One subfamily includes products of the *E2a*, E2-2, and *Heb* genes, which can bind either as homodimers or heterodimers to the μE2, μE4, and μE5 elements in the intronic enhancer. The second subfamily, the bHLH-Zip group, contains a leucine zipper region and has been considered briefly in the section above on IgH promoters. From the standpoint of IgH regulation, the best studied members of the bHLH subfamily are alternatively spliced products of the *E2a* gene, which encode the E12 and E47 proteins. The absence of E2A leads to a block of B cell development early in the pro-B stage. While these data have been interpreted to suggest that E2A is required for the progression of B cell development after commitment has occurred, such a view may need to be reconsidered. Commitment, as defined earlier, implies a time point when a cell's fate has been *irreversibly* dictated. It is becoming increasingly apparent that E2A may actually participate in the commitment of progenitors to the B lineage. This issue, the target genes regulated by E2A, and the role of bHLH proteins in T cell commitment will be discussed later in this chapter.

E2a genes are expressed ubiquitously. In some cell types, E47 and E12 can bind to their target sequences as heterodimers with other tissue-specific members of the bHLH family. No B cell-specific bHLH partners for E47 and E12 have yet been identified. How then do E2A proteins function in such a tissue-specific manner? It appears that homodimers of E47 may be uniquely found in the B lineage and that individual E47 chains in these homodimers might be covalently associated with their partners via a disulfide bridge. The B cell-specific function of E47 may also depend on a lineage-specific dephosphorylation of E47 homodimers. B cell specificity may be generated therefore by a number of means. B cells may lack other heterodimeric partners, favoring the formation of homodimers. B cells may able to specifically catalyze the cytosolic disulfide bond formation event that facilitates homodimerization or mediate a tissue-specific dephosphorylation event that is critical for function. A family of inhibitors of bHLH function known as the Id proteins might also contribute to the tissue-specific function of E2A proteins.

The Id proteins—Id1, Id2, Id3, and Id4—contain HLH domains that are homologous to those in bHLH proteins, but they lack basic regions. They can therefore sequester and functionally inactivate bHLH proteins. Id1 and Id2 are known to be expressed at high levels in certain progenitor B cell lines that do not express μ heavy chains but are not expressed in more mature B cells. Id3 has been shown to be expressed in early B lineage cells but to be downregulated in plasma cells, thus contributing to the activity of an E site in the 3′αE (discussed below). Id overexpression has also been used as a dominant negative tool that has helped reveal that bHLH proteins are important for the commitment event in both B and T cells. Id2 null mice lack peripheral lymph nodes and Peyer's patches and fail to generate NK cells. The function of Id2 in the context of lymph node biogenesis and NK cell development will be considered in more detail in Chapters 6 and 11.

While E2A proteins clearly are important transcriptional activators in the B lineage, some of the sites that they bind to in B cells can function as negative regulators of IgH expression in non-B cells. The μE4 and μE5 motifs in the intronic enhancer can bind to a zinc finger-containing protein in non-B cells called ZEB, which interferes with E2A function and transactivation, though the exact mechanism by which this interference is achieved remains unknown.

A minimal enhancer region (for the intronic enhancer) has been described consisting of the μA, μE3, and μB boxes. Occupancy of all these sites by the appropriate cognate DNA-binding proteins is necessary for B cell-specific enhancer function. The μA site is occupied by Ets-1, and another Ets family member, PU.1, occupies the μB site. The μE3 site is occupied by a bHLH-Zip protein known as TFE3. Absence of the PU.1 protein profoundly affects the commitment of cells to both lymphoid and myeloid lineages and will be discussed in more detail later in this chapter.

The intronic enhancer probably functions in conjunction with the adjacent MARs as an LCR. Formal evidence that this unit functions as an LCR, however, has yet to be obtained. This putative LCR is presumably essential for the opening of the heavy chain locus for the first step of heavy chain recombination—the joining of a *D* segment to a J_H segment—to occur. Accessibility of V_H genes might also be influenced by this LCR. However, it is likely that multiple regulatory elements cooperate to mediate transcription and accessibility.

Other Enhancers and LCRs Regulating Ig Heavy Chain Gene Activity

An enhancer/promoter region has been described upstream of the DQ_{52} segment (see Figure 2.5). This regulatory region has been hypothesized to be involved in regulating accessibility during V(D)J recombination and will be discussed in Chapter 3. A large cis-acting region that may be required during B cell development has been described as an LCR downstream of the Cα locus. Again, formal evidence of dosage-independent transgenic expression to prove that this region meets the definition of an LCR has not been obtained; evidence from transfection studies supports such a contention, and this regulatory region is now commonly described as an LCR. This 3′IgH LCR (Figure 2.8) may contribute to accessibility during class switch recombination (the evidence in support of this is actually open to question) and is likely to be important in permitting high-level transcription of the Ig heavy chain gene in terminally differentiated plasma cells.

The 3′IgH LCR consists of four functionally defined enhancer regions, some of which were originally discovered as DNase I hypersensitive sites

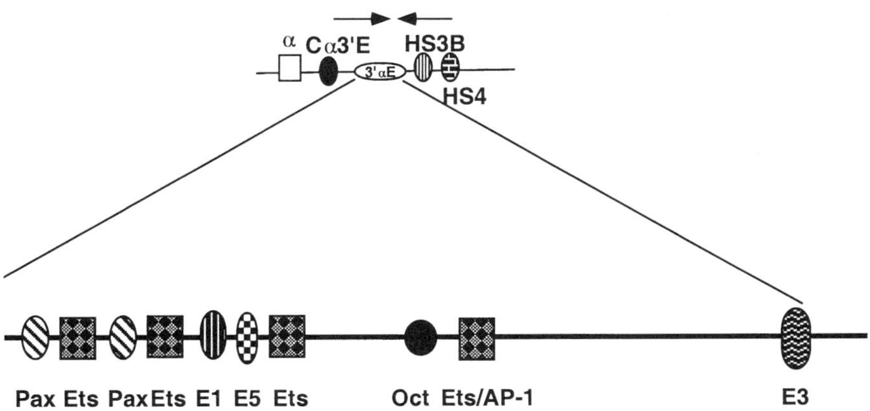

FIGURE 2.8. The 3′IgH enhancer and LCR region. This LCR is made up of four distinct enhancer regions. The 3′αE is the strongest of these four enhancers and contains a number of regulatory motifs depicted in the lower portion of the figure. See text for details.

in the course of studies on the stage-specific regulation of IgH expression. A weak enhancer, the Cα3′E (also known as HS3A), has been described 4 kb downstream of the α locus. Sixteen kilobases downstream of this locus is a relatively strong transcriptional enhancer region known as the 3′αE also known as HS1,2 because it overlaps with two DNase I hypersensitive sites. HS3B and HS4 lie further downstream. HS3B is highly homologous to the HS3A/Cα3′E sequence. Indeed, this 3′IgH LCR has a palindromic structure spread over 24 kb (the largest described palindromic region in any species) and is made up of two inverted repeats that meet head to head in the 3′αE region.

The 3′IgH LCR contains multiple binding sites for Pax-5/BSAP. Pax-5 is one of 9 members of a mammalian gene family homologous to *Drosophila*-paired rule genes. These proteins contain a 128 amino acid paired box DNA-binding domain. The sites in the HS4 region and in the 3′aE (HS1,2) region may have separate functions. Multiple Pax-5 sites are present in the HS4 region, including a particularly high-affinity site that may be especially important early in B cell differentiation, since in functional assays in transfected cell lines, it was the only active element in the pre-B stage. Pax-5 null mice have a defect in B cell development, which is more severe in fetal life and is discussed later in this chapter.

The 3′αE region itself may be required for class switching, and while it is a relatively weak transcriptional enhancer in the earlier stages of B cell differentiation, it appears to activate transcription strongly in plasmacytomas. This enhancer contains a number of well-characterized regulatory motifs, as depicted in Figure 2.8. Two Pax-5 sites are present in the 3′αE enhancer, and they appear to be important for the increased expression of Ig heavy chain genes in plasma cells. This enhancer also contains an Ets-AP-1 site, which binds to Elf-1 and AP-1 after antigen-receptor triggering. In addition, the enhancer also contains three E boxes: E1, E3 and E5. bHLH proteins presumably cannot bind to E boxes in this enhancer during some stages of B cell development, because the "dominant-negative" Id3 protein is expressed at high levels before plasma cell generation. In plasma cells, Id3 levels drop, and activation through this motif can be achieved. Numerous bHLH sites are found in the HS3A/Cα3′E, HS3B, and HS4 regions and are presumably also subject to developmental regulation. The role of this enhancer in activated and terminally differentiated B cells will be discussed in Chapter 8.

Lymphoid Progenitors and Commitment to a Lymphoid Fate

A number of different lineages can develop from a cell that may be considered a lymphoid-committed progenitor that no longer has any myeloid potential (Figure 2.9). In both humans and in mice, a progenitor cell has

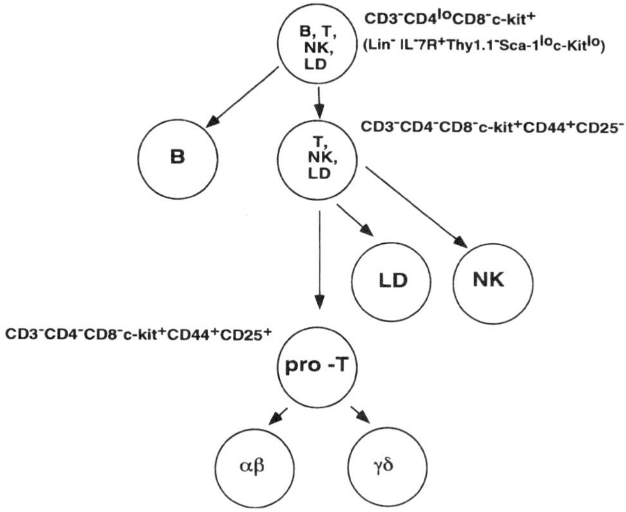

FIGURE 2.9. A common lymphoid progenitor gives rise to B cells, T cells, NK cells, and LD cells.

been described that can give rise to B cells, T cells, NK cells, and some categories of dendritic cells known as lymphoid dendritic (LD) cells but that cannot give rise to any myeloid lineage cells. What is most likely a common lymphoid progenitor cell in the mouse was first described as a $CD4^{lo}CD3^{-}$ $CD8^{-}c\text{-}kit^{+}$ cell. This cell has also been described as having a Lin^{-} $IL\text{-}7R^{+}$ $Thy\text{-}1^{-}$ $Sca\text{-}1^{lo}$ $c\text{-}kit^{lo}$ phenotype. In human bone marrow, a $CD10^{+}$ $CD34^{+}$ Lin^{-} $c\text{-}kit^{-}$ $Thy\text{-}1^{-}$ cell has been described that gives rise to B cells, NK cells, LD cells, and probably T cells as well. This precursor is assumed to represent the human common lymphoid progenitor. A slightly less multipotent murine precursor cell—presumably derived from the above $CD4^{lo}$ common lymphoid progenitor (CLP), which can give rise to T cells, NK cells, and dendritic cells, but not to B cells—has also been described. This cell does not express CD3, CD4, or CD8 (a triple-negative cell) but does express c-kit (CD117) and CD44. These progenitor cells may be found both in the bone marrow as well as in the thymus.

What specifically commits a cell to a lymphoid fate remains to be determined. Gene-targeting experiments appear to indicate that for a fetal progenitor cell to become a lymphocyte it must express the Ikaros and PU.1 transcription factors. PU.1 may be necessary for the generation of a lymphoid–myeloid-committed precursor cell, whereas Ikaros may be required for a slightly later precursor that gives rise to B cells, T cells, NK cells, and LD cells. PU.1 is an Ets family transcription factor, while Ikaros is a Kruppel-type zinc-finger protein that was discovered as a protein that bound to a regulatory sequence in the enhancer of the CD3δ gene.

As mentioned earlier, the transcriptional requirements for commitment to a lymphoid fate are distinct in fetal and adult hemopoiesis. During fetal hematopoiesis, there may be an absolute requirement for Ikaros expression to generate B, T, NK, and thymic dendritic cells. In adult lymphopoiesis, the requirement for Ikaros is slightly less stringent; other Ikaros family members may be able to partially mediate the function of Ikaros in T cells.

Ikaros and PU.1 Null Mice

A failure of development of B cells, αβT cells, Vγ3-expressing dendritic epidermal thymic γδT cells, NK cells, and certain categories of dendritic cells derived from lymphoid precursors was observed in Ikaros null mice. Essentially all lymphoid and "lymphoid-related" lineages that originate from fetal liver-derived HSCs fail to develop in the absence of Ikaros. In adult mice, the differentiation defects are more circumscribed (Figure 2.10). There is a defect in the generation of all B cells including B-1a cells (which are fetally derived). In addition, NK cells and dendritic cells are not seen. αβ T cell generation is not completely abrogated but is significantly reduced in adult Ikaros-null mice, and there is a skewing of these cells toward the CD4 subset. Vγ3-expressing dendritic epidermal γδ T cells (DETCs), intraepithelial CD8αα γδ T cells, and adult HSC-derived peripheral γδ T cells are missing or significantly reduced. $CD8^-$ intestinal γδ T cells are also reduced. However, vaginal epithelial Vγ4-bearing γδ T cells are found in adult Ikaros-null mice. These cells are generally thought to be generated from a late wave of fetal precursors. In Ikaros-null mice, these Vγ4 cells are presumably generated from adult bone marrow-derived HSCs.

The Ikaros protein exists in numerous forms derived from RNAs generated by alternative splicing. The zinc fingers fit the Kruppel C2-H2 consensus structure and come in two sets. Two C-terminal zinc fingers are found in all Ikaros proteins and are involved in homo- and hetero- dimerization. The *N*-terminal zinc finger domains contain from one to four zinc fingers in different Ikaros proteins, and these zinc fingers participate in DNA recognition. The sequence recognized by these zinc fingers contains a GGGA motif. A dominant-negative Ikaros protein (which can dimerize with other Ikaros family members but lacks the ability to bind DNA) completely blocks lymphoid development even from adult-type HSCs. This suggests that there might be related proteins whose functions are being inhibited by this dominant negative form of Ikaros. This finding initially led to the identification of an Ikaros-related lymphoid-restricted protein called Aiolos, which can heterodimerize with Ikaros. The small numbers of T cells that do develop in adult Ikaros-null mice may be driven to do so by dimers of other Ikaros-like zinc finger gene products such as Aiolos, Helios, or Daedalos. While Ikaros is expressed in HSCs, Aiolos is first expressed in lymphoid progenitors and is expressed at higher levels after the generation

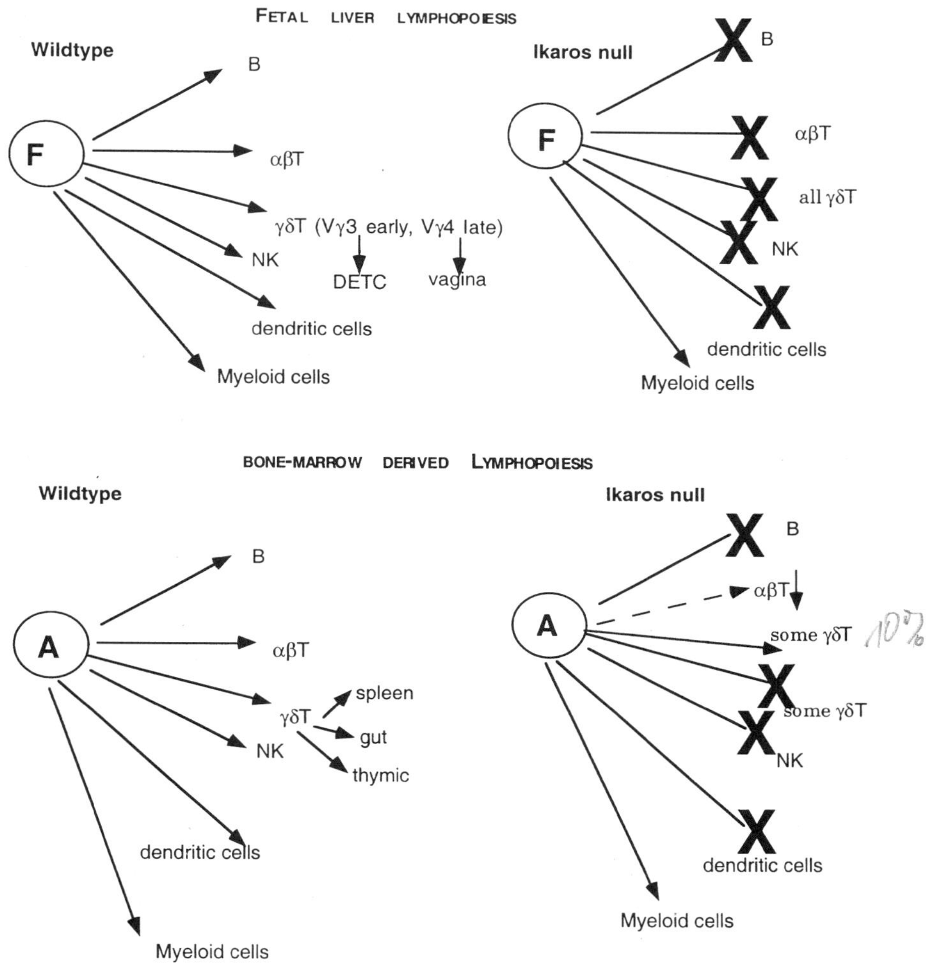

FIGURE 2.10. The Ikaros gene is absolutely required for lymphoid development in the mouse fetus but only for some populations in adult mice. DETC refers to dendritic epidermal thymic cells, a γδ T cell population. These are distinct from the dendritic cells referred to in the figure. The dendritic cells mentioned above represent populations of nonmyeloid origin (myeloid-derived dendritic cells may not depend on Ikaros for their development) and are referred to as LD cells. Adult αβ T cells are reduced to about 10% of the normal level in Ikaros-null mice (dashed line).

of pro-T or pro-B cells. Helios is expressed specifically in the T cell compartment. Binding sites for these transcriptional regulators have been identified in many lymphoid genes including *Tdt*, *Lck*, λ5, and *vPreB*.

Whether Ikaros is a transcriptional activator, or a protein that participates in gene inactivation at the chromatin level, or whether it can function both as a transcriptional activator and in a silencing capacity *in vivo* remain to be firmly established. It appears that Ikaros can function as a negative regulator of T cell activation and cell cycle progression, possibly because of its ability to repress gene expression in specialized complexes in centromeric heterochromatin. Aiolos negatively regulates B cell activation and will be considered again in Chapter 8.

PU.1 (also known as Spi 1) is an Ets-family transcription factor that is expressed specifically in hematopoietic cells. High levels of expression are seen in myeloid and B lymphoid cells. PU.1 is required in a cell-autonomous manner for the development of all myeloid and lymphoid cells but is not required for the development of erythrocytes and platelets. A rare human form of severe combined immunodeficiency known as reticular dysgenesis presents with a phenotype reminiscent of the *PU.1*-null mouse—with the total absence of myeloid and lymphoid cells but no reduction in the numbers of erythrocytes or platelets. The molecular basis of reticular dysgenesis remains to be determined. This is discussed in Chapter 12.

Stromal Signals and Lymphoid Commitment

Clearly, growth factors such as SCF, stromally expressed integrins, cytokines, and chemokines all play important roles in lymphocyte development and may participate in the process of lymphoid commitment. Stromal cells—the diverse mixture of supporting fibroblasts, macrophages, and related cells in the bone marrow—clearly play a very important role in the commitment process. The total absence of the p50 and p65 subunits of NFκB in gene-targeted mice leads to the failure of B and T cell development. When $p50^{-/-}p65^{-/-}$ fetal liver cells are transferred to recipients in the presence of wild-type fetal liver cells, the double-knockout cells can give rise to a normal complement of B cells. Presumably, this transcription factor is necessary for nonlymphoid "stromal" components to produce some regulator of lymphoid commitment. The role of "adhesion" molecules, including integrins, in lymphopoiesis will be considered in depth in Chapter 6. Survival and further differentiation of the common lymphoid progenitor might depend on signals generated by IL-7 and possibly IL-15. The common lymphoid progenitor is apparently the earliest hematopoietic progenitor to express the IL-7 receptor. The IL-2, IL-4, IL-7, IL-9, and IL-15 receptors share a polypeptide chain known as the common γ chain, or γc (see Chapter 4 for a further discussion of these receptors). γc Signals via a tyrosine kinase known as Jak3. In mice, a knockout of γc or of *Jak3* leads to a defect in B, T, and NK cell development. Information on LD cells in these knockout mice has not been obtained. In humans, the absence of the γc (in X-linked severe combined immune deficiency, or X-SCID) and of Jak3 (in an autosomal-recessive form of SCID) severely compromises T and NK cell development (Figure 2.11). These immunodeficiency states will be con-

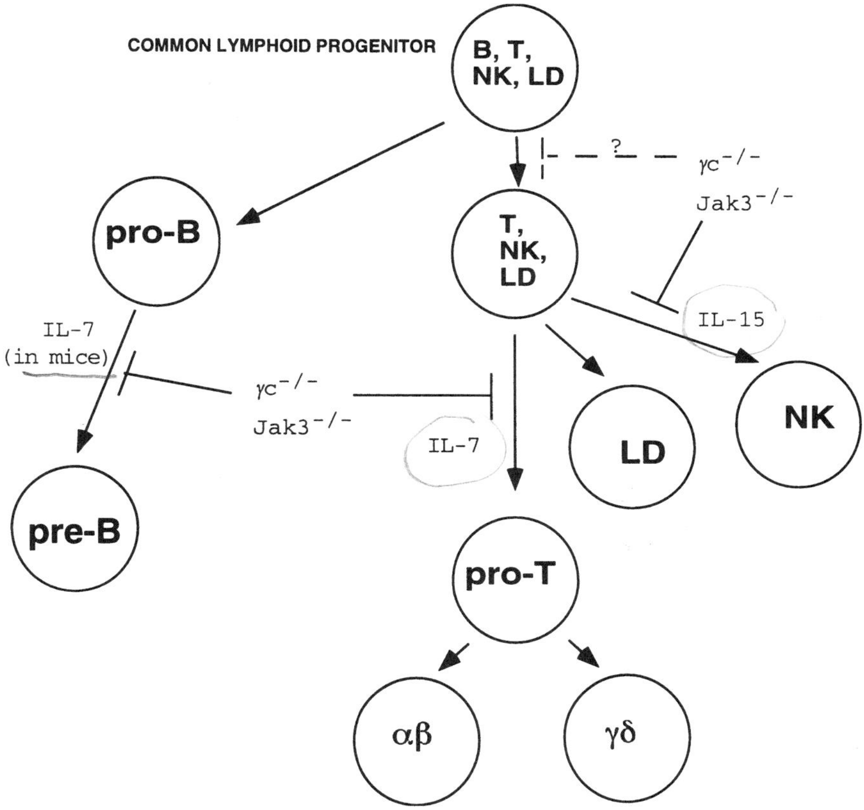

FIGURE 2.11. IL-7 and IL-15 may be required for commitment to the T and NK lineages, respectively. Commitment to the T and NK lineages may require the γc/Jak3-signaling pathway. The defect in T and NK generation is more marked in humans than in mice. IL-7 is required for the pro-B to pre-B transition in mice but may not be required for this event in humans.

sidered in more detail in Chapter 12. The defects in B and T cell development in the mouse in the absence of γc or Jak3 can be attributed to a lack of IL-7 signals. In humans, IL-7 does not appear to be critical for B cell development, though some other ligand might actually signal to human pro-B cells via a receptor that includes the IL-7 receptor α-chain but not γc. Thymic stromal lymphopoietin (TSLP) is a protein that binds to IL-7Rα and might influence B cell development.

Clearly, although the IL-7 receptor is expressed in common lymphoid progenitor cells, IL-7 signaling is not absolutely critical for the expansion or survival of this cell, since development is blocked at a later stage in IL-7 and IL-7R knockout mice. It might be critical, more so in humans, for the progenitor that gives rise to T, NK, and LD cells. However little data are available to support or refute such a possibility. The phenotypic changes seen in the absence of γc and of Jak3 might be fully explained by the requirement of IL-7 in pro-B cells and in some early T lineage cells and by the need for IL-15 in NK cell generation. An interesting connection has been made between presumed "maintenance" factors for gene expression such as Mel-18 and Bmi-1 and IL-7-dependent lymphocyte development. These proteins are mammalian homologs of the *Polycomb* gene family in *Drosophila melanogaster*, which presumably function in large protein complexes to repress the expression of homeotic genes at the level of chromatin. In vertebrates, Mel-18 and Bmi-1 may also be involved in the maintenance of the expression of selected genes including some whose products are required for IL-7 signaling.

The function of IL-7 and the IL-7 receptor will be discussed in the context of accessibility for V(D)J recombination in Chapter 3 and in terms of B and T cell development in Chapters 7 and 9. The requirement for IL-15 in NK cell generation will be considered in some depth in Chapter 11. Cytokine receptors and signaling will be broadly covered in Chapter 4.

Commitment to the B Lineage

B lineage progenitors commit to the B lineage before the initiation of Ig heavy chain gene rearrangement. The expression of *Ikaros* and *PU.1* is required for the development of common lymphoid progenitors and of bipotential B/myeloid progenitors, both of which may represent precursor cells for the B lineage. However, while gene-targeting studies suggest that these genes are necessary for development before the B lineage commitment event, they may act in concert with other gene products to orchestrate this event. The earliest demonstrable B lineage cell expresses the B220 protein (a B lineage isoform of the CD45 transmembrane tyrosine phosphatase gene), and the AA4.1 antigen on the cell surface, but contains the heavy chain locus in a germline configuration. A detailed discussion of the surface markers in the early stages of B cell development is provided in Chapter 7. In a simplified scheme, Fraction A (now subdivided into A0, A1 and A2) in the scheme of Hardy refers to the earliest committed progenitors with germline *IgH* loci, Fraction B refers to cells beginning to initiate $D\text{-}J_H$ rearrangement, and Fraction C refers to cells in which V to DJ_H rearrangement is initiated but in which light chain genes remain in germline configuration. As the earliest cells of the B lineage (in Fraction A1) differentiate further, they express the λ5 and *vpre*B surrogate light chain genes and also the genes that encode the Igα/mb-1 and Igβ/B29 proteins. These gene products play essential roles in the selection events that occur in pro-B and pre-B cells. They also turn on *Rag-1*, *Rag-2*, and *TdT*, which are lymphoid-specific components of the V(D)J -recombination machinery.

Commitment to the B lineage may require that a progenitor *not* receive signals from a Notch 1 ligand (see discussion on T cell commitment below). Notch signals may antagonize E2A function and thus prevent entry into the B lineage. The exact nature of the extracellular signals that commit a common lymphoid progenitor to a B cell fate is unclear. Numerous additional extracellular ligands are critical for early B cell development. These will be considered in detail in Chapter 7.

Based on the phenotypes of knockout mice, it has been inferred that progression through the B lineage, presumably after cells are committed to the B lineage, requires expression of the early B cell factor (*EBF*), *E2a*, and *Pax-5* genes (Figure 2.12). It is now recognized that Pax-5 is actually required for commitment to the B lineage and not just for progression after commitment has occurred.

Adult *Pax-5*$^{-/-}$ mice contain B cells that complete the D to J_H rearrangement step but do not proceed to the pre-B stage. Development is arrested between Fractions B and C in the Hardy classification. In these mice, no B cell development occurs in the fetal liver. However, when fetal liver cells are transferred to a normal host, development proceeds up to Fraction B,

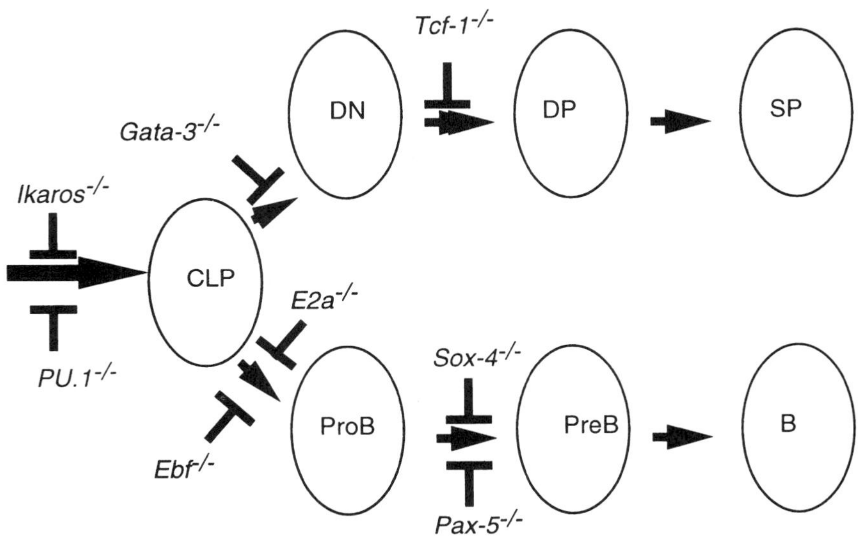

FIGURE 2.12. Transcription factors identified from gene-targeting experiments that are critical early in B and T cell ontogeny. Pax-5 is absolutely required for B cell development in fetal liver, and it may play a role in "true" commitment in the bone marrow, as discussed in the text. The T cell factor (TCF) gene is required for the progression of T cell development and will be discussed in Chapter 9.

indicating that in the fetal liver Pax-5 is apparently required in some non-B cell, which may provide signals for B cell development. Although it was initially assumed that there is no impairment of commitment to the B lineage in the bone marrow in adult *Pax-5*$^{-/-}$ mice, a careful analysis of Fraction B cells in these mice suggests that they are not "truly" committed in an irreversible Calvinistic sense. These cells are actually quite promiscuous in their ability to generate other lineages, and it may therefore be necessary to consider Pax-5 to be responsible in part for commitment to a B cell fate.

The E2A-derived E47 and E19 proteins have been discussed in some detail earlier in this chapter. Potential targets of E2A include *Ig-μ*, *Ig-κ*, *Ig-λ*, *TdT*, and *Rag-1*. E2A-derived homodimers may be essential for entry into the B lineage but may generally act in concert with another transcriptional activator known as EBF. The EBF protein is also a homodimer that is found in pro-B, pre-B, and B cells, but not in plasma cells. Homodimerization of EBF requires a structural motif that may be related to the dimerization motif found in HLH proteins. The DNA-binding motif of EBF is unique, and it recognizes variants of the palindromic nucleotide sequence 5′ATTCCCNNGGGAAT. Functionally relevant EBF sites are found in the *mb-1* promoter, in some Ig promoters, and in the promoter/enhancer region of the *λ5* and *vPreB* genes.

Inactivation of the *Ebf* and *E2a* genes individually result in very similar B lineage phenotypes—arrest at an early pro-B stage (at the boundary between Fraction A and Fraction B in the classification according to Hardy, discussed in Chapter 7) before the initiation of heavy chain gene rearrangement. These genes can functionally cooperate in inducing *λ5* expression when introduced together into nonlymphoid cells. However, the critical targets of these genes whose absence is responsible for the developmental block remain to be identified. It remains to be proven that the Fraction A precursors seen in the absence of E2A or EBF are actually *irreversibly* committed to the B lineage, even though they phenotypically appear to be early B lineage cells. Experiments to test whether cloned "B lineage type" precursors in these mice can develop into non-B cells remain to be described.

Further progression during B cell development depends on a number of other transcription factors, which will be discussed in some detail in Chapters 7 and 8.

Commitment to a T-Cell Fate

Entry of progenitor lymphoid cells into the T lineage require signals from the Notch 1 receptor (Notch signaling is discussed in Chapter 4). Triggering of Notch on progenitor cells by ligands either expressed or secreted by stromal cells can commit cells to the T lineage and prevent their developing into B cells. Although inactivation of the *Notch 1* gene results in embryonic lethality, a conditional knockout of the *Notch 1* gene has revealed a very early cell–autonomous block in T cell development while the development of other hematopoietic lineages is unaffected. It appears that commitment of a common lymphoid progenitor to a T cell fate occurs at the expense of the generation of B cells. A marked increase in thymic B cells was observed in conditional *Notch 1* knockout mice. In separate studies in which an activated form of Notch 1 was introduced into stem cells, B cell development was blocked, and enhanced ectopic T cell development was noted in the bone marrow. As discussed in Chapter 4, one of the transcriptional targets linked to Notch is the vertebrate Hes gene. An early block in T cell development has been noted in *Hes-1* null mice.

As we will discuss in Chapter 9, many γδ T cells develop in extrathymic locations. While initial commitment to the αβ T lineage (before actual rerrangement of TCR genes) can occur in the fetal liver or in the adult bone marrow, it appears that lymphoid committed cells—particularly cells that can give rise to both NK cells and T cells—home to the thymus and that specific commitment to the αβT lineage occurs primarily in the thymus. The existing data in the field are consistent with the migration to the thymus of oligopotent precursors that irrevocably commit to the T lineage in the thymus. However, commitment to the αβT lineage can occur in extrathymic sites, and the main role of the thymus, as discussed in Chapter 9, may be to provide a compartmentalization function rather than an inductive environment.

Some of our current understanding of T cell commitment was obtained by first attempting to understand T lineage-specific gene expression. The regulation of gene expression in T cells has not been as thoroughly explored as in B cells. However, the inactivation of some of the genes encoding transcription factors involved in T cell-specific expression has resulted in valuable insights being obtained about lymphocyte development in general. Ikaros was first identified as a protein involved in the T lineage-specific expression of the CD3 δ gene. GATA-3 (discussed below) and TCF (discussed in Chapter 9) were first identified by an analysis of proteins that bind to regulatory motifs in the TCRα gene enhancer.

Ikaros and PU.1 are absolutely essential for the formation of both αβ and γδ T cells that are derived from fetal liver precursors. However, in the absence of either PU.1 or of Ikaros, some αβ T cell development does occur after birth, as has been discussed earlier. A transcription factor that presumably is critical later than Ikaros and PU.1 and that is involved in T lineage commitment is GATA-3. The *Gata-3* gene is expressed in the T lineage and encodes a zinc finger-containing protein that is involved in the regulation of TCR α, β, and δ chains.

Inactivation of the *Gata-3* gene results in embryonic lethality. In studies using the $Rag^{-/-}$ blastocyst-complementation approach, the earliest recognizable T cell precursors were not seen in the absence of *Gata-3*. B cell development is normal in these mice. Although the development of NK cells and dendritic cells has not been thoroughly examined in *Gata-3* null/*Rag*$^{-/-}$ mice, it is likely that these lineages will be absent, based on the presumed lack of the triple-negative CD44^{+}CD25^{-} multipotent precursors in these mice (Figure 2.13). The commitment of cells to the T lineage likely also involves the activation of bHLH transcription factors such as the products of the *E2a* and *Heb* genes. While knockouts of individual bHLH genes have yielded less dramatic T lineage phenotypes, forced expression of the dominant-negative HLH Id3 protein has been found to abrogate the development of CD1a^{+} human T cell progenitors (which correspond to CD25^{+} murine T lineage progenitors).

The external signals that drive "oligopotent" CD44^{+}CD25^{-} cells to the T-committed CD44^{+}CD25^{+} stage remain to be identified with certainty. It is considered likely that this transition may require the generation of survival and/or expansion signals from IL-7, c-kit, and the Flk2/Flt3 ligand.

Transcription factors such as TCF and lung Kruppel like factor (LKLF), which play an important role in T cell development and survival (but not in commitment), will be discussed in Chapter 9.

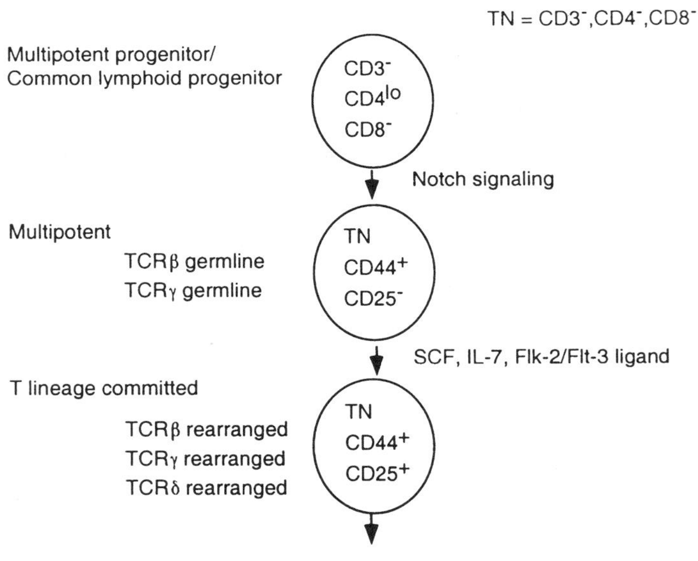

FIGURE 2.13. Commitment to the T lineage. TN (triple negative) refers to cells lacking CD3, CD4, and CD8. Notch signaling is required for commitment to the T lineage at the expense of B lineage commitment. The signals required for the transition from the "oligocommitted" CD44^{+}CD25^{-} to the T-committed CD44^{+}CD25^{+} stage are not known. IL-7, stem cell factor (SCF), and the Flk2/Flt3 ligand (FL) are candidate signal inducers at this stage.

Summary

The characteristics of HSCs have been discussed in this chapter. Some of the transcriptional regulators that are essential for the development of HSCs from ventral mesoderm have been identified, as have transcription factors that are critical for the development of cells committed to specific hematopoietic lineages. Little is known about the signaling pathways that are involved in the induction of these commitment factors. Transcripts from antigen-receptor gene segments represent some of the earliest known markers of commitment to these lineages. The inactivation in the mouse germline of transcription factors that contribute to B and T lineage-specific gene expression has led to the identification of genes that are critical for lymphocyte development. Although it is likely that all the transcriptional regulators that control lymphoid commitment have not as yet been identified, the generation of the common lymphoid progenitor capable of giving rise to T, B, NK, and LD cells requires the Ikaros and PU.1 transcription factors. Signaling via the cytokine receptor common γ chain also is critical for the generation of most lineages downstream of this progenitor. Commitment to the B lineage might require expression of the EBF, E2A, and Pax-5 transcriptional regulators. Notch 1 signaling promotes the development of T lineage cells at the expense of entry into the B lineage. *Gata-3* and *bHLH* genes may be required for T cell commitment. Some of the targets of these transcription factors are known, but the critical proteins whose absence compromises further differentiation during lymphocyte development remain to be identified. Very little information is currently available about the extracellular signals that might turn on the expression of these transcription factors during ontogeny.

Selected Reviews

Arulampalam, V., Eckhardt, L., and Petterson, S. (1997). The enhancer shift: a model to explain the developmental control of IgH gene expression in B-lineage cells. Immunology Today *11*, 549–554.

Clevers, H., and Ferrier, P. (1998). Transcriptional control during T cell development. Curr. Opin. Immunol. *10*, 166–171.

Clevers, H. C., and Grosschedl, R. (1996). Transcriptional control of lymphoid development: lessons from gene targeting. Immunology Today *17*, 336–343.

Ernst, P., and Smale, S. (1995a). Combinatorial regulation of transcription I: general aspects of transcriptional control. Immunity *2*, 311–319.

Ernst, P., and Smale, S. (1995b). Combinatorial regulation of transcription II: the immunoglobulin m heavy chain gene. Immunity *2*, 427–438.

Georgopoulos, K. (1997). Transcription factors required for lymphoid lineage commitment. Curr. Opin. Immunol. *9*, 222–227.

Glimcher, L. H., and Singh, H. (1999). Transcription factors in lymphocyte development. Cell *96*, 13–23.

Henderson, A., and Calame, K. (1998). Transcriptional regulation during B cell development. Annu. Rev. Immunol. *16*, 163–200.

Michaelson, J. S., Singh, M., and Birshtein, B. K. (1996). B cell lineage-specific activator protein (BSAP) a player at multiple stages of B cell development. J. Immunol. *156*, 2349–2351.

Morrison, S. J., Shah, N. M., and Anderson, D. (1997). Regulatory mechanisms in stem cell biology. Cell *88*, 297–298.

Nelsen, B., and Sen, R. (1992). Regulation of immunoglobulin gene transcription. Int. Rev. Cytol. *133*, 121–149.

Orkin, S. H. (1996). Development of the hematopoietic system. Curr. Opin. Genet. Dev. *6*, 597–602.

Reya, T., and Grosschedl, R. (1998). Transcriptional regulation of B cell differentiation. Curr. Opin. Immunol. *10*, 158–165.

Staudt, L. M., and Lenardo, M. J. (1991). Immunoglobulin gene transcription. Annu. Rev. Immunol. *9*, 373–398.

Weissman, I. L. (2000). Stem cells: units of development, units of regeneration, and units in evolution. Cell *100*, 157–168.

Some Relevant References

Akasaka, T., Tsuji, K., Kawahira, H., Kanno, M., Harigaya, K., Hu, L., Ebihara, Y., Nakahata, T., Tetsu, O., Taniguchi, M., and Koseki, H. (1997). The role of mel-18, a mammalian Polycomb group gene, during IL-7 dependent proliferation of lymphocyte precursors. Immunity *7*, 135–146.

Allman, D., Li, J., and Hardy, R. R. (1999). Commitment to the B lymphoid lineage ocurs before D_H–J_H recombination. J. Exp. Med. *189*, 735–740.

Bain, G., Robanus Maandag, E. C., te Reile, H. P., Feeney, A. J., Sheehy, A., Schlissel, M., Shinton, S. A., Hardy, R. R., and Murre, C. (1997). Both E12 and E47 allow commitment to the B lineage. Immunity *6*, 145–154.

Bain, G., Robanus-Maandag, E., te Riele, H., Izon, D., Amsen, D., Kruisbeek, A., Krop, I., Schlissel, M., Feeny, A. J., van Roon, M., van der Valk, M., Berns, A., and Murre, C. (1994). E2A proteins are required for proper B cell development and initiation of immunoglobulin gene rearrangements. Cell *79*, 885–892.

Brown, K. E., Guest, S. S., Smale, S. T., Hahm, K., Merkenschlager, M., and Fisher, A. G. (1997). Association of transcriptionally silent genes with Ikaros complexes at centromeric heterochromatin. Cell *91*, 845–854.

Corcoran, A. E., Riddell, A., Kroshoop, D., and Venkitaraman, A. R. (1998). Impaired immunoglobulin gene rearrangement in mice lacking the IL-7 receptor. Nature *391*, 904–907.

Cumano, A., Dieterlen-Lievre, F., and Godin, I. (1996). Lymphoid potential, probed before circulation in mouse, is restricted to caudal intraembryonic splanchnopleura. Cell *86*, 907–916.

Georgopoulos, K., Bigby, M., Wang, J.-H., Molnar, A., Wu, P., Winandy, S., and Sharpe, A. (1994). The Ikaros gene is required for the development of all lymphoid lineages. Cell *79*, 143–156.

Gstaiger, M., Georgiev, O., van Leuween, H., van der Vliet, P., and Schaffner, W. (1996). The B cell coactivator Bob-1 shows DNA sequence-dependent complex formation with Oct-1/Oct-2 factors, leading to differential promoter activation. EMBO J. *15*, 2781–2790.

Hahm, K., Cobb, B. S., McCarty, A. S., Brown, K. E., Klug, C. A., Lee, R., Akashi, K., Weissman, I., Fisher, A. G., and Smale, S. T. (1998). Helios, a T cell restricted Ikaros family member that quantitatively associates with Ikaros at centromeric heterochromatin. Genes. Dev. *12*, 782–796.

Heemskerk, M. H. M., Blom, B., Nolan, G., Stegmann, A. P. A., Bakker, A. Q., Weijer, K., Res, P. C. M., and Spits, H. (1997). Inhibition of T cell and promotion of natural killer cell development by the dominant negative helix loop helix factor Id3. J. Exp. Med. *186*, 1597–1602.

Horwitz, B. H., Scott, M. L., Cherry, S. R., Bronson, R. T., and Baltimore, D. (1997). Failure of lymphopoiesis after adoptive transfer of NFκB deficient fetal liver cells. Immunity *6*, 765–772.

Kee, B., and Paige, C. (1996). *In vitro* tracking of IL-7 responsiveness and gene responsiveness during commitment of bipotent B cell/macrophage progenitors. Curr. Biol. *6*, 1159–1169.

Kelley, C. M., Ikeda, T., Koipally, J., Avitahl, N., Wu, L., Georgopoulos, K., and Morgan, B. A. (1998). Helios, a novel dimerization partner of Ikaros expressed in the earliest hematopoietic progenitors. Curr. Biol. *23*, 508–515.

Kim, J., Sif, S., Jones, B., Jackson, A., Koipally, J., Heller, E., Winandy, S., Viel, A., Sawyer, A., Ikeda, T., Kingston, R., and Georgopoulos, K. (1999). Ikaros DNA-binding proteins direct formation of chromatin remodeling complexes in lymphocytes. Immunity *10*, 345–355.

Koipally, J., Renold, A., Kim, J., and Georgopoulos, K. (1999). Repression by Ikaros and Aiolos is mediated through histone deacetylase complexes. EMBO J. *18*, 3090–3100.

Kondo, M., Weissman, I. L., and Akashi, K. (1997). Identification of clonogenic common lymphoid progenitors in bone marrow. Cell *91*, 661–672.

Li, Y. S., Wasserman, R., Hayakawa, K., and Hardy, R. R. (1996). Identification of the earliest B lineage stage in mouse bone-marrow. Immunity *5*, 527–535.

Lin, H., and Grosschedl, R. (1995). Failure of B cell differentiation in mice lacking the transcription factor EBF. Nature *376*, 263–267.

Mebius, R. E., Rennert, P., and Weissman, I. L. (1997). Developing lymph nodes collect $CD4^+CD3^-$ $LT\beta^+$ cells that can differentiate to APC, NK cells, and follicular cells but not T or B cells. Immunity *7*, 493–504.

Medvinsky, A., and Dzierzak, E. (1996). Definitive hematopoiesis is autonomously initiated by the AGM region. Cell *86*, 897–906.

Morgan, B., Sun, L., Avitahl, N., Andrikopoulos, K., Ikeda, T., Gonzales, E., Wu, P., Neben, S., and Georgopoulos, K. (1997). Aiolos, a lymphoid restricted transcription factor that interacts with Ikaros to regulate lymphocyte differentiation. EMBO J. *16*, 2004–2013.

Mucenski, M. L., McLain, K., Kier, A. B., Swerdlow, S. H., Schreiner, C. M., Miller, T. A., Pietryga, D. W., Scott, W. J. J., and Potter, S. S. (1991). A functional c-myb gene is required for normal murine fetal hepatic hematopoiesis. Cell *65*, 677–689.

Mukouyama, Y.-S., Hara, T., Xu, M.-J., Tamura, K., Donovan, P. J., Kim, H.-J., Kogo, H., Tsuji, K., Nakahata, T., and Miyajima, A. (1998). *In vitro* expansion of murine multipotential hematopoietic progenitors from the embryonic aorta-gonad-mesonephros region. Immunity *8*, 105–114.

Nutt, S. L., Heavey, B., Rolink, A. G., and Busslinger, M. (1999). Commitment to the B lymphocyte lineage depends on the transcription factor Pax5. Nature *401*, 556–562.

Nutt, S. L., Urbanek, P., Rolink, A., and Busslinger, M. (1997). Essential functions of Pax-5 (BSAP) in pro-B cell development: difference between fetal and adult B lymphopoiesis and reduced V to DJ recombination at the IgH locus. Genes Dev. *11*, 476–491.

Okuda, T., Deursen, J. V., Hiebert, S. W., Grosveld, G., and Downing, J. R. (1996). AML1, the target of multiple chromosomal translocations in human leukemia, is essential for normal fetal liver hematopoiesis. Cell *84*, 321–330.

Osawa, M., Hanada, K.-I., Hamada, H., and Nakauchi, H. (1996). Long-term lymphohematopoietic reconstitution by a single CD34-low/negative hematopoietic stem cell. Science *273*, 242–245.

Porcher, C., Swat, W., Rockwell, K., Fujiwara, Y., Alt, F. W., and Orkin, S. H. (1996). The T cell leukemia oncoprotein SCL/tal-1 is essential for development of all hematopoietic lineages. Cell *86*, 47–57.

Pui, J. C., Allman, D., Xu, L., DeRocco, S., Karnell, F. G., Bakkour, S., Lee, J. Y., Kadesch, T., Hardy, R. R., Aster, J. C., and Pear, W. S. (1999). Notch 1 expression in early lymphopoiesis influences B versus T lineage determination. Immunity *11*, 299–308.

Radtke, F., Wilson, A., Stark, G., Bauer, M., van Meerwijk, J., MacDonald, H. R., and Aguet, M. (1999). Deficient T cell fate specification in mice with an induced inactivation of Notch 1. Immunity *10*, 547–558.

Rolink, A. G., Nutt, S. L., Melchers, F., and Busslinger, M. (1999). Long term *in vivo* reconstitution of T cell development by Pax5 deficient B-cell progenitors. Nature *401*, 603–606.

Roy, A. L., Du, H., Gregor, P. D., Novina, C. D., Martinez, E., and Roeder, R. G. (1997). Cloning of an Inr- and E-box binding protein, TFII-I, that interacts physically and functionally with USF1. EMBO J. *16*, 7091–7104.

Scott, E. S., Fisher, R. C., Olson, M. C., Kehrli, E. W., Simon, M. C., and Singh, H. (1997). PU.1 functions in a cell autonomous manner to control the differentiation of multipotential lymphoid-myeloid progenitors. Immunity *6*, 437–447.

Scott, E. W., Simon, M. C., Anastasi, J., and Singh, H. (1994). Requirement of transcription factor PU.1 in the development of multiplehematopoietic lineages. Science *265*, 1573–1577.

Sigvardsson, M., O'Riordan, M., and Grosschedl, R. (1997). EBF and E47 collaborate to induce expression of the endogenous immunoglobulin surrogate light chain genes. Immunity *7*, 25–36.

Ting, C. N., Olson, M. C., Barton, K. P., and Leiden, J. M. (1996). Transcription factor GATA-3 is required for development of the T cell lineage. Nature *384*, 474–478.

Thevenin, C., Nutt, S. L., and Busslinger, M. (1998). Early function of Pax-5 (BSAP) before the pre-B cell receptor stage of B lymphopoiesis. J. Exp. Med. *188*, 735–744.

Tomita, K., Hattori, M., Nakamura, E., Nakanishi, S., Minato, N., and Kageyama, R. (1999). The bHLH gene hes1 is essential for expansion or early T cell precursors. Genes Dev. *13*, 1203–1210.

Tsai, F. Y., Keller, G., Kuo, F. C., Weiss, M., Chen, J., Rosenblatt, M., Alt, F. W., and Orkin, S. H. (1994). An early hematopoietic defect in mice lacking the transcription factor GATA-2. Nature *371*, 221–226.

Turpen, J. B., Kelley, C. M., Mead, P. E., and Zon, L. I. (1997). Bipotential promitive-definitive hematopoietic progenitors in the vertebrate embryo. Immunity *7*, 325–334.

Urbanek, P., Wang, Z.-Q., Fetka, I., Wagner, E. F., and Busslinger, M. (1994). Complete block of B cell differentiation and altered patterning of the posterior midbrain in mice lacking Pax5/BSAP. Cell *79*, 901–912.

Wang, J. H., Nichogiannopoulou, A., Wu, L., Sun, L., Sharpe, A. H., Bigby, M., and Georgopoulos, K. (1996a). Selective defects in the development of the fetal adult lymphoid system in mice with an Ikaros null mutation. Immunity *5*, 537–549.

Wang, Q., Stacy, T., Binder, M., Marin-Padilla, M., Sharpe, A. H., and Speck, N. A. (1996b). Disruption of the CBFA2 gene causes necrosis and hemorrhaging in the central nervous system and blocks definitive hematopoiesis. Proc. Natl. Acad. Sci. USA *93*, 3444–3449.

Wang, Q., Stacy, T., Miller, J. D., Lewis, A. F., Gu, T. L., Huang, X., Bushweller, J. H., Bories, J. C., Alt, F. W., Ryan, G., Liu, P. P., Wynshaw-Boris, A., Binder, M., Marin-Padilla, M., Sharpe, A. H., and Speck, N. A. (1996c). The CBFbeta subunit is essential for CBFalpaha2 (AML1) function *in vivo*. Cell *87*, 697–708.

Wu, L., Li, C.-L., and Shortman, K. (1996). Thymic dendritic cell precursors: Relationship to the T lymphocyte lineage and phenotype of the dendritic cell progeny. J. Exp. Med. *184*, 903–911.

Yoder, M. C., Hiatt, K., Dutt, P., Mukherjee, P., Bodine, D. M., and Orlic, D. (1997). Characterization of definitive lymphohematopoietic stem cells in the day 9 murine yolk sac. Immunity *7*, 335–344.

Yokota, Y., Mansouri, A., Mori, S., Sugawara, S., Adachi, S., Nishikawa, S.-I., and Gruss, P. (1999). Development of peripheral lymphoid organs and natural killer cells depends on the helix-loop-helix inhibitor Id2. Nature *397*, 702–706.

Zhuang, Y., Soriano, P., and Weintraub, H. (1994). The helix-loop-helix gene E2A is required for B cell formation. Cell *79*, 875–884.

Zwilling, S., Dieckmann, A., Pfistrerer, P., Angel, P., and Wirth, T. (1997). Inducible expression and phosphorylation of coaxtivator BOB.1/OBF.1 in T cells. Science *277*, 221–225.

3

V(D)J Recombination and the Generation of Diversity

Lymphoid organs, unlike most other tissues, resemble diverse societies made up of innumerable distinct individuals. The antigen-binding repertoires of both B and T cells are enormous, and during development, each lymphocyte emerges with its own unique receptor. The tremendous diversity of the immune system is generated in developing B and T cells primarily by the process of gene rearrangement. The rearrangement of immunoglobulin (Ig) and T cell-receptor (TCR) genes is initially activated in pro-B and pro-T cells. Because it involves the joining of *V* (variable) region genes, *D* (diversity) segments, and *J* (junctional) gene segments, this process is known as V(D)J recombination (Figure 3.1). The rearrangement of distinct combinations of *V*, *D*, or *J* segments in different developing lymphocyte clones contributes to the process of immune-receptor diversification. A given Ig heavy chain protein may pair in different cells with one of many distinct light chains, and a given T cell-receptor (TCR) β chain may pair with numerous different TCR α chain proteins in different T cell clones. This process of heterodimeric pairing with different randomly selected partners also contributes to the process of "combinatorial" diversity. V(D)J recombination is intrinsically designed to be somewhat inaccurate—the addition or removal of bases during the joining process contributes to the generation of "junctional" diversity. As a result, rearrangement must be intimately tied to cell selection processes. Cells that make productive rearrangements are selected to survive and differentiate further, while cells that fail to do so die by default. The selection of cells that have made in-frame rearrangements at the *Ig* heavy chain locus or at the *TCR* β chain locus is driven by a signaling process that also shuts off rearrangement at the other allele. This latter phenomenon is known as allelic exclusion.

V(D)J recombination is activated more than once during lymphocyte development. This process generates *Ig* and *TCR* gene rearrangements in progenitor B and T cells. At the immature B stage, cells that encounter multivalent self-antigen are given an opportunity to reform themselves or to die. Signals delivered by the B cell receptor may reactivate the process of V(D)J recombination; self-reactive B cells that successfully make new receptors by a process known as "receptor editing" may evade death and become useful citizens. V(D)J recombination might also be reactivated in a small number of B and T cells in the periphery during an immune response; whether reactivation of *Rag* genes actually occurs in the periphery remains a subject of intense debate, and this issue will be discussed in Chapter 8.

During an immune response to T-dependent protein antigens, B cells generate further diversity in Ig variable regions by a unique process known as somatic mutation. Somatic mutation is coupled to the generation of a number of progeny of the initially activated cell, each of which may have acquired specific point substitutions in the V regions of their rearranged *Ig* genes. As a result, a number of novel B cells are generated, some of which may now have acquired receptors that can bind more tightly to the original antigen, while others may actually no longer be able to bind to antigen. Somatically mutated B cells must be stimulated by antigen to survive, and cells that can no longer see the original antigen might once again be induced to rearrange their *Ig* genes by a process akin to receptor editing. Somatic mutation represents a process of repertoire diversification that is mechanistically still poorly understood. Receptor editing, on the other hand, involves either the continued activity or the reactivation of the cellular machinery that is responsible for V(D)J recombination, a process that can now be fairly clearly described in molecular terms. The processes of receptor editing and somatic mutation will be discussed in more detail in Chapters 7 and 8. Receptor diversification in anergic T cells will be considered in Chapter 10.

In this chapter, we will first discuss the molecular mechanisms involved in V(D)J recombination. The less well understood issues of how selected loci are specifically rearranged in particular lineages and at distinct time points in differentiation will also be considered. The process of allelic exclusion will be discussed briefly in this chapter and will be revisited in the context of B and T cell differentiation in Chapters 7 and 9.

Molecular Events During V(D)J Recombination

The tremendous diversity of the repertoires of B cells, and of both αβ and γδ T cells, is generated by V(D)J recombination. In the case of antibody genes, the heavy chain locus and the light chain loci are on different chromosomes. In the heavy chain locus, a large number of *V* (variable region) genes are tandemly arranged upstream of about 20 to 30 *D* (diversity) segments. A few *J* segments are separated by the J-C intron from heavy chain constant region (C) exons. Both combinatorial and junctional mechanisms contribute to diversity. D segments are found in the Ig heavy chain locus and the *TCRβ* and *δ* loci. Two V(D)J recombinational events are involved in the joining process at these loci. D segments are not found in the Ig light

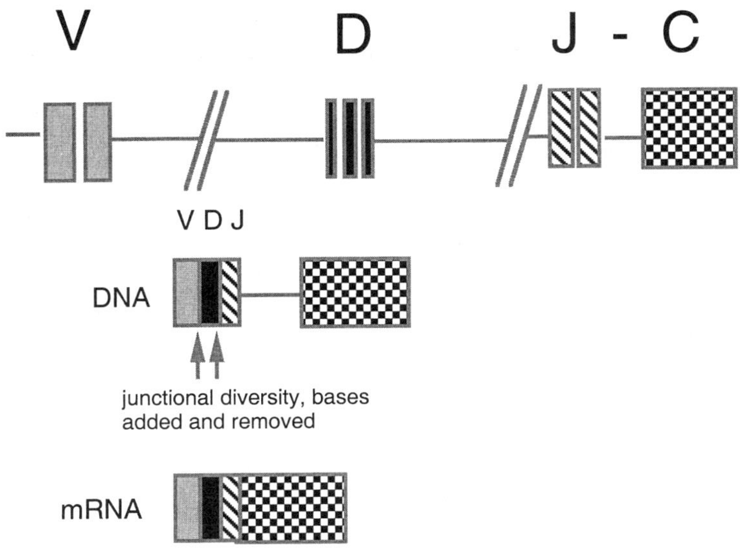

FIGURE 3.1. Schematic view of the organization of a rearranging immune receptor locus. D segments are present in the IgH chain and the *TCR β* and *δ* loci. In the *IgH* and *TCR β* loci, a D to J rearrangement occurs first. This is followed by a V to DJ recombinational event. In the case of the *TCR δ* locus, V to D rearrangement may be followed by a VD to J joining event. *Immunoglobulin light chain* loci and the *TCR α* and *γ* loci do not have D segments.

chain κ and λ loci, or in the *TCR α* and *TCR γ* loci. These individual loci will be considered in Chapters 7 and 9. A generic view of an immune receptor locus is provided in Figure 3.1.

Rearrangement is an ordered process. The first rearrangement that occurs in a cell committed to the B lineage, for instance, is a D_H to J_H rearrangement at the heavy chain locus. This event is followed by V_H to DJ_H recombinational event and sequential rearrangement at the κ and λ light chain loci.

All *Ig* and *TCR* gene segments that undergo rearrangement are flanked by recombination signal sequences (RSSs; Figure 3.2). Immediately adjacent to the coding region is a conserved palindromic heptamer (CACTGTG or its reverse complement, CACAGTG) followed by a spacer and then a relatively conserved nonamer (GGTTTTTGT or its reverse complement, ACAAAAACC). The spacer in an RSS is either approximately 12bp in length or approximately 23bp in length. RSSs with 12bp spacers recombine only with RSSs with 23bp spacers (the 12/23 spacer rule). This rule serves to ensure that V segments that may have 12-bp spacer-containing recombination signals adjacent to the 3′ end do not combine with other V segments, for instance, but do so with D segments that may have 23-bp

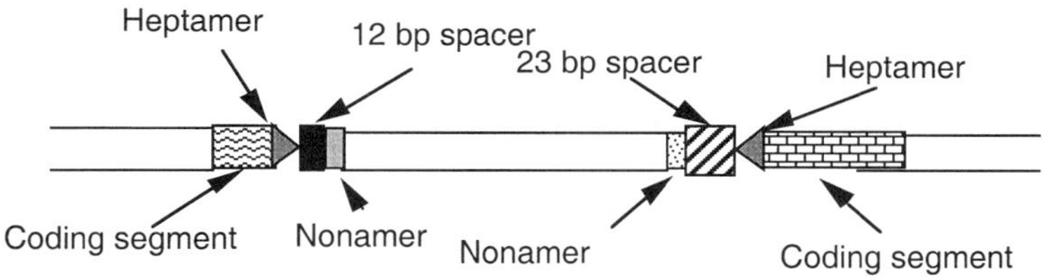

FIGURE 3.2. Recombination signal sequences are made up of a heptamer that is immediately adjacent to the coding segment (V, D, or J), and that is separated by a 12-bp or a 23-bp spacer from a nonamer sequence. Twelve base pair spacer-containing signals recombine with 23-bp spacer-containing signals.

spacer-containing RSSs at their 5′ ends (and therefore 12-bp spacer-containing RSSs at their 3′ ends).

Once an Ig or TCR locus has been made accessible, the entire process of V(D)J recombination can be divided into three steps (Figure 3.3):

1. Synapsis (bringing together the segments on the chromosome that need to be joined) and cleavage of DNA to generate a double-stranded break
2. Processing of the cut ends—primarily to generate greater diversity
3. Joining the processed ends

The division of the process into these three steps is somewhat artificial. Molecules such as RAG-1 and RAG-2, which are involved in the cutting step, may also be required for joining events.

Synapsis and Generation of a Double Stranded Break

The RAG-1 and RAG-2 proteins recognize RSSs and specifically cleave DNA between the heptamer and the coding sequences generating flush ends. RAG-1 and RAG-2 associate with one another and are both needed for recombination. In order to initiate V(D)J recombination, two selected RSSs adjacent to segments participating in a recombinational event must come together forming a large chromosomal loop between them. Synapse formation may require RAG-1/RAG-2 dimers associated with selected RSSs to come together, potentially in a tetrameric manner as shown in Figure 3.3. Apart from RAG-1 and RAG-2, high-mobility group (HMG) proteins may be involved in the formation of the initial synapse between two RSSs. The directional orientation of individual RSSs involved in a recombination event on a given chromosome determines the nature of the looping event that brings the RSSs together. If the RSSs are organized in an opposite orientation to one another (Figure 3.4), recombination will involve deletion of the intervening chromosomal segment. If the RSSs of

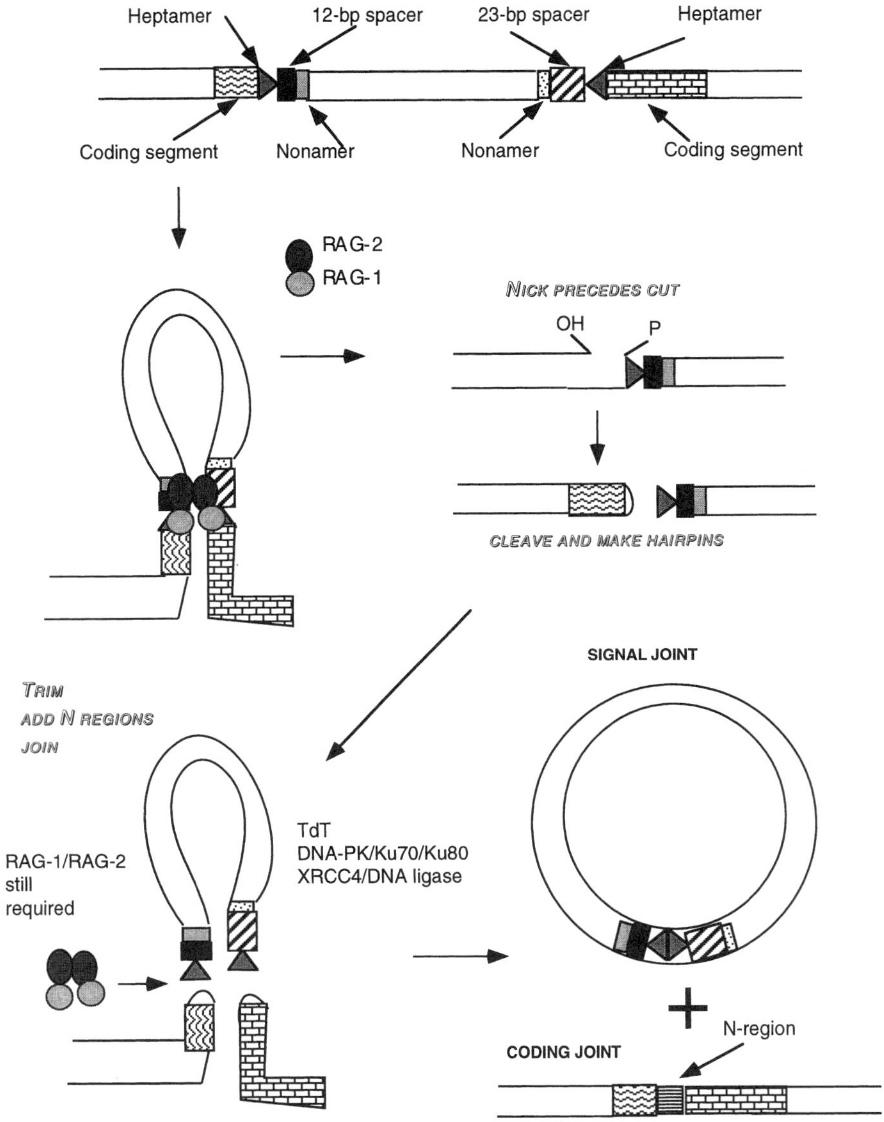

FIGURE 3.3. Current view of role of RAG-1 and RAG-2 in V(D)J recombination. DNA is cut at the heptamer-coding region boundary by RAG-1 and RAG-2. Coding ends form hairpins, while signal joints (on right) are formed between flush-cut heptamer ends. The hairpins are resolved, bases are added or removed at these ends, and these altered coding ends are then ligated together. The double-strand break repair process for coding ends requires DNA-PKc, Ku70, and Ku80, while signal joints do not require DNA-PKc but do require the Ku proteins. DNA ligase IV (in combination with XRCC4) is responsible for mediating the ligation step.

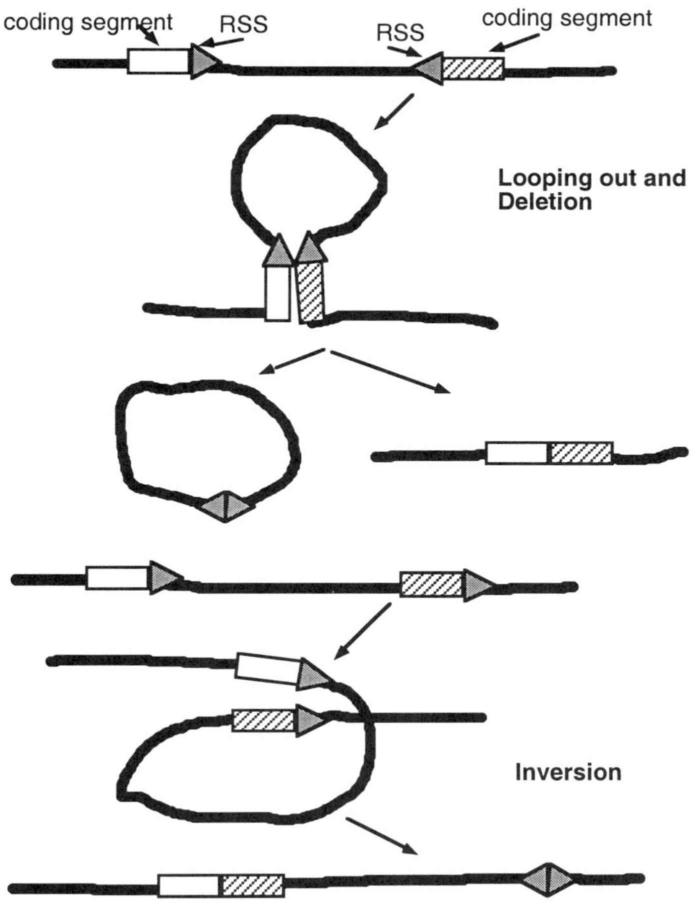

FIGURE 3.4. The relative orientation of two RSSs on a given chromosome determines whether V(D)J rearrangement will involve deletion of a chromosomal circle containing the signal joint or inversion and retention of the signal joint on the recombined chromosome.

interest are oriented in the same direction, recombination will involve the formation of an inverted structure, and there is no requirement for deletion.

RAG-1 and RAG-2 are nuclear proteins. RAG-1 dimers are involved in the actual cleavage of DNA; RAG-2 must be associated with the RAG-1 endonuclease for the latter to be functional. RAG-1 contains a homeodomain, has homology to the bacterial Hin recombinase, and participates in the binding to nonamer sequences prior to cleavage. The N-terminal region of RAG-1 is conserved through evolution and contains clusters of basic amino acids that interact with the nuclear importin, hSRP, and are required for the nuclear localization of RAG-1. There are at least seven cysteine-histidine motifs in this region including a zinc-binding RING finger (the Zn-RING motif) and another motif known as the ZnA motif. The Zn-RING and the ZnA motifs may participate in homodimerization. Three of the remaining cysteine-histidine motifs in this N-terminal region constitute a novel functional category of cysteine-rich motifs. These structures include as a consensus the sequence D/EXnHXnCXXC. They most likely serve to bind to Zn^{2+} and therefore serve a structural function. These structures closely resemble C2H2 motifs found in the N-termini of retroviral integrases and like those motifs probably participate in the generation of higher order complexes *in vivo*.

RAG-2 has a β-propeller like structure that probably serves to connect RAG-1/RAG-2 complexes to other proteins, including those required for accessibility during V(D)J recombination. The C-terminal 89 amino acids of RAG-2 are not required for the catalytic activity of RAG-1/RAG-2 complexes, but participate in the cell-cycle-linked degradation of RAG-2 (discussed below). The C-terminal fourth of RAG-2 contains a plant homeodomain finger-like motif that might contribute to the interaction of RAG-2 with chromatin proteins. This portion of RAG-2 is required in the B lineage for V to DJ_H recombination, although it appears to be dispensable for D to J_H rearrangement. It may therefore be important at the level of accessibility.

It was suggested a few years ago that the RAGs might have been derived evolutionarily from a transposable element. The shared motifs between retroviral integrases and the N-terminus of RAG-1 are striking when viewed in this context. In *in vitro* recombination experiments, it has been demonstrated that the RAG1 and RAG2 proteins can still catalyse a full transposition reaction. It is extremely likely that RAG-1 and RAG-2 were once components of a transposable element that was inserted into an ancestral antigen receptor gene about 450 million years ago around the time of the evolutionary divergence of jawless and jawed vertebrates.

It has become clear that RAG-1 and RAG-2 alone are not only sufficient to cleave DNA at heptamer-coding region junctions but that the RAG-1/RAG-2 complex is required for subsequent events during V(D)J recombination possibly because it helps to hold the coding and the signal junctions together in a synaptic complex, while other gene products complete the process of V(D)J recombination.

RAG-1 binds to nonamer sequences in RSSs and probably recruits RAG-2 to these sites. The RAG-1/RAG-2 complex may bring together RSSs containing 12-bp and 23-bp spacers, and only when this has been achieved, the complex initiates the cutting process. At the heptamer-coding sequence junction, the RAG-1/RAG-2 nuclease first makes a nick immediately 5′ to the RSS and then uses the free 3′ hydroxyl group as a nucleophile to attack the phosphodiester bond on the other strand, generating a closed hairpin structure at the coding end. A flush heptamer signal end results from the double-stranded break.

Processing of Broken DNA Ends

Blunt signal ends are joined without any processing. In contrast, the coding end hairpins are opened up, and nucleotides may be added to or removed from these ends. This results in greater immune diversity (and a high frequency of out-of frame rearrangements). Resolution of the hairpin structures might be mediated by the RAG proteins themselves or by MRE11, a protein involved in double-strand break repair that has both 3′ to 5′ exonuclease activity as well as endonuclease activity. The latter activity may be required for hairpin resolution. Opening of hairpins may also require DNA-PK, an enzyme that is required in the joining process and that is discussed in the next section. Severe combined immunodeficient (*scid*) mice carrying a mutation in DNA-PKcs, the catalytic subunit of DNA-dependent protein kinase, or mice mutant in Ku80 (a noncatalytic subunit of DNA-PK) cannot resolve hairpins efficiently. DNA-PK might therefore promote the association of the yet to be identified hairpin nuclease with closed coding ends. Very little is understood in molecular terms as to how nucleotide loss occurs at coding junctions. Presumably endo- or exo-nucleases are involved that recognize only coding/hairpin ends. It is possible that nucleotide loss occurs directly as a result of the joining mechanism. Some deletions would be expected when recombination is directed by short-sequence homologies.

Template-dependent and template-independent mechanisms exist for the addition of nucleotides to coding ends. Terminal deoxynucleotidyl transferase (TdT) is an enzyme responsible for N region addition. N regions are short stretches of bases, usually GC rich, that are found at junctions and that are added in a template-independent manner. TdT adds stretches of bases to the 3′OH ends of DNA at junctions. Mice null for Ku80 as well as *scid* mice have markedly reduced N regions. DNA-PK may serve to bring TdT in apposition to DNA-coding ends. TdT is synthesized in developing B and T cells in the bone marrow and thymus, but its expression is restricted to progenitors generated from bone marrow-derived stem cells. The absence or relative absence of TdT in lymphoid progenitors derived from fetal liver stem cells is a contributory factor to the restricted diversity of lymphocytes generated early in life. The molecular basis for the enhanced expression of TdT in bone marrow-derived B lineage progenitors (as compared with fetal liver-derived progenitors) is not well understood.

Template-dependent nucleotide addition occurs when hairpins are cleaved eccentrically, resulting in a few additional nucleotides in a single-

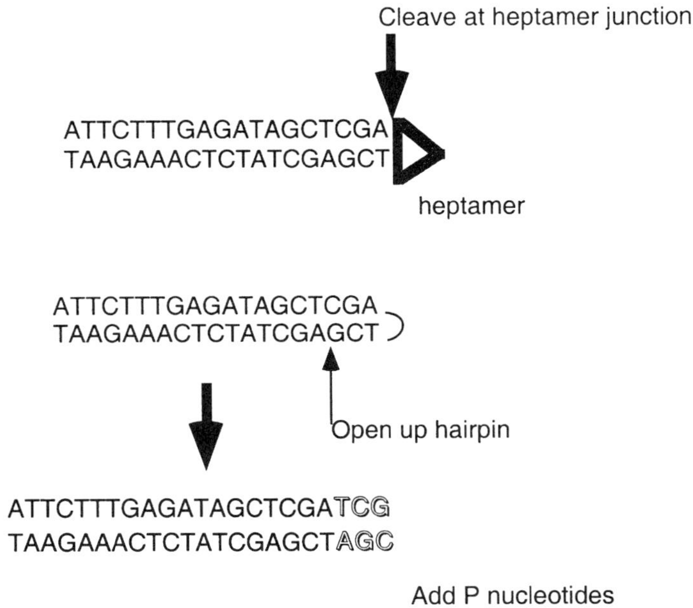

FIGURE 3.5. Templated addition of P nucleotides at a coding end depends on the site of hairpin cleavage during V(D)J recombination.

stranded extension. These extra nucleotides are retained in coding joints and are called P nucleotides (Figure 3.5).

Joining of Broken DNA Ends

Proteins that are involved in double-strand break (DSB) repair are critical for the joining process in V(D)J recombination. These genes include Ku70, Ku80, the catalytic subunit of DNA-dependent protein kinase (DNA-PKcs), XRCC4, and DNA ligase IV. Intriguingly the requirements for joining broken coding ends and broken signal ends, which apparently exist together in a synaptic complex, are distinct.

DNA-PKcs is the *scid* gene product and forms a complex with Ku70 and Ku80 (sometimes referred to as Ku86). The *scid* gene is so named because it is defective in murine severe combined immune deficiency. In the absence of DNA-PKcs, Ig and T cell-receptor rearrangement is extremely inefficient, and *scid* mice basically lack B and T cells. The DNA-PKc/Ku70/Ku80 complex is required for DSB repair. DNA-PKcs is a member of a growing family of protein and dual-protein/lipid kinases that structurally resemble the catalytic subunit of PI-3 kinase (see Chapter 4). Another important member of this family that may be required for the maintenance of chromosomal integrity during V(D)J recombination is the ataxia telangiectasia mutated (ATM) protein, discussed in Chapter 12. The Ku70 and Ku80 proteins are noncatalytic subunits of DNA-PK, and after they bind to DNA ends, they recruit and activate DNA-PKcs. What exactly DNA-PK phosphorylates to mediate its function and how exactly it participates in DNA repair is not understood at present. The possible role of DNA-PK both in promoting access of hairpins to the appropriate endonuclease as well as of resolved coding junctions to TdT has been considered in the previous section. The analysis of *scid* mice and of DNA-PKcs null mice has revealed that while Ku70 and Ku80 are required for the repair of both coding and signal joints, DNA-PKcs is not required for the repair of signal joints. Ku70 and Ku80 can apparently function during DSB repair in a DNA-PK-independent manner that remains poorly understood.

The final joining step is meditated by a DNA ligase. There are three DNA ligases in mammalian cells: DNA ligase I, DNA ligase II/III, and DNA ligase IV. A protein known as XRCC4 is essential for V(D)J recombination and can specifically associate with and activate DNA ligase IV. XRCC4 has no homology to any known protein, but it interacts with two BRCT (Breast Cancer-1 C-Terminal) motifs in a C-terminal noncatalytic domain of DNA ligase IV. BRCT motifs are found in a few proteins involved in DNA replication and repair, including the breast cancer-susceptibility gene BRCA1. It is likely that the final step in V(D)J recombination is mediated by an XRCC4/DNA ligase IV complex *in vivo*, although DNA ligase I may be able to fuse the appropriate ends *in vitro*. Knockouts of the XRCC4 and DNA ligase IV genes have revealed that the XRCC/DNA ligase IV complex is required for DSB repair and for V(D)J recombination. While Ku70, Ku80, and DNA PKcs are required for DSB repair and thus for V(D)J recombination, mice lacking XRCC4 and DNA ligase IV share these defects but also exhibit embryonic lethality. The XRCC4/DNA ligase IV complex is required in the central nervous system (CNS) during early development.

Table 3.1 lists mutations in genes required for V(D)J recombination.

Cell Cycle Control of RAG-1 and RAG-2 Activity

V(D)J recombination occurs primarily in the G0/G1 phase of the cell cycle. It has been suggested that the presence of cleaved DNA during S phase or mitosis could interfere with the faithful transmission of genetic information, and it is therefore important that V(D)J recombination be restricted to the G0/G1 phase. It has been noted that transcripts of the *Rag-1* and *Rag-2* genes are more abundant in slowly dividing or resting cells in which V(D)J recombination is ongoing, as compared with more rapidly dividing B and T cell precursors. It is possible therefore that *Rag-1* and *Rag-2* are

TABLE 3.1. Mutations in Genes Required for V(D)J Recombination

Mutation	Phenotype	Reference
Rag-1$^{-/-}$	No V(D)J recombination. No B and T cell development beyond pro-B and pro-T stages.	Mombaerts et al., 1992
Rag-2$^{-/-}$	No V(D)J recombination. No B and T cell development beyond pro-B and pro-T stages.	Shinkai et al., 1992
Scid/ DNA PKcs$^{-/-}$	Impaired V(D)J recombination; defective DSB repair. Defect in coding joint formation. Impaired B and T cell development.	Bosma et al., 1983 Taccioli et al., 1998 Gao et al., 1998a
Ku70$^{-/-}$	Impaired V(D)J recombination and DSB repair. Both coding and signal joints affected. Block in B cell development. Only partial block in T cell development. Also required for Ig class switching.	Gu et al., 1997 Ouyang et al., 1997 Manis et al., 1998
Ku80$^{-/-}$	Impaired V(D)J recombination and DSB repair. Both coding and signal joints. Block in B and T cell development.	Nussenzweig et al., 1996 Zhu et al., 1996
XRCC-4$^{-/-}$	Impaired V(D)J recombination, defective DSB repair. Block in B and T cell development. Late embryonic lethality due to defect in CNS development.	Gao et al., 1998b
DNA Ligase IV$^{-/-}$	Impaired V(D)J recombination, defective DSB repair. Block in B and T cell development. Late embryonic lethality due to defect in CNS development.	Frank et al., 1998
TdT$^{-/-}$	Lack of N regions in antigen-receptor genes.	Komori et al., 1993 Gilfillan et al., 1993

regulated at the transcriptional level in some poorly understood cell-cycle-dependent manner.

It is likely that the XRCC-4/DNA ligase IV complex is also functionally active only during the G1 phase of the cell cycle. In DSB repair induced in response to ionizing radiation, XRCC-4 is required only during G1. How exactly this activity is constrained to the G1 phase of the cell cycle is unclear, but this requirement is consistent with the thesis that V(D)J recombination occurs specifically in G0/G1.

The RAG-2 protein is degraded in a periodic manner in cycling cells. This process is now fairly well understood in molecular terms and is likely to be the major mechanism by which V(D)J recombination is restricted to the G0/G1 phase of the cell cycle. The C-terminal 89 amino acids of RAG-2 are essential for the cyclical degradation of this protein. Position 490 in the C terminal portion of RAG-2 is a threonine residue adjacent to Pro-491. Cyclin-dependent kinases (CDKs) that drive cell cycle progression are proline-directed serine/threonine kinases, and they are capable of phosphorylating this site *in vitro* and *in vivo*. Mutation of Thr-490 or Pro-491 can abrogate the periodic degradation of RAG-2. During the cell cycle, RAG-2 accumulates in G1. Degradation is initiated during the G1/S transition and persists through mitosis. The phosphorylation by CDKs of Thr 490 presumably serves as a signal for the polyubiquitination and proteasomal degradation of RAG-2. Other residues in the C-terminus of RAG-2 are also essential for the degradation process, but a precise knowledge of why they are required has not been obtained.

While a teleological reason for the G0/G1 specificity of V(D)J recombination has been considered above, it is possible that the necessity for the periodic degradation of RAG-2 is linked primarily to the process of allelic exclusion, although little evidence supports such a view. This process will be considered later in this chapter.

Tissue Specificity and Temporal Specificity of V(D)J Recombination: The Accessibility Model

V(D)J recombination is only observed in lymphocytes. A major level at which this tissue specificity is achieved is by the lymphoid specific expression of the *Rag* genes. *Rag-1* and *Rag-2* are primarily expressed in developing B and T lymphocytes, although they may be reactivated during B cell development. Since the recombination signals and the basic machinery used for V(D)J recombination in B and T cells are identical, how does a B cell know that it should rearrange Ig genes as opposed to T cell-receptor genes? What are the mechanisms by which DJ_H rearrangement in a pro-B cell is targeted before V_H to DJ_H rearrangement? Issues such as these have been conceptually dealt with in a general scheme that is known as the accessibility model. This model suggests that in a developing B cell or in a developing T cell, specific loci are "opened" at defined times during development. Locus accessibility is presumably achieved by the regulated binding of appropriate proteins to selected cis-acting elements in the vicinity of the locus that must be rearranged at a specified time during development. Obviously, these cis-acting elements must be distinct from the RSSs themselves and must be able to direct the opening, at the level of chromatin, of the appropriate segments at the requisite times. A number of very interesting questions remain to be answered. What cis-acting elements regulate V(D)J recombination at the level of accessibility, and what are the proteins that bind to these elements to open up target loci at specific times during development and in specific cells? What are the extracellular signals that functionally turn these proteins "on" and "off" during development to regulate accessibility? Is transcription functionally coupled to rearrangement?

The Temporal Order of V(D)J Recombination Events in B and T Cells and the Generation of Germline Transcripts

During B cell development, D_H to J_H rearrangement at the Ig heavy chain locus is carried out before V_H to DJ_H rearrangement. Rearrangement of light chain genes follows heavy chain gene rearrangement, and rearrangement at the Ig κ locus is generally thought to precede rearrangement at the λ locus. Similarly, in T cells, rearrangement at the T cell-receptor β chain locus first involves $D\beta$ to $J\beta$ joining followed by $V\beta$ to $DJ\beta$ rearrangement. However, for the *TCR* δ chain, rearrangement generally first proceeds from $V\delta$ to $D\delta$ followed by a $VD\delta$ to $J\delta$ event (in some cells the order is $D\delta$ to $J\delta$ followed by $V\delta$ to $DJ\delta$, but this is less frequent). In the T lineage, rearrangement at the TCR β, γ, and δ loci occur early, and *TCR* α rearrangement occurs later. Clearly, locus accessibility must be regulated in a temporally ordered manner in both T and B lineages.

Transcripts have been observed from virtually every rearranging Ig or TCR locus at the time of rearrangement. Certain observations particularly strengthen the likelihood that the correlations observed may be meaningful. The detection of sequential germline $V\gamma$ transcripts corresponding to the waves of $V\gamma$ rearrangement seen during γδ T cell ontogeny supports the notion that transcription and rearrangement may be linked events. Expression of a rearranged β chain transgene or overexpression of the *Lck* gene (a Src-family tyrosine kinase involved in pre-T-receptor and antigen-receptor signaling) have been shown to lead to allelic exclusion at the *TCR* β locus and also to contribute to the marked reduction of germline $V\beta$ transcripts. The observation that germline transcripts can be seen specifically at the time when a particular locus is being rearranged lends support to the view that transcriptional activation is required for locus accessibility. However, it cannot be ruled out that these germline transcripts merely reflect the existence of chromatin in an "open" state in the vicinity of rearranging loci; opening the locus of interest may incidentally contribute to the generation of germline transcripts. It is likely that some of the cis-acting elements used for regulating gene expression are also used to permit the V(D)J recombination machinery to have access to specific loci at precisely defined time points during development.

Cis-Acting Elements and the Regulation of V(D)J Recombination

Not surprisingly, experimental evidence supports a role for cis-acting transcriptionally relevant motifs in the regulation of V(D)J rearrangement. Mutations in chicken λ light chain promoters impair V(D)J recombination.

Inactivation of a transcriptional silencer in the chicken λ locus dramatically increases the frequency of V(D)J recombination. Similarly, mutating the octamer motif in the promoter of an unrearranged transgene containing some portions of a mouse λ light chain reduces the frequency of recombination. While some evidence suggests the importance of promoter elements and silencers in permitting V(D)J recombination in B cells, it is becoming increasingly apparent that both the cell-type specific and temporal regulation of V(D)J recombination is regulated by locus-specific enhancer elements or by LCRs that may incorporate enhancer elements as well as MARs. Sequence specific DNA binding proteins presumably bind in a highly regulated fashion to enhancer elements. This results in the recruitment of proteins with histone acetyltransferase activity and the localized acetylation of histone H3. Chromatin structure is then altered by chromatin remodeling complexes and the RAG1/RAG2 nuclease is permitted to bind and cleave an exposed RSS.

We have explored the nature of enhancers, silencers, matrix attachment regions (MARs), and locus control regions (LCRs) in Chapter 2, using the regulation of *Ig* heavy chain gene expression as a general example. In Chapters 7 to 9, we will consider the roles performed by these regulatory elements in lineage-specification in T and B cells, in class switching, and in somatic mutation. Numerous experiments have been performed to show that enhancers or LCRs not only determine in which lineage an *Ig* or *TCR* gene will be rearranged but also when this might occur.

As discussed in Chapter 2, it has been established that chromatin accessibility can be generated over relatively long distances in the absence of transcription; however, the cis-acting elements that control accessibility tend to be involved in the regulation of gene expression as well. The consensus view from both transgenic and *in vitro* approaches is that chromatin accessibility for V(D)J recombination may be achieved without a direct need for transcription. However, transcriptional enhancers may also be enhancers for recombination. In general, targeted deletion of specific *Ig* and *TCR* enhancers results in both decreased transcription and rearrangement at the locus of interest. In the case of *IgH* gene rearrangement, the intronic LCR (enhancer and flanking MARs) probably regulates accessibility of J_H segments for D_H to J_H rearrangement. A potential mechanism has been suggested to explain the initial preference for D_H to J_H rearrangement at this locus. It is possible that the D_{Q52} enhancer and promoter element that is just upstream of the D_{Q52} segment and that is in close proximity to the J_H segments (see Figure 2.5 in Chapter 2), might be critical in influencing this process. Because this enhancer/promoter is located close to the intronic enhancer, it might facilitate the D_H to J_H rearrangement process. Rearrangements involving any of the upstream *D* segments would lead to the deletion of the D_{Q52}-enhancer region, and this could "free up" the intronic enhancer to interact with upstream elements that regulate accessibility for V_H to DJ_H rearrangement.

The delineation of specific cis-acting elements that regulate accessibility is being undertaken by a number of laboratories. Gene-targeting approaches have been refined to the point where precise answers may now be expected. In early experiments in which specific sequences were replaced by a drug-resistance cassette linked to a specific promoter, the presence of this cassette was found to artefactually interfere with accessibility. Current studies involve cleaner germline excision strategies, and it is expected that a clearer picture regarding specific regulatory elements will emerge in the coming years. While dicussing B and T cell differentiation in Chapters 7 and 9, we will describe Ig light chain loci and the TCR α/δ locus in more detail and examine the role of cis-acting elements during developmental progression. However, note that in the murine κ locus, an element made up of two palindromic motifs, KI and KII, and located upstream of the J_κ cluster has been described. This element—which binds to a factor expressed in pro-B and pre-B cells—is highly homologous to a chicken "antisilencer" motif and has been demonstrated to enhance κ gene rearrangement without being involved in transcription.

Much less has been learned about how specific signals induce selected transacting factors that then contribute to the timed opening of a given locus for rearrangement. Although, as we will discuss below and in Chapters 7 and 9, signaling by the pre-B and pre-T receptors can influence locus accessibility for V(D)J recombination nevertheless very little information on the mechanism by which these receptors influence chromatin structure is presently available. The best evidence for the temporally defined regulation of accessibility by external signals has been obtained from studies examining the role of the interleukin-7 (IL-7) receptor in V(D)J rearrangement and B cell development.

IL-7 Receptor-Derived Signals Contribute to Accessibility for V(D)J Recombination

An elegant demonstration of the regulation of accessibility by extracellular signals has been obtained from studies in mice lacking the α chain of the IL-7 receptor. The IL-7 receptor contributes significantly to B and αβ T cell development and is absolutely required for the generation of γδ T cells. Signals downstream of this cytokine receptor contribute to cell survival and proliferation in developing T and B cells and also influence the process of V(D)J recombination at the level of accessibility.

In the murine *Ig* heavy chain locus. *V* segments are spread over about a megabase of DNA upstream of the D_H segments (Figure 3.6). A subset of V_H genes—the V_H*7183* family—is located close to the D_H region and is made up of about 25 related members. Most V_H segments are located much further upstream, and a large number of them belong to the V_H*J558* family,

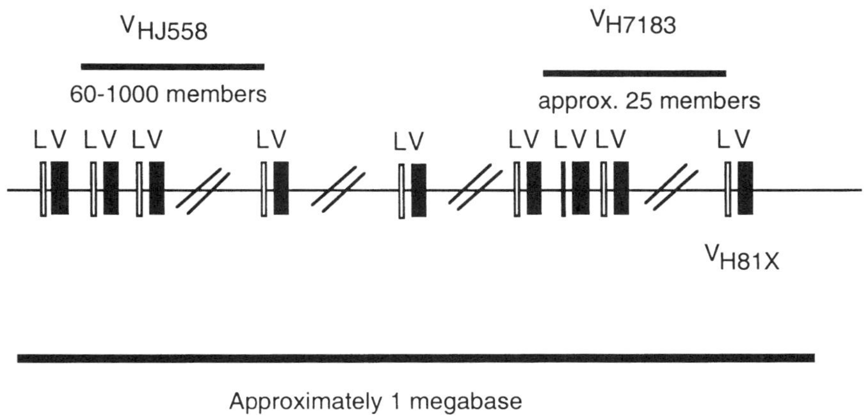

FIGURE 3.6. Distal V_H genes at the murine heavy chain locus include members of the V_{HJ558} family. Accessibility and rearrangement of these genes depend on IL-7 signaling.

which has been estimated to have from 60 to 1000 members (in contrast, the total number of functional V_H genes in humans appears to be much smaller as discussed in Chapter Two). In the absence of the IL-7 receptor α chain, D_H to J_H rearrangement is unimpaired, and proximal V_H segments of the V_{H7183} family can participate in V(D)J rearrangement. However, in these mutant mice, the expression of the *BSAP/Pax-5* transcription factor is markedly diminished. Perhaps as a consequence, chromatin accessibility is compromised, and the distal V_{HJ558} segments fail to rearrange. IL-7 or some other extracellular ligand that can interact with the IL-7 receptor may contribute to enhanced accessibility and the generation of greater immune diversity. It has yet to be demonstrated that *BSAP/Pax-5* sites actually regulate accessibility for V(D)J recombination. It has also not been formally established that the *3′IgH LCR* (which contains multiple *BSAP/Pax-5* sites) has any role to play in the accessibility of distal V_H genes, which may be more than a megabase away. If, indeed, Pax-5 is important for accessibility, additional sites might be present somewhere in the vicinity of distal V_H genes.

BSAP/Pax-5 sites are present both in the *3′IgH LCR* (see Chapter 2) as well as in the *3′Igκ* enhancer. In mice in which the *BSAP/Pax-5* gene has been knocked out, B cell development is arrested at the pro-B stage, and cells fail to make V_H to DJ_H rearrangements, although D_H to J_H rearrangement does occur. An apparent function of the IL-7 receptor α chain is to induce this transcription factor and to enhance the accessibility of distal V_H segments for V(D)J recombination. It has been suggested that the *Pax-5/BSAP* site in the *3′LCR* might be occupied in pre-B cells and that occupancy of this site could be responsible for the alteration of chromatin structure leading to V_H gene accessibility over a megabase away. Further studies must be undertaken to establish whether Pax-5/BSAP participates in the IL-7 receptor-regulated alteration of V_H accessibility. While the *BSAP/Pax-5* site in the IgH 3′LCR might function to increase chromatin accessibility, it appears that the *BSAP/Pax-5* site in the κ locus may function to restrict accessibility at that locus. Displacement of BSAP/Pax-5 by other transcriptional regulators may be required for enhanced germline transcription and rearrangement at the κ locus. An overview of developmental progression in the B lineage and of the rearrangement of the κ and λ light chain genes will be provided in Chapter 7.

IL-7R signaling is absolutely required for the induction of V(D)J recombination in λδ T cells. Signaling via the IL-7 receptor in both B and T cell progenitors can lead both to the activation of PI-3 kinase as well as the activation of STAT5 (see Chapter 4). The activation of PI-3 kinase downstream of IL-7R appears to be critical for the survival of B and T cell progenitors, but STAT5 activation may be crucial for IL-7 mediated alterations in Ig and TCR gene accessibility during B and T cell development.

Allelic Exclusion: Regulation at the Level of Recombinase Availability as well as Locus Accessibility

The specificity of the immune system is in part maintained by preventing individual lymphocytes from expressing more than one antigen receptor. The process by which lymphocytes ensure that they productively rearrange only a single allele at a given receptor locus is called allelic exclusion. Murine diploid genomes contain two *Igκ* loci and four functional *Igλ* loci. The process by which either a κ or a λ light chain, but not both, might be expressed in a given B cell is referred to as isotypic exclusion. Isotypic exclusion will also be discussed in Chapter 7.

Allelic exclusion is best understood for the *IgH* and the *TCR* β genes. At these loci, this process is mediated by signals from the pre-B and pre-T receptors, respectively, and these events will be discussed in more detail in Chapters 7 and 9. At the *Igκ* light chain locus, allelic exclusion is achieved at the level of expression and apparently does not occur in response to receptor-generated signals. Allelic excusion is not observed at the *TCR* α locus, and why this may not matter from a functional standpoint is discussed in Chapter 9.

Since allelic exclusion at the *IgH* and *TCR* β loci is mediated primarily at the level of rearrangement, there are two broad theoretical mechanisms by which this might be achieved: inhibiting recombinase activity or regulating accessibility. In the case of the B lineage, it is clear that κ gene rearrangement is carried out precisely when heavy chain rearrangement must be turned off (to mediate allelic exclusion at the latter locus). Indeed, as we will discuss in Chapter 7, the signals that shut off rearrangement at the heavy chain locus may indirectly contribute to the rearrangement process at the κ locus. Similar conclusions could be drawn about allelic exclusion at the *TCR β* locus, which occurs contemporaneously with the induction of rearrangement at the *TCR α* locus. It is unlikely that allelic exclusion at the Ig heavy chain locus or the *TCR β* locus can be mediated solely by a transient inhibition of *Rag* gene expression or RAG protein levels downstream of preantigen-receptor signaling. Allelic exclusion is probably facilitated for a very brief period by the reduction in *Rag* gene expression and the degradation of the RAG-2 protein in cells that have been induced to proliferate by signals from the pre-B or pre-T receptor. The degradation process, as discussed above, depends on the phosphorylation of Thr-490 by CDKs. This modification presumably targets RAG-2 for polyubiquitination and proteasomal degradation. It has been demonstrated that the degradation of RAG-2 is not essential for allelic exclusion mediated by a transgenic TCR *β* chain. However, transgene-induced allelic exclusion probably involves the generation of an extremely "strong" signal, exceeding the signal that would be provided by preantigen receptors that arise from the process of endogenous gene rearrangement. This CDK-dependent degradation of RAG-2 is probably important for allelic exclusion mediated during normal B and T cell development.

The major mechanism likely to be involved in allelic exclusion at the Ig heavy chain and the TCR *β* loci presumably involves the regulation of accessibility. Signals from preantigen receptors either lead to the inactivation of an accessibility factor for the Ig heavy chain locus in pre-B cells or for the TCR *β* locus in pre-T cells, or they initiate the active suppression of accessibility by some other means.

Allelic exclusion occurs at the κ locus as well as at an increasing number of cytokine loci. Expression of the IL-2 and IL-4 genes has been shown to be allelically excluded, although why this might be biologically useful is unclear. In all these cases, allelic exclusion is achieved at the level of gene expression. One κ locus in small pro-B cells is selectively demethylated, and it is this locus that undergoes rearrangement and that is expressed. The mechanism by which such selective demethylation might occur is unclear. This selection of a single allele for rearrangement is the underlying basis for allelic exclusion at the κ locus. The shutoff of *Rag* gene expression after IgM is expressed on the surface presumably contributes to the maintenance of allelic exclusion at the κ locus.

Receptor Editing/Diversification and *Rag* Gene Re-expression

As mentioned earlier, *Rag* gene expression appears to be reactivated during B cell development and possibly also in anergic T cells. *Rag* gene reexpression has been reported to occur both at the immature B cell stage as well as in germinal centers. These reexpression events will be considered in Chapters 7, 8, and 10. Immature B cells that have encountered a strong cross-linking self-antigen receive a death signal as well as a signal that is permissive for further rearrangement, particularly at the light chain locus. The potential role of an element known as the RS- or κ-deleting element in excising previously rearranged κ light chains will be described in Chapter 7. While it appears that *Rag* gene expression may be induced during the process of receptor editing, whether this involves expression at higher levels of already expressed genes or the dedifferentiation of the entire cell to a slightly less mature form in which *Rag-1* and *Rag-2* are expressed at higher levels remains to be determined.

During somatic mutation cells that acquire mutations that make them self-reactive need to be tolerized. This might occur by the induction of receptor editing, perhaps by causing activated B cells to revert to a pre-B-like phenotype. Alternatively, cells that have lost the ability to respond to selecting antigen may reactivate *Rag* gene expression as a means of generating further immune diversity. There is considerable uncertainty as to whether *Rag* genes are actually reexpressed in the periphery or whether the observed *Rag* expression in secondary lymphoid organs reflects residual *Rag* expression by less mature B cell progenitors. These issues will be discussed in Chapter 8. *Rag* gene reexpression as a means of inducing T cell anergy will also be briefly considered in Chapter 10.

V(D)J Rearrangement in Other Vertebrate Species

There are distinct differences in the organization of immune receptor gene segments and in the processes by which diversity is generated in different vertebrate species. A detailed consideration of these alternative mechanisms in species other than the mouse or human falls outside the scope of this book. In this section, we will briefly consider different diversification mechanisms of B lineage cells in a few selected vertebrates.

In chickens, sheep, and rabbits, the process of V(D)J rearrangement occurs but contributes little to the process of Ig diversity because of the limited availability of V genes for rearangement. Once IgM-bearing B cell precursors of very restricted diversity are generated in the bursa of the chicken or in the ileal Peyer's patches in sheep or rabbits, these B cell precursors colonize gut-associated lymphoid tissues (GALT), where they proliferate extensively. The preimmune repertoire is generated in these precursors by postrearrangement diversification processes. In chicken and rabbits, the diversification process involves gene conversion. In the B cell context, gene conversion refers to a process in which a template somewhere in the genome that resembles (but is distinct from) the *Ig* gene of interest is copied, and the copy is then inserted in the place of the original rearranged template.

In sheep, instead of gene conversion, an apparently constitutive, antigen-independent process of somatic hypermutation generates receptor diversity in IgM-expressing gut-associated B cells. In contrast, somatic mutation in mice and humans is an antigen-driven process, which will be discussed in Chapter 8.

Summary

Immunoglobulin and T cell receptor genes are rearranged during early B and T cell development. The process of V(D)J recombination initially requires the recognition of consensus heptamer-spacer-nonamer recombination signal sequences. Signal sequences containing 12 basepair and 23 basepair spacers are brought together by RAG-1 and RAG-2 to form a synapse. RAG-1 is involved in the actual recognition and cutting of DNA but will only do so when associated with RAG-2. RAG-2 may be involved in multiple protein-protein interaction events and probably physically links the site specific RAG-1 endonuclease to factors that mediate accessibility. Following generation of tandem double strand breaks, coding ends are generally altered by the addition or removal of nucleotides. The process of recombination is completed using components of the double strand break repair machinery, notably DNA-PK, Ku70, and Ku80, and eventually DNA ligase IV complexed with XRCC4.

V(D)J recombination involves the cell-type and developmental stage-specific generation of chromatin accessibility. The mechanisms by which accessibility is generated are poorly understood. During B and T cell development pre-antigen receptor derived signals contribute to the shut-off of rearrangement at unrearranged IgH and TCRβ chain loci. V(D)J recombination may be reactivated in immature B cells where it might contribute to receptor editing; this process may also be reactivated in anergic T cells. Ig gene rearrangement is probably not reactivated in germinal center B cells in the periphery during the course of humoral immune responses.

Selected Reviews

Gellert, M. (1996). A new view of V(D)J recombination. Genes Cells *1*, 269–275.

Gellert, M. (1992). V(D)J recombination gets a break. Trends Genet. *8*, 408–412.

Jeggo, P. A. (1997). DNA-PK: at the crossroad of biochemistry and genetics. Mutation Res. *384*, 1–14.

Oettinger, M. A. (1996). Cutting apart V(D)J recombination. Curr. Opin. Genet. *6*, 141–145.

Schatz, D. G. (1997). V(D)J recombination moves *in vitro*. Semin. Immunol. *9*, 149–159.

Sleckman, B. P., Gorman, J. R., and Alt, F. W. (1996). Accessibility control of antigen-receptor variable region gene assembly: role of cis-acting elements. Annu. Rev. Immunol. *14*, 459–481.

Tonegawa, S. (1983). Somatic generation of antibody diversity. Nature *302*, 575–581.

Thompson, C. B. (1995). New insights into V(D)J recombination and its role in the evolution of the immune system. Immunity *3*, 531–539.

Weaver, D. T. (1995). What to do at an end: DNA double strand break repair. Trends Genet. *11*, 388–392.

Some Relevant References

Agrawal, A., Eastman, Q. M., and Schatz, D. G. (1998). Transposition mediated by RAG1 and RAG2 and its implications for the evolution of the immune system. Nature *394*, 744–751.

Agrawal, A., and Schatz, D. G. (1997). RAG1 and RAG2 form a stable postcleavage synaptic complex with DNA containing signal ends in V(D)J recombination. Cell *89*, 43–53.

Bogue, M. A., Wang, C., Zhu, C., and Roth, D. B. (1997). V(D)J recombination in Ku86-deficient mice: distinct effects on coding, signal, and hybrid joint formation. Immunity *7*, 37–47.

Bosma, G. C., Custer, R. P., and Bosma, M. J. (1983). A severe combined immunodeficiency mutation in the mouse. Nature *301*, 527–530.

Corcoran, A. E., Riddell, A., Kroshoop, D., and Venkitaraman, A. R. (1998). Impaired immunoglobulin gene rearrangement in mice lacking the IL-7 receptor. Nature *391*, 904–907.

Difilippantonio, M. J., McMahan, C. J., Eastman, Q. M., Spanopoulou, E., and Schatz, D. G. (1996). RAG1 mediates signal sequence recognition and recruitment of RAG2 in V(D)J recombination. Cell *87*, 253–262.

Eastman, Q. M., Leu, T. M., and Schatz, D. G. (1996). Initiation of V(D)J recombination *in vitro* obeying the 12/23 rule. Nature *380*, 85–88.

Ferradini, L., Gu, H., De Smet, A., Rajewsky, K., Reynaud, C.-A., and Weill, J.-C. (1996). Rearrangement enhancing element upstream of the mouse immunoglobulin kappa chain J cluster. Science *271*, 1416–1420.

Frank, K. M., Sekiguchi, J. M., Seidl, K. J., Swat, W., Rathbun, G. A., Cheng, H. L., Davidson, L., Kangaloo, L., and Alt, F. W. (1998). Late embryonic lethality and impaired V(D)J recombination in mice lacking DNAligase IV. Nature *396*, 173–177.

Fugmann, S. D., Villey, I. J., Ptaszek, L. M., and Schatz, D. G. (2000). Identification of two catalytic residues in RAG1 that define a single active site within the RAG1/RAG2 protein complex. Molecular Cell *5*, 97–108.

Gao, Y., Chaudhuri, J., Zhu, C., Davidson, L., Weaver, D. T., and Alt, F. W. (1998a). A targeted DNA-PKcs-null mutation reveals DNA-PK-independent functions for Ku in V(D)J recombination. Immunity *9*, 367–376.

Gao, Y., Sun, Y., and Frank, K. M. et al. (1998b). A critical role for DNA end joining proteins in both lymphogenesis and neurogenesis. Cell *95*, 891–902.

Gilfillan, S., Dierich, A., Lemeur, M., Benoist, C., and Mathis, D. (1993). Mice lacking TdT: mature animals with an immature lymphocyte repertoire. Science *261*, 1175–1178.

Gottlieb, T. M., and Jackson, S. P. (1993). The DNA dependent protein kinase: requirement for DNA ends and association with Ku antigen. Cell *72*, 131–142.

Grawunder, U., Leu, T. M., Schatz, D. G., Werner, A., Rolink, A., Melchers, F., and Winkler, T. H. (1995). Down-regulation of RAG1 and RAG2 gene expression in pre-B cells after functional immunoglobulin heavy chain rearrangement. Immunity *3*, 601–608.

Grawunder, U., Wilm, M., Wu, X., Kulesza, P., Wilson, T. E., Mann, M., and Lieber, M. R. (1997). Activity of DNA ligase IV stimulated by complex formation with XRCC4 protein in mammalian cells. Nature *388*, 492–495.

Gu, Y., Seidl, K. J., Rathbun, G. A., Zhu, C., Manis, J. P., van der Stoep, N., Davidson, L., Cheng, H. L., Sekiguchi, J. M., Frank, K., Stanhope-Baker, P., Schlissel, M. S., Roth, D. B., and Alt, F. W. (1997). Growth retardation and leaky SCID phenotype of Ku70 deficient mice. Immunity *7*, 653–665.

Han, S., Zheng, B., Schatz, D. B., Spanopoulou, E., and Kelsoe, G. (1996). Neoteny in lymphocytes: rag 1 and rag 2 expression in germinal center B cells. Science *274*, 2094–2097.

Hiom, K., and Gellert, M. (1997). A stable RAG-1-RAG2-DNA complex that is active in VDJ cleavage. Cell *88*, 65–72.

Hiom, K., Melek, M., and Gellert, M. (1998). DNA transposition by the RAG1 and RAG2 proteins: a possible source of oncogenic translocations. Cell *94*, 463–470.

Hiramatsu, R., Akagi, K., Matsuoka, M., Sakumi, K., Nakamura, H., Kingsbury, L., David, C., Hardy, R. R., Yamamura, K.-I., and Sakano, H. (1995). The 3′ enhancer region determines the B/T specificity and pro-B/pre-B specificity of immunoglobulin VK-JK joining. Cell *83*, 1113–1123.

Komori, T., Okada, A., Stewart, V., and Alt, F. W. (1993). Lack of N regions in antigen receptor variable region genes of TdT-deficient lymphocytes. Science *261*, 1171–1175.

Kwon, J., Imbalzano, A. N., Mathews, A., and Oettinger, M. A. (1998). Accessibility of nucleosomal DNA to V(D)J cleavage is modulated by RSS positioning and HMG1. Mol. Cell *2*, 829–839.

Li, Z., Dordai, D. I., Lee, J., and Desiderio, S. (1996). A conserved degradation signal regulates RAG-2 accumulation during cell division and links VDJ recombination to the cell cycle. Immunity *5*, 575–589.

Li, Z., Otevrel, T., Gao, Y., Cheng, H.-L., Seed, B., Stamato, T. D., Taccioli, G. E., and Alt, F. W. (1995). The XRCC4 gene encodes a novel protein involved in DNA double-strand break repair and VDJ recombination. Cell *83*, 1079–1090.

Lin, W. C., and Desiderio, S. (1993). Regulation of V(D)J recombination activator protein RAG-2 by phosphorylation. Science *260*, 953–959.

Manis, J. P., Gu, Y., Lansford, R., Sonoda, E., Ferrini, R., Davidson, L., Rajewsky, K., and Alt, F. W. (1998). Ku 70 is required for late B cell development and immunoglobulin heavy chain class switching. J. Exp. Med. *187*, 2081–2089.

McBlane, J. F., van Gent, D. C., Ramsden, D. A., Romeo, C., Cuomo, C. A., Gellert, M., and Oettinger, M. A. (1995). Cleavage at VDJ recombination signal requires only RAG1 and RAG2 proteins and occurs in two steps. Cell *83*, 387–395.

McMurry, M. T., and Krangel, M. S. (2000). A role for histone acetylation in the developmental regulation of V(D)J recombination. Science *287*, 495–498.

Melek, M., Gellert, M., and van Gent, D. C. (1998). Rejoining of DNA by the RAG1 and RAG2 proteins. Science *280*, 301–303.

Mombaerts, P., Iacomini, J., Johnson, R. S., Herrup, K., Tonegawa, S., and Papaioannou, V. E. (1992). Rag-1 deficient mice have no mature T and B lymphocytes. Cell *68*, 869–877.

Monroe, R. J., Seidl, K. J., Gaertner, F., Han, S., Chen, F., Sekiguchi, J., Wang, J., Ferrini, R., Davidson, L., Kelsoe, G., and Alt, F. W. (1999a). RAG2:GFP knockin mice reveal novel aspects of RAG2 expression in primary and peripheral lymphoid tissues. Immunity *11*, 201–212.

Monroe, R. J., Sleckman, B. P., Monroe, B. C., Khor, B., Claypool, S., Ferrini, R., Davidson, L., and Alt, F. W. (1999b). Developmental regulation of TCRδ locus accessibility and expression by the TCRδ enhancer. Immunity *10*, 503–514.

Mostoslavsky, R., Singh, N., Kirillov, A., Pelanda, R., Cedar, H., Chess, A., and Bergman, Y. (1998). Kappa chain monoallelic demethylation and the establishment of allelic exclusion. Genes Dev. *12*, 1801–1811.

Nussenzweig, A., Chen, C., daCosta Soares, V., Sanchez, M., Sokol, K., Nussenzweig, M. C., and Li, G. C. (1996). Requirement for Ku80 in growth and immunoglobulin V(D)J recombination. Nature *382*, 581–585.

Oettinger, M. A., Schatz, D. G., Gorka, C., and Baltimore, D. (1990). RAG-1 and RAG-2, adjacent genes that synergistically activate V(D)J recombination. Science *248*, 1517–1523.

Ouyang, H., Nussenzweig, A., Kurimasa, A., da Costa Soares, V., Li, X., Cordon-Cardo, C., Li, W. H., Cheong, N., Nussenzweig, M., Illiakis, G., Chen, D., and Li, G. C. (1997). Ku 70 is required for DNA repair but not for T cell antigen receptor gene recombination *in vivo*. J. Exp. Med. *186*, 921–929.

Paull, T. T., and Gellert, M. (1998). The 3′ to 5′ exonuclease activity of Mre11 facilitates repair of double stranded breaks. Mol. Cell *1*, 969–979.

Ramsden, D. A., Paull, T. T., and Gellert, M. (1997). Cell free V(D)J recombination. Nature *388*, 488–491.

Reynaud, C. A., Garcia, C., Hein, W., and Weill, J.-C. (1995). Hypermutation generating the sheep immunoglobulin repertoire is an antigen-independent process. Cell *80*, 115–126.

Roman, C. A. J., Cherry, S. R., and Baltimore, D. (1997). Complementation of V(D)J recombination deficiency in Rag-1–/– B cells reveals a requirement for novel lements in the N-terminus of RAG-1. Immunity *7*, 13–24.

Schatz, D., Oettinger, M., and Baltimore, D. (1989). The V(D)J recombination activating gene RAG-1. Cell *59*, 1035–1048.

Scherer, D. C., Brockman, J. A., Bendall, H. H., Zhang, G. M., Ballard, D. W., and Oltz, E. M. (1996). Corepression of Rel-A and c-Rel inhibits immunoglobulin κ gene transcription and rearrangement in precursor B lymphocytes. Immunity *5*, 563–574.

Shaffer, A. L., Peng, A., and Schlissel, M. S. (1997). *In vivo* occupancy of the κ light chain enhancers in primary pro- and pre-B cells: A model for κ locus activation. Immunity *6*, 131–143.

Shinkai, Y., Rathbun, G., Lam, K. P., Oltz, E. M., Stewart, V., Mendelsohn, M., Charron, J., Datta, M., Young, F., Stall, A. M., and Alt, F. W. (1992). RAG-2 deficient mice lack mature lymphocytes owing to inability to initiate V(D)J rearrangement. Cell *68*, 855–867.

Spanopoulou, E., Zaitseva, F., Wang, F. H., Santagata, S., Baltimore, D., and Panayotou, G. (1996). The homeodomain region of Rag-1 reveals the parallel mechanisms of bacterial and V(D)J recombination. Cell *87*, 263–276.

Stanhope-Baker, P., Hudson, K. M., Shaffer, A. L., Constantinuescu, A., and Schlissel, M. S. (1996). Cell-type specific chromatin structure determines the targeting of VDJ recombinase activity *in vitro*. Cell *85*, 887–897.

Taccioli, G. E., Amatucci, A. G., Beamish, H. J. D. G., Xiang, X. H., Torres Arzayus, M. I., Priestley, A., Jackson, S. P., Rothstein, A. M., Jeggo, P. A., and Herrera, V. L. M. (1998). Targeted disruption of the catalytic subunit of the DNA-PK gene in mice congers severe combined immunodeficiency and radiosensitivity. Immunity *9*, 355–366.

Taccioli, G. E., Rathbun, G., Oltz, E., Stamato, T., Jeggo, P. A., and Alt, F. W. (1993). Impairment of V(D)J recombination in double strand break repair mutants. Science *260*, 207–210.

van der Stoep, N., Gorman, J. R., and Alt, F. W. (1998). Reevaluation of 3′E kappa function in stage and lineage specific rearrangement and somatic hypermutation. Immunity *8*, 743–750.

van Gent, D. C., Mizuuchi, K., and Gellert, M. (1996). Similarities between initiation of V(D)J recombination and retroviral integration. Science *271*, 1592–1594.

Yaneva, M., Kowalewski, T., and Lieber, M. R. (1997). Interaction of DNA dependent protein kinase with DNA and with Ku: biochemical and atomic-force microscopy studies. EMBO J. *16*, 5098–5112.

Yu, W., Nagaoka, H., Jankovic, M., Misulovin, Z., Suh, H., Rolink, A., Melchers, F., Meffre, E., and Nussenzweig, M. C. (1999). Continued RAG expression in late stages of B cell development and no apparent re-induction after immunization. Nature *400*, 682–686.

Zhu, C., Bogue, M. A., Lim, D. S., Hasty, P., and Roth, D. B. (1996). Ku-86 deficient mice exhibit severe combined immunodeficiency and defective processing of V(D)J recombination intermediates. Cell *86*, 379–389.

4
Pathways Mediating Signal Transduction and Cell Cycle Progression in Lymphocytes

Every event in lymphocyte development is determined by signals. Most of these signals are delivered by surface receptors responding to external cues. Some developmental transitions may be driven by receptors that do not require extracellular ligands but generate constitutive, or "tonic," signals to drive cells past specific checkpoints. Most developmental signals lead either to the post-translational modification of key cytoplasmic proteins, or the regulation of gene expression, or both. The initiation of a signaling pathway might lead to one or more of a number of events. These include the induction of survival factors, the initiation of proliferation, the activation of a differentiation program, or, in certain circumstances, the triggering of cell cycle arrest or of cell death.

In this chapter, we will initially consider general mechanisms and pathways involved in the transmission of signals from the membrane to the nucleus. How these various signals mesh with the machinery for protein synthesis and for cell cycle progression will also be discussed. We will then address issues that relate more specifically to lymphocyte signaling. These include signaling mediated by antigen receptors and coreceptors, the two-signal hypothesis for lymphocyte activation, and the critical role played by costimulatory receptors. Some general features of cytokine signaling will also be described. Signals that lead to the induction of apoptosis will be discussed in Chapter 5. While the role played by signaling molecules in lymphocyte development will be touched upon in this chapter, the developmental consequences of specific defects in signal transduction in individual lineages will be considered in more detail in the chapters on B, T, and NK cell development (Chapters 7 to 11). Table 4.1 provides an outline of the topics discussed in this chapter.

Some specialized themes relating to the integration of signals during lymphocyte activation will be more fully considered in Chapters 8 and 10. In a mature T cell, for instance, ligation of the T cell receptor (TCR) with the concomitant and optimal activation of a costimulatory receptor, the CD28 molecule, leads to T cell proliferation and activation. However, triggering of the TCR without costimulation, or in conjunction with a "negative" cos-

TABLE 4.1. Signal transduction in lymphocytes

A general introduction to signaling pathways
A brief overview of receptors that initiate signal transduction
Kinases and phosphatases involved in cellular signaling
 A minimalist view of how protein kinases function
 Some protein–protein and protein–lipid interaction domains that are frequently found in signaling molecules
 Tyrosine kinases frequently participate in signal initiation
 Nonreceptor tyrosine kinases involved in antigen-receptor signaling
 Protein phosphatases
Cytosolic signaling pathways and downstream nuclear events
 Signals from the cell surface may activate monomeric GTP-binding proteins that in turn regulate mitogen-activated protein kinase (MAPK) pathways
 The ERK MAP kinase
 Cytoskeletal reorganization mediated by the Rho family proteins, Cdc42, Rac, and Rho
 The JNK/SAPK pathway
 Rap-1, a small G protein that may negatively regulate signal transduction
 The phosphatidylinositol 3-OH kinase pathway
 Phospholipase C, phosphoinositide hydrolysis, and calcium signaling
 The protein kinase C family
 The NFκB/Rel family of transcription factors
 The IκB family of inhibitors
 The induced degradation of IκB
 Heterotrimeric G proteins and protein kinase A-mediated signals
 Ceramide signaling
 Notch signaling
 Wnt signaling
 Signals from the cell surface that target protein synthesis
 The regulation of G1 progression by extracellular signals

Antigen receptors, coreceptors, and costimulators
General features of antigen-receptor signaling
 Some aspects of antigen receptor structure of relevance to signaling
 Initiation of signal transduction: crosslinking versus allosteric change
 Coreceptors facilitate antigen-receptor activation: defining coreceptors and costimulators
 Adaptor proteins involved in antigen-receptor signaling
Signal transduction by the B cell receptor
 The CR2/CD19/TAPA-1 coreceptor complex
 Negative regulation of BCR signaling by CD22
 Inhibition of B cell signaling by FcγRIIb
 CD5 as a negative regulator of B-1 cell signaling
 Other mechanisms by which B cell signaling may be attenuated
 CD40 is a member of the TNFR family and a critical costimulator of B cells
 BlyS/BAFF may also provide activating signals to B cells
Signal transduction in T cells
 Signal transduction via the T cell receptor: the CD4 and CD8 coreceptors
 Early events in TCR-mediated activation
 Immunological synapses and the merging of lipid rafts during T cell activation
 The two-signal hypothesis and the need for costimulators
 Signal transduction via CD28 and CTLA-4
 Costimulation as a signal-amplification mechanism
 ICOS is an inducible costimulator on T cells
 SLAM is a costimulatory ligand and receptor that may trigger B and T cell responses
 Other costimulatory receptors

TABLE 4.1. (*Cont.*)

An overview of cytokine signaling
The Jak-STAT pathway for Type I and Type II cytokine receptors
Specificity in Jak-STAT signaling
Inhibitors of cytokine signaling—the SOCS proteins
Signaling by selected Type I cytokine receptors
Receptors that share γc
The IL-7 receptor
The IL-2 receptor
The IL-15 receptor
The IL-4 and IL-13 receptors
The IL-5 receptors and members of the βc family
The IL-6 receptor and the gp130 subfamily
The IL-12 receptor
Type II cytokine receptors: Interferon receptors and IL-10 signaling
Type III cytokine receptors of the TNFR family
Type IV cytokine receptors: IL-1 and IL-18 signaling

timulatory signal, may lead to the induction of anergy or nonresponsiveness. Issues such as these will be introduced in this chapter but considered in more detail in Chapters 8 and 10. Signal transduction pathways are quite complex. To complicate matters for those of us hoping for cut and dried answers, different pathways often talk to one another. In addition, there are often differences in the mechanisms by which a given pathway may be initiated by different receptors and in distinct cell types. In other words, keeping up with signal transduction can be intimidating, especially for the uninitiated. An attempt has been made to present pathways in a linear fashion as far as possible, but please keep in mind that signaling is probably never linear and that most of the schemes presented focus only on a selected view of the overall landscape.

A General Introduction to Signaling Pathways

A Brief Overview of Receptors That Initiate Signal Transduction

Signal transduction is usually initiated at the cell surface by the binding of extracellular ligands to integral membrane proteins that function as receptors (Figure 4.1). A special category of membrane-permeable ligands may initiate signal transduction, not at the cell surface, but by directly interacting with nuclear receptors. While the majority of receptors of direct importance in lymphocyte development are cell surface receptors, nuclear receptors also participate in the process. Some signals may also be initiated and transduced purely from intracellular locations. Examples of this latter

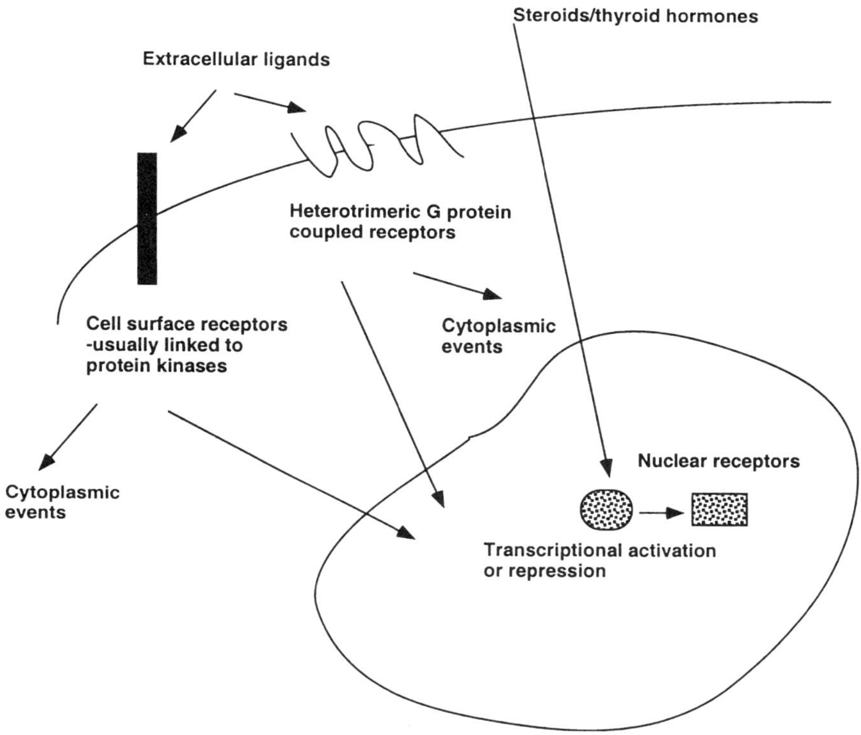

FIGURE 4.1. Signal transduction might be initiated from the cell surface by extracellular ligands or by membrane-permeable ligands that interact with nuclear receptors.

kind of signaling include the checkpoint pathways that are transduced in cells following DNA damage or when chromosomes fail to segregate appropriately. Most of the issues discussed in this chapter reflect the initiation of signals from the cell surface.

Signal transduction pathways generally consist of a number of sequential steps that may culminate in nuclear or non-nuclear events. The existence of multiple steps in a pathway offers the opportunity for multiple branch points and therefore for a single initial molecular event to lead to the functional modification or activation of a large number of targets. The existence of multiple intermediate steps also allows for signal amplification, although this could well be achieved without the need for many steps in the pathway.

Some signaling pathways may be initiated from cell surface receptors that may themselves be protein kinases or that may be physically linked to protein kinases. Most protein kinases in vertebrates either phosphorylate proteins on tyrosine residues (protein tyrosine kinases) or on serine or threonine residues (serine/thronine kinases). In the immune system, a large number of receptors physically associate with cytosolic protein tyrosine kinases, some of which snuggle up to the inner face of the plasma membrane. Crosslinking of receptors that belong to this category (which include antigen receptors and the majority of cytokine receptors) leads to the activation of the linked tyrosine kinase and the subsequent initiation of signaling pathways. A number of receptors of importance in the immune system are not known to be directly linked to tyrosine kinases. However, ligation of these receptors leads to the activation of downstream pathways because the cytoplasmic tails of these receptors are induced to recruit critical signaling effectors that mediate their function without necessarily inducing a specific phosphorylation or dephosphorylation event. Although receptors such as the CD40 molecule, or those for cytokines such as tumor necrosis factor (TNF) and interleukin-1 (IL-1), may not be directly associated with protein kinases; nonetheless, they initiate a number of signaling pathways that depend on the activation of protein kinases.

A distinct category of receptors, often referred to as serpentine receptors, is made up of proteins that contain seven transmembrane domains and that are coupled to heterotrimeric G-proteins (that bind guanosine triphosphate or GTP and guanosine diphosphate or GDP). While receptors of this category exist in virtually every cell, from the viewpoint of lymphocyte development, a particularly important group of serpentine receptors are those that mediate responses to chemokines.

Nuclear receptors are basically transcription factors that contain a ligand-binding site for membrane-permeable mediators such as glucocorticoids or vitamin D. This category of ligand will bind to nuclear receptors and induce a conformational change, which then permits these proteins to function as specific transcriptional activators or repressors. Glucocorticoids may participate in T cell development, as will be discussed in Chapter 9.

Signaling via cell surface receptors may be transduced in many locations within the cell. Signal transduction pathways may target cytosolic proteins; they may lead to the post-translational modification of membrane proteins, which may be other signaling receptors or ion-channels; and they might also influence nuclear function, generally by regulating gene expression.

Kinases and Phosphatases Involved in Cellular Signaling

Protein phosphorylation regulates almost every aspect of cellular function. In lymphocytes, proteins may either be phosphorylated on tyrosine residues by tyrosine kinases or on serine or threonine residues by serine/threonine kinases. Certain kinases can phosphorylate both serine/threonine residues as well as tyrosine residues and are known as dual-specificity kinases. These phosphate residues may be removed in turn by phosphatases, which may be tyrosine phosphatases, or serine/threonine phosphatases, or dual-specificity phosphatases. Regulation of cellular events may therefore often be achieved either by modulating the activity of a specific protein kinase or by influencing the activity of a protein phosphatase.

Although most of the phosphorylation events that we consider involve proteins as substrates, keep in mind that kinases and phosphatases that modify lipids are also critical players in signaling. In this section, we will primarily consider kinases involved in signal initiation. Other kinases will be discussed in subsequent sections.

In bacteria, regulatory phosphorylation generally involves the modification of histidine residues. Phosphorylation of histidine, arginine, and lysine resisues have also been described in vertebrates, but the significance of these relatively rare modifications is unclear at this time.

Kinases that phosphorylate lipids are also key regulators of signal transduction. Most of the enzymes initially identified as lipid kinases have been shown to be dual lipid/protein kinases. We will discuss an important member of this family of enzymes, phosphatidyl inositol 3-OH kinase, in a later section.

A Minimalist View of How Protein Kinases Function

It is worth considering, initially at a conceptual level, some aspects of kinase structure that contribute to function. An obvious key feature of all kinases (about 400 have been characterized in terms of primary structure, and crystal structures have been obtained for about a dozen) is a well-conserved catalytic domain. Kinases generally contain regulatory regions that allow the enzyme to be turned on and off. In addition, the enzyme may have sequences that target it to specific locations within the cell and may also contain domains that help it interact with its potential targets.

The catalytic domains of most tyrosine and serine/threonine kinases have broad similarities in structure. Some understanding of how a protein kinase works is important from a lymphocyte development viewpoint. Apart from the biological significance of the enzymes, kinases with altered catalytic domains have frequently been employed as tools to examine the role of signaling pathways during lymphoid ontogeny. The kinase domain includes an adenosine triphosphate (ATP) binding site adjacent to the cleft into which a specific substrate might slide. Protein kinases contain a structure called an activation segment, which is part of the catalytic domain and which functionally occludes the substrate binding site. Phosphorylation of a conserved residue in the activation segment dramatically alters the disposition of this segment and "opens up" the catalytic site (Figure 4.2) so that the substrate may be properly accessed. In many tyrosine kinases, a specific tyrosine residue in the activation segment represents an autophosphorylation site. In many, but not all, serine/threonine kinases, the equivalent site contains a critical serine or threonine residue, which may be autophosphorylated. The corresponding site in certain kinases, such as MAP mitogen activated protein kinases and cyclin-dependent kinases (CDKs), cannot be

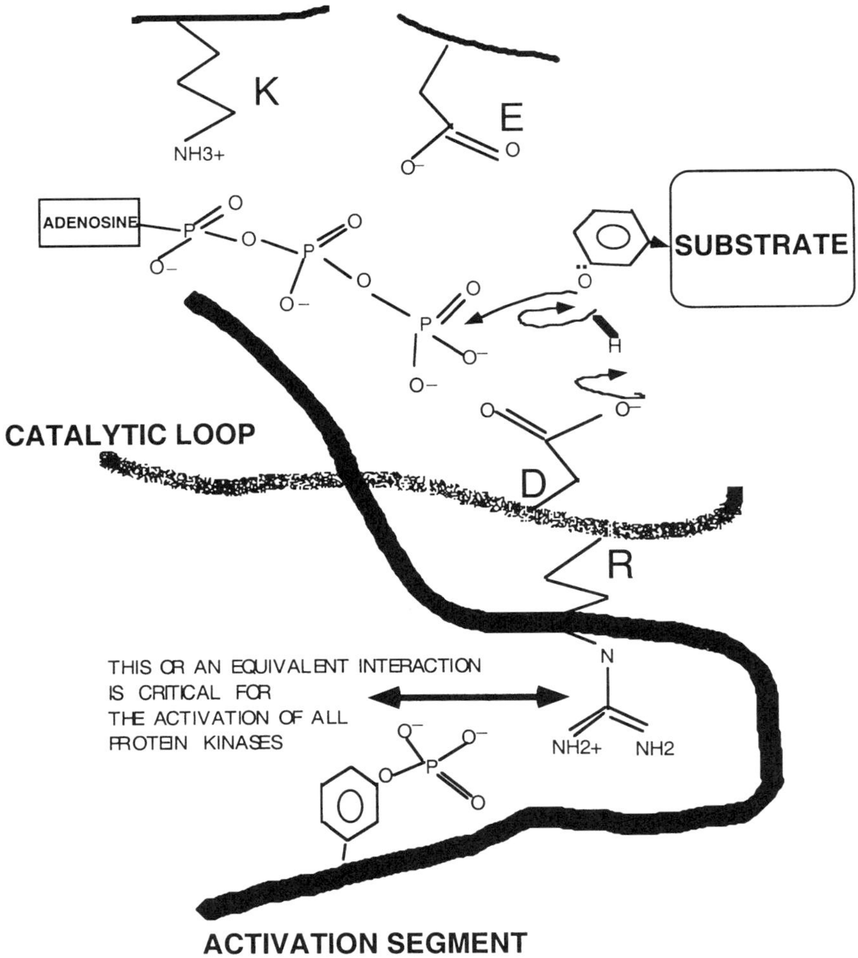

FIGURE 4.2. Phosphorylation of a critical residue (shown here as a tyrosine) in the activation segment of a kinase allows this segment to free the catalytic loop so that it can appropriately interact with the substrate and ATP. Most protein kinases contain adjacent R (arginine) and D (aspartate) residues in the catalytic loop. The arginine interacts with the phosphorylated residue in the activation segment, and the aspartate is essential for catalysis. Kinases are inactivated if a critical lysine (K) in the ATP binding site is mutated. This lysine is localized to the ATP-binding site because it forms a salt bridge with a defined glutamic acid residue (E). An additional aspartate residue (not shown) also contributes to an interaction with Mg^{++} ions, which bind to the β and γ phosphates of the ATP.

autophosphorylated. In these enzymes, a critical residue in the activation segment must be phosphorylated "in trans" by other regulatory enzymes. The phosphorylation of this critical residue of most protein kinases is a key event in signal transduction and in the initiation of many pathways in development. This modified side chain in the activation segment acquires the ability to interact with a critical arginine in the catalytic loop. This arginine residue is adjacent to and immediately precedes an invariant aspartate in most protein kinases (forming an arginine-aspartate (RD) motif). This latter aspartate side chain is absolutely required for catalysis. It presumably functions as a base that deprotonates a hydroxyl group (on a tyrosine, serine, or threonine residue) on the substrate. The resulting phenolate (or alcoholate in the case of serine/threonine substrates) ion may then attack the γ phosphate of ATP (Figure 4.2). A critical lysine residue in the ATP-binding site is frequently mutated to yield a dead kinase. This lysine residue is held in place by a salt bridge that it forms with an equally important glutamic acid residue. If the dead kinase retains a high affinity for substrates, it may, when introduced into lymphocytes, compete with the endogenous wild-type kinase for substrates and thus function as a dominant inhibitor. If the mutated kinase does not have a high affinity for substrates, it is unlikely to function in a dominant-negative fashion even if expressed at high levels in transfected cells or in transgenic mice.

Many kinases contain regulatory domains. These portions of the protein typically fold "over" the kinase domain, keeping the enzyme in a generally inactive mode. The binding of specific ligands to the regulatory domain may alter the conformation of the enzyme, freeing up the catalytic domain and thus permitting at least partial activation. In the case of some protein kinase C family members, for instance, the binding of phosphatidyserine, diacylglycerol, and Ca^{++} to the regulatory domain may be essential for activation. Protein kinase A is activated by the binding of cAMP to the regulatory domain, and calmodulin-dependent kinases may be activated by the binding of Ca^{++}/calmodulin complexes to their regulatory domains. Sometimes, as we shall see for some enzymes, protein–protein interactions might contribute to kinase activation.

Most kinases typically contain structural motifs that tell it where to go in a cell. Myristylation signals (discussed below) might direct a kinase to the inner face of the plasma membrane or to other cellular membranes. A nuclear localization signal might direct specific kinases to the nucleus and so on. Localization within a cell will obviously contribute to the modification of specific targets or substrates, by placing the enzyme in the right neighborhood.

The catalytic domain contributes in a major way, as discussed above, to substrate specificity. However, the kinase may also contain specific protein–protein or protein–lipid interaction domains that allow it to physically interact with specific targets.

Some Protein–Protein and Protein–Lipid Interaction Domains that Are Frequently Found in Signaling Molecules

A number of well-defined protein modules participate in protein–protein and protein–lipid interactions, particularly during signal transduction. We will consider a few of the modules that are frequently considered when discussing signal transduction mediated by tyrosine kinases.

The Src homology 2 (SH2) domain is a module of about 100 amino acids and is found in a large number of signaling molecules. It was originally discovered in Src-family kinases. A phosphotyrosine-binding pocket in the SH2 domain recognizes short peptide sequences within proteins. In these peptide stretches, the N-terminal residue is a tyrosine, while the three amino acids C-terminal to this residue also contribute to the recognition by a specific SH2 domain. The SH2 domains of Src-family kinases recognize tetraphosphopeptides that fit a YP-X-X-I/L consensus (where X is usually negatively charged). The tyrosines in a motif known as the immunoreceptor tyrosine-based activation motif (ITAM), which is found in the cytoplasmic tails of proteins involved in antigen-receptor signaling, fit this consensus. (ITAMs will be considered in more detail later in this chapter.) SH2 domains of other signaling molecules recognize distinct phosphotyrosinylated consensus sequences.

A distinct protein module found in a number of signaling molecules is known as a phosphotyrosine-binding domain (PTB). This type of domain also recognizes phosphotyrosine-containing peptides, but unlike SH2 domain-binding peptides, these peptides contain phosphotyrosine as the C-terminal residue in the consensus binding site.

Another important domain found in numerous signaling molecules but first identified in Src-family kinases is the SH3 (Src homology 3) domain. SH3 domains are also about 100 amino acids in length and fold to generate distinct pockets for the binding of proline rich peptides. A general consensus for SH3 ligands is (R/K)XXpXXP.

SH3 and SH2 domains are involved in both intramolecular and intermolecular associations. In Src-family kinases, a phosphorylated C-terminal negative regulatory tyrosine residue is intramolecularly bound to the SH2 domain. Dephosphorylation of this tyrosine residue by the CD45 tyrosine phosphatase is critical to the activation of Src-family kinases. The protein–protein associations mediated by the SH2 and SH3 domains may facilitate contacts between these enzymes and potential regulators and specific substrates.

A specific domain that may be involved in protein–inositol phospholipid, protein–protein interactions is known as a Pleckstrin homology (PH) domain. PH domains are found in a very large number of signaling molecules, and they generally bind to inositol polyphosphates or phosphatidylinositol phospholipids, such as phosphatidyl inositol (3,4,5) trisphosphate (PI-3,4,5P), which are phosphorylated in the D-3 position of the inositol ring. For some proteins, the PH domain may contribute to membrane attachment. Specific protein–protein interactions may also be mediated by this domain but are less well understood. Some proteins involved in vesicular transport contain a zinc-coordinating Ring finger domain, which contains eight cysteine residues, which can bind phosphatidyl inositol 3-phosphate (PI3-P). This domain is called a FYVE domain, named after the first four proteins discovered to contain such a domain.

Many other conserved domains and motifs are present in signaling proteins, but the "ligands" that they recognize are in general not understood as clearly as the specific structural motifs seen by the domains discussed above.

Tyrosine Kinases Frequently Participate in Signal Initiation

The initiation of signal transduction often depends on the activation of a protein tyrosine kinase. This category of kinases may be subdivided broadly into receptor tyrosine kinases and nonreceptor tyrosine kinases. Receptor tyrosine kinases are transmembrane proteins whose extracellular domains bind ligands and whose cytoplasmic tails contain one or more tyrosine kinase domains. While the best studied receptor tyrosine kinases are the insulin receptor, the epidermal growth factor (EGF) receptor, and the platelet devised growth factor (PDGF) receptor, from the viewpoint of lymphocyte development, critical receptor tyrosine kinases are c-Kit (CD117), which binds SCF (stem cell factor), and the Flk2/Flt3 kinase, which is triggered by the Flt ligand (FL).

Nonreceptor tyrosine kinases are synthesized on cytosolic ribosomes and may localize either in the cytosol or the nucleus. They are often attached to the cytosolic face of the plasma membrane via a lipid modification. There are a number of families of nonreceptor tyrosine kinases, many of which include members that are critical for lymphocyte signaling and lymphocyte development. These include Src-family kinases, Syk family kinases, Tec family kinases, and Jak kinases (Figure 4.3).

Nonreceptor Tyrosine Kinases Involved in Antigen-Receptor Signaling

Three nonreceptor tyrosine kinase families are active participants in pre-antigen receptor and antigen receptor signaling. The three families are the Src-family, the Syk family, and the Tec family. Key structural features of these tyrosine kinase families are summarized in Figure 4.4. The developmental roles of some of these kinases are summarized in Table 4.2. The Jak family of kinases is critical for cytokine signaling and will be discussed later in this chapter.

Src-family kinases are the best studied cytosolic tyrosine kinases. Members of this family are related to *c-src*, the cellular homolog of the *v-src* protooncogene encoded by the Rous sarcoma virus. Relevant members in lymphocytes include Lck, Fyn, Lyn, Blk, and Fgr. These kinases possess

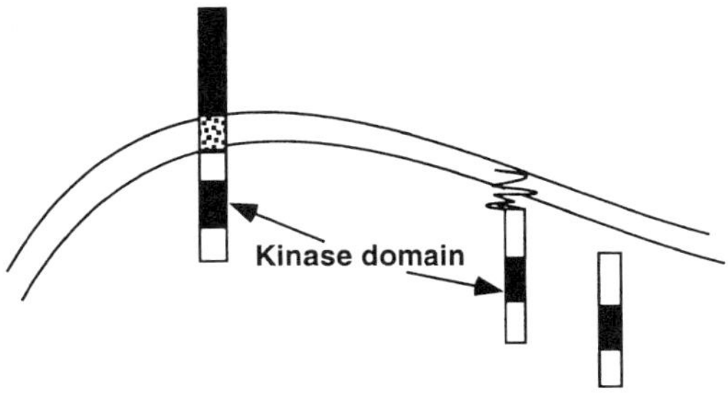

RECEPTOR TYROSINE KINASES	NON-RECEPTOR TYROSINE KINASES				
	Src family	*Syk family*	*Abl family*	*Tec family*	*Jak family*
C-Kit	Fyn	Zap70	Abl	Btk	Jak1
Fk2/Flt3	Lck	Syk	Arg	Itk	Jak2
	Lyn			Rlk	Jak3
	Blk				Tyk2
	Fgr				

FIGURE 4.3. Protein tyrosine kinase families. Members that may be relevant from a lymphocyte development standpoint are listed.

a number of interesting structural features that are intrinsic to their function. The C-terminal Src homology 1 (SH1) region is the kinase domain. A regulatory tyrosine residue in the activation segment is a target for autophosphorylation, which in turn leads to increased substrate accessibility. A tyrosine residue in the penultimate position just adjacent to the C-terminus serves a negative regulatory function. The c-Src kinase, Csk, phosphorylates this tyrosine residue (corresponding to tyrosine 527 in c-Src). This phosphotyrosine residue forms an intramolecular association with the SH2 domain of the same molecule, rendering the kinase domain inaccessible to substrate and ATP. Tyrosine phosphatases of the CD45 family dephosphorylate this critical negative regulatory tyrosine residue and thus contribute to the activation of Src-family kinases. Antigen-receptor signaling requires that a functional CD45 molecule be present presumably to ensure that Src-family kinases are dephosphorylated and available for further activation following receptor ligation.

SH2 and SH3 domains were first recognized in Src-family kinases. The Src homology 4 region (SH4) at the N-terminus is also known as the membrane localization domain. A glycine residue in postion 2 serves as a signal for myristylation. Myristylation at the amino-termini of Src-family kinases (following removal of the initiator methionine) allows for membrane localization and is essential for the function of these kinases. In some Src-family kinases, signals in the SH4 domain also contribute to palmitylation of these molecules (often on a cysteine residue that is immediately adjacent ot the N-terminal glycine), which probably further stabilizes their membrane association.

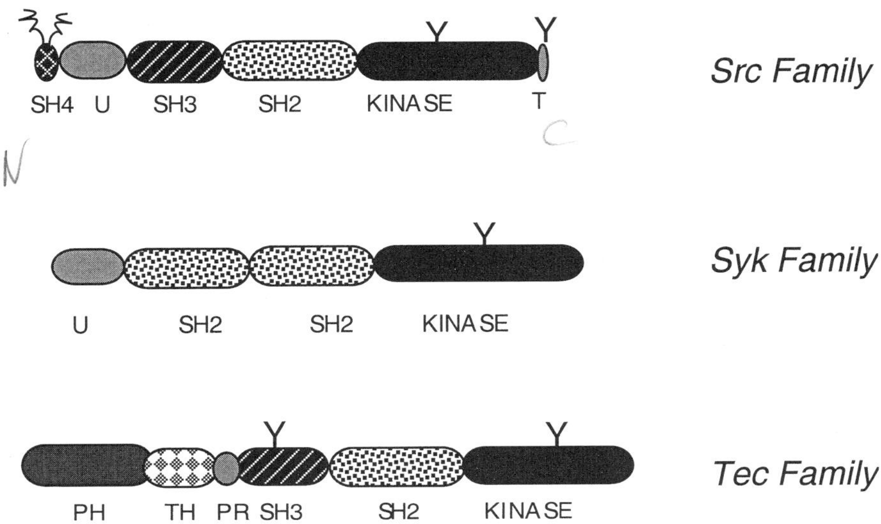

FIGURE 4.4. Nonreceptor tyrosine kinases belonging to three families are involved in antigen-receptor signaling. Src-family kinases contain an SH2 and SH3 domain and a motif, sometimes called an SH4 domain, that facilitates the covalent addition of lipids, including myristic acid, on glycine at position 2 (which is the N-terminal residue after removal of methionine) and a palmitate residue at position 3. Src-family kinases typically possess a C-terminal tyrosine residue in the tail (T) that is phosphorylated by Csk and serves a negative regulatory function. Syk and zeta associated protein-70 (Zap-70) lack an SH3 domain and have tandem SH2 domains. They contain neither a negative regulatory tyrosine residue of the sort found in Src-family kinases nor a myristylation signal. Btk, Itk, Rlk, and Tec are members of the Tec family. Tec kinases lack a myristylation signal and a C-terminal negative regulatory tyrosine residue. They generally contain a PH domain and a Tec homology (TH) domain. PR refers to a short proline-rich region between the TH and SH3 domains. U refers to a unique region. Rlk lacks a PH domain but is targeted to membranes by palmitoylation of a cysteine-string N-terminal domain.

Syk and zeta associated protein-70 (Zap-70) are members of a distinct family of nonreceptor tyrosine kinases, often referred to as Syk kinases. Syk is expressed in B, T, and myeloid cells and functions downstream of many receptors including the B cell receptor, the IgE receptor, and the T cell receptor. Zap-70 is a closely related kinase expressed primarily in T and NK cells. Syk family kinases possess no myristylation consensus sequences and contain no SH3 domains. They contain two tandem SH2 domains that can effectively bind to closely spaced phosphotyrosine residues in ITAMs. Activation of Syk family kinases depends on their recruitment to ITAM motifs. Since tyrosine-phosphorylated ITAM peptides can activate Syk, it is likely that recruitment results in allosteric modification and activation of Syk family kinases. Src-family kinases contribute to Syk activation by

TABLE 4.2. A summary of the developmental roles of tyrosine kinases linked to antigen-receptor signaling

Kinase	Developmental role	Reference
Csk$^{-/-}$	Early embyonic lethal; Src-family kinases activated.	Imamoto and Soriano, 1993 Nada et al., 1993
Src-family		
Lck$^{-/-}$	Partial block in T cell positive selection I; T cell positive selection II is blocked; defective T cell activation.	Molina et al., 1992 Wen et al., 1995
Fyn$^{-/-}$	CNS defects; mild defect in SP thymocyte activation; mild reduction in IgG3 responses; impaired IL-5 mediated B cell proliferation.	Appleby et al., 1992 Stein et al., 1992 Appleby et al., 1995
Lck$^{-/-}$/*Fyn*$^{-/-}$	T cell positive selection I impaired.	Groves et al., 1996 van Oers et al., 1996
Fyn$^{-/-}$/*Fak*$^{+/-}$	Loss of DP cells (variable penetrance).	Kanazawa et al., 1996
Lyn$^{-/-}$	Alterations in B cell proliferation; defects in B cell maintenance and maturation; increase in B1 cells; enhanced BCR-mediated ERK activation.	Hibbs et al., 1995 Nishizumi et al., 1995 Wang et al., 1996 Chan et al., 1997
Syk Family		
Syk$^{-/-}$	Block at pro-B to pre-B transition; Peripheral maturation of B cells impaired.	Cheng et al., 1995 Turner et al., 1995
Zap-70$^{-/-}$	T cell-positive selection II is defective.	Negishi et al., 1995 Wiest et al., 1997
Syk$^{-/-}$/*Zap-70*$^{-/-}$	Abrogation of T cell-positive selection I.	Cheng et al., 1997
Tec Family		
Btk R28C (*Xid*) and *Btk*$^{-/-}$	Maintenance defect in peripheral B cells; T-independent responses impaired; No B-1 cells; IgM and IgG3 decreased.	Scher et al., 1975 Khan et al., 1995 Kerner et al., 1995 Hendriks et al., 1996
Itk$^{-/-}$	Maintenance defect in peripheral T cells; Reduced T cell proliferation.	Liao and Littman, 1995 Liu et al., 1998
Itk$^{-/-}$/*Fyn*$^{-/-}$	Similar to *Itk*$^{-/-}$ but proliferative defect more marked.	Liao et al., 1997
Itk$^{-/-}$/*Rlk*$^{-/-}$	Defective T cell proliferation, cytokine production, and adaptive immune responses. Defect in PLC-γ activation, calcium signaling, and MAP kinase activation.	Schaeffer et al., 1999

phosphorylating ITAMs and may also transphosphorylate the Syk family kinase directly. This category of kinases is essential for preantigen receptor and antigen receptor-mediated signals in T and B cells.

A third family of tyrosine kinases is involved in antigen receptor-mediated signal transduction and lymphocyte development. The Tec family includes Btk, Itk (sometimes known as Tsk or Emt), and Rlk, which are homologous to a kinase known as Tec (which is expressed in many non-lymphoid cells, and at low levels in lymphoid cells). Btk was identified as the defective gene in X-linked agammaglobulinemia. This disease is also known as Bruton's disease, hence the kinase that is defective in these patients is referred to as Bruton's tyrosine kinase (Btk). Btk is expressed in B cells and myeloid cells, while Itk is primarily expressed in T cells. Structurally, Tec kinases are distinct from Src-family kinases, although they also have similar SH3, SH2, and SH1 domains. Members of the Tec family lack a myristylation signal and do not have a negative regulatory tyrosine residue equivalent to the c-Src 527Y residue. Most of these kinases possess an N-terminal PH domain followed by a Tec homology (TH) domain. Between the TH domain and the SH3 domain is a proline-rich stretch that contains consensus SH3 domain-binding motifs. Structural studies have revealed that the proline-rich region of Itk is involved in an intramolecular association with the SH3 domain, and this probably constrains the enzyme from being fully activated (Figure 4.5). The SH3 domain of Btk contains an autophosphorylation site at residue Y223. The binding of specific SH3 ligands or possibly the autophosphorylation of Tec kinases may permit their appropriate activation, potentially by relieving the inhibitory intramolecular SH3-proline peptide interaction.

A point mutation in the PH domain (R28C) completely abrogates the function of Btk. Mice with this Xid mutation have a defect in B cell proliferation and development. The phenotype of the Xid mouse is identical to that of Btk-null mice. B cell development is unimpaired in the bone marrow, but mature peripheral B cells die prematurely. The developmental defect in these mice will be considered in more detail in Chapter 7. The PH domain of Btk probably serves primarily to mediate the recruitment of Btk to PIP3 in the membrane (generated by the activity of phosphatidylinositol 3-OH kinase). However, this domain has also been shown to be involved in the association of this Btk with protein kinase C β and with TFII-I (also known as BAP-135), a transcription factor critical for transcription from initiator sequences. The C-terminal portion of the PH domain also interacts with the $\beta\gamma$ subunits of trimeric G proteins.

Tec kinases may be activated by a number of different types of cell surface signals. These include the antigen receptors on B and T cells, the IL-5 receptor, and certain serpentine receptors that signal via Gq-containing trimeric G proteins. Activation of Tec kinases downstream of antigen receptors involves the critical phosphorylation of a tyrosine in the activation segment of the catalytic domain by a Src-family kinase. Optimal

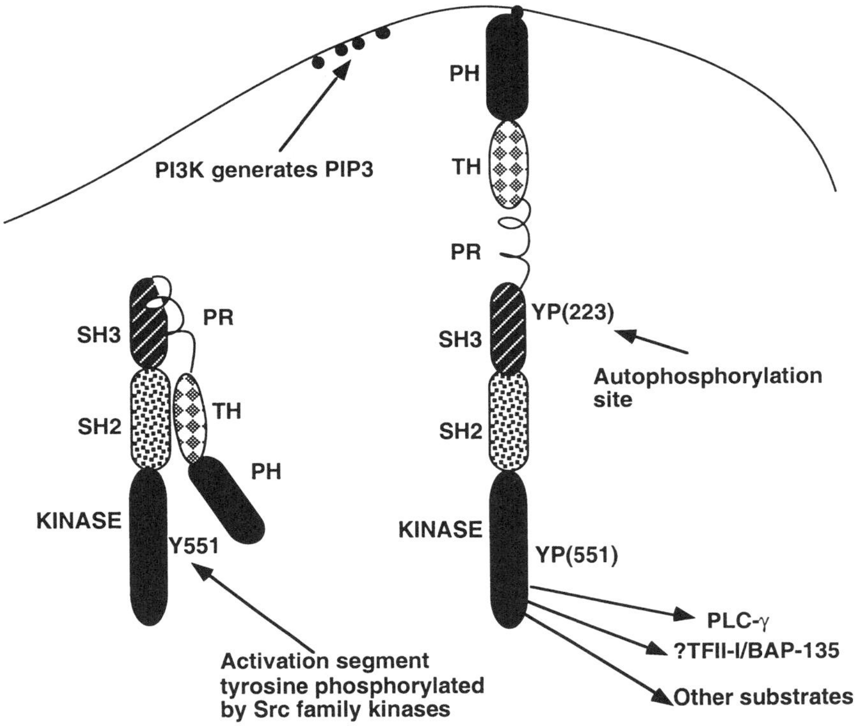

FIGURE 4.5. The activation of Btk involves both Src-family kinases and PI3K. Phosphorylation of Y551 in the activation segment by Src-family kinases is followed by autophosphorylation of Y223. This latter event or possibly the binding of a regulatory protein to the SH3 domain may release the autoinhibitory intramolecular proline-rich (PR)–SH3 interaction. Complete activation may require the PH domain-mediated recruitment of Btk to PIP3 in the membrane generated by PI3K.

activation of Btk also requires the generation of PIP3 in the membrane by phosphatidylinositol 3-OH kinase (PI3K).

Tec family kinases such as Btk, Itk, and Rlk are critical for sustaining calcium signaling in activated B and T cells. A critical substrate of Btk in this regard is phospholipase C-γ (discussed below). Apart from this phopsholipase, other potential targets of Btk have been identified including TFII-I/BAP-135. This protein binds to the core promoter of a number of genes that lack TATA boxes but contain initiator sites.

A distinct member of the Tec family has been identified in T cells called Rlk (or Txk). Like the other Tec kinases, Rlk lacks a C-terminal regulatory tyrosine residue but does contain a proline-rich stretch, SH2 and SH3 domains, and is activated by Src-family kinases. Unlike the other Tec kinases, Rlk lacks a PH domain and is activated independently of PI3 kinase. It does contain a cysteine-string motif that is palmitoylated and thus contributes to membrane localization. In the absence of both Itk and Rlk, PLC-γ activation and calcium signaling downstream of the TCR are severely impaired.

Protein Phosphatases

Protein phosphorylation and signal generation represent a fine balance between the activities of kinases and phosphatases. Phosphatases, like kinases, may be integral membrane proteins with cytosolic catalytic domains (receptor type, though these may be expressed either on the cell surface or sometimes remain in the endoplasmic reticulum membrane), or they may be cytosolic/nuclear proteins. Phosphatases may also be categorized as serine/threonine phosphatases, tyrosine phosphatases, or dual-specificity phosphatases, which can remove phosphates from both serine/threonine as well as tyrosine residues.

We will briefly discuss a few phosphatases that are of particular interest from the standpoint of lymphocyte development and that may be encountered in the context of signaling pathways discussed in this chapter. The CD45 family of receptor tyrosine phosphatases and the SHP-1 and SHP-2 cytosolic tyrosine phosphatases play important roles in lymphocyte signaling and development. Calcineurin is an important serine/threonine phosphatase that participates in lymphocyte signaling. Dual-specificity phosphatases such as PAC-1 are involved in the inactivation of MAP kinases. Cdc25A, B, and C are also dual-specificity phosphatases that regulate the activity of cyclin-dependent kinases. These dual specificity phophatases will be briefly considered in the sections on MAP kinases and cell cycle regulation.

The CD45 protein tyrosine phosphatase is abundantly expressed on the surface of most hematopoietic cells. This receptor tyrosine phosphatase contains an extracellular region, a transmembrane domain, and a cytosolic tail (Figure 4.6). This tail region contains tandem protein tyrosine phosphatase (PTP) catalytic domains. The extracellular region is composed of a heavily glycosylated cysteine-free N-terminal region and a membrane proximal cysteine-rich region, which contains three fibronectin III type repeat sequences. The N-terminal cysteine-free region contains multiple helix-breaking amino acids (such as proline and glycines) and probably has an extended rod-like structure. This putative rod-like region contains multiple N- and O-glycosylation sites. The region that includes all the O-glycosylation sites is encoded by exons 3–8, of which exons 4, 5, and 6 are alternatively spliced in different lymphoid subpopulations, giving rise to multiple isoforms of CD45. These isoforms contain varying numbers of glycosylation sites and therefore differ considerably in length. All CD45RA forms contain a segment encoded by exon 4, CD45RB forms

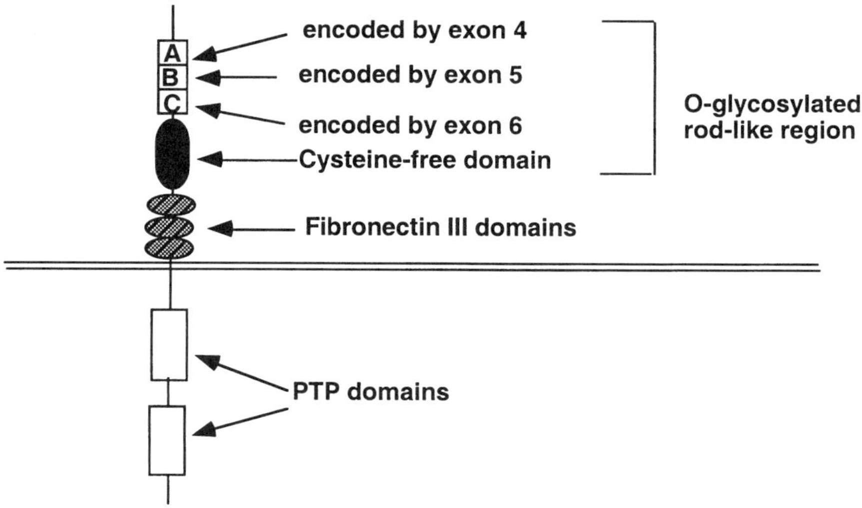

FIGURE 4.6. The CD45 receptor-type protein tyrosine phosphatase. The catalytic protein tyrosine phosphatase PTP domains are located on the cytosolic face of the membrane.

require exon 5, and CD45RC forms require exon 6. CD45RA, RB, and RC forms are expressed abundantly on naive T cells but not on activated and memory T cells. The form that lacks all the variable exons is called CD45RO and is expressed at relatively high levels on activated and memory T cells. The long form of CD45, which includes all the variable exons, is expressed on B cells (and a few natural killer, NK, cells) and is generally referred to as B220, which is the name of the monoclonal antibody used to detect this protein.

It is generally believed that the major function of CD45 is to dephosphorylate the negative regulatory tyrosine at the C-terminus (corresponding to position 527 in c-Src) of Src-family kinases. This residue is phosphorylated by a tyrosine kinase known as Csk (C-Src kinase). This phosphotyrosine residue binds to the SH2 domain of the Src-family kinase. This intramolecular interaction presumably occludes the kinase domain. The dephosphorylation of this negative regulatory residue best explains why CD45 is required for the activation of Src-family kinases in B and T cells and for signal transduction via both the B and T cell receptors. The function of CD45 is probably far more complex than outlined above. This phosphatase may dephosphorylate not only the negative regulatory tyrosine residue at the C-terminus but also the tyrosine residue in the activation segment (at the autophosphorylation site).

There are nine different isoforms of CD45. It is likely that different isoforms function in distinct, but poorly understood ways, at different stages of development. Since these isoforms are structurally distinct in their extra-

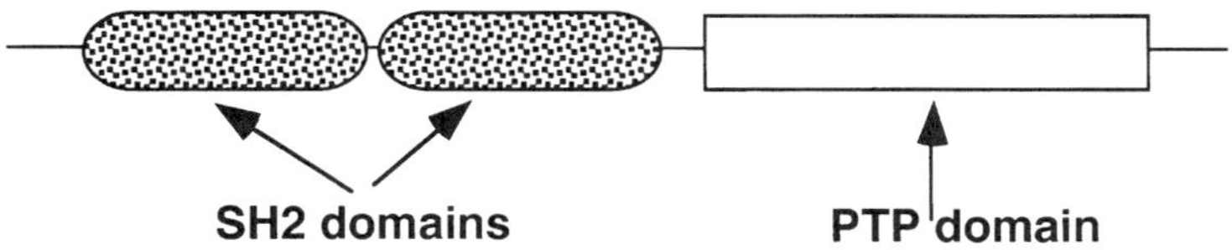

FIGURE 4.7. A schematic view of the structure of the SHP-1 and SHP-2 cytosolic tyrosine phosphatases.

cellular portions, it is assumed that ligands influence the functions of these receptor tyrosine phosphatases. Evidence for functionally significant ligands of CD45 remains to be obtained. Although CD22—an α2,6 sialic acid-binding lectin on B cells—might bind to CD45 and other sialylated glycoproteins, it is unclear whether the association of CD22 with CD45 is of any functional significance. The developmental consequences of the absence of CD45 are discussed in Chapters 7 and 9.

SHP-1 and SHP-2 (SH2 containing PTP-1 and 2) are cytosolic protein tyrosine phosphatases that are important negative regulators of signaling in lymphoid and myeloid cells. SHP-1 is also sometimes called hematopoietic cell phospatase (HCP) or PTP-1C. These phosphatases contain tandem SH2 domains and a C-terminal catalytic PTP domain (Figure 4.7).

SHP-1 is recruited to phosphorylated tyrosine-containing motifs called immunoreceptor tyrosine-based inhibition motifs (ITIMs) in the cytoplasmic tails of inhibitory receptors such as CD22 in B cells, and Ly49 C-type lectins and killer inhibitory receptors (KIRs) in NK cells. These motifs are tyrosine phosphorylated by Src-family tyrosine kinases such as Lyn, following which, SHP-1 may bind and be brought in proximity to its targets. Dephosphorylation mediated by SHP-1 and SHP-2 generally results in the dampening of signal transduction. We will discuss the role of SHP-1 in CD22 signaling and the putative role of SHP-2 in cytotoxic T lymphocyte antigen-4 (CTLA-4)-mediated negative signaling later in this chapter. The role of SHP-1 in negative signaling in NK cells is considered in Chapter 11.

Naturally occurring mutations in SHP-1 are found in *motheaten* mice. *Me/me* mice have a frameshift mutation in the early part of the SHP-1 coding sequence and are virtually null for this phosphatase. These mice die at 2 to 3 weeks because of severe abnormalities in myeloid cells. *Motheaten* viable (me^v/me^v) mice express two defective SHP-1 proteins, which have only 10% to 20% of the activity of the wild-type phosphatase. These mice die around 9 to 12 weeks after birth, also because of myeloid abnormalities. The developmental consequences of a defect in SHP-1 in B and T cells will be addressed in Chapters 7 and 9.

A small subfamily of tyrosine phosphatases that contain proline-glutamic acid-derine-threonine (PEST) domains may participate in the dephosphorylation of Src-family kinases downstream of the B cell receptor (BCR) and the TCR. PEST domains are enriched in proline, glutamic acid, serine, and threonine, and they usually contribute to enhanced protein degradation. One member of this family, proline enriched phosphatase (PEP) is expressed mainly in lymphocytes and may participate in the deactivation of Src-family kinases downstream of antigen-receptor activation. A distinct and unrelated T cell protein tyrosine phosphatase (TC-PTP) may also be required downstream of the TCR.

An important phosphatase in lymphocyte signaling that does not target phosphorylated proteins is the SH2 domain containing inositol-5 polyphosphate phosphatase (SHIP). SHIP removes the phosphate from the D5 position of the inositol ring in inositol polyphosphates and from inositol-containing phospholipids. SHIP also binds to ITIM motifs. SHIP modulates the consequences of PI-3 kinase signaling, particularly in B cells. The best studied receptor that is known to recruit SHIP is the FcγRIIb receptor in B cells. This receptor will be discussed later in the context of the negative regulation of B cell signaling. A distinct phosphatase that serves a similar function to SHIP in many cell types is the tumor suppressor protein tyrosine phosphatase with homology to tensin (PTEN). It is now recognized that PTEN is also an inositol phosphatase that can antagonize the anti-apoptotic function of PI-3 kinase.

A calcium- and calmodulin-dependent serine/threonine phosphatase that is of considerable importance in lymphocyte signaling is calcineurin. Calcineurin, also known as PP2B, is made up of two subunits, calcineurin A, which is the larger catalytic subunit, and calcineurin B, which is smaller, resembles many other Ca^{++} binding proteins, and serves a regulatory function. In quiescent T cells, when intracellular Ca^{++} concentrations are low, an autoinhibitory domain of subunit A binds and inactivates the catalytic domain. When calcium levels rise after T cell activation, Ca^{++} calmodulin complexes bind to a specific domain on the A subunit and free the catalytic site from autoinhibition. Activated calcineurin dephosphorylates a number of proteins including nuclear factor of activated T cells (NFAT), an important transcription factor in T and B cells. Dephosphorylated NFAT, as discussed later, can be translocated to the nucleus. Some evidence indicates that activated calcineurin may also be translocated into the nucleus with NFAT. Calcineurin is the target for two widely used immunosuppressive compounds. Cyclosporin A complexes with a protein known as cyclophilin and, this complex can associate with calcineurin B and inactivate it. Similarly, FK-506 is a drug that complexes with a cytosolic protein known as FKBP12. These latter complexes can also inactivate calcineurin.

Cytosolic Signaling Pathways and Downstream Nuclear Events

Although the field of signal transduction has seen explosive growth in the last two decades, much remains to be learned as to how these pathways mesh with the mechanisms involved in cell cycle progression, cell cycle arrest, the induction of cell death, or cellular differentiation. While connections between signaling and the induction of cell death will be discussed in Chapter 5, in this chapter we will explore signaling pathways in general and also examine how some of these may be linked to cell cycle progression. After gaining an overview of the various signaling cascades in the following section, we will review how specific receptors such as antigen receptors, costimulatory receptors, and cytokine receptors generate signals that engage these general pathways.

Signals From the Cell Surface May Activate Monomeric GTP-Binding Proteins That in Turn Regulate Mitogen-Activated Protein Kinase Pathways

A recurring theme in signal transduction initiated downstream of a number of receptors is the activation of monomeric GTP-binding proteins that function as molecular switches. These switches may initiate cytosolic protein kinase cascades, which frequently culminate in the phosphorylation and activation of specific transcription factors.

The Ras protein is an example of a monomeric or small G protein (as opposed to the slightly larger heterotrimeric G proteins, which will be discussed in a separate section). Monomeric G proteins have a binding site for guanine nucleotides. When bound to GDP, these proteins exist in a different conformation from the GTP-bound form. Signals from cell surface receptors lead to the activation of regulatory proteins known as guanine nucleotide exchange factors (GEFs), which facilitate the conversion of the inactive GDP-bound regulator to the activated GTP-bound form. In the GTP-bound form, a surface of the small G protein is exposed, which can interact with and activate downstream targets (generally known as effectors) of this regulator. The GTPase activity of most small G proteins is limited but can be greatly enhanced by enzymes known as GTPase-activating proteins (GAPs).

More than 60 different monomeric G proteins have been described in vertebrates. These are divided, on the basis of sequence homology, into five families, listed in Table 4.3.

Ras is an oncogene that is frequently mutated in tumors to yield a dominantly activated protein. There are six Ras proteins in vertebrates. Of these, K-Ras, H-Ras, and N-Ras may be involved in signal transduction downstream of tyrosine kinases and serpentine receptors. R-Ras 1 and 2 have been shown to participate in the regulation of integrin signaling, a topic that will not be discussed in detail in this book. Other than the six Ras proteins, other members of the family such as Rap1 and Rap2 under certain circumstances may functionally oppose Ras. Because of the potential for functional redundancy among Ras proteins, studies on the role of these proteins in lymphocyte development have often depended on the use of activating mutations of Ras or of dominant negative forms of Ras that will be discussed below.

TABLE 4.3. Small G protein families

Family	Selected members	General functions
Ras	H-Ras, K-Ras, N-Ras R-Ras 1, 2 (TC21), Rap1A/ Rap 1B, Rap2A/Rap2B	Cell Growth Development
Rho	cdc42, Rac1, Rac2, RhoA RhoB, RhoC, RhoE, RhoG Tc10	Regulation of the actin cytoskeleton
Ran	Ran	Nuclear transport
Rab	Approximately 40 Rabs	Vesicular transport

Monomeric G proteins such as Ras share the following structural features:

1. Well-defined regions are involved in nucleotide binding and for intrinsic GTPase activity. Mutations that convert the glycine in position 12 to a valine or to other residues are transforming because they result in reduced GTPase activity and therefore persistence of Ras-GTP. Similarly, a mutation affecting codon 61, altering a glutamine to a leucine, reduces GTPase activity, results in an activated Ras-GTP molecule, and is transforming. Engineered codon 12 and codon 61 mutants have been used as tools to examine the consequences of constitutive Ras activation on lymphocyte development.

2. These proteins contain a domain that interact with GEFs that facilitate the coversion of the G protein from a GDP-bound form to a GTP-bound form. Mutations converting residue 17 to an asparagine (N17-Ras, N17-Rho, and so on) result in the small G protein binding very tightly to its GEF. Engineered N-17 mutants have been used to probe the role of small G proteins in lymphocyte development.

3. Monomeric G proteins contain an "effector" domain, which, in the GTP-bound molecule, changes shape and can facilitate the association with and the activation of downstream molecules and pathways.

4. Many G proteins contain a C-terminal C-A-A-X box (where X is the C-terminal amino acid and A is an amino acid with an aliphatic R group). This motif facilitates prenylation of G proteins, which in turn allows these proteins to associate with the inner face of the plasma membrane. Prenylation involves the formation of a stable thioether of the cysteine residue in the C-A-A-X box with either a 15-carbon farnesyl moiety or a 20-carbon geranylgeranyl moiety. After prenylation, the protein is translocated to a membrane compartment where the A-A-X tripeptide is removed and the COOH group of the now C-terminal prenylcysteine residue is methylated. Since Ras seems to mediate its function in part by bringing other proteins to the membrane, prenylation and membrane

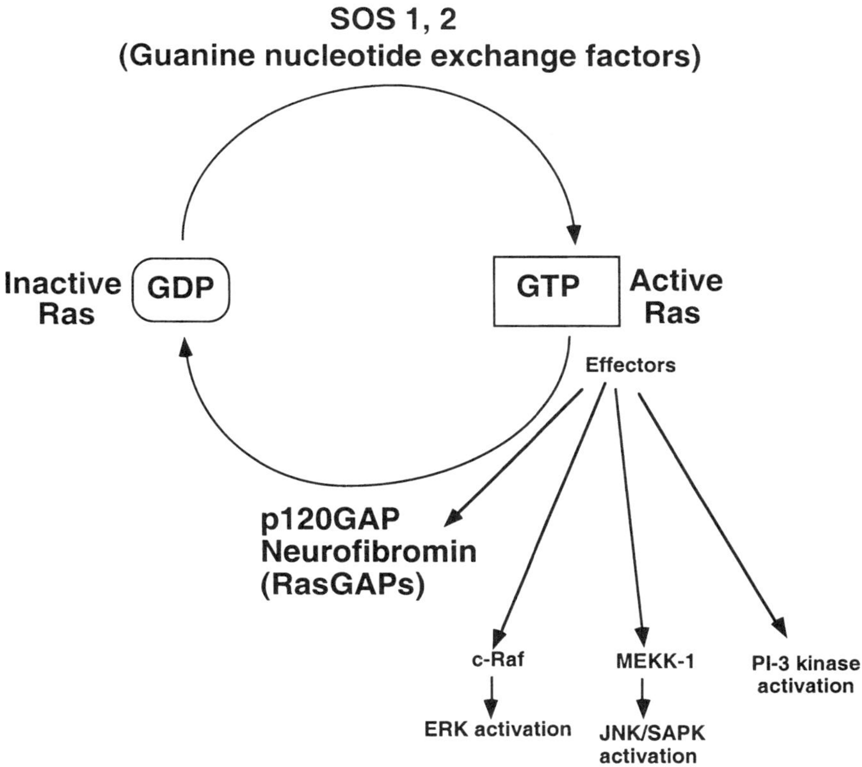

FIGURE 4.8. The Ras cycle. Inactive GDP-bound Ras may be converted to the GTP-bound state either by the activation of a guanine nucleotide exchange factor (such as mSOS) or by the inhibition of a RasGAP such as p120RasGAP or neurofibromin. Ras effectors include, but are not restricted to, c-Raf and other members of the Raf family, MEKK1, PI-3 kinase, and the neurofibromin and p120 RasGAP proteins. c-Raf is a MAP kinase kinase kinase (MAPKKK) for the ERK pathway. MEKK1 is a MAPKKK for the JNK/SAPK pathway.

localization are critical for the function of Ras. Ras is also palmitylated at an internal cysteine.

The Ras cycle (Figure 4.8) provides an example of a typical monomeric G protein regulatory cycle. Input signals regulate the G protein by either enhancing the activity of a specific GEF or by inhibiting the activity of a GAP. The GTP-bound G protein then can bind to effectors, one of which in many cases is a serine-threonine kinase, which activates a specific mitogen-activated protein kinase (MAPK) cascade. The MAPK at the end of the cascade may then be translocated to the nucleus, where it may phosphorylate and generally activate specific transcription factors.

Ras activation downstream of tyrosine kinases depends largely on the activation of a GEF known as mSOS (the murine homolog of the Drosophila Son of Sevenless protein). The recruitment to the membrane and activation of mSOS via tyrosine phosphorylation will be described when we dicuss antigen-receptor signaling later in this chapter. Other exchange factors may participate in Ras activation when cells are triggered by trimeric G protein-coupled receptors.

Two RasGAPs may participate in the Ras cycle in lymphocytes. These are p120RasGAP and neurofibromin. The latter is the protein product of a tumor-suppressor gene mutated in patients with type I neurofibromatosis. Protein kinase C-mediated phosphorylation of p120RasGAP may contribute to the inactivation of GAP activity and the activation of Ras. In lymphocyte signaling, Ras activation probably depends on both an increase in exchange factor activity as well as an inhibition of GAP activity. When Ras is in the GTP-bound state, its effector domain changes shape particularly in two patches known as the Switch 1 and Switch 2 regions. A number of proteins bind to these switch motifs, and these are generally assumed to be effectors of Ras and include p120RasGAP and neurofibromin. Proteins that bind to the effector domain of GTP-Ras may be activated because they are recruited to the membrane in association with this isoprenylated G protein. An important effector of Ras is c-Raf, which is a serine-threonine kinase. Ras-GTP binds to Raf and contributes to the activation of the latter by bringing Raf to the membrane. Other members of the Raf family such as B-Raf and A-Raf may also be activated by Ras-GTP. Membrane-localized Raf may be "fully" activated by a poorly defined Raf kinase. c-Raf is a MAP kinase kinase kinase (MAPKKK) for the ERK (extracellular signal-regulated kinase) MAPK pathway. Other effectors of Ras include MEKK1 (MAP and ERK kinase kinase 1), a MAPKKK for the JNK/SAPK (Jun N-terminal kinase/stress-activated protein kinase) pathway, and PI3 Kinase (PI3K, phosphatidylinositol 3-OH kinase). Although Ras may activate the JNK pathway via MEKK1 activation, it is likely that JNK activation in lymphocytes is mediated primarily by other monomeric G proteins, (discussed below). The pathways that lead to PI3 Kinase activation will be discussed at length in a separate section. Although both p120 RasGAP and neurofibromin are RasGAPs that bind to Ras-GTP and help convert it to Ras-GDP, both these proteins also probably function as effectors of Ras. p120 RasGAP may modulate the activation of Rho family proteins, and neurofibromin might be a regulator of cAMP (cyclic adenosine monophosphate)-dependent protein kinase activity.

A typical sequence of events that leads from tyrosine kinase activation to the activation of a monomeric G protein and a MAP kinase cascade is depicted in Figure 4.9.

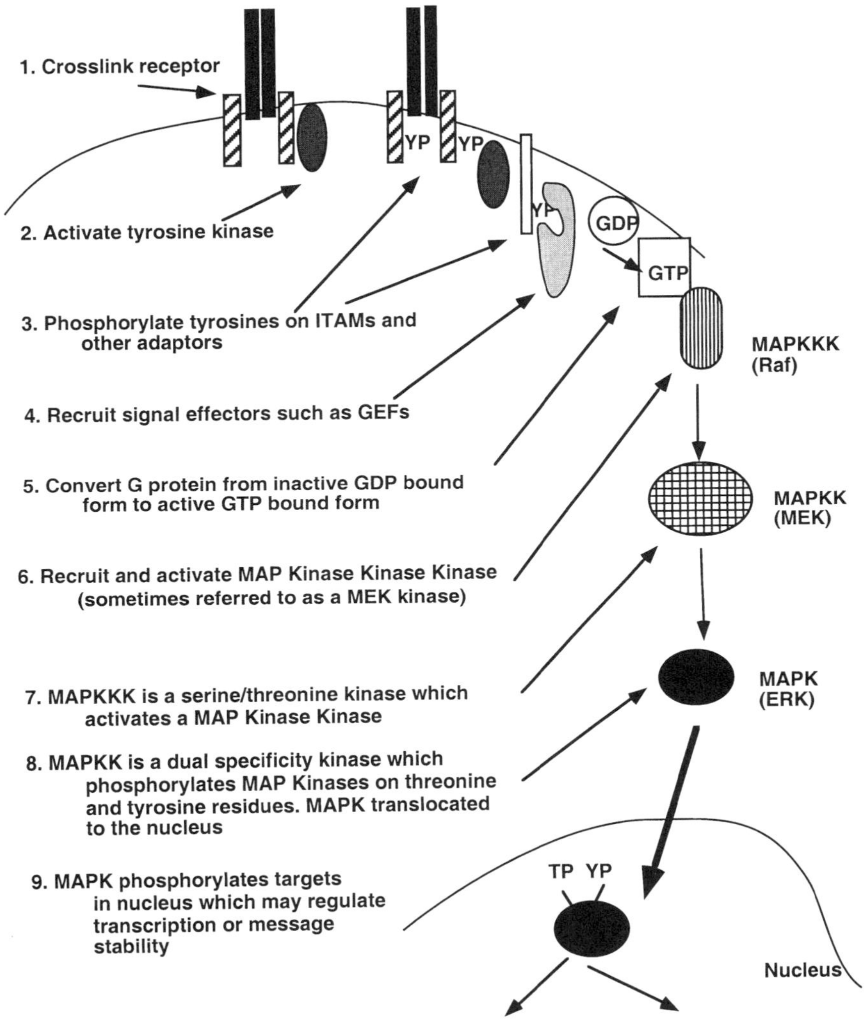

FIGURE 4.9. A prototype signaling pathway leading from tyrosine kinase activation to the induction of a MAP kinase. This sequence of events is based, to a large extent, on our understanding of how the Ras/Raf/ERK pathway is turned on. In the Ras/Raf/ERK pathway, the MAPKKK is Raf, the MAPKK is MEK1/2, and the MAPK is ERK1 or ERK2.

The ERK MAP Kinase

The major MAP kinase turned on by Ras is ERK (in some cells Ras can contribute to JNK/SAPK activation). The predominant way in which the ERK1 and ERK2 MAP kinases are activated is via the activation of Ras and Raf, but other Ras-independent pathways can also contribute to ERK activation. Raf phosphorylates a critical serine residue on the activation segment of MAPKKs known as MEK1 and MEK2 (for MAP and ERK kinase 1 and 2). MEK1 and MEK2 are dual-specificity kinases. This family of kinases phosphorylate threonine and tyrosine residues, which are found in a critical tripeptide (discussed below) in the activation segment of MAP kinases.

MAPKs such as ERK, JNK, and p38 share the following features:

1. They are proline-directed serine/threonine kinases. Their target proteins are phosphorylated on serine or threonine residues, which are immediately followed by a proline residue.
2. They contain a critical TXY tripeptide in the activation segment, which must be phosphorylated by a dual-specificity MAPKK. For ERKs, this tripeptide is TEY (threonine-glutamic acid-tyrosine). For JNKs the tripeptide is TPY (threonine-proline-tyrosine). For p38 family members this target sequence is TGY (threonine-glycine-tyrosine).
3. When activated, MAPKs are generally translocated from the cytosol to the nucleus.

Nuclear ERK phosphorylates a number of targets. For instance, an Ets family transcription factor, Elk-1, is activated by ERK and transcriptionally activates the *c-Fos* gene, and in some cell types, probably the cyclin D1 gene as well. ERK may also phosphorylate and activate pp90Rsk, a kinase that phosphorylates the S6 protein in 40S ribosomes and that may be required for efficient protein translation. In germinal center B cells, ERK may phosphorylate the Bcl-6 transcription factor and thus facilitate its degradation as a prelude to the induction of further differentiation. Some of the steps in the major vertebrate MAPK pathways are listed in Figure 4.10. We will describe the activation of the JNK/SAPK pathway in more detail in the next section.

Cytoskeletal Reorganization Mediated by the Rho Family Proteins: Cdc42, Rac, and Rho

A number of Rho family members participate in the regulation of the actin cytoskeleton, and some members of this family, in particular Rac and Cdc42, may contribute to the activation of JNK/SAPK by tyrosine kinase-derived signals.

Reorganization of the cytoskeleton is required for a lymphocyte to move, for the generation of polarity when T cells interact with B cells or other antigen-presenting cells (APCs), and for cytokinesis. Rho, Rac, and Cdc42 each participate in a distinct pattern of actin cytoskeletal reorganization. Rac is required for membrane ruffling, which involves the formation of sheet-like structures known as lamellipodia. Rho is required for the formation of actin stress fibers. Cdc42 was identified first in budding yeast,

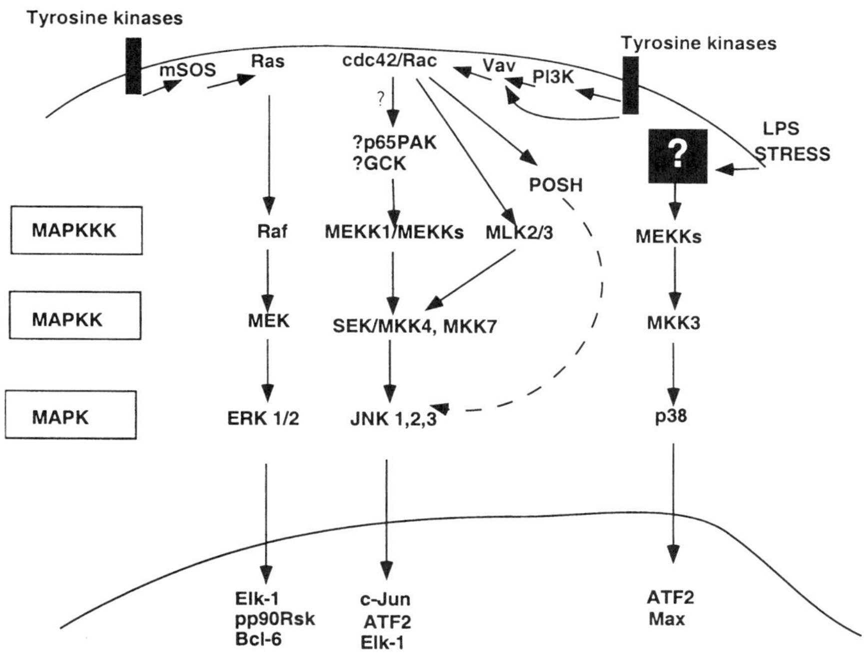

FIGURE 4.10. A comparative view of vertebrate MAPK pathways. Although only signal initiation from tyrosine kinases is depicted for the ERK and JNK/SAPK pathways, many other types of receptors activate these pathways, and in different ways. The JNK/SAPK pathway will be discussed in the next section.

where it participates in bud site selection. The vertebrate homolog of Cdc42 is required for the formation of actin-containing microspikes called filopodia. Cdc42 is also required for the polarization of T cells when they form "synapses" with APCs.

A large number of GEFs and GAPs can regulate the activation of Rho family GTPases. Vav is a critical GEF for Rac activation, and TIAM-1 (T lymphoma invasion and metastasis protein-1) is one of a number of GEFs that may participate in the activation of Rho. These GEFs typically contain a Dbl homology domain that possesses catalytic activity and a PH domain that binds PI-(3,4,5) P3, which is a product pf PI3K activity. Activation of these GEFs by tyrosine kinase-linked receptors may require both its recruitment to the membrane following PI3K activation as well as tyrosine phosphorylation of the GEF, which is probably facilitated by membrane localization.

Rho-GTP binds to a number of Rho effectors that probably participate in actin reorganization. Among these effectors are four serine/threonine kinases, a protein kinase C-like enzyme called PKN, two related kinases, p164 Rho-kinase (RhoK or ROKα), and p160 Rho-associated coiled–coil-containing protein kinase (p160ROCK), and the Citron kinase. All these kinases associate with Rho (but not with cdc42 or Rac) in a GTP-dependent manner. Association with Rho-GTP also contributes to an increase in kinase activity. P164 RhoK can phosphorylate a subunit of myosin light chain (MLC) phosphatase, which also binds to Rho, and this phosphorylation leads to a decrease in MLC phosphatase activity. MLC is phosphorylated by an MLC kinase and dephosphorylated by an MLC phosphatase. Rho activation leads to an overall increase in MLC phosphorylation, and this is probably a critical event in the actin bundling that leads to stress fiber formation.

The Citron kinase activated by Rho has been shown to be involved in cytokinesis, the process by which cells actually split apart during mitosis. Rho-GTP also contributes to cell cycle progression in fibroblasts, presumably via the induction of *c-fos* transcription mediated by a transcription factor known as SRF (serum response factor). The molecular basis for this function of Rho is relatively poorly understood. Unlike Rac and Cdc42, Rho does not contribute to the activation of the JNK/SAPK pathway. Rho can be specifically inactivated when it is ADP ribosylated by a bacterial toxin called C3. C3 has been used as a tool to examine Rho function in lymphocytes, and (as will be discussed in Chapter 9) studies using such an approach suggest that Rho might have a role in the generation of survival and proliferative signals in lymphocytes as well. Rho also plays an important role in integrin-mediated lymphocyte adhesion. Rho is activated by chemokine signaling initiated via serpentine receptors and may subsequently participate in adhesion and migration events.

Cdc42-GTP probably accumulates in specific locations in the plasma membrane and may thus be critical for the polarization of cells and for the formation of microspikes or filopodia. T cell polarization at the site of contact with APCs requires Cdc42. Cdc42 interacts with the Wiskott-Aldrich syndrome protein (WASP). In this X-linked disease, cytoskeletal abnormalities in platelets and lymphocytes have been described, and lymphocyte activation is also defective. WAS cells have defects in microvilli and probably have defects in "synapse" formation between lymphocytes and APCs. WASP, like many Cdc42- and Rac-binding proteins, contains a consensus 18-amino acid stretch called a CRIB (Cdc42/Rac interactive binding) motif. WASP is expressed only in hematopoietic cells, contains a number of proline-rich stretches, and interacts with SH3-binding proteins such as the adaptor protein Nck. WASP probably interacts with actin as well, although the structural basis for this is not well understood.

Vav is an important regulator of B and T cell development and activation, and these functions will be discussed in Chapters 7 and 9. A number of putative effectors have also been described for Rac, many of which also bind to Cdc42. One of the effectors that has been reported to bind to both Cdc42 and Rac is the p70 S6 kinase, which will be discussed in the section on signal transduction and translational regulation. Effectors such as p65PAK, MLK2/3, and POSH will be discussed in the next section when we discuss the activation of the JNK pathway.

The JNK/SAPK Pathway

The JNK/SAPK MAP kinase has been implicated in the induction of apoptosis in some cells, while it is critical for cell cycle progression in others. There are three different JNK genes in vertebrates, which encode JNK1, JNK2, and JNK3. Alternative splicing of these three genes yield 10 distinct proteins. JNK1 and JNK2 are expressed in lymphocytes. The best studied substrate of JNK/SAPK is the transcription factor, c-Jun. c-Jun is phosphorylated on Ser-63 and Ser-73 in its N-terminal activation domain, and phosphorylation increases its transcriptional activity. Other substrates of JNK/SAPK include the Elk-1 and ATF2 transcriptional regulators. As we will discuss later, ERK and JNK/SAPK cooperate in the induction of activity of Jun/Fos heterodimers, which are known as AP-1 (activation protein-1). In the context of lymphocyte development, most of the available evidence suggests that JNK is required for negative selection in T cells (see Chapter 9). JNK2 may be required for Th1 cell differentiation. B and T cell development is not impaired in *Jnk-2*$^{-/-}$ mice, but activated T cells fail to differentiate into activated effector Th1 cells, because JNK2 activity is apparently recquired downstream of the IL-12 receptor to induce IFN-γ secretion. Many different receptors can activate the JNK/SAPK pathway. In lymphoid cells, although JNK may be activated by antigen receptor signaling, the magnitude of activation by these receptors is relatively weak. More potent triggering of the JNK pathway is achieved by CD28 in T cells and by CD40 signaling in B cells.

Cdc42 and Rac may participate in the activation of JNK/SAPK by tyrosine kinases. As mentioned earlier, Vav is activated by tyrosine phosphorylation and by recruitment to PI(3,4,5)-P3. Vav is a GEF for Rac and will facilitate the generation of Rac-GTP. Rac-GTP may recruit effectors that activate the JNK pathway. In T cells, both calcineurin and a specific member of the PKC family, PKCθ, have been shown to contribute to JNK activation. It is believed that PKCθ may activate Rac, perhaps by inactivating a specific RacGAP. How exactly calcineurin activates the JNK pathway is unclear.

Many different downstream effectors of Rac and Cdc42 have been postulated to be involved in the activation of the JNK/SAPK pathway. Many of these effectors contain CRIB motifs. p65PAK is a serine/threonine kinase that is homologous to the *S. cerevesiae* Ste 20 kinase, which is upstream of MAPKKK in yeast. Another mammalian Ste20 homolog that has not been linked to Cdc 42 and Rac but that might regulate JNK activation is a serine/threonine kinase expressed at high levels in germinal centers known as germinal center kinase (GCK). It has been suggested that Cdc42 or Rac may activate p65PAK, which in turn might activate MEKK1 or other MEKKs that are MAPKKKs (see Figure 4.10). One of these MAPKKKs may then activate the subsequent steps in the pathway. Limited evidence suggests that p65PAK or GCK is actually involved in the activation of the JNK pathway *in vivo*. More likely candidates for Cdc42 and Rac-dependent activation of JNK are the mixed lineage kinases MLK2 and MLK3. These MAPKKKs contain a CRIB motif and can directly phosphorylate SEK1/MKK4, or a related protein known as MKK7 (which is expressed at high levels in T cells), which are MAP kinase kinases (MAPKKs) for the JNK/SAPK pathway. Another CRIB motif containing protein called POSH (plethora of SH3s) associates with Rac and has been shown to potently activate the JNK/SAPK pathway and the nuclear translocation of NFκB. POSH contains four SH3 domains and is believed to serve as a scaffold for JNK and NFκB activation.

In some cells, Ras-GTP may contribute to the activation of the JNK/SAPK pathway by interacting with and activating the MAPKKK, MEKK1. Potent activation of JNK/SAPK may, however, also result from pathways that do not activate Ras, Cdc42, or Rac. JNK activation by receptors such as CD40, TNF, and Fas probably does not require monomeric G proteins and will be discussed in other sections of this chapter as well as in Chapter 5. The known functions of small G proteins and components of MAPK pathways in lymphocyte development is summarized in Table 4.4.

Rap-1 Is a Small G Protein that May Negatively Regulate Signal Transduction

Rap1, a small G protein of the Ras family, has been implicated as a negative regulator of Ras signaling. Rap1 was originally identified as a negative regulator of Ras-induced transformation called K-Rev. Rap1 is transiently activated following T cell receptor ligation. It may be activated folowing the recruitment of a GEF for Rap1, C3G, to the membrane by the Cbl and CrkL adaptor proteins (described later in this chapter). Rap1 may also be activated by calcium transients, though the mechanism by which a calcium flux leads to the activation of this small GTPase remains unknown. Rap1 may be a negative regulator of the Ras pathway in some cells, including lymphocytes, perhaps because of its ability to compete with RasGTP for c-Raf. In other cells, Rap1 may actually interact with and activate B-Raf and thus function as a positive regulator of the ERK pathway.

The Phosphatidylinositol 3-OH Kinase Pathway

Phosphatidylinositol (PI) is an abundant component of the inner leaflet of the plasma membrane that is converted enzymatically into phosphatidylinositol-4 phosphate (PI4-P) and then into phosphatidylinositol (4,5) bisphosphate (PIP2). PIP2 serves as a substrate for hydrolysis by PI-specific phospholipases, as will be discussed in the next section. PI4,5-P2 can also

Table 4.4. Small G proteins and MAPK pathways in lymphocyte development

	Phenotype	References
Activated Ras targeted to the T lineage	Enhancement of $CD4^-8^-$ to $CD4^+8^+$ transition; expansion of $CD4^+8^+$ cells.	Swat et al., 1996
Dominant negative Ras or MEK1 targeted to the T lineage	Positive selection at $CD4^+8^+$ stage abrogated; negative selection unaffected.	Swan et al., 1995 Alberola-Ila et al., 1995, 1996
Dominant negative Ras or MEK1 targeted to the B lineage	Early block in B cell development, at or soon after lineage commitment.	Iritani et al., 1997
Jnk-1$^{-/-}$	T cell hyperproliferation, defective activation induced cell death (AICD), increased nuclear NFATc1 (nuclear factor of activated T cells, type c1), and preferential differentiation into Th2 cells.	Dong et al., 1998
Jnk-2$^{-/-}$	B and T cell development unaffected. Th1 cell differentiation impaired.	Yang et al., 1998
Sek-1$^{-/-}$	Lympadenopathy, polyclonal B and T cell activation.	Swat et al., 1998
Vav$^{-/-}$	Defective positive selection of thymocytes; defective antigen-receptor–mediated proliferation of T and B cells; defective actin cap formation.	Fischer et al., 1995 Tarakhovsky et al., 1995; Zhang et al., 1995; Turner et al., 1997; Holsinger et al., 1998
Inactive Rho in T lineage	Survival of pro-T cells compromised.	Henning et al., 1997 Galandrini et al., 1997

be phosphorylated by phosphatidylinositol 3-OH kinase (PI3K) in the 3 position of the inositol ring to yield an important second messenger, phosphatidylinositol (3,4,5) trisphosphate (PIP3) (Figure 4.11). PIP3 cannot diffuse into the cytosol but remains in the membrane, where it primarily provides binding sites for PH domain-containing proteins. PI3K may also phosphorylate PI4-P to yield PI3,4-P2. By facilitating the relocalization and activation of numerous PH domain proteins including TEC family kinases (such as Btk and Itk) of phospholipase C-γ, and anti-apoptotic proteins such as Akt, the PI3K pathway influences many different cellular events. These include mitogenesis, preventing apoptosis, reorganizing actin filaments, and influencing vesicular transport.

A major form of PI3K that is involved in signal transduction is made up of two subunits, p110 and p85. The regulatory subunit, p85, contains and SH2 domain, and the catalytic subunit is called p110. Since its substrate,

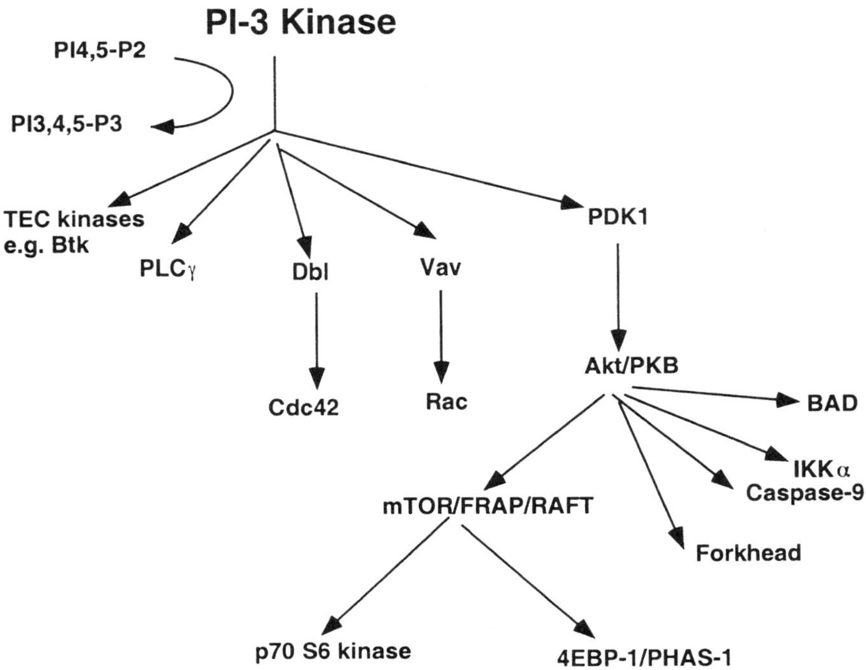

FIGURE 4.11. Some of the pathways that are influenced by PI3 kinase. A number of molecules that contain PH domains are activated by 3-phosphoinositides such as PI3,4,5-P3. The antiapoptotic function of PI3 kinase and Akt are considered in more detail in Chapter 5.

PI4,5-P2, is in the membrane, an important component of the activaton of PI3K involves its recruitment to the membrane. The SH2 domain of p85 specifically recognizes phosphorylated YMXP motifs. Following the activation of signaling in B and T cells, as will be discussed later in this chapter, the appropriate motifs in proteins such as CD19 in B cells and CD28 in T cells may recruit p85 to the membrane and thus activate PI3K. PI3K is also an effector of the Ras pathway, as discussed earlier, and Ras-GTP may also contribute to the activation of this kinase.

PI3K is the prototypic member of a growing family of dual-lipid/protein kinases. Members of this family include mTOR/FRAP (mammalian target of Rapamycin), a kinase that is inhibited by the immunosuppressant rapamycin. Interestingly, mTOR may be a direct downstream protein target of PI3K. The role of PI3K and activated mTOR in regulating translational initiation will be discussed later is this chapter. Other mamalian dual-lipid/protein kinases that are structurally related to PI3K include ATM (ataxia telangiectasia mutated). ATR (ataxia telangiectasia related), and the catalytic subunit of DNA-PK, which is the *scid* gene product. Defects in

ATM lead to immunodeficiency and lymphomas, as discussed in Chapter 12. We have already discussed the function of DNA-PK in Chapter 3.

The major lipid products of PI3K signaling are PI3,4,5-P3 and PI3,4-P2. These lipid products may influence the function of a number of PH domain-containing molecules by helping localize them to the membrane, often to the vicinity of other activating enzymes. Whether or not occupancy of the PH domain can allosterically influence the function of recruited enzymes is unclear. The levels of PIP3 in the cell are not only dictated by the activity of PI3K, but are also held in balance by the activity of a negative regulator, an SH2 domain-containing inositiel phosphatase, SHIP, which is recruited to certain phosphorylated ITIM-containing receptors. PI3K may activate a phosphoinositide-dependent kinase called PDK1 (3-phosphoinositide-dependent kinase-1), which contributes to the PI3K-dependent activation of an antiapoptotic kinase called Akt. In Chapter 5, we will examine the role of Akt in cell survival. PDK1 also phosphorylates a critical Thr residue in the activation loop of certain protein kinase C (PKC) isozymes, most notably PKCζ, contributing to their activation. In the next section, we will discuss in detail how PI3K sustains calcium signaling and therefore cellular proliferation and differentiation, through its influences on PLCγ and on TEC kinases.

Phospholipase C, Phosphoinositide Hydrolysis, and Calcium Signaling

The activation of PI-specific phospholipases that can cleave PI4,5-P2 leads to the the generation of two important second messengers, inositol triphosphate (IP3), and diacyl glycerol (DAG). Inositol triphosphate binds to a ligand-gated calcium channel in the endoplasmic reticulum membrane, the IP3 receptor, and this results in a rapid release of calcium into the cytosol. The sharp initial increase in cytosolic and nuclear calcium is often followed by a sustained increase of lower magnitude, which in combination with the initial flux leads to the activation of a number of calcium-dependent signaling events. Some of these events will be discussed in this section. Both DAG and calcium contribute to the activation of members of the PKC family. The PKC family will be discussed in the next section.

Many receptors, including antigen receptors, can activate members of the phospholipase C (PLC) family, which in turn can cleave PI4,5-P2 into DAG and IP3. Some trimeric G protein-coupled serpentine receptors may activate phospholipase C-β (PLC-β), whereas tyrosine kinase signaling may result in the activation of PLC-γ1 or PLC-γ2. PLC-γ1 and γ2 contain an N-terminal PH domain followed by tandem SH2 domains. In unstimulated cells, these enzymes are largely cytosolic, but following the induction of tyrosine kinase activity, they may be recruited to the membrane in a PI3K-dependent manner and also via PY-SH2 domain interactions. Activation of PLC-γ requires that it be recruited to the membrane to bring it close to its substrate, PI4,5-P2, and to also bring it in proximity to Syk and Tec family tyrosine kinases. Tyrosine phosphorylation of PLC-γ, particularly by Tec kinases, appears to be necessary for the optimal activation of this enzyme (Figure 4.12).

PI3,4,5-P3 therefore contributes in a major way to the activation of PLC-γ in lymphocytes. This contribution occurs directly, as well as indirectly via the activation of TEC kinases. PLC activation results in an initial spike in IP3 and cytosolic calcium, followed by a sustained but less marked increase of both these functionally linked mediators. It has been suggested that the

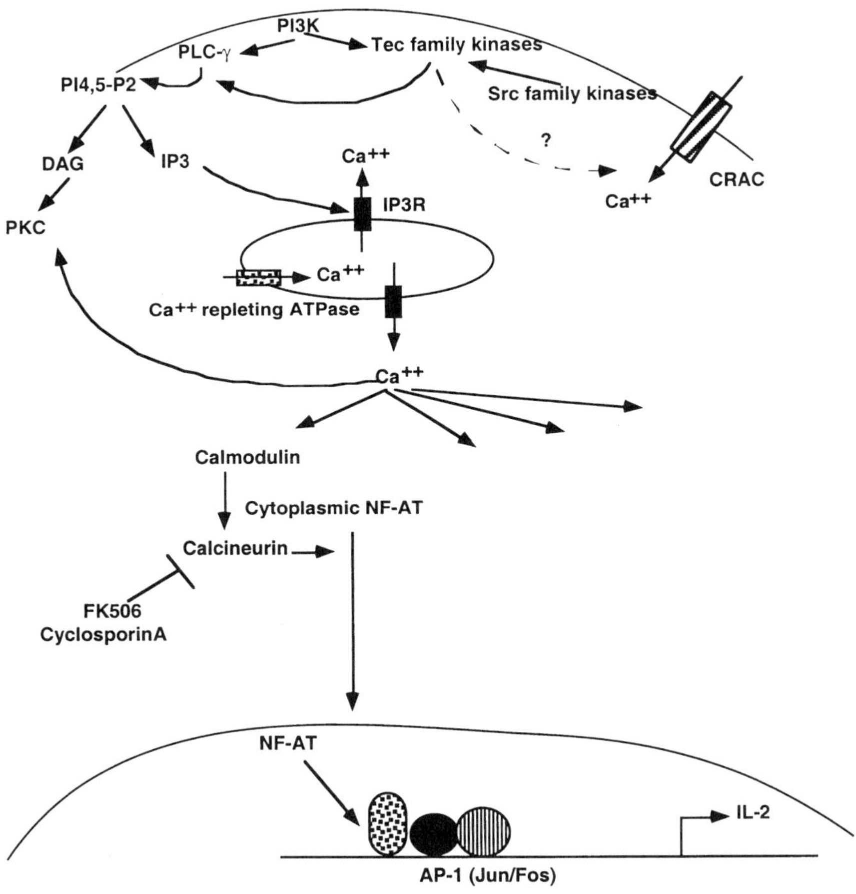

FIGURE 4.12. An overview of calcium signaling and NFAT activation. Activation of PLC-γ involves recruitment via SH2 domains to membrane adaptors as well as association of the PH domain of this enzyme with PI3,4,5-P3. Activation also involves phosphorylation by both Syk family and Tec kinases. PLC-γ generates IP3 and DAG from PIP2. IP3 production is critical for the initial Ca^{++} spike and probably also for capacitative calcium entry via CRAC channels.

initial large rise in cytosolic calcium is linked to NF-κB and JNK activation, while the lower but sustained wave of calcium is repsonsible for the activation of NFAT. The sustained increase in calcium depends, at least in part, on the generation of PI3,4,5-P3 and on the activity of Tec kinases. The levels of intracellular calcium stores are limited and are rapidly depleted in the initial spike of a calcium flux. Influx of extracellular calcium is required to maintain a sustained calcium signal. In lymphocytes, sustained (or capacitative) entry of calcium from the exterior occurs via calcium release-activated channels (CRAC) channels. The activity of these channels is regulated by a subset of intracellular calcium stores known as CRAC control stores. The initial spike of IP3 depletes CRAC control stores, and this leads to the opening of the CRAC channels. The calcium stores may be repleted by a refilling Ca^{++} ATPase in the ER and mitochondrial membranes. PI3,4,5-P3 and TEC kinases may contribute directly to the opening of CRAC channels, but it is more likely that they do so indirectly, by continuing to activate PLC-γ and thus sustaining IP3 levels. These IP3 levels ensure that CRAC control stores can continue to be sufficiently depleted to keep the CRAC channels open.

What ends the calcium flux? In B cells considerable evidence indicates that SHIP, recruited by the FcγRIIb1 regulator, is a major contributor to this tailing-off process. SHIP cleaves inositol phosphates in the D-5 position and thus negates the sustaining role played by PI3,4-5-P3 in both PLC-γ activation as well as in Tec kinase activation. It is likely that similar SHIP-dependent processes operate in T cells as well.

Apart from participating in the activation of protein kinase C, an increase in cytosolic calcium contributes to the initiation of a number of signaling events, some of which are mediated by Ca^{++}/calmodulin complexes. Ca^{++}/calmodulin can activate a range of cellular kinases (CaM kinases), and can also, as we have briefly discussed earlier, contribute to the activation of calcineurin, a serine/threonine phosphatase. Calcineurin targets many regulatory proteins including the nuclear factor of activated T cells (NFAT) family of transcription factors. Many other cellular proteins possess calcium-binding sites and may be functionally altered after they complex with cytosolic calcium. Cytosolic and nuclear calcium spikes are involved in the activation of cytotoxic T lymphocytes (CTLs) and NK cells, are required for cellular proliferation, can regulate gene expression, and may in certain circumstances accelerate apoptosis, in part by inducing a critical Ca^{++}-dependent endonuclease.

NFAT was initially discovered in the context of the regulation of IL-2 transcription. NFAT proteins bind to consensus sites in promoters or enhancers that are usually adjacent to an AP-1 site that binds Jun/Fos heterodimers. NFAT proteins contain a C-terminal Rel homology domain that binds DNA and an N-terminal regulatory region that includes a transactivation domain, a calcineurin-binding segment, a serine-rich regulatory domain, a nuclear localization signal (NLS), and an NLS-masking domain.

Some investigators refer to AP-1 components of NFAT complexes as NFATn (because they are nuclear proteins) and to the Rel homology proteins (which can be detected in the cytosol) as NFATc proteins. For convenience, in this book we refer to the Rel homology components of NFAT complexes as NFAT proteins, and the individual members are called NFATc1, NFATc2, NFATc3, and NFATc4 (the different nomenclatures used are summarized in Chapter 10 in Table 10.1).

There is some uncertainty as to the actual kinases that are involved in the phosphorylation of NFAT proteins. These proteins are probably phosphorylated initially in the serine-rich region by unidentified kinases that may in turn be activated by MEKK1 (a mediator of JNK activation) or by glycogen synthase kinase 3β (GSK-3β), an enzyme that is a major player in Wnt signaling (discussed later in this chapter). Phosphorylation of the serine-rich region might facilitate the interaction of NFAT proteins with casein kinase 1α, which phosphorylates serine/threonine residues in the NLS-masking domain. For some NFAT proteins, JNK/SAPK may directly participate in the phosphorylation event. Constitutive phosphorylation of this domain contributes to its ability to mask the NLS on NFAT and thus prevent NFAT from shuttling into the nucleus.

Following the induction of a calcium flux, Ca^{++}/calmodulin complexes bind to calcineurin. Activated calcineurin binds to the calcineurin-binding domain of NFAT proteins and mediates the dephosphorylation of the residues on the NLS masking domain. This results in the "unmasking" of the NLS and the subsequent nuclear translocation of NFAT (Figure 4.13). As mentioned earlier, in the section on phosphatases, cyclosporin A–cyclophilin complexes and FK506-FKBP12 complexes both bind and inactivate calcineurin B and thus prevent the dephosphorylation and subsequent nuclear translocation of NFAT proteins. The different NFAT proteins and the phenotypes of mice lacking specific NFAT proteins or combinations thereof are described in Chapter 10.

The Protein Kinase C Family

The PKC family is made up of at least 11 phospholipid-dependent serine/threonine kinases (Figure 4.14). PKC was originally descrbed as a calcium-activated, phospholipid-dependent cytosolic serine/threonine kinase. This description now holds true only for a subset of PKCs known as conventional PKCs. PKC subfamilies exist that either do not require calcium (novel PKCs) or that require neither calcium nor DAG for optimal activation (atypical PKCs). Specific PKC isozymes play important roles in lymphocyte development and activation, but the study of the function of individual isozymes is in its infancy.

All PKCs are single-chain proteins that contain an N-terminal regulatory region and a C-terminal catalytic domain. The N-terminal region contains membrane-targeting domains and a pseudosubstrate motif. The membrane-

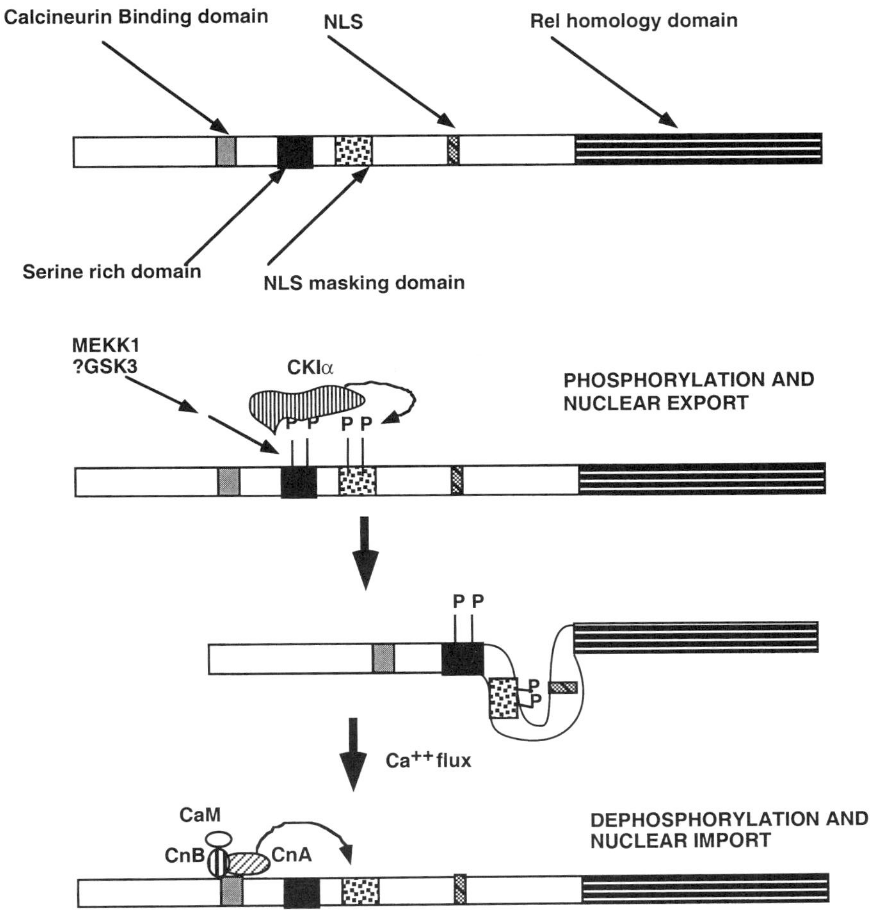

FIGURE 4.13. Mechanisms involved in nuclear shuttling of NFAT proteins. Phosphorylation of the serine-rich domain by unidentified serine/threonine kinases (which might be activated by other kinases such as MEKK1 or GSK3β) permits the recruitment and activation of casein kinase 1α. This kinase directly phosphorylates sites in the NLS-masking domain and prevents NLS accessibility. Following a calcium flux, calcineurin binds as a complex with Ca^{++}/calmodulin to the calcineurin-binding site and dephosphorylates NFAT, thus facilitating its nuclear import.

targeting domains help ensure that the interactions of these enzymes with their substrates are specific and of high affinity. Two cysteine-rich stretches make up the C1 domain. This domain can bind DAG or phorbol esters. In atypical PKCs, the C1 domain contains only a single cysteine-rich motif, but this domain does not bind DAG. The C2 domain binds phosphatidylserine. In conventional PKCs, this domain also binds Ca^{++}. The binding of Ca^{++} leads to an allosteric change and consequently an increase in affinity of this

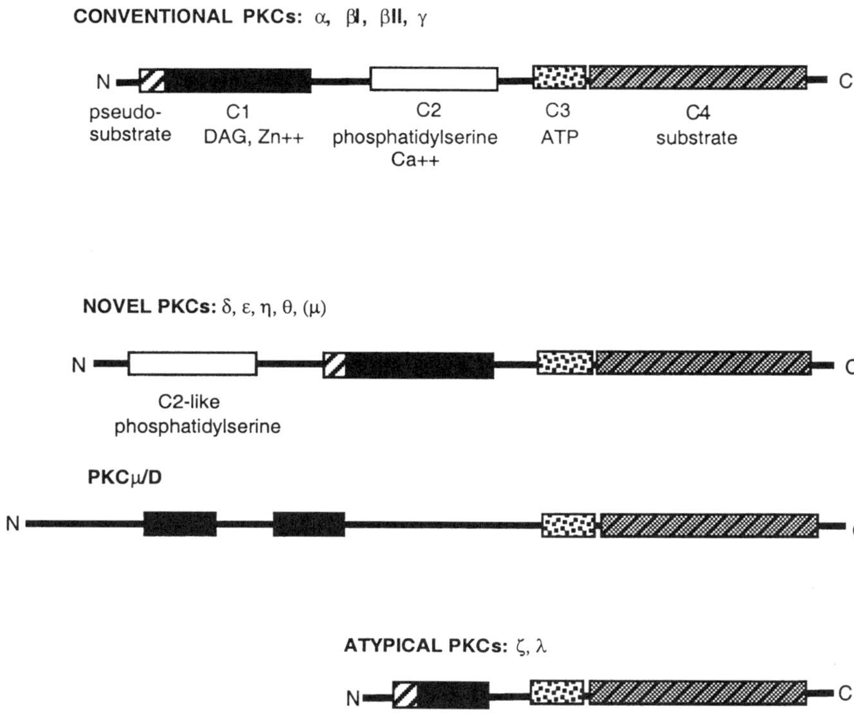

FIGURE 4.14. Members of the PKC family. All PKCs are activated in a phospholipid-dependent manner. Conventional PKCs respond to both DAG and Ca^{++}, whereas novel PKCs do not require Ca^{++} to be optimally activated. Atypical PKCs neither require Ca^{++} nor DAG for activation. PKCμ is sometimes categorized separately. It lacks a pseudosubstrate motif, and its catalytic domain is more similar to calmodulin-dependent kinases.

domain for phosphatidylserine. In novel PKCs, the C2 domain lacks critical aspartate residues involved in binding to Ca^{++}. The pseudosubstrate motif in the *N*-terminus interacts with the substrate binding site in the C-terminus and thus maintains protein kinase C in an inactive state. The C-terminal catalytic domain is made up of the C3 region, which contains an ATP-binding site, and the C4 region, which is the substrate-binding lobe. The activation segment in the catalytic domain of all PKC isozymes contains a critical threonine, which is not a site for autophosphorylation but must be modified by some other kinase for PKC to be activated. The only PKC-activating kinase identified so far is PDK1 (which lies downstream of PI3K), which activates PKCζ.

Conventional PKCs (cPKCs) include PKCα, PKCβI, PKCβII, and PKCγ. They require phosphatidylserine, DAG, and Ca^{++} for maximal activation. Both PKCβI and PKCβII are generated by alternative splicing at the PKCβ locus. This locus has been shown to be critical for B cell development. In the absence of PKCβ, B cells fail to proliferate in response to antigen-receptor ligation, T-independent responses are defective, and B-1 cells are not seen. This phenotype is similar to that observed in Btk-deficient or CD45-deficient mice, and it is possible that PKCβ and Btk are functionally linked.

Another subfamily of PKCs are referred to as novel PKCs (nPKCs). These include PKCδ, PKCε, PKCη, PKCθ and PKCμ. These enzymes are regulated by DAG and phosphatidylserine. This subgroup of kinases is structurally similar in many ways to cPKCs, but the C2 domain does not bind Ca^{++}. PKCμ (sometimes referred to as PKCD in rodents) is often categorized separately, based on its distinct structural features. It lacks a pseudosubstrate domain, and its catalytic domain is more similar to the corresponding domain of calmodulin-activated kinases than of other PKCs. PKCθ is an important player in T cell receptor-mediated signaling and translocates in T cells to the site of T cell-APC contact. PKCθ, PKCε, and PKCα all participate in the activation of the ERK pathway in antigen receptor and CD28-activated T cells, but only PKCθ particpates in the induction of JNK activity. This might reflect the specific role that this PKC isozyme might play in the activation of Rac, perhaps by inactivating a specific RacGAP. PKCμ may negatively regulate antigen-receptor–mediated signaling in B cells. This enzyme has been shown to be physically associated with the B cell receptor. Both Syk and PLC-γ may be substrates of PKCμ. It has been suggested that this PKC family member may downregulate the ability of Syk to phosphorylate PLC-γ, thus attenuating signaling via the BCR.

Atypical PKCs (aPKCs) include PKCζ, PKCι, and PKCλ. These kinases do not require Ca^{++} or DAG for their activation but do bind acidic phospholipids. In some cells, 3-phosphoinositides may activate PKCζ. Although PKCζ is expressed in lymphocytes, the function of atypical PKCs during lymphocyte development has not been systematically explored.

The NFκB/Rel Family of Transcription Factors

NFκB was originally described as a DNA-binding protein that associated with a specific site in the intronic immunoglobulin κ light chain enhancer. This factor was noted to be constitutively active in B cells. It could be induced by the exposure of pre-B cells and nonlymphoid cells to lipopolysaccharide (LPS) or phorbol esters (which activate PKC). Further studies revealed that this induction did not require the new synthesis of protein but depended on the release of NFκB from a cytosolic inhibitor. NFκB is the prototypic example of a transcriptional regulator that is retained in the cytosol in quiescent cells and that migrates to the nucleus in response to a wide range of signals. The molecular basis for this translocation event is understood in a broad sense, though many details remain to be worked out.

The interest in NFκB today extends far beyond the κ enhancer and B lymphocytes. Members of the NFκB/Rel family participate in stress and inflammatory responses, in the expression of numerous cellular and viral genes, and in developmental events in organisms that range from *Drosophila* to humans. In vertebrates, NFκB members are required during early hematopoiesis and also during many aspects of lymphocyte development. Signals for the induction of NFκB activity are delivered by many receptors including those specific for antigens, costimulatory ligands, and cytokines. NFκB is also required in many stress situations and may be induced by bacterial and viral pathogens, by agents that induce DNA damage, and by reactive oxygen intermediates generated during inflammation. Both during normal development, as well as in certain stress situations, a unifying functional role played by NFκB proteins might be the induction of genes that mediate cell survival.

The NFκB/Rel family in vertebrates includes NFκB1 (p50/p105), NFκB2 (p52/p100), RelA (p65), RelB, and c-Rel. All Rel family proteins contain a Rel homology domain, an approximately 300-amino acid strectch that includes sequences involved in DNA binding, nuclear localization, and dimerization. NFκB proteins bind to cognate κB sites in a range of promoters and enhancers, either as heterodimers or homodimers. The "original" description of NFκB referred to a p50/p65 heterodimer.

NFκB1 (p50) and NFκB2 (p52) contain little more than a Rel homology domain and do not contain a protein domain that can mediate transactivation (Figure 4.15). These proteins are derived by proteolytic cleavage of larger precursors (p105 is the precursor of p50, and p100 gives rise to p52). The C-terminal halves of p105 and p100 contain ankyrin repeats that help retain these proteins in the cytosol by masking nuclear localization signals on p50 and p52. These C-terminal portions resemble the inhibitory IκB proteins discussed below and probably provide another level of control during NFκB induction. A glycine-rich region in the middle of p100 and p105 directs an unidentified protease to cleave these proteins a little beyond this recognition signal. The C-terminal fragments of these proteins are then ubiquitinated and degraded in proteasomes. RelA (p65), RelB, and c-Rel all contain N-terminal Rel homology domains and C-terminal transactivation domains. A p50/p50 homodimer therefore cannot be a transcriptional activator, but a p50/p65 heterodimer, for example, can activate transcription.

Homodimers of p50 and p52 function as transcriptional repressors. Repression is an active process and does not depend merely on competition for DNA binding with transcriptionally active forms of NFκB. The Bcl-3 protein, discussed below, may participate in the repression process. Homo- and hetero-dimers that contain at least one transactivating chain (p65, RelB, or c-Rel) activate transcription, but distinct activating NFκB species bind most efficiently to subtly different κB sites. Pre-B cells express mainly

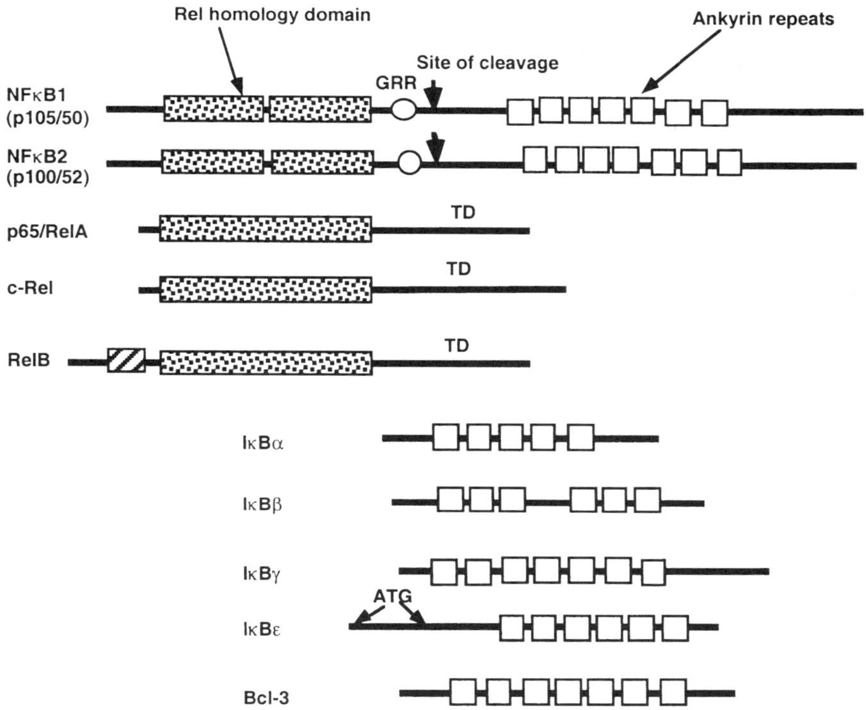

FIGURE 4.15. Vertebrate NFκB/Rel and IκB family members. p50 and p52 are produced by proteolytic cleavage of larger precursors. All NFκB family proteins contain a Rel homology domain. p65, c-Rel, and RelB also contain C-terminal transactivation domains (TD). The striped box at the N-terminus of Rel B represents a leucine zipper region. Ankyrin repeats in IκB family members are required for the binding of these proteins to NFκB dimers. Two different translational products are generated from IκBε transcripts.

p50/p65 dimers, immature and mature B cells primarily contain p50/c-Rel dimers, and plasma cells mainly express p52/RelB dimers.

The IκB Family of Inhibitors

In the cytosol, NFκB homo- and hetero-dimers are tightly associated with IκB (inhibitor of NFκB) proteins. All IκB proteins contain multiple ankyrin repeats. IκB prevents NFκB from entering the nucleus and activating transcription. When signals are received from the cell surface, specific kinases that initiate the phosphorylation of IκB are activated. These kinases phosphorylate IκB on serine 32 and serine 36. The phosphorylation of IκB leads to the ubiquitination of this protein and its subsequent degradation in 26S proteasomes. Following the degradation of IκB, nuclear localization signals on NFκB are no longer obscured, allowing this protein to migrate into the nucleus and to contribute to the transcriptional activation of specific targets (see Figure 4.16, later). The two members of this family that associate with most of the p50/p65 and p50/c-Rel complexes in the cytosol are IκBα and IκBβ.

IκBα and IκBβ contain an N-terminal regulatory domain, a set of ankyrin repeats, and a C-terminal PEST domain. The N-terminal domain contains two critical serine residues that are phosphorylated as a prelude to the induction of rapid degradation. The ankyrin repeats bind to NFκB and obscure the nuclear localization signal. The C-terminal PEST domain regulates the basal turnover of these proteins (as opposed to the induced rapid degradation controlled via the N-terminus). The primary function of IκBα is to regulate a rapid and transient increase in NFκB activity in response to specific triggers. The response is transient because of an interesting feedback effect. The IκBα promoter contains NFκB sites. Induction of degradation of IκB results in the activation of NFκB and consequently the increased synthesis of IκBα, thus completing the loop.

The persistent activation of NFκB that is seen after activation of cells by IL-1, bacterial LPS, or human T cell leukemia virus HTLV Tax-1, as well as the constitutive activation of this protein that is seen in mature B cells, may depend on the unique activity of IκBβ. The transcription of this inhibitor is not dependent on NFκB. It appears that both IκBα and IκBβ are basally phosphorylated on C-terminal serines by casein kinase II. However, the basal phosphorylation of IκBβ is regulated, and this might depend on the activity of a specific phosphatase. When cells are activated, the rapid degradation of both IκBa and IκBβ ensues. Newly synthesized IκBβ may be largely unphosphorylated, perhaps because of the activation of the above mentioned specific but as yet unidentified phosphatase. Unphosphorylated IκBβ can bind to NFκB in a unique way, ensuring that the nuclear localization signal of this transcription factor is not obscured. This association also competitively prevents IκBα from binding NFκB. The complex of NFκB and unphosphorylated IκBβ apparently migrates to the nucleus in a transcriptionally active form. Unphosphorylated IκBβ therefore functions as a chaperone mediating the persistent activation of NFκB.

Very little is known about the specific role played by IκBγ. This protein is encoded by a message that is initiated from a second promoter in the p105 gene, which transcribes exons corresponding to the C-terminal half of p105 (which contains multiple ankyrin repeats). The IκBγ protein has only been identified in mature B cell lines and appears to specifically interact with p50 or p52 homodimers. The biological significance of this potential inhibitor of transcriptional repressors remains to be elucidated.

IκBε appears to be a specific inhibitor of p65/p65 and p-65/c-Rel complexes. This inhibitor is also degraded in response to the phosphoryla-

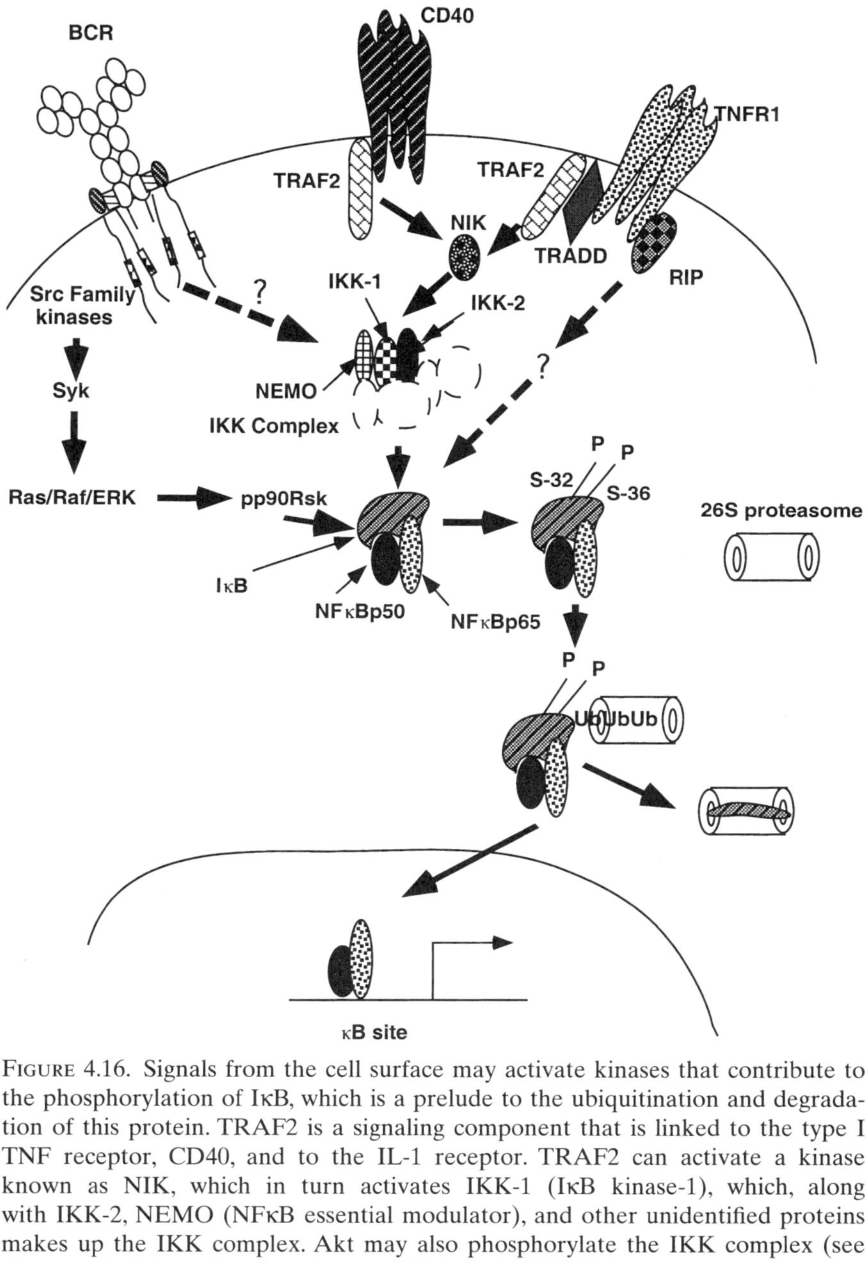

FIGURE 4.16. Signals from the cell surface may activate kinases that contribute to the phosphorylation of IκB, which is a prelude to the ubiquitination and degradation of this protein. TRAF2 is a signaling component that is linked to the type I TNF receptor, CD40, and to the IL-1 receptor. TRAF2 can activate a kinase known as NIK, which in turn activates IKK-1 (IκB kinase-1), which, along with IKK-2, NEMO (NFκB essential modulator), and other unidentified proteins makes up the IKK complex. Akt may also phosphorylate the IKK complex (see Chapter 5). Other kinases such as pp90Rsk and MEKK1 have also been linked to IκB phosphorylation.

tion of N-terminal serine residues but may regulate the expression of very specific genes that depend on p65 and c-Rel-containing NFκB complexes.

The final member of the IκB family that we will consider here is Bcl-3. Bcl-3 was discovered as the gene on chromosome 19 that was linked to the Ig heavy chain locus in patients with chronic lymphocytic leukemia who carry at (14:19) translocation. Although Bcl-3 structurally resembles the other IκB proteins, it has a number of very distinct properties. Unlike IκBα and IκBβ, which are cytosolic proteins, Bcl-3 is a nuclear protein. It binds specifically to p50 and p52 homodimers, and when combined with these repressors, it converts them into transcriptional activators. Unlike the other IκB proteins, which generally repress transcription, Bcl-3 activates specific transcriptional events.

The Induced Degradation of IκB

The kinase or kinases that phosphorylate IκB proteins on critical serine residues in the N-terminal region (serines 32 and 36 in IκBα) have long been recognized as key players in the activation of NFκB. Considerable progress has been made in the identification of IκB kinases and toward understanding the steps leading to the activation of these enzymes.

An important point that has been considered in examining kinases that are relevant has been the ability of these kinases to phosphorylate IκBα specifically on serine-32 and serine-36. Very few kinases have met this stringent criterion of specificity. One particular kinase complex that is activated downstream of receptors such as the TNF Receptor type I TNFRI, CD40, and the IL-1R (all of which signal via the TNFR2-associated Factor, TRAF2, protein, which will be discussed later) is now known as the IκB kinase (IKK) complex. This large multisubunit complex contains at least two catalytic subunits with serine/threonine kinase activity that are individually capable of modifying serine-32 and serine-36 on IκBα. These two subunits, generally referred to as IKK-1 and IKK-2 (or sometimes IKK-α and -β) cannot, however, phosphorylate IκB in quiescent cells. In activated cells, TRAF-2 interacts with a serine/threonine kinase known as NFκB-inducing kinase (NIK) and activates it in a poorly understood fashion. NIK in turn interacts with the IKK-1 and IKK-2 kinases, which are components of a large complex, and these kinases then acquire the ability to phosphorylate the critical serine residues in the N-terminal portion of IκB. Recent evidence suggests that NIK is not critical for IKK phosphorylation and that a distinct kinase called NAK (NFκB activating kinase) may be a key player in this process.

Apart from kinase domains, IKK-1 and IKK-2 each contain a leucine zipper and a helix-loop-helix (HLH) motif. IKK-1 was originally called CHUK (conserved HLH ubiquitous kinase). Another essential component of the IKK complex has been identified that contains a leucine zipper. This protein, known as NEMO (NFκB essential modulator), is absolutely required for the activity of this kinase complex. NEMO associates directly with IKK-2 and might regulate its activity.

Once IκB is phosphorylated on serine-32 and serine-36, it becomes a target for ubiquitination and subsequent proteasomal degradation. While it appears that TRAF2-linked receptors might sequentially activate NIK and the IKK complex, how other receptors target IκB proteins for degradation is unclear. Kinases such as MEKK1 (an activator of the JNK/SAPK pathway) and pp90Rsk, and S6 kinase activated by the ERK MAP kinase, have been implicated in the activation of NFκB. Very few kinases actually target serine-32 and serine-36 of IκBα specifically. Other than the IKK complex enzymes, one of the few kinases that does so is pp90Rsk. A number of receptors including antigen receptors on B and T cells might activate PKC and Ras and thereby initiate the degradation of IκB by pp90Rsk. Some receptors might trigger the activation of the IKK complex via kinases other than NIK. Some receptor tyrosine kinases activate PI3K and Akt, and Akt in turn may phosphorylate the IKK complex and thus contribute to NFκB ativation.

A brief summary of the phenotypes of knockouts that affect the NFκB and IκB proteins is provided in Table 4.5.

As may be inferred from Table, 4.5, NFκB and IκB proteins are critical for a number of developmental events, many of which impact directly or directly on lymphocyte ontogeny. NFκB1, RelA, and c-Rel are required for lymphocyte survival and proliferation and also participate in isotype switching in activated B lymphocytes. NFκB1 mediates survival of resting B cells (in G0), whereas c-Rel-containing complexes are required both for survival and cell cycle progression during the G1 phase after lymphocytes are triggered either via their antigen receptors or by other mitogens. NFκB/Rel proteins probably regulate the expression of distinct target genes when they mediate survival as opposed to cell cycle progression. Although studies in the WEHI 231 B lymphoma have suggested that NFκB may regulate the expression of *c-myc*, studies on primary B lymphocytes suggest that *c-myc* might not be downstream of NFκB.

NFκB2, Bcl-3, and RelB are required mainly in nonlymphoid cells, possibly downstream of receptors of the TNF receptor family (see section on "Signaling by Type III Cytokine Receptors" toward the end of this chapter and also Chapter 6). RelB is required for the proper development of dendritic cells and of the thymic medulla. It is also required for proliferative responses in B cells. NFκB2 and Bcl-3 probably work together and are required for the formation of follicular dendritic cell (FDC) networks and B cell follicles and for the survival and activation of specialized macrophage populations such as the metallophilic macrophages of the splenic marginal zone. These factors are critical for the proper organization of secondary lymphoid organs, a topic that will be discussed in Chapter 6. A common functional role played by many NFκB/Rel family dimers, both in lymphoid or nonlymphoid cells, may be in the protection of cells from apoptosis.

TABLE 4.5. Developmental roles of NFκB and IκB proteins

Knockout/ Transgenic line	Phenotype	Reference
NFκB1 (p50)$^{-/-}$	Defective B cell proliferation in response to LPS; normal response to CD40L and mIg ligation; defects in B cell maturation; defects in isotype switching (IgG3 and IgE reduced); defective T cell proliferation	Sha et al., 1995 Snapper et al., 1996b
NFκB2 (p52)$^{-/-}$	Reduction in B cell numbers; defects in humoral immune responses, germinal center formation, and splenic micro-architecture; lack FDC networks and marginal zone metallophilic macrophages	Franzoso et al., 1998 Caamano et al., 1998
NFκB1$^{-/-}$/ *NFκB2*$^{-/-}$	Defective B cell and osteoclast development	Franzoso et al., 1997b
RelA (p65)$^{-/-}$	Death at day 16 of embryogenesis due to massive liver degeneration; rescued B and T cells have defects in proliferation; defects in isotype switching to IgG1 and IgA	Beg et al., 1995b Doi et al., 1997
c-Rel$^{-/-}$	Defective B cell proliferation in response to LPS, CD40, and anti-mIg; defects in isotype switching (IgG1 and IgE reduced); defective T cell proliferation and cytokine generation	Kontgen et al., 1995
RelB$^{-/-}$	Reduced thymic epithelial and dendritic cells; myeloid hyperplasia; inflammatory infiltrate in many organs; defective B cell proliferation	Weih et al., 1995, 1997b Burkly et al., 1995 Snapper et al., 1996a
NFκB1$^{-/-}$/*RelA*$^{-/-}$	Mice die at day 12 of embryogenesis due to massive liver necrosis; stem cell rescue studies reveal a failure of lymphopoiesis due to defect in stromal cells	Horwitz et al., 1997
RelB$^{-/-}$/*NFκB1*$^{-/-}$	Defects in B cell proliferation, thymic medullary development, and dendritic cells, as in individual KOs; inflammation more severe.	Weih et al., 1997a
Bcl-3$^{-/-}$	Defects in germinal center formation, splenic and lymph node microarchitecture; humoral immune responses defective; phenotype resembles NFκB2 KO.	Franzoso et al., 1997a Schwarz et al., 1997
IκBα$^{/-}$	Mice die soon after birth of runting, enhanced granulopoiesis, skin abnormalities, atrophy of spleen and thymus; constitutive activation of NFκB	Beg et al., 1995a Klement et al., 1996
IκBα$^{/-}$/*NFκB1*$^{-/-}$	Onset of lethal phenotype of IκBα$^{-/-}$ mouse is delayed	Beg et al., 1995a
NFκB1 ΔC/ΔC (p105$^{-/-}$*)*	Enhanced proliferation of B cells; impaired cytokine production by macro-phages; chronic inflammation and suceptibility to bacterial infections	Ishikawa et al., 1998
NFκB2 ΔC/ΔC (p100$^{-/-}$*)*	Increased lymphocyte proliferation; gastric hyperplasia	Ishikawa et al., 1997
IκBα transgene overexpressed in T lineage	Defective T cell proliferation; decreased single positive CD8^{+} T cells	Boothby et al., 1997 Esslinger et al., 1997

Heterotrimeric G Proteins and Protein Kinase A-Mediated Signals

Serpentine (seven transmembrane/heptahelical) receptors are linked to heterotrimeric G proteins. The α subunits of these G proteins are GTPases. When in the GDP-bound state, Gα proteins associate with βγ heterodimers (which functionally behave like monomers), and the trimeric complex associates with serpentine receptors. Ligand binding results in a conformational change in the receptor, which results in a decrease in the affinity of Gα for GDP. As a result, the GDP is exchanged for GTP, and Gα is activated and dissociates both from the receptor and from Gβγ. The intrinsic GTPase activity of Gα results in the hydrolysis of the bound GTP to GDP and the reassembly of the heterotrimeric Gαβγ complexes and their reassociation with receptors.

While the classical example of a serpentine receptor is the β-adrenergic receptor, from a lymphocyte development standpoint, the most important receptors of this class are the receptors for chemokines. Once trimeric G proteins are activated, both the Gα-GTP and Gβγ subunits can contribute to the induction of downstream pathways. These include the activation of enzymes, the regulation of ion channels, and the facilitation of vesicular transport. Hetrotrimeric G proteins are often linked to the catalytic activation of a number of enzymes. These include adenyl cyclase, phosphodiesterases, phospholipase C-β, and Tec family tyrosine kinases such as Btk.

The activation of adenyl cyclase by trimeric G proteins leads to the activation of cAMP-regulated protein kinase isoforms known as PKAs. PKAs are made up of two regulatory subunits and two catalytic subunits. Very little is understood about the circumstances in which the activation of PKA in lymphocytes is important. However, the activation of PKA in lymphocytes generally antagonizes antigen-receptor signaling. Phosphorylation of c-Raf by PKA can inhibit the association of Raf and Ras and thereby antagonize ERK activation. Significantly, both PKA (type I) and Raf have been shown to be physically associated with the T cell receptor. The physiological significance of PKA activation in lymphocytes and the receptors that may be involved in this process remain largely unexplored issues.

Ceramide Signaling

Ceramide signaling is discussed briefly for completeness, although its importance in lymphocyte development remains a matter of conjecture. Lipid second messengers are critical for many signaling events. We have already discussed how DAG and PI(3,4,5)P participate in signal transduction. It is likely that, in many cells and for many receptors, signaling events that depend on the generation of ceramide from sphingomyelin are likely to be as important as the activation of protein kinase C by DAG.

Sphingomyelin is a phospholipid that is preferentially found in the outer leaflet of the plasma membrane and is cleaved by acidic or neutral sphingomyelinases (which have different pH optima) into ceramide and phosphocholine. Activation of acidic sphingomyelinases has been linked to receptors such as TNFRI, Fas, IL-1R, and the NGF (nerve growth factor) receptor. Sphingomyelinase activity may also be indirectly induced after a considerable lag phase by the B cell receptor and by CD28. It is possible that ceramide generation is part of an inhibitory signaling cascade in lymphocytes. An important target of ceramide is the kinase suppressor of Ras, which is a ceramide-activated serine/threonine kinase that may inactivate Ras signaling in certain cell types. Ceramide can also activate a distinct kinase that activates the Ras pathway, a specific ceramide-activated phosphatase, Vav (which is an exchange factor for Rac), and PKCζ.

Notch Signaling

An important signaling pathway known to be involved in cell fate determination in *Drosophila* and *Caenorhabditis elegans* is linked to the Notch receptor. There is growing evidence for the importance of this pathway in many aspects of vertebrate development including the development of T and B lymphocytes. Two broad messages are generated from the study of Notch in invertebrates that might be worth keeping in mind. Notch signaling is often linked to the maintenance of cells in an undifferentiated state. Since, in general, proliferating cells do not differentiate, the ability of Notch to maintain cells in the undifferentiated state may be linked to its ability to induce or potentiate signals involved in cellular proliferation. Notch is also involved in a type of signaling during differentiation called lateral specification or lateral inhibition. This is a phenomenon in which neighboring cells start out being identical; the stochastic induction of some change in a subset of these cells allows them to induce neighbors to acquire different fates, resulting in the generation of nonequivalent neighbors. Lateral inhibition is easily distinguished from other forms of signaling between cells that may influence development. Signaling in the thymus, between an antigen-presenting cell and a T cell for instance, might represent a kind of "inductive signaling" where the cells that signal to each other are nonequivalent. It is unclear whether lateral specification is important in lymphocyte development; intutitively, it might be assumed to be more likely to be involved in thymic epithelial development than in thymocyte development, for instance. However, it is possible that in the developing thymus thymocytes do talk to each other and thus influence cell fate decisions via lateral signaling.

Notch is an integral membrane protein with a large number (34 in most vertebrate Notch proteins) of extracellular EGF-like repeats, three extracellular cysteine-rich repeats, six intracellular ankyrin/cdc10 repeats and a PEST region near the C-terminus (Figure 4.17). The EGF-like repeats are involved in ligand binding, and the ankyrin repeats facilitate the interaction of the cytoplasmic tail of Notch with C promoter binding factor 1 CBF1/RBPJκ1, a transcription factor that is the vertebrate homolog of the *Suppressor of Hairless* gene in *Drosophila*.

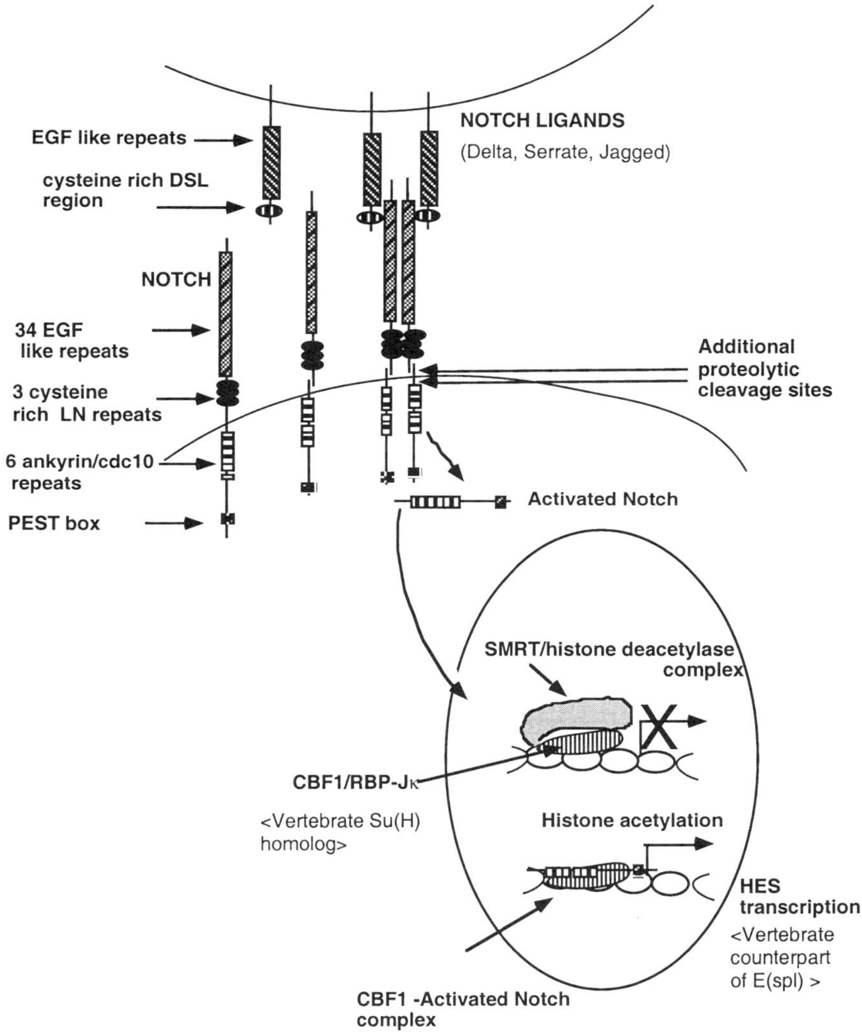

FIGURE 4.17. A nuclear translocation model for Notch signaling. In this model, Notch is cleaved in a stepwise fashion, the final cleavages depending on its interaction with its ligand. Activated Notch is translocated to the nucleus, where it converts CBF1 from a repressor into an activator.

There are four Notch proteins in the mouse, Notch1, Notch2, Notch3 and Int-3/Notch4. Of these, Notch2 is expressed at particularly high levels on splenic B cells. Notch was discovered in vertebrates at the translocation breakpoint of a T cell lymphoma, which brought the TCR β locus in apposition to a truncated but active form of Notch including largely the transmembrane domain and the cytoplasmic tail region; this is known as TAN-1.

Although TAN-1 contains the hydrophobic membrane anchor sequence of Notch, it lacks the N-terminal signal peptide and most of the extracellular domain sequences. TAN-1 therefore cannot be translocated into the endoplasmic reticulum and is presumably synthesized on cytosolic ribososmes and is subsequently translocated to the nucleus.

All the known Notch ligands in invertebrates and vertebrates are integral membrane proteins expressed on the plasma membrane. Vertebrate homologs of the *Drosophila* Notch ligands, delta and serrate, and the *C. elegans* ligand, Lag-2, have been identified. A mammalian serrate-like protein is called Jagged. All Notch ligands contain multiple extracellular EGF repeats and also a cysteine-rich motif known as the DSL (for delta, serrate and lag-2) region. The DSL motif on the ligand and some of the EGF motifs on Notch are required for the ligand–receptor interaction and subsequent signaling.

The Notch protein is cleaved in its extracellular domain by a constitutively active protease. The extracellular portion and the rest of the cleaved protein remain noncovalently associated on the cell surface. When Notch and its ligand interact, the receptor aggregates and may induce two further proteolytic cleavages in quick succession. The first cleavage is extracellular and may depend on the protease encoded by the *Kuzbanian* gene in *Drosophila*. The subsequent cleavage event occurs in a part of the cytosolic domain in close proximity to the plasma membrane, releasing a fragment referred to as "activated Notch," which enters the nucleus and activates transcription (Figure 4.17). One model suggests that activated Notch regulates the activity of a transcription factor known as CBF1/RBPJκ/KBF. This nuclear protein was originally thought to be involved in Igκ light chain recombination but has turned out to be a transcriptional regulator. CBF1 is the vertebrate homolog of the *Drosophila suppressor of hairless* (Su(H) gene. In *Drosophila*, Su(H) regulates the expression of genes of the enhancer of split (E(spl)) complex, which encode bHLH repressors of the *achaete-scute* complex genes. Notch signaling in Drosophila therefore downregulates the expression of *achaete scute* genes and consequently of *delta*, which encodes a ligand for Notch.

An alternative model for the Notch pathway envisages the binding of a protein called Deltex to the ankyrin repeat regions of Notch following ligand-induced aggregation. This model postulates that, in noninduced cells, the cytoplasmic tail of Notch physically associates with CBF1. Ligation of Notch apparently results in the displacement of CBF1 by Deltex, thus facilitating the entry of CBF1 into the nucleus. In versions of this model, activated Notch may also migrate into the nucleus and cooperate with CBF (Figure 4.18). Deltex contains a proline-rich stretch that can bind to an adaptor protein called Grb2 (discussed later in this chapter) that contains an SH3 domain and two SH2 domains. Deltex has been described to mediate notch signaling in a CBF-independent manner, possibly because it can activate the Ras pathway via Grb2 and SOS.

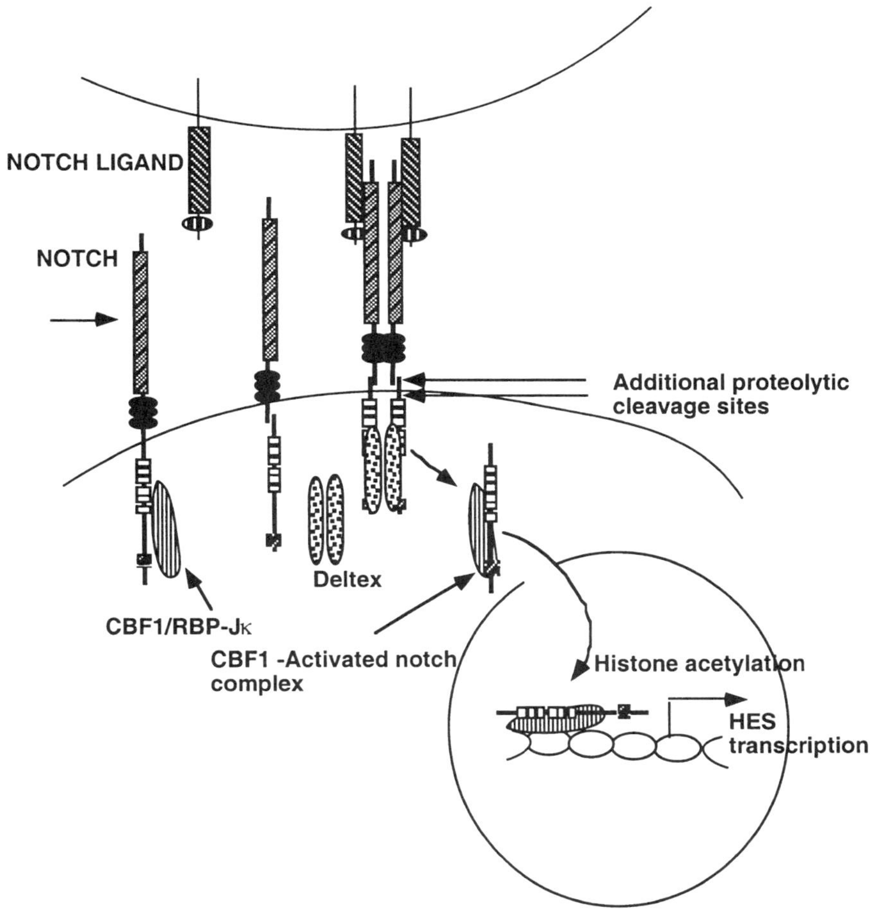

FIGURE 4.18. An alternative CBF1 displacement model for Notch signaling. Deltex is an activator of Notch signaling and might displace CBF1 from the cytoplasmic tail of Notch. Deltex can bind Grb2 and may contribute to the activation of other pathways.

In mammalian cells, CBF1 functions as a repressor in the absence of Notch sigaling and associates with activated Notch to mediate transcriptional activation. Before induction, CBF1 is a part of a repressor complex that contains the SMRT (silencing mediator of retinoid and thyroid hormone receptors) protein and the histone deacetylase HDAC-1. Histone deacetylation contributes to repression of target-gene expression, the mammalian targets including a homologue of the hairy and enhancer of split complex genes known as *Hes-1*. This CBF1/SMRT/HDAC-1 repressor complex is disrupted when activated Notch binds to CBF1 and (presumably) promotes histone acetylation and transcriptional activation of Hes genes.

Apart from studies that have revealed the ability of Notch to influence cell fate in *Drosophila* and *C. elegans*, this pathway has been shown to inhibit myogenic, neural, and myeloid differentiation in vertebrate cell culture systems.

As discussed in Chapter 2, Notch 1 signaling can influence the entry of lymphoid progenitors into the T cell compartment at the expense of B cell generation. Interestingly, an early block in T cell development is also observed in *Hes-1*$^{-/-}$ mice. Constitutively activated Notch contributes to the generation of leukemias involving immature T cells. Studies using activated Notch targeted to the T lineage in mice suggest that this pathway may contribute to the development of αβ T cells (as opposed to γδ T cells), perhaps by helping mediate positive selection I (discussed in Chapter 9). Notch 1 may have an antiapoptotic role in double-positive T cells. Notch signaling might also influence lineage commitment during positive selection II, directing $CD4^+8^+$ T cells toward the $CD8^+$ lineage. In this context, HES-1 has been demonstrated to be involved in the silencing of the CD4 locus.

Notch signaling has also been postulated to influence B cell development. It has been shown that Notch signaling (via Deltex and independent of CBF1) suppresses the activity of the bHLH protein E47, a product of the E2A gene, which is required for B cell development. Notch might inhibit B cell development, maintaining precursors in an undifferentiated and cycling state.

Wnt Signaling

Wnt proteins have been demonstrated to be critical in developmental pathways in *C. elegans*, *Drosphila*, and the mouse. They participate in embryonic induction events, in the generation of cell polarity, and in a number of cell fate decisions. Although there is no direct evidence for the involvement of a secreted Wnt protein in lymphocyte development, transcription factors that are targets of Wnt signaling were originally discovered in the context of lymphocyte development. Given the importance of the TCF/LEF-1 family of proteins in lymphocyte development, it is reasonable to assume that the upstream component of the Wnt pathway might participate in lymphoid development as well.

Wnt proteins are secreted and bind to serpentine receptors of the Frizzled family (Figure 4.19). Although Frizzled looks like a G-protein coupled receptor, and some circumstantial evidence suggests a role for trimeric G proteins in Wnt signaling, no genetic or biochemical evidence as yet unequivocally places G-proteins in this pathway. Frizzled presumably talks to Disheveled (Dsh), a cytosolic phosphoprotein that in a poorly understood way negatively regulates a kinase known as glycogen synthase Kinase-3 (GSK-3). Dsh may be hyperphosphorylated in response to Wnt

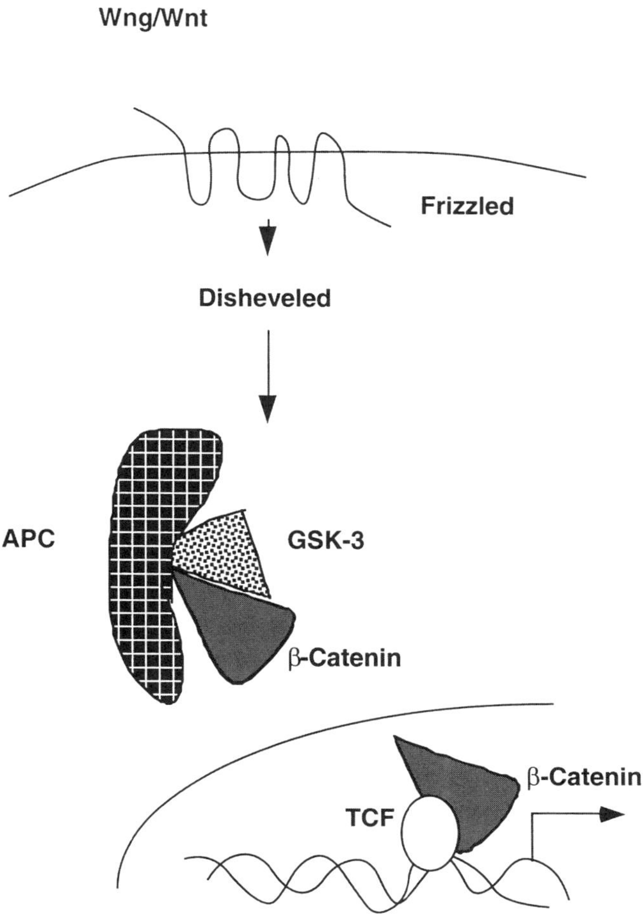

FIGURE 4.19. An overview of Wnt signaling. Frizzled is a serpentine receptor that inactivates GSK-3β and thus stabilizes β-catenin. β-catenin facilitates the transcriptional activation of target genes by TCF/LEF-1 proteins.

signals, but it is unclear whether phosphorylation of Dsh is critical for its function.

GSK-3β is somewhat unusual. It is a kinase that is constitutively active in nonstimulated cells and is inactivated in cells that have been triggered by a number of mitogens. Phosphorylation of Ser-9 of GSK-3β by pp90Rsk or by Akt may reduce its catalytic activity. Whether a kinase is interposed between Dsh and GSK-3β is unclear, although it has been suggested that in mammalian cells PKC may be induced downstream of Frizzled and that this kinase may mediate the inactivation of GSK-3β.

GSK-3β forms a complex with a protein known as β-catenin and with the APC protein (the product of the *adenomatosis polyposis coli* gene, involved in the transformation of colonic cells). Inactivated GSK-3β stabilizes β-catenin, presumably because it is not aggressively targeted for degradation by the ubiquitin-proteasomal pathway when it is underphosphorylated. GSK-3β might actually phosphorylate APC, and APC might be responsible for regulating the degradation of β-catenin. Stabilization of β-catenin facilitates its accumulation in the cell and its migration into the nucleus, where it associates with high mobility group (HMG) box-containing transcription factors of the TCF/LEF-1 family.

TCF and LEF-1 are required for early T cell development, as discussed in Chapter 9, and for the development of some NK cells (see Chapter 11). TCF-1 is active in a β-catenin independent manner in the context of the *TCR-α* enhancer. However TCF activation may require β-catenin in the context of other genes. Wnt 1 and Wnt 4 are known to be expressed in the thymus, and it is known that early B lineage precursors may proliferate and differentiate when exposed to Wnt proteins.

Signals From the Cell Surface That Target Protein Synthesis

Signals from the cell surface must be integrated to facilitate cell cycle progression but also must be molecularly coordinated to facilitate cell growth, which is usually a necessary prelude to cell cycle progression. The study of the mechanism of action of the immunosuppressive compound, rapamycin, has contributed to our understanding of how signals are integrated within the cell to enhance protein translation.

Translation of capped mRNAs requires that a set of proteins bind to the cap and contribute to the initiation of translation. The eukaryotic initiation factor (eIF)-4E is a subunit of the eIF-4F complex that binds to the cap and recruits eIF-4G and eIF-4A. In quiescent cells, eIF-4E exists in a nonphosphorylated state and binds to a protein known as 4E-binding protein 1 (4E-BP1, or PHAS-1), which prevents eIF-4E from associating with eIF-4A and eIF-4G. In this situation, translation of capped mRNAs is drastically inhibited. Once a cell is triggered by the appropriate mitogens, a cascade of events is set in place, primarily by the activation of PI3K (Figure 4.20). The generation of PI3,4,5P3 in the membrane facilitates the recruitment and activation of PDK1 and Akt as discussed earlier. Activated Akt can phosphorylate and activate the mammalian target of rapamycin (mTOR/FRAP/RAFT), which is a rapamycin-sensitive member of the extended PI3K family. Catalytically active mTOR may in turn phosphorylate 4E-BP1 and thus cause it to release eIF-4E, which is then free to associate with the other eIF-4F proteins and initiate cap-dependent translation. Translation is further facilitated by the phosphorylation of eIF-4E by a kinase that lies downstream of the ERK MAP kinase.

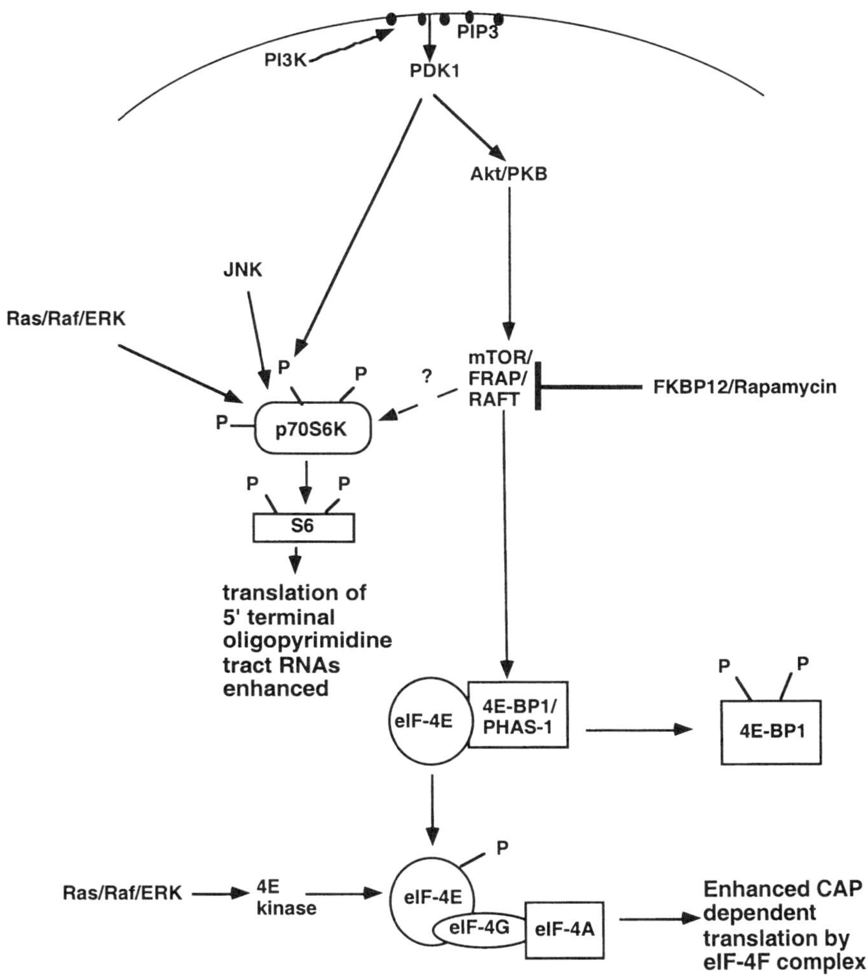

FIGURE 4.20. Signals from a number of pathways are integrated to regulate protein translation.

A second general mechanism by which translation may be regulated by extracellular signals depends on the convergence of a number of signaling pathways on a specific kinase that phosphorylates the ribosomal S6 protein. This enzyme is known as the p70S6 kinase (p70S6K). It contains an autoinhibitory domain whose repressive activity may be negated by proline-directed kinases such as ERK and JNK/SAPK. p70S6K may also be phosphorylated on sites in its kinase domain by PDK1 (which is downstream of PI3K) as well as by the rapamycin-sensitive mTOR kinase.

The Regulation of Gl Progression by Extracellular Signals

A resting or G0 cell may be induced to move out of this quiescent stage by signals that increase the overall efficiency of transcription and translation in the cell. We have just considered how translation may be enhanced by a number of signaling pathways. External signals during G1 may drive a cell to progress further in the cell cycle. In G1, the cell needs to be driven into the S phase of the cell cycle, during which DNA synthesis occurs as a prelude to mitosis (Figure 4.21).

External signals that drive proliferation do so primarily by contributing to the activation of specific cyclin-dependent Kinases (CDKs), that are critically required during the G1 phase of the cell cycle. CDKs are proline-directed serine-threonine kinases and, as their name implies, are active only when they are complexed with cyclins. Cyclins are small proteins that are noncatalytic regulatory subunits of CDKs but that are degraded in a cyclical fashion during the cell cycle. A number of different types of cyclins are required at distinct stages of the cell cycle. The cyclins that are critical in

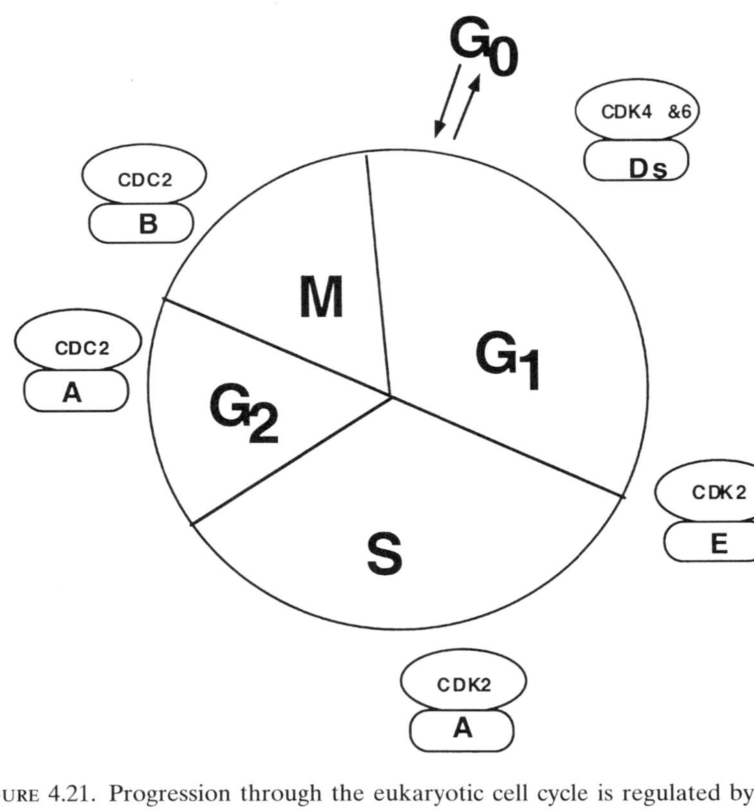

FIGURE 4.21. Progression through the eukaryotic cell cycle is regulated by cyclin-dependent kinases, which are active only in complex with a cyclin partner.

the early to mid G1 phase are D-type cyclins. In lymphocytes, cyclin D2 and cyclin D3 are the members of this class, which are induced in G1 and associate with and activate CDK4 and CDK6. Late in G1, the cell cycle is driven by CDK2/cyclin E complexes, and CDK2/cyclin A complexes help drive cells through the S phase. Mitosis depends on the catalytic activity of CDC2-containing complexes.

The activity of CDKs may be regulated in the following ways:

1. They are activated, as mentioned above, when they complex with cyclins.

2. The activation segment (sometimes referred to as the T-loop) in CDK domains may be phosphorylated on a critical threonine residue, Thr-160 in CDK2, by a CDK-activating Kinase (CAK). There are some indications that a distinct regulatory kinase other than CAK may be induced in G1. Phosphorylation of the activating segment is required for the full activation of most kinases, as discussed earlier.

3. CDKs may be maintained in an inactive configuration when they are phosphorylated on conserved threonine and tyrosine residues in the region of the ATP-binding site (Thr-14, and Tyr-15 in CDK2). The Wee-1 and Mik-1 protein kinases are responsible for these inhibitory modifications. These phosphate residues are removed by CDC25 phosphatases, which are positive regulators of the cell cycle. CDC25A is active at the G1-S transition, and CDC25C is required at the G2-M transition. CDC25 molecules may themselves be phosphorylated and catalytically activated by CDKs, thus generating a positive feedback loop for CDK activation.

4. The activities of CDKs may also be regulated by CDK inhibitors that negatively regulate kinase activity. These inhibitors associate with CDKs and can inactivate them either by displacement of the regulatory cyclin or by sterically limiting access to substrate. There are two major families of CDK inhibitors. The p16 family (sometimes referred to as the INK4 family (inhibitors of CDK4), includes p15, p16, p18, and p19. The members of this family only inactivate CDK4 and CDK6, kinases that drive cells through G1 and that are both dependent on D-type cyclins. The p21 family includes p21, p27, and p57, and members of this family inhibit a wide range of cyclin-CDK complexes. p27 has been co-crystallised as part of a cyclin-CDK-CDKI complex and has been shown to occlude the catalytic cleft of the CDK, thus preventing access to substrate.

Extracellular signals may contribute to G1 progression in a number of ways. Activation of the Ras/ERK pathway leads to the transcriptional induction of D-type cyclins (this is known for cyclin D1, which is not expressed in lymphocytes). Sufficiently high levels of these proteins may displace p16 family CDKIs from CDK4 and CDK6 and thus contribute to the activation of these G1 kinases. Induction of *c-myc* is believed to contribute to the transcriptional activation of CDC25A, a phosphatase that dephosphorylates the inhibitory phosphates on Thr-14 and Tyr-15 on CDKs.

Studies performed on a B cell line led to the conclusion that NFκB might be involved in the induction of *c-myc*. It has not been possible to make such a conclusion based on the study of B cells from mice lacking NFκB1 and c-Rel. Although a constitutively active kinase known as CAK, whose catalytic subunit is itself a CDK-type kinase known as CDK7, is considered to be responsible for phosphorylating the activation loop threonine on CDKs, no evidence indicates that CAK is responsible for the phosphorylation of CDK4 or CDK6 in G1. A yet to be characterized G1 specific T-loop (activation segment) kinase probably is induced or activated early in the cell cycle.

Once CDK4/6 are catalytically activated, they contribute to the phosphorylation of the retinoblastoma protein (Rb). Rb contains a pocket to which many proteins are bound, the most prominent being a set of transcription factors belonging to the E2F family. Phosphorylation of Rb leads to the release of bound E2F factors, which contribute to the induction of a number of genes required for DNA synthesis and other aspects of S phase progression. There are least six known E2F factors in vertebrates: E2F-1, E2F-2, E2F-3, E2F-4, E2F-5 and E2F-6. Each E2F protein is associated with a heterodimeric partner known as DP (there are actually two DPs, DP-1 and DP-2). While E2F-2, E2F-3, and E2F-4 probably contribute to S phase progression by transcriptionally inducing some of the genes that are involved in DNA synthesis, E2F-1 may play a part in driving cells down apoptotic pathways. Activation of CDK4 and CDK6 and G1 progression therefore either drives a cell through the G1/S transition and inexorably through the cell cycle, or in certain circumstances it may actually set up a suicide pathway.

Apart from the activation of CDK4 and CDK6 as described above, further progression through the cell cycle may also depend on other intracellular changes driven by external signals. These include the induction of the degradation of the p27 CDK inhibitor during the G0 to G1 transition in activated B and T cells. In IL-2 stimulated T cells, this degradation of p27 depends on a rapamycin-sensitive pathway. How exactly this pathway impacts upon the proteasomal degradation of p27 is not well understood. Whether this is somehow linked to the rapamycin-sensitive, mTOR-regulated, translational control mechanisms discussed in the previous section is unclear.

Antigen Receptors, Coreceptors, and Costimulators

In this section, we will initially discuss issues of relevance to antigen-receptor signaling in general and will then examine how antigens specifically trigger B and T cells. Some of the shared structural features of B and T cell receptors will be described, and we will examine the mechanisms by which antigen-receptor signaling is initiated, focusing on the role of receptor crosslinking. We will distinguish between coreceptors and costimulatory receptors and describe some of the adaptor molecules that are critical for signal initiation in B and T cells.

General Features of Antigen-Receptor Signaling

Some Aspects of Antigen-Receptor Structure of Relevance to Signaling

The antigen receptors in B and T lymphocytes (BCR and TCR) are broadly similar in structure and are believed to initiate signaling in similar ways. The antigen-binding chains of these receptors do not directly contact cytosolic signaling molecules but interact with accessory proteins that are also anchored in the plasma membrane and whose cytoplasmic tails contain motifs known as ITAMs. ITAMs contain the following consensus sequence:

$$\mathrm{YxxL/Ix_{6\text{–}8}\ YxxL/I}$$

These motifs are found in proteins associated with a number of signaling receptors in the immune system. When tyrosine residues in "YxxL/I" tetrapeptides are phosphorylated, they may be recognized by specific SH2 domains present in certain proteins. Phosphorylated ITAMs therefore can recruit molecules that contain specific SH2 domains to activated antigen receptors.

The TCR is made up of either α and β chains (in $\alpha\beta$ T cells) or γ and δ chains (in $\gamma\delta$ T cells). These antigen-binding transmembrane polypeptides are linked to one another by a disulfide bridge. Each heterodimer is associated with the CD3 complex made up of integral membrane proteins that contain ITAMs in their cytoplasmic tails (Figure 4.22). The pre-T receptor is generated only in $\alpha\beta$ lineage cells before α chain rearrangement and will be discussed in detail in Chapter 9. This receptor consists of the β chain linked to a surrogate α chain known as preTα, and the noncovalently associated CD3 complex. The CD3 complex consists of CD3γ, CD3δ, CD3ε, and ζ chains (the latter are generally not called CD3 proteins, but they do belong to this complex). ζ Chains generally exist as disulfide-linked homodimers and contain three ITAMs in their cytoplasmic tail regions; a small proportion of TCRs may be associated with ζ–η heterodimers. The η chain is the product of an alternatively spliced version of the ζ chain gene. The γ, δ, and ε chains of the CD3 complex each contain a single cytoplasmic ITAM.

B lineage preantigen receptors (the pro-B receptor and the pre-B receptor) as well as the B cell antigen receptor, share very similar structures. The antigen receptor on B cells is made up of membrane immunoglobulins (IgM and IgD in naive B cells; IgG, IgA, or IgE in some activated and memory B cells) associated with a disulfied-linked heterodimer made up of two integral membrane glycoproteins, Igα and Igβ. These latter proteins each contain an ITAM in their cytoplasmic tails. The pre-B receptor lacks conventional light chains but contains pre-B specific surrogate light chains. Pro-

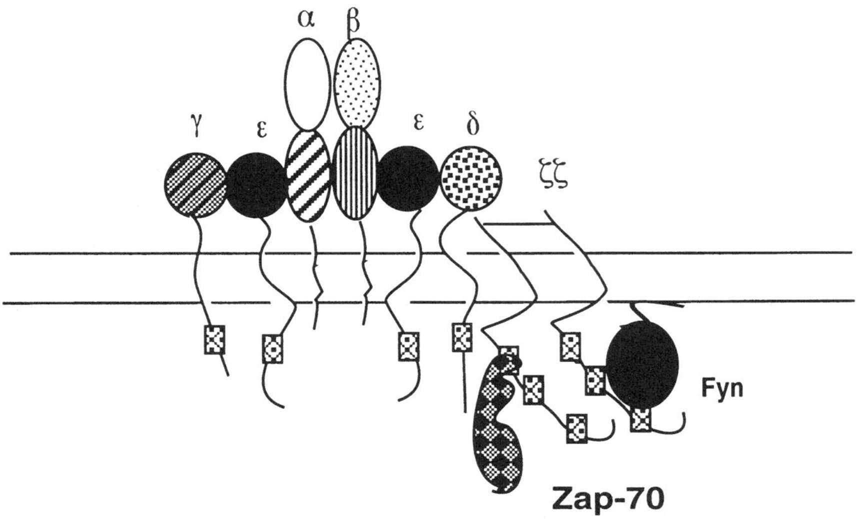

FIGURE 4.22. A schematic view of the antigen receptor on T cells, prior to its interaction with a cognate antigen. Fyn is a myristylated Src-family kinase that associates with the receptor, although the details of this interaction are poorly understood. Zap-70 is a member of the Syk family that is recruited to phosphorylated ITAMs. Some "constitutive" association of Zap-70 with the TCR in resting B cells may reflect low-level signaling by self-MHC (major histocompatibility complex) ligands that are responsible for the generation of survival signals.

B and Pre-B receptors will be discussed in greater detail in Chapter 7. The cytoplasmic tails of the μ and δ heavy chains contain only three amino acids (KVK), which basically function as a "stop transfer" sequence during the process of translocation into the ER and do not present a significant surface for the association of these receptors with cytosolic signal transducers. The ITAMs in Igα and Igβ are critical for signal transduction) by the B cell receptor (Figure 4.23).

Although antigen receptors may potentially initiate signals in multiple ways, the major mechanism by which signals are generated is the activation of tyrosine phosphorylation. Src-family kinases associate with the antigen receptor even in quiescent cells. The cytoplasmic tails of coreceptor proteins (discussed below) also bind to Src-family kinases. Some activation of antigen receptors occurs even in naive B lymphocytes and might represent "tonic" signaling. In T cells, it is clear that the "tickling" of naïve TCRs is mediated by self-MHC molecules presenting self-peptides. In B cells, it remains to be established if signaling in naïve cells is self-antigen driven or "tonic," an issue that will be discussed in Chapter 7. These low-level "tonic" or self-antigen–driven signals might contribute to the just detectable but

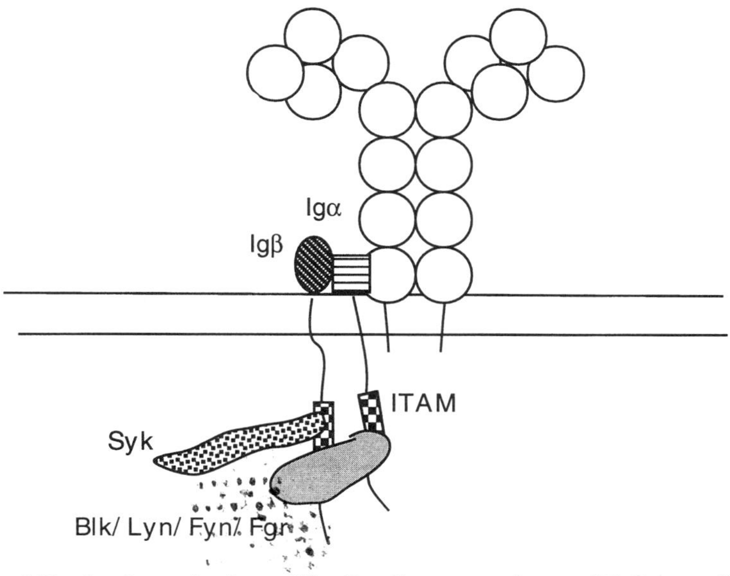

FIGURE 4.23. A schematic view of the B cell receptor. Igα and Tgβ form disulfide linked heterodimers that are noncovalently associated with membrane Ig H-L tetramers. A single Igα-β heterodimer may actually be associated with one Ig tetramer. Blk, Lyn, Fyn, Fgr, and Lck are Src-family kinases that associate with the B cell receptor. These kinases are myristylated (not shown) and associated with the inner face of the plasma membrane. Their N-termini are involved in the interaction with the spacer portion of the ITAM. Syk probably associates via its tandem SH2 domains with constitutively phosphorylated tyrosine residues in the ITAMs.

constitutive phosphorylation of tyrosine residues in ITAMs. This phosphorylation probably permits the minimal, but detectable, recruitment to the receptor, in "resting" lymphocytes, of members of the Syk/Zap-70 family of tyrosine kinases.

Src-family kinases (Blk, Lyn, Lck, Fyn, Fgr) associate with the inner face of the plasma membrane via N-terminal lipid modifications (discussed in the section entitled "A General Introduction to Signaling Pathways") and are loosely associated with antigen receptors in nonstimulated B and T cells. In the case of the BCR, this association involves an interaction between the unique N-terminal region of the Src-family kinase with a loop region between the tyrosine residues of the ITAM motif in the cytoplasmic tails of Igα and Igβ. In a similar fashion, Lck associates with the cytoplasmic tails of CD4 and CD8 co-receptors in T cells. Fyn probably associates with the ITAMs of CD3 chains, particularly ζ and ε, although this interaction might be indirect.

Initiation of Signal Transduction: Crosslinking Versus Allosteric Change

How does occupancy of an antigen receptor lead to activation? An allosteric change in the receptor could be theoretically postulated to occur following receptor ligation, but little evidence supports this. The major mechanisms by which antigen receptors are activated follow:

1. Receptor crosslinking *or*
2. The coligation of transmembrane proteins (known as coreceptors) with the antigen receptor. Coreceptors will be discussed below. They can contribute to signal initiation but may also influence receptor clustering.

The clustering of receptor molecules presumably brings signaling molecules on the cytosolic face of the receptor into close apposition. This close physical contact is assumed to lead to the catalytic activation of quiescent enzymes. In the case of the BCR, for instance, a monosaccharide cannot trigger the antigen receptor (and is therefore nonimmunogenic), whereas a polysaccharide can facilitate B cell activation (Figure 4.24).

While there has never been a conceptual problem explaining how a multivalent antigen might induce clustering of the BCR, it has been unclear how major histocompatibility complex (MHC)–peptide complexes induce TCR clustering. It has generally been assumed that the TCR complex on T cells is monomeric and that MHC–peptide complexes on antigen-presenting cells (APCs) are also monomeric. How do monomeric ligands induce clustering? Moreover, since an APC may present numerous different peptides at any given time, it is likely that the actual number of specific MHC–peptide complexes on a given APC that would be recognized by the TCR on a single T cell is relatively small. It has been estimated that this number might range from approximately 10 specific peptide-MHC complexes to a few hundred. In addition, the affinity of TCRs for their ligands is relatively modest, estimates being typically in the low micromolar range. Given this modest affinity, it is difficult to understand how a low-affinity

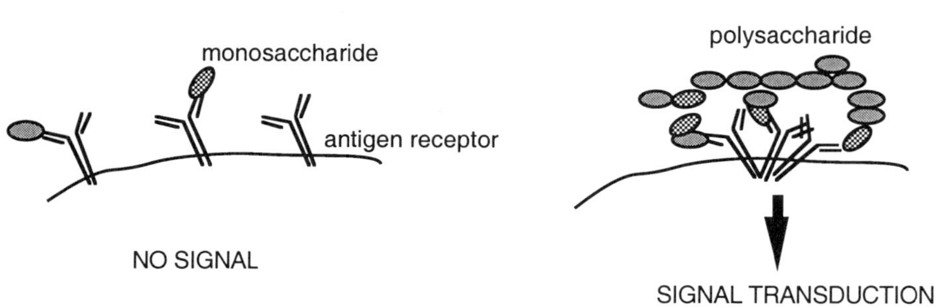

FIGURE 4.24. Crosslinking of the B cell receptor is one way in which signal transduction may be initiated.

monomeric recognition of one TCR on a T cell for a single specific MHC-peptide on the APC could result in the kind of sustained signal transduction that would be required to induce gene activation in a T cell. We will return to this issue below when we discuss the formation of immunological synapses.

Growing evidence indicates that MHC–peptide complexes may actually functionally exist as multimers and that they might be capable of inducing TCR aggregation. The crystal structure of MHC class II molecules suggests that they may exist as dimers of α–β heterodimers. While MHC molecules do change conformation when they bind peptides, it is unclear whether this change in conformation results in homotypic aggregation of identical MHC–peptide complexes in the plane of the APC membrane. Light-scattering studies examining soluble forms of the TCR and cognate-soluble MHC–peptide complexes indicate that stable aggregates may form involving up to six MHC–peptide complexes and six TCRs. For MHC oligomerization to be meaningful, oligomers must exist between identical MHC molecules occupied by the same peptide. This assumes that, when peptide occupancy alters MHC conformation, this conformational alteration facilitates the formation of homotypic aggregates in the plane of the membrane. No hard evidence supports such an assumption.

It is now recognized that CD4 molecules can form dimers because of self-association and that MHC class II molecules have two spatially separated sites that are required for CD4 association. CD4 molecules may therefore not only be brought in close proximity to the TCR by MHC molecules, but they may also facilitate the formation of higher order oligomers of the TCR bridged by CD4 dimers.

How exactly crosslinking or receptor aggregation leads to activation is not very well understood. It is assumed that tyrosine kinases that are separated from other similar molecules are not efficiently phosphorylated on critical tyrosines (such as the activation loop tyrosine residue). Bringing tyrosine kinases into close proximity to one another may facilitate transphosphorylation events that permit full activation. These intermolecular phosphorylation events are usually considered to represent "autophosphorylation." CD45, as discussed earlier (see section entitled "A General Introduction to Signaling Pathways"), dephosphorylates critical negative regulatory tyrosine residues in Src-family kinases and thus permits proper activation of these kinases. Although it is sometimes suggested that antigen-receptor crosslinking leads to the catalytic activation of loosely associated CD45 molecules and thus permits the turning on of antigen receptor linked Src-family kinases, little evidence supports such a role for CD45. Truncated CD45 molecules that lack extracellular domains can support antigen-receptor signaling. This tyrosine phosphatase is probably constitutively active and does not require regulation by antigen-receptor crosslinking.

While crosslinking of receptor molecules can lead to activation, a second way in which receptor triggering can be achieved is by the process of coligation with a coreceptor. An antigen may draw a receptor and coreceptor into close proximity and thus initiate signal transduction.

Coreceptors Facilitate Antigen-Receptor Activation: Defining Coreceptors and Costimulators

The portion of the antigen described as an epitope contacts the highly variable surface generated by complementarity determining regions of antigen receptors. Less diverse portions of the antigen (nonvariable portions of MHC molecules for T cells and complement components or Fc portions of antibodies present in immune complexes for B cells) contact receptors physically associated with the antigen receptors. These molecules are known as coreceptors. A *coreceptor* is defined as a signaling receptor that directly associates with the very same antigen that is seen by the antigen receptor. As a result, an antigen may physically bring together an antigen receptor and a coreceptor. Signaling molecules associated with the coreceptor may "turn on" the antigen receptor. As discussed above, bivalency or multivalency of the coreceptor might also contribute to receptor clustering and thus facilitate signal generation.

The major coreceptor in B cells is CR2/CD21, which is a receptor for a cleaved complement component, C3d. An Fc receptor on B cells, FcγRIIb, also essentially functions as a coreceptor, although it delivers a negative signal. CD22 is a sialic acid-binding lectin that can cluster with the BCR and regulate BCR signaling. It is sometimes called a coreceptor, although it does not contact antigen and should not strictly be classified as such. This is really a semantic distinction. The coreceptors for the TCR are either CD4 or CD8 molecules that associate with invariant portions of MHC class II or class I molecules respectively.

Coreceptors facilitate signal initiation by the antigen receptor. A separate category of signaling receptors that may cooperate with antigen receptors (but don't really "turn them on") are known as costimulatory receptors. Some molecules originally thought to be costimulatory receptors have turned out to be inhibitors of antigen-receptor–induced signals. *Costimulatory* (and "coinhibitory") *receptors*, by definition, are cell surface proteins that may or may not physically associate with antigen receptors, but that generate signals integrated within lymphocytes with signals from antigen receptors. The ligands that trigger these receptors are known as *costimulatory ligands*. We will discuss costimulatory receptors in the context of B and T cell signaling a little later in this chapter. Although cytokines may also be described as costimulators, we will consider them separately.

Adaptor Proteins Involved in Antigen-Receptor Signaling

A broad outline of the initial events that characterize antigen-receptor–mediated signal transduction is provided in Figure 4.25. A key initial event is activation of receptor-associated tyrosine kinase activity.

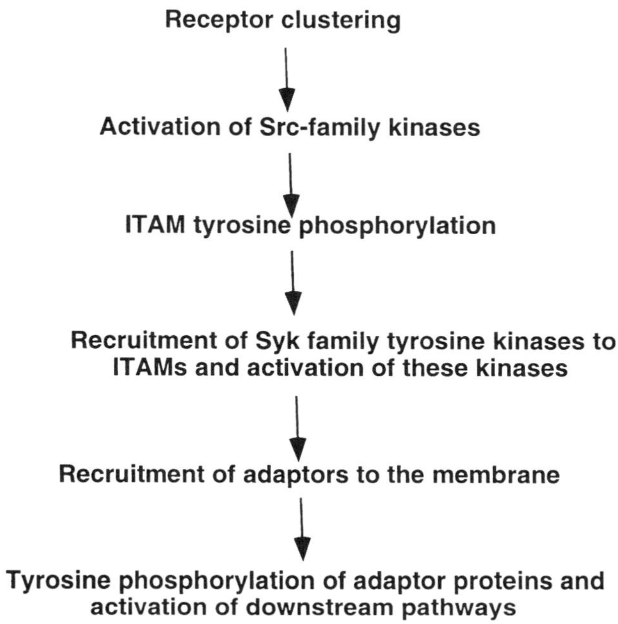

FIGURE 4.25. Initial events in antigen-receptor–mediated signaling.

Occasionally, tyrosine phosphorylation of a target protein may result in a conformational alteration that could, for instance, result in a change in catalytic activity. Alternatively, tyrosine phosphorylation of one protein may generate a novel recruitment site for another protein that contains an appropriate SH2 or PTB domain. Proteins that primarily serve as agents for the recruitment of other cytosolic proteins during signal transduction are known as adaptors. Antigen-receptor–mediated signaling involves a number of cycles of catalytic activation and recruitment. Recruitment sites may also be created by enzymes that are not tyrosine kinases. PI3 Kinase, as discussed earlier, also generates novel phosphoinositide motifs in cell membranes for protein recruitment and consequent activation.

The ITAM-containing chains in antigen receptors are, in effect, transmembrane adaptor proteins. One of the key functions of these proteins is to recruit and activate Syk family kinases. These kinases in turn phosphorylate other adaptors. In T cells, two adaptor proteins appear to be immediate targets of Zap-70 activity when TCR signaling is initiated. One protein is a transmembrane adaptor known as LAT (linker for activation of T cells), and the other is a cytosolic protein known as SLP-76 (SH2 domain containing leukocyte phosphoprotein of 76 kd).

LAT (Figure 4.26) is a type III transmembrane protein that contains a very short N-terminal extracellular extension and a long cytoplasmic tail

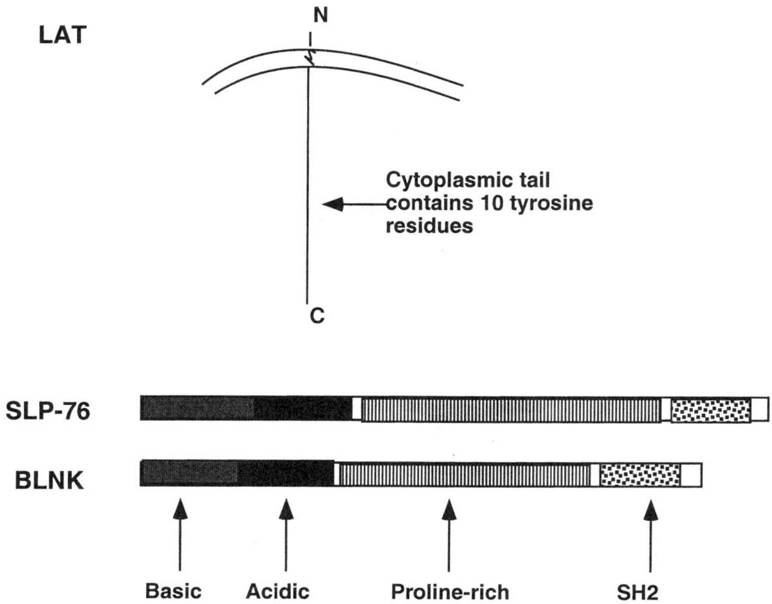

FIGURE 4.26. Adaptors involved in early signaling events downstream of lymphocyte antigen receptors. Strictly speaking, ITAM-containing chains also function as adaptors. LAT is a transmembrane protein with multiple tyrosine residues in its cytoplasmic tails. SLP-76 and BLNK serve similar functions in T and B cells, respectively. They contain multiple YxxP motifs, proline-rich regions, and an SH2 domain.

with 10 tyrosine residues, 5 of which contribute to consensus (YxNx) sites for the binding of the SH2 domain of Grb2, an adaptor that will be discussed below. LAT also contains two juxtamembrane cysteine residues that are palmitolylated and facilitate the migration of LAT to lipid rafts (discussed below). LAT and SLP-76 are expressed in T cells but not in B cells. B cells contain a homolog of SLP-76 known as BLNK (for B cell linker protein) or SLP-65. Both SLP-76 and BLNK have N-terminal basic and acidic regions followed by a proline-rich region in the central portion of the molecule and a C-terminal SH2 domain (Figure 4.26). The N-terminal portion of SLP-76 contains three YxxP motifs (which are target sequences for phosphorylation by Syk family kinases), and BLNK contains six YxxP motifs. The phosphorylated tyrosines at the N-terminus of SLP-76 can bind the SH2 domains of Vav (an exchange factor for Rac) and of another adaptor protein called Nck. The proline-rich regions of these molecules contain multiple consensus sites for association with SH3 domains.

An adaptor briefly mentioned earlier is Nck. Nck contains an SH2 domain as well as an SH3 domain. In activated T cells, the SH2 domain of

Nck is recruited to tyrosine-phosphorylated SLP-76 (and to BLNK in B cells). The SH3 domain of Nck can associate with WASP, as described earlier, and may also associate with a serine-threonine kinase called Pak (p21 associated kinase). TCR triggering leads to the formation of a SLP-76-Vav-Nck complex, and this complex is required for the activation of Pak and for the polarization of F-actin. SLP-76-dependent activation of small G proteins and of Vav, Pak, and WASP appear to be crucial for TCR-dependent cytoskeletal reorganization, which in turn contributes to TCR capping and provides a positive feedback loop for efficient TCR signaling.

In the absence of LAT or of SLP-76, T cell development is abrogated at the pre-TCR checkpoint. BLNK is required at the pre-BCR checkpoint in human B cells. While LAT, SLP-76, and BLNK represent adaptors that may have very specialized functions linked to antigen-receptor signaling, a number of other adaptors that participate in antigen-receptor signaling are also used by numerous other signaling receptors in many different tissues. Grb2 is an adaptor in which an SH2 domain is sandwiched between two SH3 domains (Figure 4.27). The SH2 domain of Grb2 may associate with other tyrosine phosphorylated adaptors such as Shc, LAT, and SLP-76. The SH3 domains of Grb2 can bind effectors of signaling that contain proline-rich stretches. One particularly important Grb2-associated protein is mSOS, the GEF for Ras. In most cells, Grb2 SH3 domains are constitutively associated with SOS. Recruitment of Grb2 via its SH2 domain to tyrosine-phosphorylated adaptors at the plasma membrane helps bring SOS to the membrane and close to its target, which is Ras-GDP. Grb2 may also form complexes with proteins other than SOS such as the Cbl adaptor protein, which will be discussed below.

Another adaptor that may be recruited by T and B cell receptors is Shc. The p52 form of Shc contains an N-terminal PTB domain and a C-terminal SH2 domain. A collagen-like stretch in the portion of the protein between these two domains contains tyrosine residues, which are phosphorylated during antigen-receptor signaling. Shc can be recruited to phosphorylated ITAM motifs, and this may be particularly important in the context of BCR and TCR signaling. Tyrosine phosphorylation of Shc leads to the formation of a complex with Grb2-SOS, with two Grb2 molecules binding via their SH2 domains to one molecule of Shc. Shc-Grb2-SOS complexes may thus contribute to Ras activation. Two other adaptors that we will consider later in the chapter in the context of the negative regulation of antigen receptor signaling are CrkL and Cbl. CrkL contains an N-terminal SH2 domain and two C-terminal SH3 domains.

Cbl is a large protein that contains an N-terminal variant SH2 domain, a RING finger domain, a proline-rich stretch that contains 17 distinct PxxP motifs capable of binding to SH3 domains, and a C-terminal leucine zipper motif (Figure 4.27).

Cbl may also form complexes with Grb2, and these complexes might be recruited to the membrane by Shc. Cbl may mediate both positive and neg-

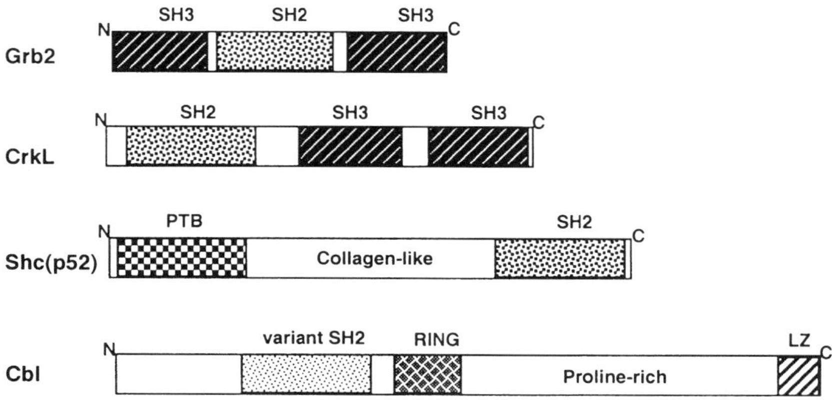

FIGURE 4.27. Schematic view of the structure of selected signaling adaptors that are involved in signaling downstream of the antigen receptors. The SH2 domain in Cbl is atypical. The RING domain of Cbl is required for the activation of an E2 ubiquitin conjugating enzyme. There are three Cbl proteins: c-Cbl, Cbl-b and Cbl-3. See text for details.

ative signaling events in lymphocytes. It may serve as an adaptor for the recruitment and activation of PI3K downstream of the BCR and the TCR (although the cytoplasmic tails of CD19 and CD28 might be the primary "adaptors" for this process). It may also be involved in the attenuation of tyrosine kinase-mediated signaling. The RING domain of Cbl may contribute to the function of this protein as an E2-dependent ubiquitin protein ligase (an E3), which facilitates the proteasomal degradation of receptor and nonreceptor tyrosine kinases.

Genetic evidence from *C. elegans* suggest that Cbl (Sli-1 in worms) might be a negative regulator of the Ras pathway in that organism. Growing evidence indicates that Cbl might actually be a negative regulator of the Ras pathway in B and T cells (and this might depend on its participation in the degradation of tyrosine kinases). One way in which Cbl might function is to recruit the SH2 and SH3 domain-containing adaptor CrkL. CrkL in turn may be complexed via an SH3–proline peptide interaction with C3G, which is a GEF for Rap-1A. Rap-1A-GTP may be an antagonist of Ras-GTP because it competes for and somehow sequesters Raf. The existing data addressing this issue are somewhat contradictory. In some cell types, it has been suggested that Cbl/CrkL/CrG/Rap1-A contributes to the activation of B-Raf and of PKA. There are three Cbl proteins that have been described, c-Cbl, Cbl-b, and Cbl-3. All three proteins can function as adaptors and as E2-dependent ubiquitin protein ligases. However these adaptors may have other functions in addition to their roles in protein degradation. Cbl-b is particularly important from the viewpoint of lymphocyte signaling. In the absence of Cbl-b, B cells exhibit hairtrigger responses to antigen receptor and CD40 ligation and T cells also respond in an exaggerated fashion to TCR triggering. These T cells do not require CD28 mediated costimulation to proliferate when their antigen receptors are ligated. Activation of Vav, the JNK pathway, and PI-3 kinase occur readily in *cbl-b*$^{-/-}$ mice in response to TCR ligation in the absence of costimulation.

A number of other adaptor proteins have been described in lymphocytes whose biochemical functions are still being elucidated. These include Grb2-like proteins such as Grap/Gads and GrpL, FYB/SLAP (Fyn T binding protein/SLP-76 associated protein), and SKAP-55 (Src-kinase-associated protein of 55 kDa).

Signal Transduction by the B cell Receptor

B cells may be activated by multivalent T-independent antigens such as polysaccharides, as well as by T-dependent protein antigens. Protein antigens may be monovalent or paucivalent (if they exist as oligomers or if they are multimerized as components of immune complexes). While B cell activation by protein antigens does depend in large part on signals induced by T cell ligands such as CD40L, and cytokines such as IL-4, there is still a need for signals to be delivered via the BCR. This signal may be generated either by a paucivalent form of the protein antigen that can crosslink the BCR or by the recruitment of a coreceptor that facilitates BCR signaling.

A truly monomeric protein may be unable to trigger the BCR. Such a protein might be endocytosed by antigen-specific B cells without any signals being generated by the BCR. This B cell may then present specific MHC–peptide complexes to activated T cells (if T cells have been activated in the host animal). However, as discussed in Chapter 8, in the absence of survival signals delivered via the BCR, a B cell that attempts to collaborate with a T cell is repulsed and summarily executed as the result of a FasL–Fas interaction.

Signaling via the BCR is therefore initiated either by multivalent antigens bearing repeated determinants such as microbial polysaccharides, or by protein antigens that are arrayed in multivalent form, or by antigens that are within immune complexes and that are therefore able to recruit a critical coreceptor, CD21, to BCR complexes. Receptor clustering leads to the activation of BCR-associated Src-family kinases, which include Lyn, Fyn, Blk, Lck, and Fgr. A major consequence of Src-family kinase activation is the phosphorylation of ITAM motifs in Igα and Igβ. These phosphorylated ITAMs facilitate the recruitment and activation of Syk, as discussed earlier (see section entitled, "A General Introduction to Signaling Pathways"). While Lyn does participate prominently in BCR-mediated phosphorylation of Igα, Syk, and Shc, the "positive" influences mediated by this Src-family kinase are not unique and can also presumably be initiated by other Src-family kinases. The negative regulation of B cell signaling, however, does depend on the activity of Lyn, as will be discussed below.

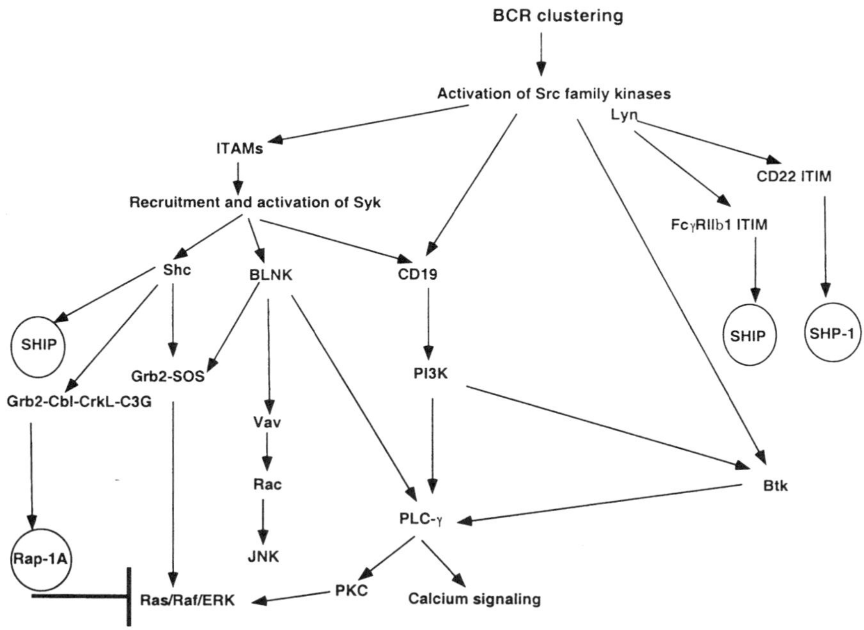

FIGURE 4.28. A summary of selected events in BCR signaling. Circled molecules represent negative regulators. Although not shown in the figure, Btk activation may also require recruitment to BLNK. Details are in the text.

Some of the signaling pathways dowstream of the antigen receptor are listed in Figure 4.28.

A large number of the pathways downstream of the B cell receptor depend on the activation of Syk. Syk phosphorylates BLNK and creates sites for the recruitment of PLCγ, Vav, Grb2, and Nck (an adaptor protein whose function is less well understood) via their respective SH2 domains. Modification of BLNK is critical for the activation of PLC-γ and consequently for the activation of many PKC isozymes and of calcium dependent signaling pathways. Since Grb2-containing complexes, including Grb2-SOS, can be recruited to tyrosine-phosphorylated BLNK, it is likely that BLNK might participate in Ras activation. It appears, based on the introduction of a dominant interfering mutant, that BLNK is not absolutely required for Ras activation and that Ras can be activated in different ways in B cells, as will be discussed below. BLNK can recruit Vav, which, as discussed, is an exchange factor for Rac. BLNK may therefore participate in the activation of the JNK pathway. Btk may associate with tyrosine phosphorylated BLNK and thus be brought in proximity to the membrane and to PLCγ. BLNK is presumably recruited to the membrane, but it is unclear how this actually occurs.

One of the ways in which the Ras pathway might be activated downstream of the BCR is believed to depend on the phosphorylation of ITAMs by Src-family kinases and of the Shc adaptor by Syk. Although the *in vivo* evidence for this is sketchy, it is believed that the SH2 domain of Shc associates with phospho-ITAMs, allowing this adaptor to be a target of Syk phosphorylation. The SH2 domain of Grb2 associates with phosphotyrosine motifs in the collagen-like stretch in Shc, and thus an ITAM-Shc-Grb2-mSOS complex is formed at the membrane. There are two other ways in which the Ras pathway might be activated in B cells. One mechanism involves PKC activation: PKC might inhibit the activity of p120RasGAP, as discussed earlier. Another way in which the Ras pathway might be activated in B cells is via the CD19-dependent activation of PI3K. The link between PI3K and Ras has been discussed earlier this chapter, and we will briefly consider CD19 signaling below.

The activation of PI3 kinase depends on the recruitment of the p85 subunit of PI3K to YMXM motifs in the cytoplasmic tail of CD19, which are probably phosphorylated by Lyn and other Src-family kinases downstream of the BCR. The activation of PI3K facilitates the recruitment of Btk to the membrane via its PH domain (see discussion in the section entitled, "A General Introduction to Signaling Pathways"). The optimal activation of Btk requires both Src-family kinases as well as PI3K to be activated. Although Btk probably contributes to the phosphotyrosinylation of a number of targets including TFII-I and PLC-γ, the latter is probably the most critical known substrate of this tyrosine kinase. Sustained elevations of calcium during signaling depend on the activity of Btk and PLC-γ. (This issue has been considered in more detail in the preceding section.)

The CR2/CD19/TAPA-1 Coreceptor Complex

Antigens may directly interact with complement via the alternative pathway or may do so indirectly after forming antigen–antibody complexes that can activate and bind to complement components. Complement-coated T-dependent as well as T-independent antigens can signal 10- to 100-fold more efficiently than "naked" antigens. Mice express two receptors on B cells and follicular dendritic cells (FDCs) that bind activated complement C3 and C4 fragments and that can function as coreceptors in the B lineage. These two murine complement receptor proteins CD21 (CR2, complement receptor type 2) and CD35 (CR1, complement receptor type I) are products of the same locus in mice and of different genes in humans. In the mouse, CD21 is made up of 15 short consensus repeats (repeated units each of about 60 to 70 amino acids), a transmembrane domain, and a cytoplasmic tail. C3b may covalently attach itself to antigen or to antigen–antibody complexes and may be cleaved to yield iC3b or C3dg, both of which may be recognized by CR2. In humans, this receptor also mediates the entry of the Epstein-Barr virus into B cells. CD35 contains all the domains in

CD21, and in addition has an N-terminal extension that contains binding sites for C3b and C4b.

In both mice and humans, CD21 forms a complex with CD19, an integral membrane protein that can initiate signal transduction, and a tetraspan protein (with four transmembrane domains) known as CD81 (Figure 4.29). An interferon-inducible protein called Leu-13 has also been described in this complex.

Complement-coated antigens bring together the coreceptor and the antigen receptor. There are a number of reasons why CR2 is such an effective coreceptor. The CD19 component transduces signals. It is phosphorylated by Lyn and Fyn, and possibly also by Syk, and can recruit and activate PI3K and also contribute to the recruitment and activation of Vav. As a result, CD19 qualitatively contributes to the activation of PI3K (and thus of Akt, Btk, PDK1, and so on), activates Rac and the JNK/SAPK pathway, and can indirectly (via PI3K) contribute to the activation of the Ras pathway. Coligation of CD19 with the BCR has been shown to augment the activation of MAP kinase pathways both qualitatively and quantitatively.

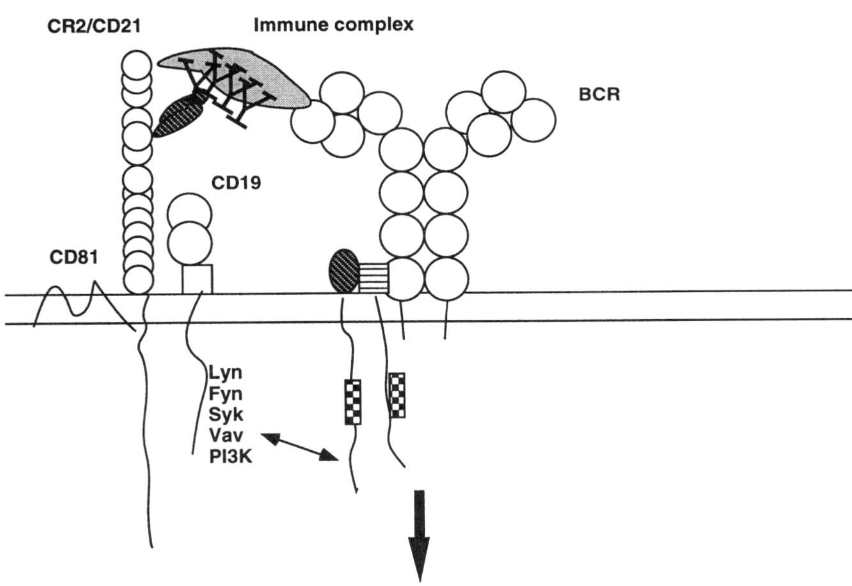

FIGURE 4.29. The CD21/CD19 complex functions as a coreceptor for the BCR. CD21 is also referred to as CR2, or complement receptor type II. Other proteins in the complex include CD81 and Leu13. Antigens that are complexed with degradation products of C3b can bring the BCR and CR2 coreceptor together. CD19 may be phosphorylated by Src and Syk family kinases and may recruit and activate PI3K and Vav.

A second simple way in which this coreceptor might work, particularly if it exists in the membrane as a dimer, is to facilitate B cell receptor clustering by low-valency antigens that are coated with activated C3 fragments. Antigens that would intrinsically be unable to crosslink the BCR might be able to do so efficiently if they were coated with C3d.

Negative Regulation of BCR Signaling by CD22

B cell signaling may be modulated by CD22, which is a sialic acid-binding lectin expressed on the surface of B lymphocytes (Figure 4.30). Multiple extracellular Ig-like domains encompass the carbohydrate binding site, and the cytoplasmic tail of CD22 contains motifs known as ITIMs (immunoreceptor tyrosine inhibitory motif). These ITIMs are probably phosphorylated by activated Lyn. Syk has also been shown to be associated with CD22 and might phosphorylate ITIMs as well, although Lyn is probably the critical kinase for inhibitory signaling. The phosphorylated ITIMs can bind to the SH2 domain of SHP-1 which is a tyrosine phosphatase and can poten-

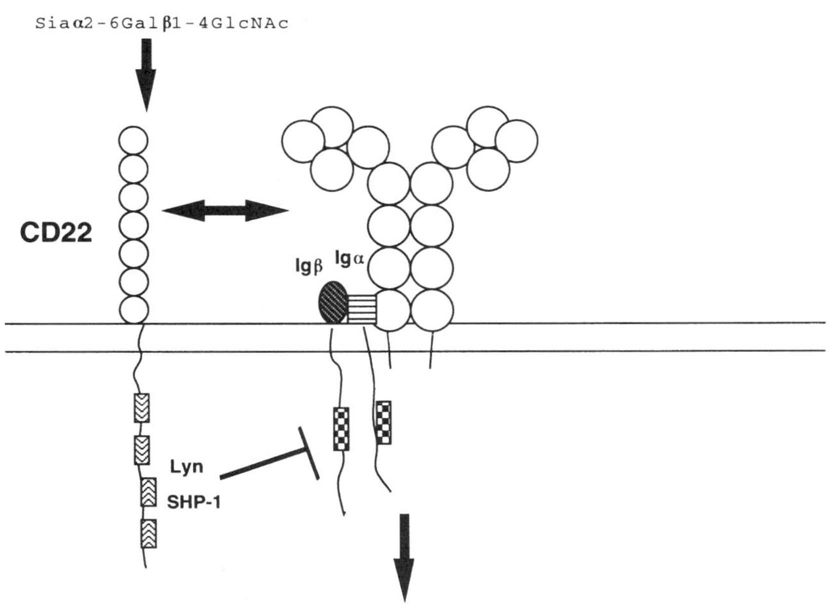

FIGURE 4.30. CD22 negatively regulates BCR-mediated signals. Proximity of CD22 to the BCR is likely to attenuate signaling because of the activity of SHP-1 associated with its cytoplasmic tail. This sialic acid lectin may be "drawn" away from the B cell receptor during T–B collaboration. Alternatively it may be more avidly brought into BCR complexes when they are heavily clustered. It is also possible that the association of CD22 with the BCR depends on the regulation of α2,6 siallytransferase and neuraminidase activity during B cell activation.

tially dephosphorylate substrates that are phosphorylated as a consequence of B cell receptor signaling.

At least three models might explain how CD22 negatively regulates B cell signaling. CD22 binds terminal α2-6 sialic acid moieties found in many glycoproteins in B and T cells, including membrane IgM. This lectin activity allows it to associate with the B cell receptor. In one model, CD22 associates with the antigen receptor in B cells most readily prior to the initiation of T-B collaboration. Since the cytoplasmic tail of CD22 may be linked to SHP-1, B cell signaling is likely to be attenuated under these circumstances. This model suggests that, when B and T cells are in contact with each other, the high density of CD22 ligands on T cells can draw this lectin away from the receptor and dampen negative regulation. Evidence for this model is quite indirect. It is unclear why ligands on T cells would compete more avidly for CD22 than α2-6 sialylated glycoproteins on B cells.

Another model suggests that relatively little CD22 is associated with the BCR in quiescent cells. However, once the BCR is triggered, receptor clustering greatly increases the local density of α2-6 sialic acid on membrane IgM, and this leads to the recruitment of CD22 and the subsequent negative regulation of signaling.

A separate model suggests that the regulation of glycosylation and deglycosylation of the BCR is of critical importance in orchestrating signal intensity. In such a model, neuraminidases could theoretically promote B cell activation by preventing CD22 from readily associating with the BCR. It has been demonstrated that when B cells are activated the expression of a B cell-specific α2,6 sialyltransferase is enhanced at the level of transcription. This sialyltransferase was originally identified as a presumed CD antigen, CDw75, but it is now known that new sialyl determinants are generated by the upregulation of this B cell-specific sialyltransferase. Enhanced sialylation of the BCR in activated B cells may facilitate the association of CD22 with the BCR and thus attenuate signaling via this receptor.

CD72 is a Negative Regulator of BCR Signaling

As discussed above, CD22 negatively regulates B cell signaling via the ITIM dependent recruitment of SHP-1. Tyrosine phosphorylation of the ITIM is required for recruitment of SHP-1 and this modification is probably mediated by Lyn. BCR hyperreactivity and an increase in B-1 cells are features observed in $CD22^{-/-}$ mice, in viable motheaten mice (which harbor a defect in SHP-1), and in $Lyn^{-/-}$ mice. Another knockout mouse strain with a broadly similar phenotype is the $CD72^{-/-}$ mouse. CD72 is a type II integral membrane glycoprotein which is a member of the C-type lectin superfamily. This protein exists as a disulfide linked homodimer on most cells of the B lineage (it is not expressed on plasma cells) and functions as a negative regulator of both pre-BCR and BCR signaling. In the absence of CD72, pre-B cell numbers in the bone marrow are enhanced but mature B cells in the bone-marrow (often called recirculating bone marrow B cells) are reduced. Peripheral B-1 cell numbers are enhanced but there is a relative decrease in B-2 cells. B cell proliferation in response to both LPS and antigen receptor crosslinking is enhanced. The cytoplasmic tail of CD72 contains two ITIM motifs, one of which has been shown to be capable of recruiting SHP-1. Although CD72 and CD22 appear to mediate broadly similar functions, CD72 may be a more significant negative regulator of signaling at the pre-B stage than CD22. The physiological relevant ligand for CD72 is not known.

Inhibition of B Cell Signaling by FcγRIIb

Immune complexes generated in conditions of IgG antibody excess have the ability to bind both to the antigen receptor as well as to a specific Fc receptor on B cells known as FcγRIIb. Immune complexes will form a bridge between this molecule and the antigen receptor, allowing it to meet the definition of a co-receptor—but with a difference: This receptor inhibits signals from the B cell receptor (Figure 4.31).

The FcγRIIb receptor contains an ITIM in its cytoplasmic tail. The tyrosine residue of the ITIM is phosphorylated by Lyn, and the phospho-ITIM recruits an SH2 domain containing inositol polyphosphate-5 phosphatase (SHIP). SHIP is phosphorylated on tyrosine residues following B cell activation and itself dephosphorylates inositol phosphate moieties in the 5 position. Since it is located near the membrane in its active form, it is in the right place to dephosphorylate PI3,4,5-P3 which it converts to PI3,4-P2. In a sense, SHIP serves as a functional anatagonist of PI3K. In the absence of PI3,4,5-P3, Btk and PLC-γ are not activated, and as a result, many aspects of signaling are severely compromised in these B cells.

CD5 as a Negative Regulator of B-1 Cell Signaling

We will discuss B-1 cells in detail in Chapters 7 and 8. For the sake of completeness, we mention the CD5 molecule, which is expressed in T cells and in the B-1 subset of B cells. CD5 can also associate with the BCR in B-1 cells and may function as a negative regulator of BCR signaling. CD5 is a type I membrane protein, and although some signaling molecules such as calcium-calmodulin–dependent kinase IV have been detected in association with CD5, exactly how this molecule generates inhibitory signals is very poorly understood.

Other Mechanisms by Which B Cell Signaling May Be Attenuated

Tyrosine phosphorylation of Shc can contribute to the activation of the Ras pathway, as discussed earlier. However, tyrosine phosphorylated Shc may also bind to SHIP and bring this inositol polyphosphate 5-phosphatase to the

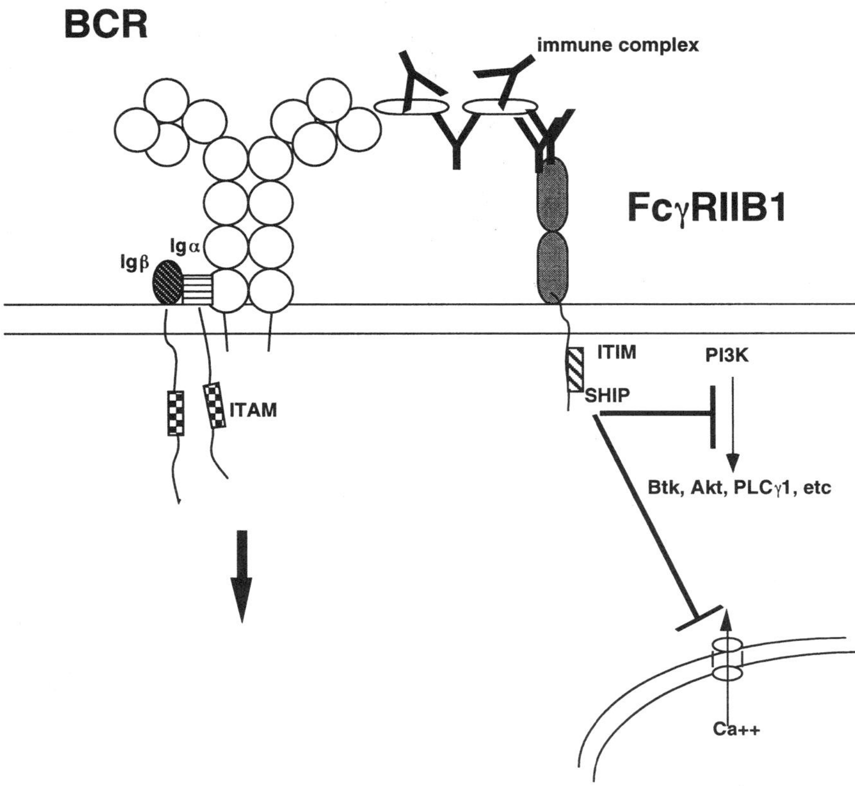

FIGURE 4.31. Negative regulation of B cell signaling by SHIP (SH2 domain containing inositol-5-phosphatase). Since SHIP can remove phosphate residues from the 5′ position of the inositol ring, it can convert PI3,4,5-P3 to PI3,4-P2 and consequently compromise the activation of Btk, PLCγ, and calcium signaling.

membrane. How exactly this process is regulated is unclear, but Shc–SHIP complexes may provide an important mechanism for signal attenuation.

The B cell receptor has also been shown to be complexed with an atypical member of the PKC family, PKCμ (see "A General Introduction to Signaling Pathways," earlier in this chapter). PKCμ might phosphorylate and inactivate Syk and thereby attenuate antigen-receptor signaling.

Other negative regulators of BCR signaling include the Aiolos zinc finger protein (discussed in Chapter 8) and Cbl-b (discussed earlier in this chapter). Signaling via the BCR and CD40 is enhanced in mice that lack Aiolos as well as in mice that lack Cbl-b. These proteins alter BCR signaling thresholds in poorly understood ways. In the absence of Aiolos, signaling via Src-family kinases and Btk is enhanced. Cbl-b may be an E2 dependent ubiquitin ligase that facilitates the degradation of activated Src-family kinases or of other signaling molecules such as Vav.

CD40 is a Member of the TNFR Family and a Critical Costimulator of B Cells

CD40 is a member of the TNFR family, and there are many similarities in the pathways downstream of TNFR2, CD40, and CD30 (Figure 4.32). TNF is the prototypic member of the Type III "cytokine" family (discussed in the latter portion of this chapter), which is made up of trimeric ligands that may be cell-bound or secreted and that activate signaling by a structurally related family of receptors (the TNFR family). While a few members of this family of ligands are actually cytokines, many cell surface ligands that belong to this family—the CD40 ligand and the Fas ligand being prominent examples—are not considered to be cytokines. Signaling via CD40 is absolutely essential for T-dependent activation of B cells. Activated T cells express a transmembrane glycoprotein known as gp39 or the CD40 ligand (CD40L/CD154). This homotrimeric glycoprotein binds to CD40, which is expressed on the surface of B cells (as well as of dendritic cells, macrophages, and some epithelial cells). Signaling via CD40 is required for T-dependent B cell proliferation, class switching, and somatic mutation. Signals from this receptor also contribute to the upregulation of the expression in B cells of Fas, ICAM-1, CD23, and of lymphotoxin-α. In a human disease known as the X-linked hyper-IgM syndrome, mutations are found in the gp39 gene. Patients with this disease present with B cells that cannot class switch, as well as other defects in B and T cell activation (discussed in Chapter 12).

CD40 signaling leads to the activation of a number of signaling pathways. These include the JNK/SAPK and the NFκB pathways, the activation of Src-family kinases, and of PLC-γ2. Ligation of CD40 leads to the trimerization of this receptor in the plane of the membrane and the subsequent activation of signals via proteins of the TRAF (TNFR-associated factor) family, which associate with CD40. TRAF2, TRAF3, TRAF5, and TRAF6 can directly bind to the cytoplasmic tails of CD40 trimers. TRAF2 has a C-terminal TRAF domain (divided into TRAF-N and TRAFC regions), which is involved in the association with CD40, and an N-terminal RING finger domain, which facilitates its interaction with a serine-threonine kinase known as NIK (Fig. 4.32). TRAF2, TRAF5, and TRAF6 can activate NFκB, but the function of TRAF3 remains poorly understood. All three TRAFs mentioned above can presumably associate with and activate NIK. NIK in turn associates with the IκB kinase complex, which can directly phosphorylate IκB leading to its degradation and the subsequent nuclear translocation of NFκB. The pathway leading from TRAF proteins to the initiation of NFκB degradation has been described in detail in the section entitled "A General Introduction to Signaling Pathways."

Although a number of TRAFs might participate in the activation of NFκB, we now know from gene inactivation studies that TRAF2 is uniquely required for the activation of the JNK/SAPK pathway by CD40. It is

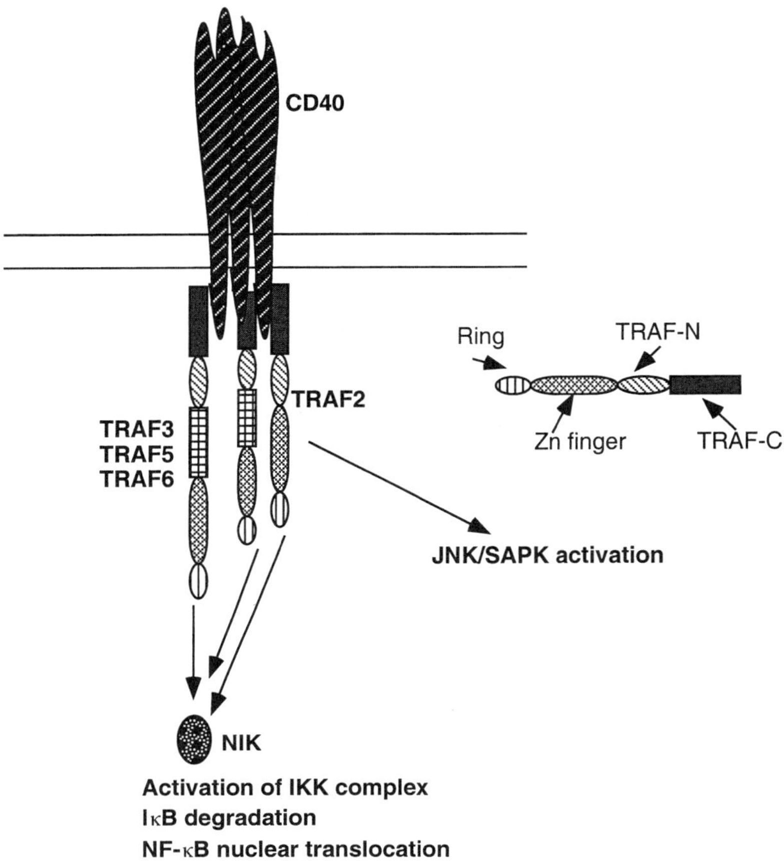

FIGURE 4.32. CD40 signaling. TRAF2 is absolutely required for the activation of the JNK/SAPK pathway. TRAF2, TRAF5, and TRAF6 may all be involved in the activation fo the NFκB pathway. The Jak3 kinase may also associate with CD40 but is not shown in this figure.

unclear whether small G proteins such as Cdc42 or Rac participate in this process, and most of the events that lead from TRAF2 to JNK/SAPK remain to be elucidated.

Constitutive association of the Jak3 tyrosine kinase with CD40 has been described in a number of human B cell lines. This association depends on a proline-rich stretch of eight amino acids in the cytoplasmic tail of CD40 that resembles the Box 1 motif seen in cytokine receptors (discussed later, in the section entitled, "A Overview of Cytokine Signaling"). Some of the events that occur downstream of CD40 ligation—in particular, the induction of CD23, ICAM-1 and lymphotoxin-α expression—may require this associated Jak3 activity.

BlyS/BAFF May Also Provide Activating Signals to B Cells

A recently discovered member of the TNF family has been discovered that is released by proteolytic cleavage (by a member of the furin family of proprotein convertases) from T cells, dendritic cells, and monocytes. This ligand has been called BlyS (for B lymphocyte stimulator) and BAFF (B cell-activating factor of the TNF family). Receptors for this ligand have been found on B cells, and BAFF/BlyS can induce the proliferation of anti-IgM-triggered B cells and enhance Ig secretion. This TNF family member is likely to emerge as an important costimulator of B cells. It may be synthesized by marginal zone macrophages and could well play a crucial role in the generation of marginal zone B cells.

Signal Transduction in T Cells

Activation of the T cell receptor by a cognate MHC–peptide complex leads to its activation via processes that are broadly similar to those that occur downstream of the B cell receptor. Many issues that relate to TCR signaling will be addressed in Chapter 10. In this section, we will initially discuss how signals are initiated by the TCR and its coreceptors CD4 and CD8. We will then discuss the two-signal hypothesis and examine how CTLA-4 and CD28 help set thresholds for T cell activation. We will examine how Signal 1 and Signal 2 are integrated. We will briefly discuss other costimulatory receptors, focusing primarily on SLAM. Issues such as TH1 versus TH2 activation and the phenomenon of anergy are discussed in Chapter 10.

Signal Transduction Via the T Cell Receptor: The CD4 and CD8 Coreceptors

We have already discussed how MHC–peptide complexes might initiate clustering of the TCR and have considered how coreceptors may be critical for the initiation of signaling via this receptor. The CD4 coreceptor is an integral membrane protein that contains four extracellular Ig-like domains and generally exists as a monomer; at higher protein concentrations, it forms noncovalently associated dimers. Since it can associate both with the TCR as well as MHC class II molecules, CD4 can increase the avidity of TCR-MHC class II interactions. We have pointed out earlier that CD4 may well influence clustering of the TCR. Its cytoplasmic tail associates with Lck, and when it is brought close to the TCR, it may also contribute to ITAM phosphorylation and the activation of Zap-70 or Syk.

CD8 is structurally quite distinct from CD4. It exists largely as a disulfide-linked heterodimer of CD8α and β chains, each of which is encoded by a distinct gene. Each CD8 chain contains an N-terminal Ig-like domain and a heavily O-glycosylated extended polypeptide region that leads up to the transmembrane domain. The cytoplasmic tail of the CD8 α chain binds Lck, although the β chain does not. CD8 also forms αα homodimers, but αβ heterodimers constitute the major CD8 species on $CD8^+$ αβ (referring here to the type of TCR) T cells. There is a marked reduction in $CD8^+$ cells in mice in which the CD8 β chain gene has been knocked out, suggesting that the β chain plays a significant developmental role. Why exactly β chains are so important is not totally clear. It has been suggested that the kinase activity of Lck bound to the α chain may be modified by the associated β chain. Although CD8 αα and αβ dimers bind MHC class I equally well, the αβ heterodimers may more significantly influence the binding of the TCR to MHC molecules. Presumably, αβ heterodimers bind more efficiently to the TCR than αα homodimers.

The only available crystal structures for CD8 (and also for a CD8–HLA-A2 complex) have been obtained for CD8 αα crystals. Interestingly, a flexible loop in the α3 domain of MHC class I molecules is sandwiched between the CDR1- and CDR3-like loops from both CD8 chains in the dimer. An analysis of this structure suggests that a CD8 dimer may be "monovalent." That is, it is unlikely to bind more than a single MHC class I molecule. However, some evidence suggests that CD8 dimers form higher order multimers and thus facilitate the clustering of the TCR when it sees antigen.

Early Events in TCR-Mediated Activation

Once the TCR is clustered, or coligated with its CD4 or CD8 coreceptors, signaling is probably initiated by the Lck and Fyn Src-family tyrosine kinases (Fig. 4.33). Most studies suggest that Lck must be activated to optimally activate Syk and Zap-70, and that the recruitment of these latter kinases to phosphorylated ITAM motifs is a critical event. As described earlier, a conformational alteration of these kinases probably mediates their catalytic activation.

T cells express both Syk and Zap-70. Once Syk family kinases are activated, they modify critical tyrosines on adaptor proteins and also generate some recruitment sites on themselves by autophosphorylation of tyrosine residues. It is likely that the phosphorylation of adaptors rather than the recruitment sites on the Syk family kinases themselves are crucial for the initiation of signaling cascades. The two most critical adaptor proteins early in T cell activation are SLP-76 and LAT. Phosphorylated LAT can recruit PLC-γ1, Grb-2 containing complexes, and the p85 subunit of PI3K, all via SH2-PY interactions. SLP-76 can also directly bind the SH2 domains of Grb2, Vav, Nck, and probably PLC-γ1 as well. This adaptor is absolutely required for the activation of the Ras pathway and of PLC-γ1. It appears that LAT might also be required for the activation of these very same pathways. It is believed that tyrosine-phosphorylated LAT nucleates a signaling complex that includes SLP-76. This complex recruits a number of key signaling molecules to the membrane. LAT and SLP-76 activate the

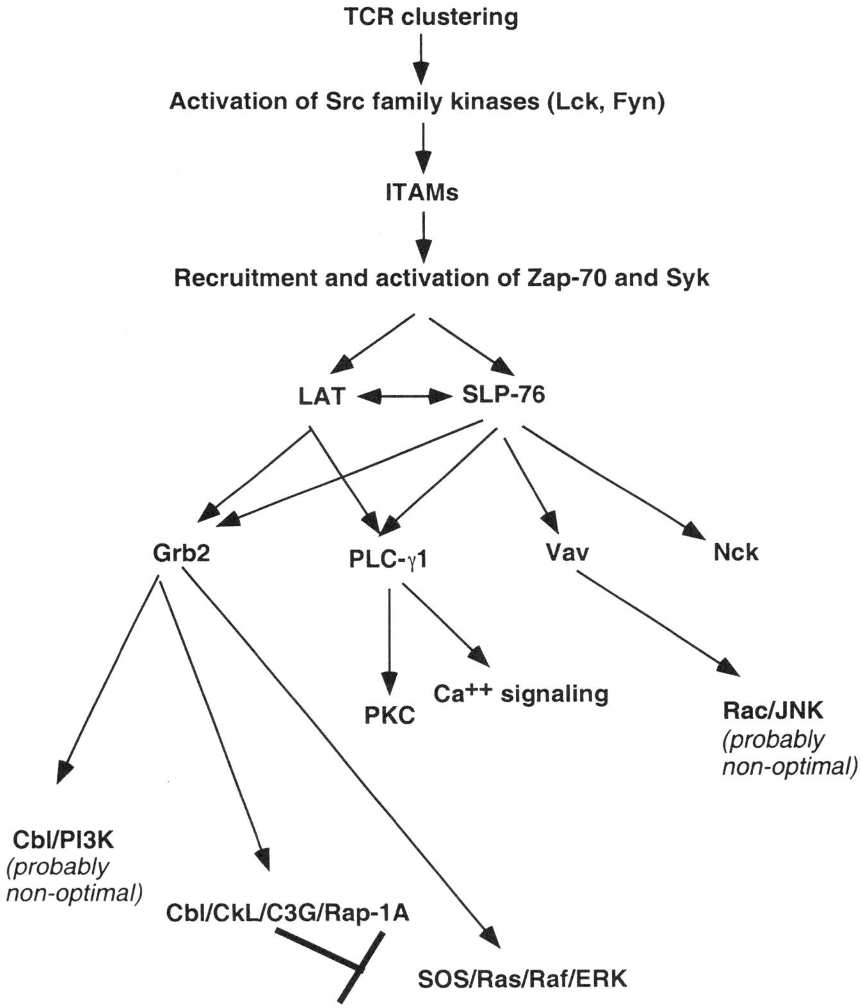

FIGURE 4.33. Some of the signals that are generated by the TCR. Details in the text.

Ras/Raf/ERK pathway and PLC-γ1 as mentioned above but also might activate PI3K and possibly Vav/Rac/JNK signaling as well. LAT has consensus sites for the recruitment of the SH2 domains of Grb2 and PLC-γ1 but not for the p85 subunit of PI3K. It is likely that PI3K is recruited indirectly to the complex, possibly bound to Grb2-Cbl complexes (p85 can associate with tyrosine phosphorylated Cbl). It is likely that activation of PI3K and of the JNK pathway via the LAT/SLP-76 complex may not be optimal; certainly, proper JNK activation appears to need costimulatory signals.

Immunological Synapses and the Merging of Lipid Rafts During T Cell Activation

When a T cell and an APC come into contact, they form a tight synapse. From the known physical dimensions of the TCR and of MHC molecules, it is known that a gap of about 15 nm separates a T cell membrane from that of an APC. While the initial association of a T cell and a relevant APC or target cell may be stabilized by interactions between adhesion molecules, this does not lead to stable synapse formation unless the TCR is triggered. There are likely to be only a small number of MHC–peptide complexes specific for a given TCR on an APC, and the known affinity of the TCR for cognate MHC–peptide complexes is relatively low (on the order of 10^{-4} to 10^{-6} M). The interaction of the TCR with MHC ligands per se is not "sticky" enough to stabilize the formation of synapses. However, TCR signaling does orchestrate a remarkable cytoskeletal rearrangement event that contributes to the formation of a highly organized synaptic junction between T cells and APCs. This leads to the generation of an outer ring-like structure that surrounds an inner contact region, which is separated by an approximately 15-nm intermembrane gap from APCs or target cells (Figure 4.34). Apart from the TCR and the CD4 or CD8 coreceptors, other T cell proteins that are small enough to fit in this gap and that are concentrated in this contact region include the CD2 adhesion molecule and the CD28 costimulatory receptor. Molecules that are too large to fit in this region include the β-integrin, LFA-1, and the CD45 tyrosine phosphatase.

CD2 is an adhesion molecule present on T cells and NK cells that can associate with a glycosylphosphatidylinositol-linked protein on APCs and target cells. The ligand for CD2 in rodents is CD48, and the ligand in human cells is a related molecule known as CD58. The cytoplasmic tail of CD2 contains three proline-rich SH3-binding motifs and associates with a protein known as CD2AP. CD2AP contains three SH3 domains, one of which has a very high affinity for a proline-rich stretch at the C-terminus of CD2. Following TCR signaling, the association of CD2AP with CD2 is enhanced, and in some poorly understood way, CD2AP contributes to cytoskeletal reorganization and the formation of the ordered ring structure that characterizes immunological synapse generation. The clustered outer ring of LFA-1 and the inner circle of CD2 contribute in a major way to maintaining the adhesive interactions between a T cell and APCs.

During lymphocyte activation, signals are initiated from specialized membrane microdomains known as lipid rafts, DIG (detergent-insoluble glycolipid) domains, DRMs (detergent-resistant membranes) or GEMs (glycolipid-enriched microdomains/ganglioside enriched membranes). GPI (glycosyl phosphatidyl inositol)-anchored proteins as well as acylated signaling molecules such as Src-family kinases, and monomeric and

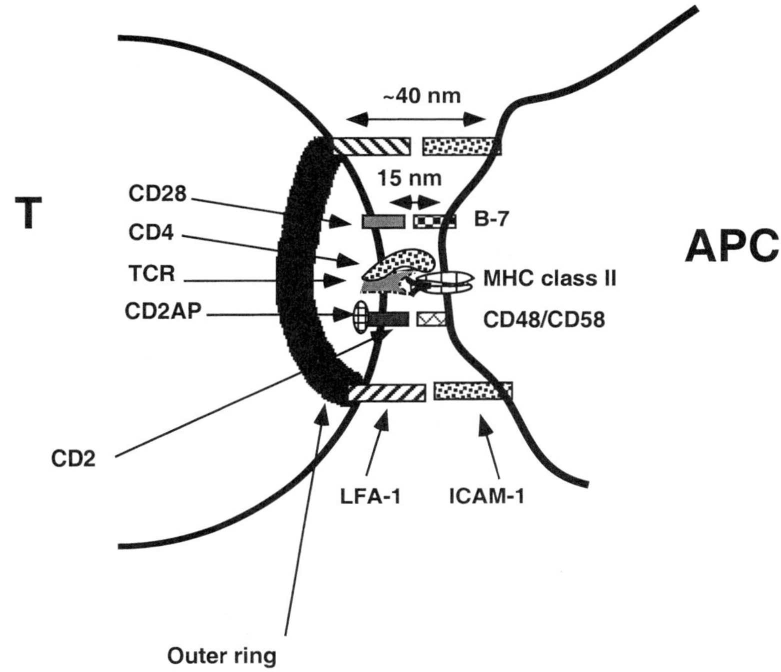

FIGURE 4.34. Features of an immunological synapse. Smaller proteins concentrate in the region of an intermembrane cleft, which is about 15 nm wide. These include the TCR-MHC pair, CD4 or CD8, CD2 and its ligand, and CD28 and B7. CD2AP participates in synapse formation The outer ring surrounding this contact region contains larger proteins such as LFA-1, which associates with intercellular adhesion molecule-1 (ICAM-1).

heterotrimeric G proteins, are generally found in these sphingolipid- and cholesterol-rich microdomains, which are resistant to nonionic detergents such as Triton X-100 (an arylpolyethoxyethanol). Lipid rafts are approximately 70 nm in diameter and are related to well-characterized cave-like invaginations of the plasma membrane known as caveoli that have long been ultrastructurally recognized and studied in epithelial cells. These regions are rich in PI-4,5,-P2. GPI anchors and lipid modifications such as palmitoylation and farnesylation faciliate the partitioning of proteins into these microdomains. PIP3 also presumably accumulates in the rafts, drawing in Tec family kinases and other PH domain-containing proteins into these specialized microdomains. During T cell activation, lipid rafts fuse with one another and converge in the region of immunological synapses; the contact surface in these synapses represents the aggregation of a multitude of rafts.

The Two-Signal Hypothesis and the Need for Costimulators

Very sophisticated signaling thresholds are utilized to ensure that T cells are activated only when the immune system requires to respond—at times of "danger." The two-signal hypothesis for lymphocyte activation (Figure 4.35) as it was first enunciated suggests that a signal from the TCR–coreceptor complex (Signal 1 in this parlance) is insufficient to fully activate a T cell. A second signal, termed *Signal 2*, is required for a T cell to be properly activated.

The best studied costimulators are the B7-1 (CD80) and B7-2 (CD86) proteins expressed on activated APCs including B lymphocytes. Resting B cells and macrophages express barely detectable levels of both B7 ligands. Dendritic cells in the spleen and peripheral blood express constitutive low levels of B7 proteins. Following activation of B cells, macrophages, and dendritic cells, the kinetics of induction of B7-1 and B7-2 are quite distinct. B7-2 induction can be observed at 6 hours, and this peaks around 18 to 24 hours after stimulation. B7-1 induction is first observed about 24 hours after stimulation, and peak expression is seen 48 to 72 hours after stimulation.

Two costimulatory receptors on T cells recognize B7-1 and B7-2. CD28 can, in many circumstances, provide Signal 2, whereas CTLA-4 (cytoxtoxic T lymphocyte antigen-4; CD152), which is structurally similar to CD28, has an approximately 10-fold higher affinity for B7 ligands but produces inhibitory signals. Both CD28 and CTLA-4 are Ig superfamily members, which exist as disulfide-linked homodimers. CD28 is constitutively expressed on the surface of T cells, whereas CTLA-4 expression at detectable levels on the cell surface is first seen 24 hours after T cell activation, and these levels peak 48 to 72 hours after activation. Although this induction pattern initially suggested that CTLA-4 might function primarily to terminate T cell responses, it is now known that most CTLA-4 in non-triggered T cells is present in the Golgi and the trans-Golgi network. Very soon after TCR signaling, there appear to be barely detectable but functionally significant focal increases of cell surface CTLA-4, which has been translocated from the Golgi to the sites of T cell–APC interaction. While CTLA-4 may well be critical for the termination of T cell responses, growing evidence suggests that CTLA-4 may also help set thresholds for T cell signaling in the earlier stages of T cell activation.

In circumstances in which costimulatory ligands are not expressed at high levels on antigen-presenting cells, Signal 1 is generated via the T cell receptor, but the low levels of B7-1 and B7-2 cannot induce Signal 2 via CD28. Signal 1 induces focal increases in CTLA-4 on the responding T cell, and this molecule is preferentially ligated when B-7 levels are low. Not only is Signal 2 not generated, but inhibitory signals via CTLA-4 may modulate signaling pathways in ways that will be discussed below (Figure 4.36). This may lead to the generation of T cell anergy. Anergic T cells cannot respond

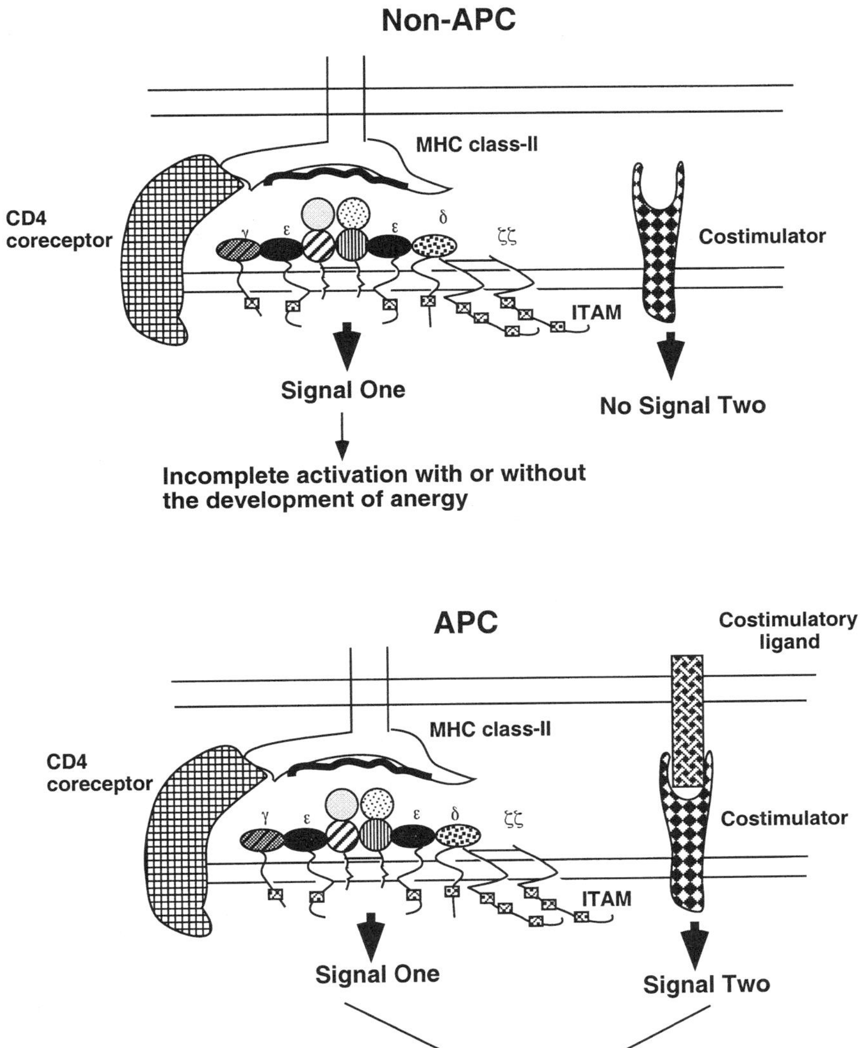

FIGURE 4.35. The two-signal hypothesis for T cell activation. Signal 1 from the TCR is insufficient for full activation. A combination of Signal 1 and Signal 2 leads to full activation. The ligand that induces Signal 2 is termed a costimulator. Note: MHC class-II is rarely expressed on non-APCs.

to further antigenic challenge, and this process will be discussed in Chapter 10.

It is unclear when exactly anergy might be induced *in vivo*. However, T cell triggering in the absence of inflammation leads to the preferential trig-

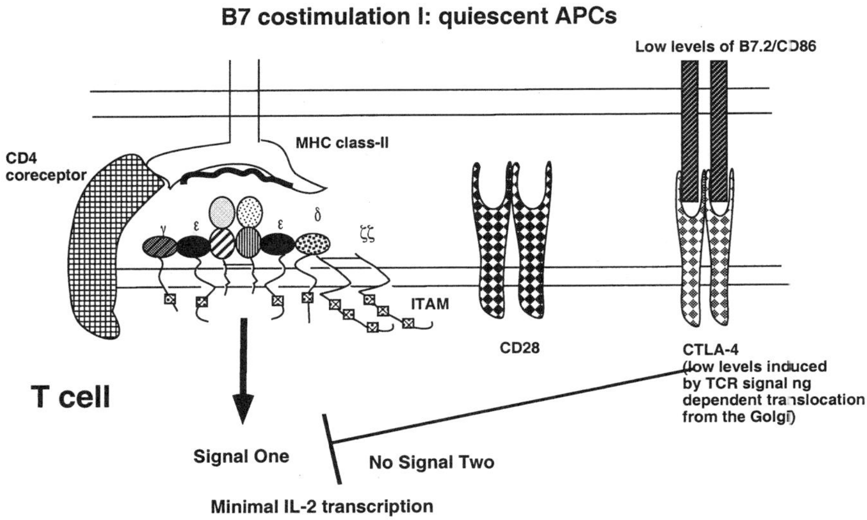

FIGURE 4.36. Quiescent APCs express low levels of B-7 ligands, which preferentially trigger CTLA-4 in T cells. CTLA-4 translocation to the cell surface may be induced by Signal 1. Because of its higher affinity for B-7, CTLA-4 may preferentially signal when APCs express low levels of B-7.

gering of CTLA-4, which essentially contributes to the setting of a threshold for signaling; T cells can be triggered only when APCs express relatively high levels of B-7 molecules. During the course of an infection, antigen-presenting cells may express high levels of B7-1 and B7-2. These levels of B7 are sufficient to activate CD28. CD28 signaling can overcome the negative signals from CTLA-4 and provide Signal 2 (Figure 4.37). The combination of Signal 1 and Signal 2 contributes to the optimal activation of T cells.

The combination of signals from the TCR, CD28, and the IL-2 receptor contributes to the transcriptional induction of high levels of CTLA-4 on T cells that have been efficiently activated. At this point, it is important to downregulate TCR-mediated signaling. The enhanced levels of CTLA-4 on activated T cells contribute to a tilt toward the transduction of negative signals via CTLA-4, which can begin to counter the growth-stimulating effects of CD28 as well as TCR signaling (Figure 4.38). In *Ctla-4*$^{-/-}$ mice, the uncontrolled activation of T cells leads to massive lymphoproliferation with early lethality.

Signal Transduction via CD28 and CTLA-4

The requirement for costimulation has been extensively analyzed at the molecular level in terms of the requirements for the induction of IL-2 transcription. Transcriptional induction of IL-2 requires nuclear factor of

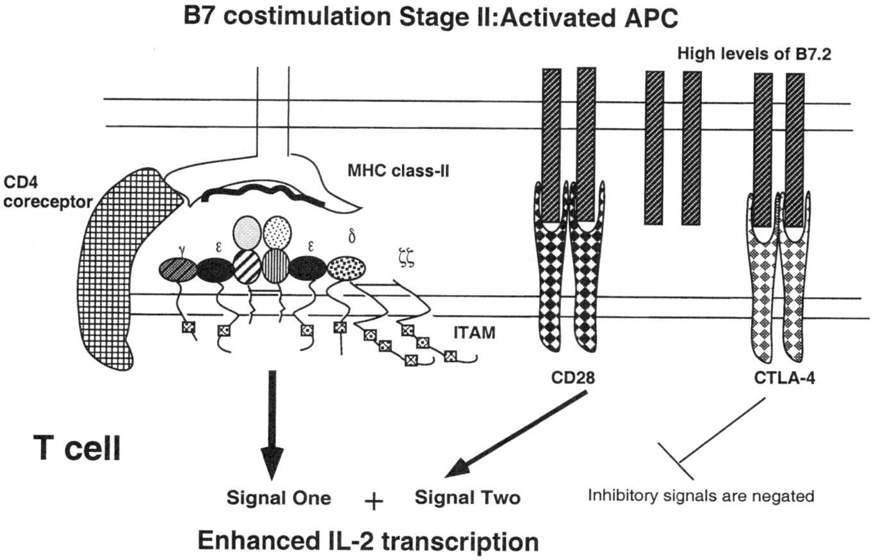

FIGURE 4.37. Activated APCs express high levels of B7 and thus permit the activation of CD28. CD28 signaling facilitates the generation of Signal 2 and negates inhibitory signals from CTLA-4.

activated T cells (NFAT), AP-1, NFκB/Rel factors (which bind to a site often referred to as the CD28 response element), and other transcriptional activators. In addition, transcriptional repressors such as Nil-2a (a protein of the ZEB family) contribute to the regulation of IL-2 transcription.

Triggering of CD28 contributes to the activation of PI3K, of JNK, and also probably of the Ras/Raf/ERK pathway. Ras activation is noted when cells are triggered using anti-CD28 antibodies but not in response to B-7 (which presumably also triggers CTLA-4 and shuts off Ras activation). The cytoplasmic tail of CD28 contains a YMXM motif, which, when tyrosine phosphorylated, can recruit the p85 subunit of PI3K. PI3K activation may in turn contribute to the activation of Vav, a GEF for Rac (which contains a PH domain and is recruited to the membrane in part by PIP3). In principle, the generation of Rac-GTP may contribute to the activation of the JNK pathway, which is not optimally turned on in response to Signal 1. JNK activation is, as discussed, critical for the phosphorylation of specific serine residues in the N-terminal portion of c-Jun, and therefore the activation of AP-1. AP-1 and nuclear NFAT contribute cooperatively to the induction of IL-2 transcription (see Chapter 10). The induction of c-Fos is presumably largely dependent on Signal 1, which activates the Ras/Raf/ERK pathway and thus contributes to the phosphorylation of Elk-1, a transcriptional inducer of the *c-fos* gene. It is likely that CD28 not only activates

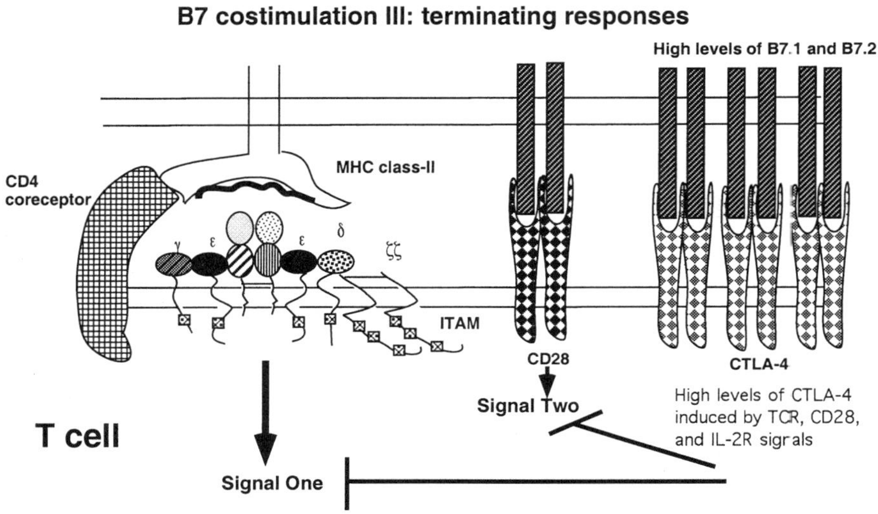

FIGURE 4.38. Signals via the TCR, CD28, and the IL-2R contribute to the transcriptional induction of high levels of CTLA-4. Enhanced CTLA-4 signals can counter Signals 1 and 2 and thus contribute to the termination of T cell responses.

JNK but also contributes to the magnitude and duration of ERK activation as well.

The identity of the nonreceptor tyrosine kinase or kinases that potentially phosphorylate the cytoplasmic tail of CD28 remains to be determined. The TEC family kinase Itk functions downstream of the TCR in a manner similar to that of Btk in B cells and contributes to capacitative calcium signaling. Itk also associates with CD28 but apparently functions in this context to negatively modulate the amplitude of CD28 signaling in a poorly understood fashion.

An additional, though less well understood function, of CD28 signaling is the ability of this costimulatory receptor to influence the turnover or displacement of a repressor of IL-2 transcription known as Nil-2a (discussed in Chapter 10). Exactly how CD28 influences Nil-2a turnover is unclear.

CTLA-4 ligation inhibits the activation of both the ERK and JNK pathways. It therefore counters the effects of both Signal 1 and Signal 2. This regulatory event occurs somewhere downstream of T cell receptor ζ chain phosphorylation and the activation of Zap-70. It has been suggested that CTLA-4 recruits and activates SHP-2 (also known as SYP or PTP1D). SHP-2 might presumably dephosphorylate a critical Zap-70 substrate such as LAT or SLP-76; studies examining this issue in more detail remain to be performed. CTLA-4 has also been reported to be capable of activating PI3K, but the functional significance of this event is unclear.

Costimulation as a Signal-Amplification Mechanism

An alternative view to the signal integration model described above has been propagated to explain what costimulation is all about. It has been suggested that costimulation serves as a signal amplification mechanism and that this is achieved by initiating the directional actin-based transport of lipid rafts with associated signaling molecules to the area of cell–cell contact, also known as the immunological synapse.

Lipid rafts are also known as DIG domains. They correspond to the microscopic invaginations known as caveolae described by electron microscopy in many nonlymphoid cell types. Lipid rafts are rich in glycosylphosphatidylinositol-linked membrane proteins and in Src-family kinases. These areas are cholesterol rich and can contribute to increased membrane rigidity and thereby increase the affinity of receptor–ligand interactions. By excluding tyrosine phosphatases, they may contribute to enhanced tyrosine phosphorylation as well.

ICOS Is an Inducible Costimulator on T Cells

While CD28 is expressed constitutively on most T cells, a distinct protein that is structurally similar to CD28, but is expressed on T cells only after TCR activation, has been described. This protein has been detected on tonsillar T cells close to B cells in the apical light zones of germinal centers. This protein, inducible co-stimulator (ICOS), provides costimulatory signals to activated T cells and can induce high-level secretion of IL-10. IL-10 can contribute to the terminal differentiation of B cells into memory cells and plasma cells, and it is thought that ICOS may prove an important player in T–B collaboration, particularly in germinal centers.

The ligand for ICOS is, not surprisingly, structurally related to B7-1 and B7-2, and is known as B7h (for B7 homolog) or B7RP-1 (for B7 related protein-1). B7h is constitutively expressed in B cells and macrophages but its expression can be induced in non-lymphoid tissues in response to TNF. It is possible that B7h-ICOS interactions participate not just in APC-T cell interactions during secondary immune responses, but that in an inflammatory context, B7h on non-lymphoid target cells may augment the effector functions of previously activated T cells that have been induced in lymph nodes to express ICOS.

SLAM is a Costimulatory Ligand and Receptor that Triggers T and B cell Responses

A type I membrane glycoprotein called signaling lymphocyte activation molecule (SLAM) is a potentially important costimulator that is believed to be involved in the bidirectional triggering of T and B lymphocytes during T–B collaboration. SLAM acts as its own ligand. Anti-SLAM monoclonal antibodies in conjunction with TCR triggering can induce resting T cells to proliferate and to activate cytokines such as γ-interferon. Anti-SLAM antibodies can modulate the phenotype of antigen-specific Th2 clones, redirecting them to a Th1 or Th0 phenotype (discussed in Chapter 10). Although anti-SLAM antibodies alone do not trigger resting T cells, they can induce the proliferation of human $CD45RO^{hi}$ memory T cells.

A soluble form of the extracellular domain of SLAM, as well as SLAM expressed on transfected cells, can enhance B cell proliferation, differentiation, and antibody secretion in conjunction with other stimuli such as the CD40 ligand. Anti-SLAM monoclonal antibodies alone do not induce B cell proliferation.

The study of the biological role of SLAM is still being elucidated. A very small SH2 domain-containing protein called SLAM-associated protein (SAP) has been discovered to be the defective gene in the X-linked lymphoproliferative syndrome (XLP) (see Chapter 12). This protein associates with the cytoplasmic tail of SLAM and may block the recruitment of SH2 domain-containing signaling molecules to this portion of SLAM during signal transduction. This suggests that SLAM is an important component of T and B cell activation that remains to be studied in detail.

Other Costimulatory Receptors

Elsewhere in this chapter, we have briefly considered a few other costimulatory receptors that may influence T cell activation. While considering immunological synapse formation, we have discussed the important roles played by CD2 and its ligand (CD58/LFA-3 in humans and CD48 in mice). Another important costimulatory receptor, considered briefly in that section, that is involved both in adhesion as well as in signaling is the β2 integrin, LFA-1 (CD11a), which binds to ICAM-1 on APCs. Integrins are discussed in Chapter 6.

A number of TNFR family members including 4-1BB, OX40/CD134, RANK (receptor activator of NFκB), CD27, and CD30 may have important costimulatory roles. RANK is the receptor for OPGL (osteoprotegerin ligand), also known as TRANCE (TNF related activation induced cytokine), and this receptor-ligand pair will be considered in more depth in Chapter 6 in the context of lymph node biogenesis. When exactly during immune responses specific costimulatory receptors other than CD28 are biologically relevant is not well understood. 4-1BB is a costimulatory receptor expressed primarily on activated $CD4^+$ and $CD8^+$ T cells. Its ligand, 4-1BBL, is expressed at low levels on activated B cells and dendritic cells. 4-1BB is probably not required during the initiation of immune responses but may contribute to IL-2 generation and T cell proliferation after T cells are activated.

The OX40L-OX40 receptor-ligand pair may be required for the generation of optimal $CD4^+$ T cell responses. Mice that lack the OX40L have no defects in humoral immune responses or delayed type hypersensitivity, but exhibit a defect in contact hypersensitivity, a phenomenon that involves the activation of $CD4^+$ T cells by dendritic cells. The proliferation of $CD4^+$ T cells is impaired in mice that lack OX40. It is likely that this receptor-ligand pair participates in the activation of some $CD4^+$ T cells by dendritic cells.

CD27 is a TNFR family member whose ligand is CD70. CD27 is expressed on T cells and some B cells and its expression is upregulated following lymphocyte activation. CD70 is a TNF family member that is expressed transiently on activated T and B cells, but not on dendritic cells. CD27-CD70 interactions may enhance T cell proliferation, TNF secretion, and the generation of $CD8^+$ effector T cells. However, the physiological role of CD27 remains poorly understood.

CD30 is a TNFR family member that is expressed at high levels on Hodgkin-Reed-Sternberg cells (HRS), a defining cell type in Hodgkin's lymphoma. HRS cells are malignant B lymphocytes that do not express surface Ig because their antigen receptor genes are somatically mutated. CD30 is normally found on some developing medullary thymocytes and may be of importance in negative selection.

An Overview of Cytokine Signaling

Cytokines are generally defined as soluble biological response modifiers that mediate cell–cell communication. They include a range of molecules, some of which are also categorized as interleukins, colony-stimulating factors, and interferons. Chemotactic cytokines, or chemokines, will be discussed separately in Chapter 6. Signal transduction via cytokine receptors is too vast a subject to discuss adequately in a brief section of a chapter on signal transduction. This section is an overview of the signaling strategies used by a range of cytokines and addresses the issue of how specificity is achieved when considering cytokines that utilize the same kinases or transcriptional regulators to transduce signals.

Most cytokine receptors are Type I glycoproteins, which lack intrinsic protein kinase activity. The majority of cytokine receptors are physically linked either directly, or indirectly during signal transduction, to nonreceptor tyrosine or serine-threonine kinases. Some receptors exist in the nonactivated state as monomers, while many exist as hetero-oligomers. Receptors of the TNFR family are induced to form homotrimers. Cytokine receptors can be classified into four major groups, which each share some broad features and modes of downstream signaling. Hormones that are generally not classified as cytokines, such as growth hormone and prolactin, have receptors that belong to the Type I cytokine receptor family. Some immune ligands that are not cytokines, such as CD40L, CD30L, and FasL, trigger receptors that belong to the Type III cytokine family.

Type I cytokines all share a similar four-α-helical bundle structure. As seen in Table 4.6, these cytokines may be categorized as short chain or long

TABLE 4.6. A broad classification of cytokine receptors

1. Type I cytokine receptors
   1.1. Receptors for short chain cytokines
      a. Receptors that share γc: IL-2, IL-4, IL-7, IL-9, and IL-15
      b. Receptors that share IL-4Rα: IL-4, IL-13
      c. Receptors that share βc: IL-3, IL-5, GM-CSF
   1.2. Receptors for long chain cytokines
      a. Receptors that share gp130: IL-6, IL-11, OSM, CNTF, LIF, CT-1
      b. Receptors with a gp130-like component: IL-12, leptin
      c. Homodimeric receptors: growth hormone, prolactin, erythropoietin, thrombopoietin
2. Type II cytokine receptors of the interferon family
   Receptors for Type I interferons (IFN- α and β) Type II interferons (IFN-γ) and the IL-10R
3. Type III cytokine receptors of the TNFR family
   TNFR Type I, TNFR Type II, CD40, CD30, CD27, NGFR, Fas, LTβR
4. Type IV cytokine receptors of the IL-1R family
   IL-1R, IL-18R
5. Other cytokine receptors
   CD117/c-Kit and CSF-1R are receptor tyrosine kinases

chain cytokines. Type I cytokine receptors (often described as members of the hematopoietin receptor superfamily) also share some structural features (Figure 4.39). Their extracellular domains contain four positionally conserved cysteine residues and a WSXWS motif. The membrane-proximal regions of the cytoplasmic tails of many, but not all, of these receptors contain two conserved regions known as Box 1 and Box 2. The Box 1 motif is found in the first 20 amino acids of the cytoplasmic tail and may contain two or three positionally conserved proline residues. The Box 2 regions generally begin with a cluster of hydrophobic amino acids and end with one or two charged residues. More significant similarities may be observed when the Box 2 regions of specific receptor chains are compared. Between the Box 1 and Box 2 regions, there is a conserved tryptophan residue. The Box 1 and Box 2 regions as well as the conserved trptophan residue are all required for Jak activation, perhaps because they are needed for Jak binding.

Many cytokine receptors share specific receptor chains. A common γ chain known as γc is shared by the IL-2, IL-4, IL-7, IL-9, and IL-15 receptors. A common β chain (βc) is shared by the IL-3, IL-5, and GM-CSF receptors, and the gp130 protein is a common subunit of the receptors for IL-6, IL-11, oncostain M (OSM), ciliary neurotrophic factor (CNTF), leukemia inhibitory factor (LIF), and CT-1. The IL-2Rβ chain is also a component of the IL-15R. The IL-4Rα chain forms heterodimers with γc as well as with IL-13Rα1 and IL-13Rα2.

There are broad similarities in the signal transduction strategies utilized by Type I and Type II cytokine receptors. Both these types of receptors use

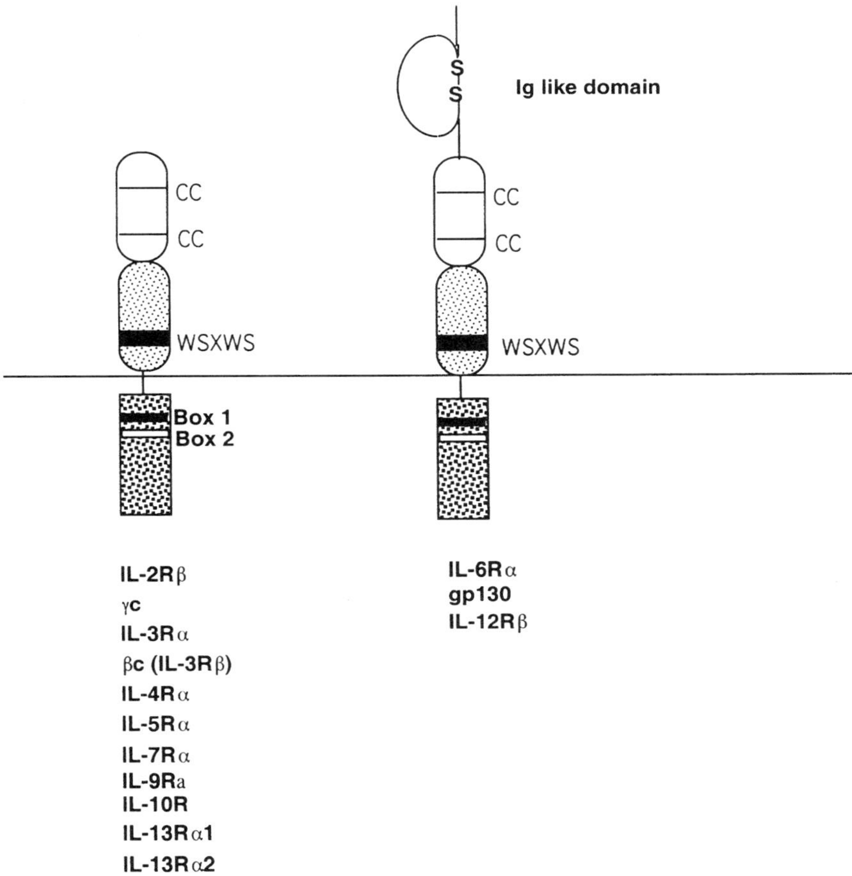

FIGURE 4.39. Type I cytokine receptor chains (sometimes referred to as members of the hematopoietin receptor family) contain conserved structural motifs. These include four positionally conserved extracellular cysteines and a WSXWS motif. Some of these receptors share broadly similar intracellular Box 1 and Box 2 sequences. A small number of receptor chains have an N-terminal Ig-like domain. The IL-2Rα chain and the IL-15Rα chain are structurally unrelated to all the chains listed above.

a very simple signaling pathway involving nonreceptor tyrosine kinases known as Jaks (for just another kinase or Janus kinases) and transcription factors that start out in association with these receptors but that are induced by signaling to migrate to the nucleus. The transcription factors are known as STATs (for signal transducers and activators of transcription), and these each contain an SH2 domain. However, individual Type I cytokine receptors often also activate other signaling pathways in addition to the activation of STATs.

Type III and Type IV cytokine receptors do not in general use Jak–STAT signaling but there are broad similarities in the type of signaling pathways that are induced downstream of these receptors. We have already considered some of the signaling pathways downstream of TNF receptors (examples of Type III cytokine receptors). The CD40 receptor, discussed above, and the Fas receptor discussed in Chapter 5, represent examples of receptors for noncytokine ligands that nonetheless belong to the Type III cytokine receptor family. Critical roles played by members of the TNFR family in the biogenesis of secondary lymphoid organs will be discussed in Chapters 7 and 8.

The Jak–STAT Pathway for Type I and Type II Cytokine Receptors

The Jak–STAT pathway is remarkably simple in comparison to most signal transduction pathways (Figure 4.40). The cytoplasmic tail of many "nonactivated" cytokine receptors are associated with Jaks. Receptor clustering induces Jak autophosphorylation and activation, which is followed by tyrosine phosphorylation of critical tyrosine residues in the cytoplasmic tail of the receptor. Specific phosphotyrosine residues in the cytoplasmic tail of the receptor recruit STATs via the SH2 domains of the latter. The STAT protein is phosphorylated on tyrosine residues by the same Jak. This leads to the release of STATs and their dimerization by PY-SH2 interactions. STAT homo or heterodimers are translocated to the nucleus, where they bind to specific regulatory sites in cytokine target genes.

There are a total of four Jak kinases (Jak1, Jak2, Jak3, and Tyk2) and seven STAT proteins (STAT1, 2, 3, 4, 5a, 5b, and 6). Jak kinases contain a number of homology domains numbered JH1 to JH7, but the only domain whose function is known is the JH1, or kinase, domain. The JH2 domain is a pseudokinase domain, whose function remains unknown. It may serve to inhibit kinase function; however, mutations in this domain have been found in human patients with severe combined immunodeficiency disease (SCID) with nonfunctional Jak3 (see Chapter 12), suggesting a functional role for this domain. The JH4 domain structurally resembles SH2 domains but has not been shown to bind phosphotyrosine-containing peptides. The N-terminal JH6 and JH7 domains contribute to the association of Jaks with the cytoplasmic tail of the receptor.

STATs were originally discovered as proteins that bound to the promoters of interferon-inducible genes. STATs contain a C-terminal SH2 domain, which is close to the site of tyrosine phosphorylation, and an N-terminal dimerization domain, which facilitates the dimerization process that is mainly driven by C-terminal SH2-PY interactions. In addition, these proteins contain a central DNA-binding domain (Figure 4.41). STAT homo- and heterodimers tend to bind semipalindromic motifs in the promoters of

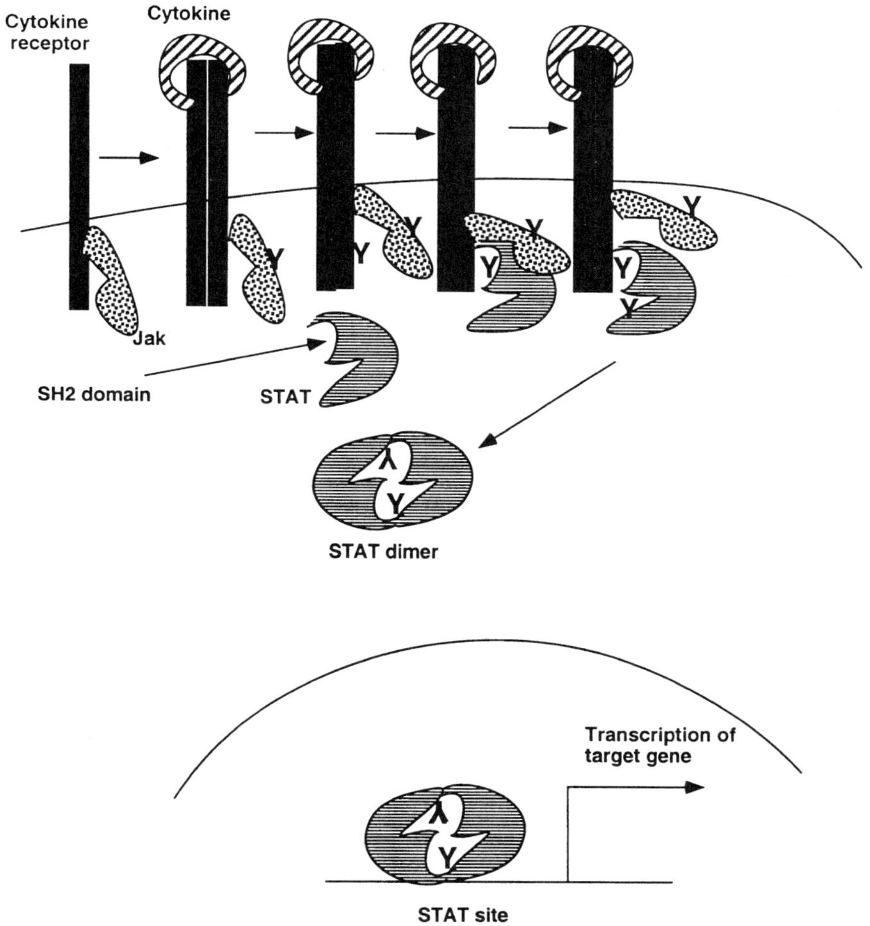

FIGURE 4.40. An overview of Jak–STAT signaling. Ligation of cytokine receptors leads to Jak autophosphorylation and activation. Specific tyrosine residues on the tail of the receptor are phosphorylated and recruit STATs via PY-SH2 interactions. STATs are phosphorylated by Jaks and dimerize. STAT dimers are translocated to the nucleus and activate transcription.

target genes, which are generically referred to as GAS sequences (for IFN-γ activated sites) even if they bind to STATs activated by a range of cytokines. GAS motifs generally fit the following consensus: TTCNNN(N)CAA. Distinct STATs bind to different GAS motifs.

Interferon α and β signaling leads to the formation of a STAT1/STAT2/p48 complex (p48 is not a STAT), which binds to a nonpalindromic site in the promoters of target genes, is distinct from a GAS motif, and is called an interferon signaling response element (ISRE).

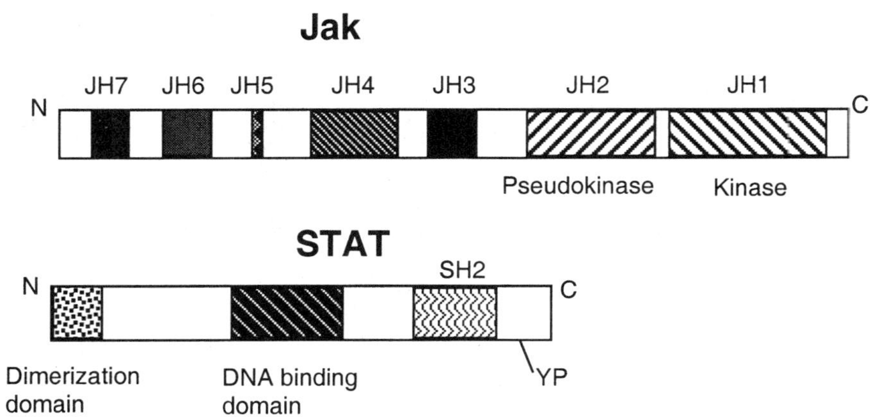

FIGURE 4.41. Structural features of Jaks and STATs. The only Jak homology region whose function is understood is the JH1, or kinase, domain. STATs contain a C-terminal site for tyrosine phosphorylation distal to an SH2 domain. An additional dimerization domain is located in the N-terminal portion. The DNA-binding domains of different STATs are specific for distinct but related motifs.

Specificity in Jak/STAT Signaling

In Table 4.7, Jaks and STATs that are linked to selected Type I and Type II cytokine receptors are listed. Given that many cytokine receptors share signaling chains and are often associated with the same subsets of Jaks, how is specificity in signaling achieved? One major contributor to specificity is the existence on different receptor chains of specific tyrosine residues for the docking of signaling molecules that include, but are not restricted to, STATs. Different docking sites can therefore recruit different STATs or other SH2-containing signaling molecules such as Src-family kinases, Syk family kinases, the p85 subunit of PI-3 kinase, and so on.

Not surprisingly, knockouts of specific *Jak* and STAT genes often compromise lymphocyte development in very specific ways. Many of these phenotypes will be considered in the chapters on B and T cell development and are summarized in Table 4.8.

Inhibitors of Cytokine Signaling: The SOCS Proteins

A family of SH2 domain-containing proteins known as suppressors of cytokine signaling (SOCS) have been described. There are eight SOCS proteins (SOCS1-7 and CIS), and of these, SOCS1 and SOCS3 appear to be the most efficient inhibitors of cytokine signaling. SOCS proteins inhibit cytokines that trigger Jak-STAT signaling primarily by binding to the activation loops of Jaks and inhibiting their catalytic activity. SOCS proteins may be important modulators of cytokine signaling. In the absence of

TABLE 4.7. Jaks and STATs linked to selected Type I and Type II cytokine receptors

Receptors	Jaks	STATs
Type I		
IL-2R, IL-7R, IL-9R, IL-15R (γcRs)	Jak1, Jak3	STAT5a, STAT5b, STAT3
IL-4R (γcR)	Jak1, Jak3	STAT6
IL-13R	Jak1, Jak2, Tyk2	STAT6
IL-6R, IL-11R, OSMR, CNTFR LIFR, CT-1R (gp130Rs)	Jak1, Jak2, Tyk2	STAT3
IL-12R	Jak2, Tyk2	STAT4
Type II		
IFN-α, IFN-β	Jak1, Tyk2	STAT1, STAT2
IL-10	Jak1, Tyk2	STAT3
IFN-γ	Jak1, Jak2	STAT1

TABLE 4.8. Phenotypes of selected Jak and STAT knockout mice

Knockout	Phenotype	Reference
Jak1$^{-/-}$	No biological responses to all Type II cytokines. No responses to γc cytokines and gp130 cytokines.	Rodig et al., 1998
Jak2$^{-/-}$	Not required for lymphoid differentiation. Embryonic lethal; no definitive erythropoiesis. Fetal liver precursors do not respond to erythropoietin, thrombopoietin, IL-3, GM-CSF, and IFN-γ.	Parganas et al., 1998
Jak3$^{-/-}$	Severe combined immunodeficiency resembling X-linked SCID	Thomis et al., 1995; Nosaka et al., 1995; Park et al., 1995
Stat1$^{-/-}$	Defective responses to IFN-α, IFN-β, and IFN-γ	Meraz et al., 1996; Durbin et al., 1996
Stat3$^{-/-}$	Embryonic lethal.	Takeda et al., 1997
Stat4$^{-/-}$	Defective Th1 development.	Kaplan et al., 1996b; Thierfelder et al., 1996
Stat5a$^{-/-}$	Defective NK cell proliferation and cytolytic activity.	Imada et al., 1998
Stat5b$^{-/-}$	Defective IL-2-induced expression of IL-2R.	Nakajima et al., 1997
Stat 5a/5b$^{-/-}$	Defective IL-2-induced cell cycle progression.	Teglund et al., 1998; Moriggl et al., 1999
Stat6$^{-/-}$	Defective Th2 cell development.	Kaplan et al., 1996a; Shimoda et al., 1996; Takeda et al., 1996

SOCS1, mice die within 2 to 3 weeks after birth, death occurs in a T cell and interferon-γ dependent manner. SOCS1 may have two critical functions. In T cells, it may regulate T cell differentiation or activation to restrain the development of activated T cells secreting IFN-γ. In addition, SOCS1 may function in both lymphoid and nonlymphoid cells to modulate IFN-γ signaling and to restrict the production of inflammatory cytokines such as TNF.

Signaling by Selected Type I Cytokine Receptors

The following sections discuss some aspects of signal transduction downstream of a few type I cytokine receptors considered to be particularly important from the viewpoint of lymphocyte development. The focus will be primarily on the structures of these receptors and the signaling pathways downstream of them. The IL-7 receptor plays an important role in lymphopoiesis for both B and T cells. The IL-2 receptor is one of the best studied cytokine receptors, and it is critical for activation-induced cell death; it is also an important autocrine factor for T cell proliferation. IL-4 and IL-12 play key roles in the differentiation of Th2 and Th1 cells respectively. IL-6 and IL-13 have also been postulated to be involved in Th2 cell differentiation, and signaling via their receptors will also be briefly discussed. IL-15 is critical for the development of NK cells, NT cells, and intraepithelial γδ T cells. IL-15 also contributes to lymphocyte proliferation in the periphery and the homing of lymphocytes to peripheral lymph nodes. While phenotypes of cytokine receptor and cytokine knockouts will be summarized in Table 4.9, a more detailed discussion of the roles played by these molecules in B, T, and NK cell development will be found in Chapters 7 to 11.

Receptors That Share γc

As mentioned before, the IL-2, IL-4, IL-7, IL-9, and IL-15 receptors share γc. We will initially discuss the IL-7, IL-2, and IL-15 receptors. The IL-4 receptor shares a chain with the IL-13 receptor (which does not belong to the shared γc category) and will be discussed separately.

The IL-7 Receptor

IL-7 is a cytokine that plays a critical role in lymphopoiesis. It is secreted by stromal cells in the bone marrow and by cortical epithelial cells in the thymus. It is a major player in the survival and the induction of proliferation of progenitor T and B lymphocytes. It can also influence V(D)J recombination in the B lineage and is absolutely required for the rearrangement of the TCRγ chain and therefore for the development of the γδ lineage of T cells.

The IL-7 receptor (Figure 4.42) plays a critical role in both B and T cell development. This receptor is made up of two chains, IL-7Rα and γc. The common γ chain (γc) was originally identified as the IL-2Rγ chain, and

TABLE 4.9. Selected Cytokine and Cytokine Receptor Knockouts That Influence Lymphocyte Development

Cytokine/receptor	Phenotype	References
IL-2$^{-/-}$	Lymphoproliferation, autoimmunity.	Schorle et al., 1991 Kundig et al., 1993 Sadlack et al., 1993
IL-2Rα$^{-/-}$	Lymphoproliferation, autoimmunity.	Willerford et al., 1995
IL-2Rβ$^{-/-}$	Lymphoproliferation, autoimmunity. Absence of NK cells, NT cells, and intraepithelial lymphocytes (IELs).	Suzuki et al., 1995 Suzuki et al., 1997a Ohteki et al., 1997
γc$^{-/-}$	Marked impairment of B, T, and NK cell development.	Cao et al., 1995 DiSanto et al., 1995
IL-4$^{-/-}$	Impaired Th2 responses.	Kuhn et al., 1991 Kopf et al., 1993
IL-4Rα$^{-/-}$	Impaired Th2 responses.	Noben-Trauth et al., 1997 Barner et al., 1998
IL-5$^{-/-}$	Transient defect in B-1 lymphocyte generation. Reduction in eosinophil counts.	Kopf et al., 1996
IL-5Rα$^{-/-}$	Defective B-1 cell development	Yoshida et al., 1996a
IL-6$^{-/-}$	Defective antibody production and acute phase responses. Defects in hematopoiesis Other defects involving innate immunity, chemokine induction, susceptibility to infection, etc. Defects in Th1 induction in response to some pathogens. Probably no *in vivo* role in Th1 or Th2 induction.	Kopf et al., 1994 Bernad et al., 1994 Romani et al., 1996 Suzuki et al., 1997b Kopf et al., 1998
gp130$^{-/-}$	Embryonic or postnatal lethality.	Yoshida et al., 1996b Kawasaki et al., 1997
IL-7$^{-/-}$	Marked impairment of B and T cell development.	von Freeden-Jeffry et al., 1995
IL-7Rα$^{-/-}$	Marked impairment of B and T cell development.	Peschon et al., 1994
IL-10$^{-/-}$	Inflammatory bowel disease.	Kuhn et al., 1993
IL-12 p35$^{-/-}$ *IL-12 p40*$^{-/-}$	Impaired Th1 responses.	Magram et al., 1996
IL-13$^{-/-}$	Impaired Th2 responses. Reduced IgE levels.	McKenzie et al., 1998
IL-15Rα$^{-/-}$	Deficient NK cells, NT cells, γδ IELs; defective lymphocyte proliferation and homing to lymph nodes. Reduced CD8 T cells.	Lodolce et al., 1998
IL-18$^{-/-}$	Defective NK cell activation and Th1 cell development.	Takeda et al., 1998

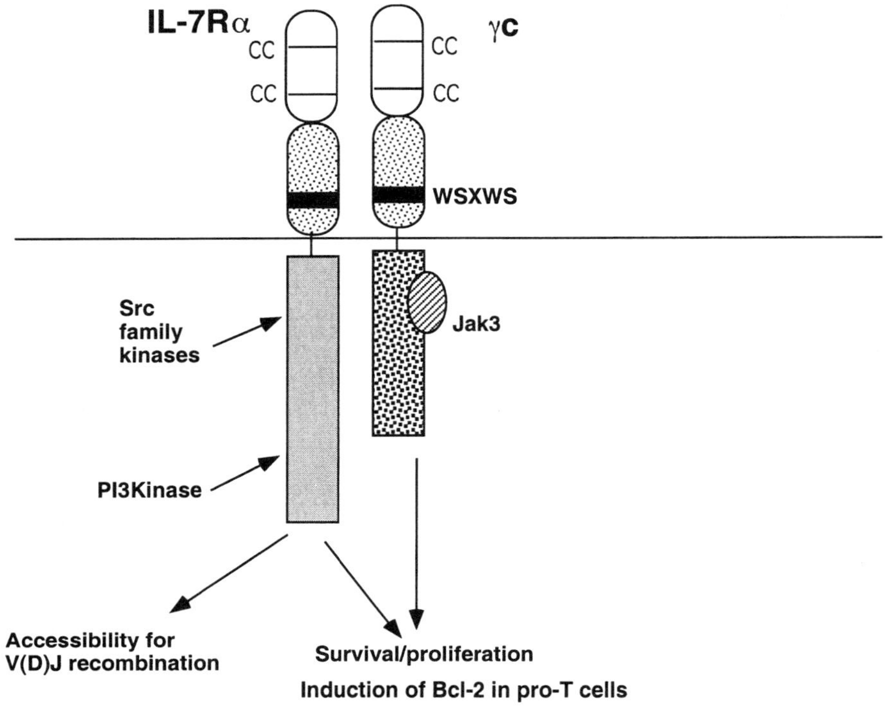

FIGURE 4.42. The IL-7 receptor is required for lymphopoiesis. It consists of the IL-7Rα chain and γc, both of which are members of the Type I cytokine/helmatopoietin receptor family. Bcl-2 may be induced by IL-7 in pro-T cells but not in murine B cell progenitors. IL-7 does not influence the development of B cells in humans.

like the IL-7Rα chain, structurally fits the prototype of a Type I cytokine/hematopoietin receptor chain. Now γc is known to be a signaling component of the IL-2, IL-4, IL-7, IL-9, and IL-15 receptors.

The γc chain is X-linked, and in humans, mutation of this gene leads to X-linked SCID (X-SCID) (see Chapter 12). This disease is characterized by a severe deficiency of T and NK cells. In mice, γc and Jak3 knockouts result in marked lymphopenia, which results from the impairment of T, B, and NK cell development.

$IL\text{-}7^{-/-}$ and $IL\text{-}7R\alpha^{-/-}$ mice have major defects in B and T cell development, but NK cell development is not compromised. The defect in NK cell development seen in $\gamma c^{-/-}$ mice reflects the absence of IL-15R signaling. In mice, IL-7R signaling is crucial for both B and T cell development, but in humans, B cell development may not require IL-7 or the IL-7R. In the absence of IL-7, proliferation of pro-T and pro-B cells is compromised, but some αβ T cell development and B cell development does occur. In contrast, γδ T cell development is very severely compromised, and few, if any, γδ T cells are seen in these knockout animals. The lymphopenia seen in *IL-7Rα*$^{-/-}$ mice is more profound than that seen in *IL-7*$^{-/-}$ mice. The IL-7Rα chain can bind a second cytokine, thymic stroma-derived lymphopoietin (TSLP), but the role, if any, of TSLP in lymphocyte development is unclear.

Signal transduction by the IL-7R leads to the activation of Jak1 (which associates with IL-7Rα), Jak3 (which associates with γc), STAT5, and also of other downstream molecules. The p85 subunit of PI-3K is recruited to a tryosine (Y449) in the cytoplasmic tail of IL-7Rα. The activation of PI-3 kinase and Akt contributes to lymphocyte survival and proliferation. STAT5 influences V(D)J recombination and lymphocyte differentiation. In the T lineage, the IL-7R contributes to the induction of Bcl-2 in pro-T cells and is required for the survival of T cell progenitors. IL-7R signaling also contributes to the activation of the α4β1 integrin, increasing the affinity of this molecule for fibronectin. The IL-7R may also provide accessibility signals for the rearrangement of the TCR γ gene in γδ T cells. In B cells, IL-7 contributes to the expansion of pro-B cells but apparently does so in a Bcl-2-independent manner. IL-7R signaling may also be required to make distal V_H genes accessible for V(D)J recombination at the H-chain locus, as discussed in Chapter 3. Mel-18 and Bmi-1 are mammalian homologs of *Drosophila* polycomb proteins. These polycomb group proteins influence cell cycle progression downstream of IL-7R signaling. These proteins presumably function as transcriptional repressors of the p16 CDK inhibitor and of the p19Arf tumor suppressor protein, which, along with p16, is encoded by the mouse *ink4a* locus.

The IL-2 Receptor

The IL-2 receptor (Figure 4.43) is a trimer made up of an inducible α chain (IL-2Rα), which is expressed on activated T cells, and two other chains, IL-2Rβ and γc, which are expressed constitutively as a heterodimer by naive T cells. The heterotrimeric complex has a high affinity for IL-2 (Kd ~ 1×10^{-11} M). IL-2Rα (CD25, Tac, or T activation antigen) is a 55-kd protein and is therefore also referred to as p55. This integral membrane protein is not a member of the Type I cytokine receptor superfamily, has two unique extracellular domains known as "sushi" domains (which were originally described in some complement proteins and coagulation factors), and a short cytoplasmic tail. This chain is primarily involved in ligand binding. It has a lower affinity (Kd ~ 1×10^{-8} M) for IL-2 than IL-2Rβ (Kd ~ 1×10^{-9} M), but in combination with IL-2Rβ and γc, it generates a very high-affinity receptor complex.

Based on its molecular size, IL-2Rβ is often referred to as p70 or p75. Like γc, which is slightly smaller, IL-2Rβ is a member of the Type I cytokine receptor family. This subunit is also a component of the IL-15 receptor. IL-2 has both positive and negative effects on the immune system. This cytokine can trigger T cell proliferation, although this function can also be

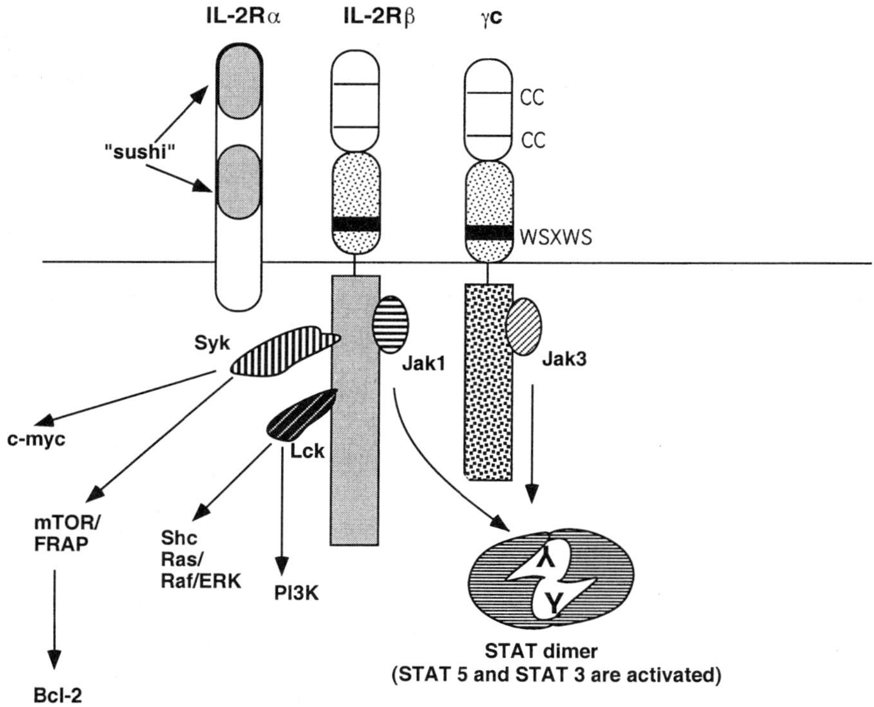

FIGURE 4.43. The IL-2 receptor is made up of three subunits. IL-2Rβ (p75) and γc are expressed together in naive T cells and constitute a low-affinity receptor for IL-2. Both these chains are structurally related to other members of the Type I cytokine receptor family. The IL-2Rα chain (p55, CD25) is induced in activated T cells and combines with the other chains to form a high-affinity receptor for IL-2. The β chain and γc are responsible for signal transduction, which involves the activation of Jak1, Jak3, Src-family, and Syk family tyrosine kinases.

mediated by other cytokines, since T cell development and proliferation are not significantly impaired in *IL-2*$^{-/-}$, *IL-2Rα*$^{-/-}$, and *IL-2Rβ*$^{-/-}$ mice. However, IL-2 appears to be absolutely required for activation-induced cell death, which is discussed in greater detail in Chapters 5 and 10. Lymphoproliferation and the expansion of activated, autoreactive T cells are noted in *IL-2Rα*$^{-/-}$ and *IL-2Rβ*$^{-/-}$ mice.

IL-2 signaling is mediated by IL2-Rβ and γc. Signaling involves the activation of a number of nonreceptor tyrosine kinases. IL-2Rβ crosslinking can activate Jak1, and γc crosslinking activates Jak3. These Jak kinases probably cooperate in the activation of STAT5 and STAT3. IL-2Rβ also activates Src-family and Syk family kinases. Lck phosphorylates critical tyrosine residues in the cytoplasmic tail of the IL-2Rβ chain. Other Src-family kinases such as Fyn and Lyn also probably mediate the same event. These phosphotyrosines participate in the recruitment of p85 and the activation of PI3 kinase and also in the recruitment of Shc and the activation of the Ras/Raf/ERK pathway. The PI3 kinase pathway, as discussed earlier in this chapter, may contribute to the activation of Akt, and the Ras pathway may, among other things, contribute to the induction of c-Fos. Although IL-2Rβ lacks ITAM motifs, it can recruit and activate Syk and Zap-70. Syk family kinases contribute to the induction of Bcl-2 in a rapamycin-sensitive manner, suggesting that activation of mTOR/FRAP may be involved in the induction of Bcl-2. Syk also contributes to the induction of c-myc and therefore presumably to the activation of Cdc25A, a phosphatase that contributes, as discussed earlier, to CDK activation. One of the less well-understood consequences of IL-2 signaling is the reduced expression of FLIP (Flice Inhibitory Protein), a competitive inhibitor of caspase-8 recruitment to the Fas/FADD complex (discussed in Chapter 5). This inhibition of FLIP expression might be critical for the induction of activation-induced cell death.

The IL-15 Receptor

IL-15 is a member of the four α-helical bundle family of cytokines. It is secreted by a range of cell types including B cells, macrophages, a variety of epithelial and endothelial cells, and stromal cells. Unlike its close relative IL-2, IL-15 is not produced by activated T cells. IL-15 is absolutely required for the development of NK cells, NT cells, and intraepithelial γδ T cells. It also contributes to the activation of cytotoxic T lymphocytes (CTLs). It can also induce T cell proliferation and function as a chemotactic factor for T cells.

The IL-15 receptor is a trimeric complex that is structurally very similar to the IL-2 receptor (Figure 4.44). These two receptors have two subunits in common. The IL-15 receptor is made up of a unique α chain, which has a very high affinity for IL-15 and associates with IL-2Rβ and γc. IL-15Rα is structurally related to IL-2Rα and like IL-2Rα is not a member of the Type I cytokine receptor superfamily. It contains a single "sushi" domain at the N-terminus, and its cytoplasmic tail is not involved in signal transduction. IL-15Rα binds IL-15 (with a 1000-fold higher affinity than of IL-2Rα for IL-2), while signaling is mediated by the β-γc complex (which is capable of binding independently not only to IL-2 but also to IL-15 with low affinity). While IL-2Rα is expressed mainly by activated lymphocytes, IL-15Rα is expressed almost ubiquitously. It may be synthesized at high levels by a number of cell types in response to viral infections. In these situations, it triggers the proliferation of NK cells and might also trigger IL-2-like responses in some T cells, primarily via the β–γc complex. Disruption of any one of the genes that encode IL-15Rα, or IL-2Rβ, or γc leads to a marked impairment of NK cell development, of NT cell development, and of intraepithelial γδ T cells.

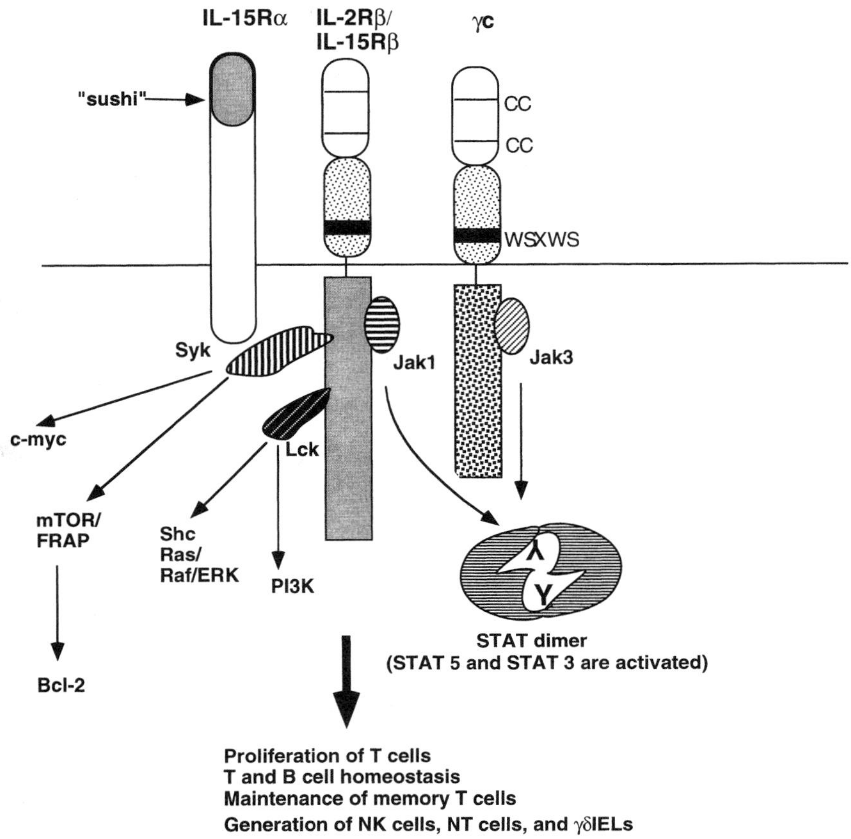

FIGURE 4.44. The IL-15 receptor structurally resembles the IL-2 receptor, uses identical chains for signaling, but has very distinct functions *in vivo*.

Although T and B cell development in *IL-15Rα*$^{-/-}$ mice is grossly normal, these mice present with a marked lymphopenia because lymphocyte proliferation is decreased and lymphocytes have a defect in homing to peripheral lymph nodes. It is possible that IL-15 is particularly important for the recirculation and maintenance of memory T cells. It has been suggested that IL-15 is expressed ubiquitously to maintain memory lymphocytes in peripheral tissue sites.

The *in vivo* functions of IL-2 and IL-15 are very different; in the absence of IL-2, mice present with lymphoproliferation, while in the absence of IL-15Rα, profound lymphopenia is observed. This is particularly surprising, since the cytoplasmic tail of IL-15Rα has been reported not to be required for signal transduction. Why do receptors that utilize identical chains have such different *in vivo* functions? The way IL-15 and IL-2 assemble their receptor complexes may be quantitatively different, given that the affinity of IL-15 for IL-15Rα is about a thousand times higher than the affinity of IL-2 for IL-2Rα. This intriguing puzzle remains to be resolved.

The IL-4 and IL-13 Receptors

IL-4 and IL-13 are both four-α-helical bundle Type I cytokines that have overlapping functions. These cytokines are secreted primarily by activated T cells, mainly Th2 cells, although IL-4 is also secreted by mast cells and natural T (NT) cells. IL-4 is required for the IgE class switch in B cells. IL-4 also inhibits class switching to IgG2a and IgG3, both of which can be induced in mice by IFN-γ. IL-4–deficient mice do not make IgE. IL-4 also triggers the growth and differentiation of T cells. Signaling via the IL-4 and IL-13 receptors is important for the development of Th2 cells. It has been suggested that there might be a hierarchical order to this process, with IL-13 signaling being required for the generation of IL-4, which is in turn required for the generation of other Th2 cytokines. Such a hierarchy remains to be established.

The significance of distinct types of receptor complexes and the specificity of IL-4 and IL-13 signaling is at present poorly understood. One type of IL-4 and IL-13 receptor is a heterodimer consisting of the IL-4Rα chain and γc. IL-4Rα can bind to both IL-4 and IL-13. Two other categories of IL-4/IL-13 receptor have been described. These receptors contain either the IL-13Rα1 or the IL-13Rα2 chain, each of which is usually complexed with the IL-4Rα chain to form distinct heterodimeric IL-4/IL-13 receptors. While the IL-13Rα1 and α2 chains can individually bind to IL-13, they cannot bind IL-4. IL-13R*α*2 may actually function as an inhibitor of IL-13 receptor signaling, though the mechanism by which this inhibition of signaling occurs is unclear. IL-4Rα, IL-13Rα1, and IL-13Rα2 are all members of the Type I cytokine receptor family of proteins, depicted in Figure 4.39. IL-4 induces the activation of Jak1, which is linked to IL-4Rα, and Jak3, which is linked to γc. Signaling by IL-4/IL-13 receptors involves the activation of STAT6. IL-13 can activate Jak1 and Jak3 as well as Tyk2, suggesting some form of receptor specificity for IL-13 that remains to be explained in terms of receptor structure. In addition, the IL-4 receptor can also also induce rapid phosphorylation of the insulin receptor substrate-1 and -2 proteins (IRS-1 and IRS-2). Tyrosine phosphorylarion of these large docking proteins permits the recruitment of SH2-containing signaling molecules. The activation of PI3 kinase by IL-4 is mediated by the recruitment of the p85 subunit of this enzyme to tyrosine-phosphorylated IRS-1 and IRS-2.

The IL-5 Receptor and Receptors of the βc Subfamily

IL-5 is a four-α-helical bundle cytokine secreted as a homodimer by activated T cells (mainly Th2 cells) and activated mast cells. IL-5 was originally characterized as a B cell growth factor, a factor that facilitates the generation of IgA responses, and as a factor that promotes the development of eosinophils. IL-5–deficient mice, however, have no obvious defects in the generation of antibody responses, including mucosal IgA responses, or in the development of conventional B cells (B-2) cells. These mice exhibit a transient defect in B-1 cell generation. At the age of 2 weeks, *IL-5*$^{-/-}$ mice have marked reductions in the number of $CD5^+$ peritoneal B-1a cells (while B-1b cell numbers are apprently unaffected). By the age of 6 to 8 weeks, however, $CD5^+$ peritoneal B-1 cell numbers return to normal levels in *IL-5*$^{-/-}$ mice.

The IL-5 receptor is made up of an α chain, which constitutes a low-affinity receptor, and a common β chain (βc), which is responsible for signal generation and is also a component of the IL-3 and GM-CSF receptors. Both these chains are members of the hematopoietin/type I cytokine receptor superfamily (see Figure 4.39). The IL-5 Rα chain is not expressed in naive B cells but can be induced upon activation of B cells.

IL-5Rα and βc combine to form a high-affinity receptor. Crosslinking of the receptor leads to tyrosine phosphorylation of βc and the tyrosine phosphorylation and activation of Jak1, Jak2, and Btk. Jak1 and Jak2 may be involved in the activation of STAT5. Other proteins that are tyrosine phosphorylated in IL-5–triggered B cells include the p85 subunit of PI3K, Vav, HS1, and Shc. Activation of the IL-5R therefore presumably leads to the activation of a number of pathways including the PI3K pathway, the Ras pathway, and the activation of Rho family GTPases. The activation of Btk might contribute to capacitative calcium signaling and the activation of PKC.

The IL-6 Receptor and the gp130 Subfamily

One subfamily of Type I cytokines is defined by receptors that share the gp130 subunit. These cytokines include IL-6, IL-11, leukemia inhibitory factor (LIF), oncostatin M (OSM), ciliary neurotrophic factor (CNTF), and cardiotrophin-1 (CT-1). Many members of this family of cytokines have functions beyond lymphocyte development and the immune system. IL-11 might be involved in the maintenance and proliferation of early B lineage progenitors but is not absolutely required for this function. *IL-11Rα* knockout mice have no discernible phenotype other than female sterility. LIF-deficient mice have defects in blastocyst implantation and therefore exhibit female sterility and also present with postnatal growth retardation. They also exhibit defects in hematopoiesis and thymocyte proliferation.

IL-6 is also a four-α-helical bundle cytokine that is believed to be secreted as a homodimer. It is produced primarily by nonlymphoid cells such as macrophages and fibroblasts, generally in response to IL-1, but can also be produced by activated T cells. IL-6–deficient mice exhibit numerous defects in immune function and cannot effectively generate an acute phase response. From the standpoint of lymphocyte development, it is of interest to note that IL-6–deficient mice have defects in hematopoiesis, in the formation of germinal centers, and have decreased IgG2a and IgG2b responses and a slight increase in IgE production. The partial defect in germinal center formation reflects the reduced local production of the complement factor C3 (see Chapter 8).

The participation of IL-6 in Th1 and Th2 cell generation is quite controversial. Although IL-6 may influence Th2 development *in vitro*, the examination of $IL\text{-}6^{-/-}$ mice initially suggested that *in vivo* Th1 cell development may be impaired when these mice are challenged with certain pathogens. More detailed analyses suggest that $IL\text{-}6^{-/-}$ mice show no bias toward either Th1 or Th2 cell development. Alterations in isotype switching in the absence of IL-6 may reflect the reduced local production of C3 rather than any alteration at the helper T cell level.

gp130 is the prototypic member of the gp130 subfamily of Type I cytokine receptor chains. IL-6Rα is also a member of this small subfamily of proteins. These receptors contain the paired cysteine and WSXWS motifs found in type I cytokine receptors but in addition also contain an N-terminal Ig-like domain. The cytoplasmic tail of IL-6Rα contains two tyrosine-containing Box 3 motifs, which are required for signaling. Signaling leads to the activation of STAT1 and STAT3.

The IL-12 Receptor

IL-12 is a disulfide-linked heterodimer that is made up of a 35-kd (p35) and a 40-kd (p40) subunit. It is secreted by activated macrophages and dendritic cells and induces IFN-γ production by T cells and by NK cells. It is an important inducer of Th1 differentiation. The p35 subunit is structurally quite distinct from the p40 protein and is synthesized by a large variety of cell types, whereas p40 is synthesized largely by activated macrophages and dendritic cells. p35 is homologous to other α-helical cytokines such as IL-6 and G-CSF, whereas p40 is structurally unrelated to all other cytokines but is actually homologous to the extracellular domains of Type I cytokine/hematopoietin receptors. p40 most closely resembles the extracellular domains of IL-6Rα and CNTFR.

p35–p40 heterodimers can activate the IL-12 receptor. Homodimers of p40 may bind to the IL-12 receptor and function as antagonists of IL-12 signaling. Some evidence, supported by studies on $p40^{-/-}$ mice, indicates that p40 homodimers may also function in an agonist role. Both $p40^{-/-}$ and $p35^{-/-}$ mice are defective in generating Th1 responses and producing IFN-γ. While both p35 and p40 are required for these responses and for resisting intracellular pathogens, the inability to respond to pathogens such as *Listeria monocytogenes* and *Cryptococus neoformans* is more severe in $p40^{-/-}$ mice. This may be because in $p35^{-/-}$ mice p40 homodimers are still made and can initiate some IL-12–type responses.

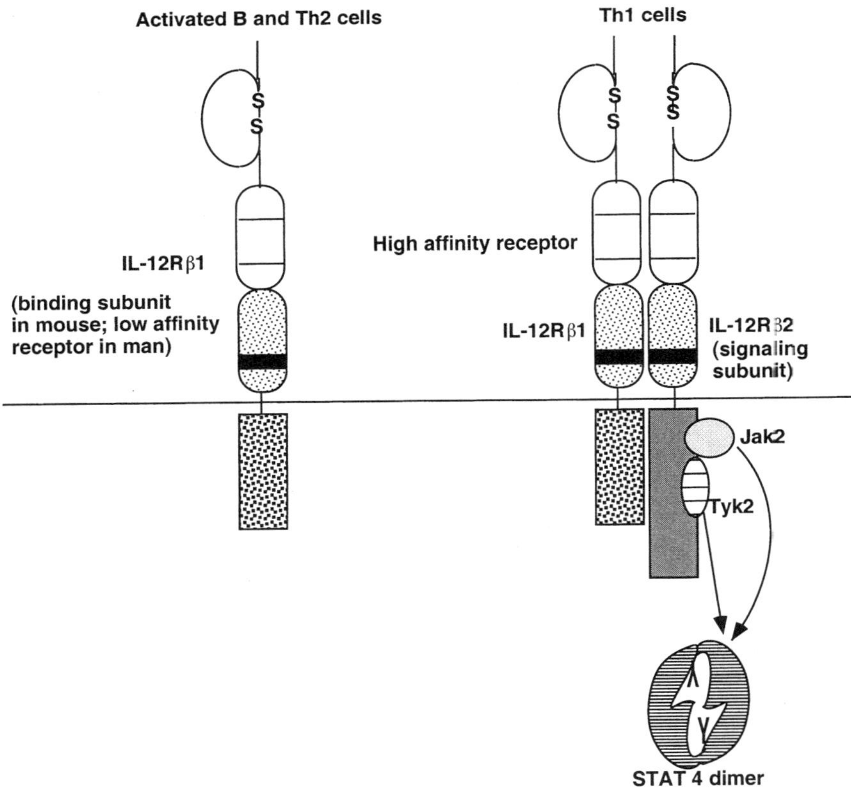

FIGURE 4.45. The IL-12 R is a heterodimer made up of a binding subunit, IL-12Rβ1, and a signaling subunit, IL-12Rβ2. IL-12Rβ2 expression is regulated by other cytokines and contributes to the Th1 versus Th2 switch. IL-4 may downregulate IL-12Rβ2 expression, whereas IFN-γ can induce it. The functional IL-12 R heterodimer is expressed on Th1 cells, but only IL-12Rβ1 is expressed on Th2 cells. IL-12 is the only cytokine that activates STAT4.

The IL-12 receptor is expressed primarily on activated T and NK cells and is made up of two chains: IL-12Rβ1 and IL-12Rβ2 (Figure 4.45). Both these chains are members of the Type I cytokine receptor family and are members of the gp130 subgroup of receptors (see Figure 4.39). IL-12Rβ1 in the mouse contributes mainly to the binding of IL-12, while IL-12Rβ2 contributes mainly to signal transduction. Receptor ligation leads to the activation of Jak2 and Tyk2 followed by the translocation of STAT4 to the nucleus. IL-12 is the only cytokine receptor that activates STAT4.

While Th1 cells express both IL-12Rβ1 and IL-12Rβ2, only IL-12Rβ1 is expressed on activated B cells and Th2 cells. Expression of IL-12Rβ2 is required for IL-12 responsiveness. Regulation of the expression of IL-12Rβ2 is an important way in which the differentiation of Th1 and Th2 cells may be modulated. The gene encoding this receptor is not expressed in naive T cells, but is transcriptionally induced by TCR signaling. The expression of this gene is then modulated by cytokines such as IL-4 and IFN-γ. IL-4 can inhibit the expression of IL-12Rβ2 in cells that are committing to Th2 development. IFN-γ can override this IL-4–mediated inhibition of IL-12Rβ2 expression and enhance the expression of this receptor chain on cells that are committed to a Th1 fate.

Type II Cytokine Receptors: The Interferon Receptors and IL-10 Signaling

Type II cytokine receptors of the interferon family share many structural similarities with Type I cytokine receptors and utilize broadly similar signaling strategies. Type II cytokines are not major players in lymphocyte development. In this section, we will briefly discuss signaling by IFN-γ and by IL-10.

The name interferon (IFN) was originally given to secreted antiviral proteins produced by cells that were responding to viral insults. Type I interferons are produced primarily by viral infection of cells and include the IFNα family of proteins and IFNβ. There are 17 related proteins, each encoded by a separate gene, that make up the IFNα family. These interferons are secreted mainly by granulocytes. There is a single IFNβ protein that is made mainly by fibroblasts. Type I interferons are not of direct relevance to lymphocyte development and will not be considered further.

Type II interferon, also known as IFNγ, is made only by T lymphocytes and NK cells as part of innate or acquired immune responses. Its expression is induced by IL-12 and IL-18. IFNγ is a homodimeric glycoprotein and functions primarily as an immune modulator. It binds to a heterodimeric receptor that is expressed on a very wide range of cell types. This receptor consists of an α chain whose cytoplasmic tail contains a membrane proximal Jak1-binding site and a C-terminal tyrosine-containing motif which when phosphorylated can bind STAT1. The second chain of the receptor, IFNγRβ, contains a membrane proximal Jak2-binding site. The membrane-proximal Jak-binding sites are similar to the Box 1 and Box 2 motifs in Type I cytokine receptors.

IFNγRα knockout mice have no obvious defects in lymphocyte development, although they are susceptible to a number of pathogens. IFNγ may directly or indirectly modulate class switching toward IgG2a and IgG3, and by contributing to the induction of IL-12Rβ2 in Th1 cells may also positively influence Th1 development.

IL-10 signaling is discussed with Type II cytokines even though IL-10 is a member of the four-α-helical bundle family, and one of its receptor chains, IL-10R (now also referred to as IL-10Rα), is also a member of the Type I cytokine/hematopoietin receptor family. This chain contributes to ligand binding but does not participate in signal transduction. The second receptor chain, IL-10Rβ, is a member of the Type II cytokine receptor family.

IL-10 is secreted by Th2 cells, macrophages, and keratinocytes, and the IL-10 receptor is expressed primarily on macrophages. IL-10 primarily inhibits cytokine production by macrophages and contributes to the reduced expression on macrophages of the B7-1 and B7-2 costimulators and of MHC class II. As a result, IL-10 might inhibit the activation of Th1-type T cell responses. IL-10 knockout mice do not have significant defects in lymphocyte development but present with an inflammatory bowel disease-like syndrome that might result from the uncontrolled activation of macrophages and inflammatory T cells. IL-10 signaling in macrophages involves Jak–STAT signaling and the activation of STAT1 and STAT3.

Signaling by Type III Cytokine Receptors (Receptors for the Trimeric Ligand Family)

A broad family of related receptors respond that to a number of structurally related trimeric ligands is sometimes referred to as the TNF receptor superfamily. In general, these ligands are synthesized as Type II transmembrane proteins (whose C-termini are extracellularly oriented) which are expressed on the cell surface as trimers. Lymphotoxin α is an exception and is secreted. Not all of these ligands are cytokines. While TNF and lymphotoxin belong to the Type III cytokine family, many related receptors respond to noncytokine cell-bound ligands, which are involved in proliferation, differentiation, and death. We have already encountered and discussed CD40L–CD40 signaling earlier in this chapter in the context of signal transduction in B cells. In Chapter 5 we will discuss Fas ligand–Fas signaling in the context of apoptosis. Other receptors of this superfamily include the low-affinity nerve growth factor receptor, CD27, CD30, OX-40, and 4-1BB.

In Chapter 6, when we discuss signals involved in the generation of secondary lymphoid organs, we will consider the receptors for TNF and lymphotoxin and closely related ligands. A subfamily of five receptors and six ligands of the TNF/TNFR family will be considered in more depth in Chapter 6. These molecules are major players in the generation of secondary lymphoid organs and the formation of follicular dendritic cell (FDC) networks and germinal centers.

TNF (often referred to as TNF-α in older nomenclatures) is a major mediator of inflammation and can induce cell death and proliferation. TNF is synthesized both by LPS-activated macrophages as part of innate immunity as well as by activated T cells. NK cells, and mast cells also produce this cytokine. TNF can stimulate macrophages and neutrophils, induce chemokine secretion, activate the endothelium, recruit leukocytes to sites of inflammation, induce angiogenesis, and is an endogenous pyrogen. Its major role in inflammation, fever, and clearing intracellular microbes will not be discussed here. TNF is initially synthesized as a trimeric nonglycosylated Type II membrane protein. It is cleaved by a cellular metalloprotease called TNFα-converting enzyme (TACE) and is released as a soluble homotrimer. The relative importance of membrane-associated and cleaved TNF in biological responses remains somewhat controversial, and both forms are likely to be biologically active. TNF can activate two distinct receptors, TNFRI (p55) and TNFRII (p75). Both these receptors may also be triggered by lymphotoxin-α (see Chapter 6). TNF does not bind to or activate the lymphotoxin β receptor (LTβR).

Knockouts of a number of ligands and receptors of the TNF family have revealed fairly complex phenotypes. Some of these will be discussed in Chapter 6 in the context of the biogenesis of lymph nodes and other secondary lymphoid organs. In brief, signaling via the LTβR is essential for the generation of lymph nodes and proper splenic architecture, whereas signaling via TNFRI in conjunction with LTβR signaling contributes to the formation of the FDC network and of germinal centers. Knockout mice have failed to reveal any function of TNFRII in lymphocyte development. In this section, we will briefly consider signal transduction by TNFRI. The osteoprotegerin ligand (OPGL) will be considered in more detail in Chapter 6. OPGL is required for the development of all lymph nodes but is not essential for proper splenic architecture and has a limited role in Peyer's patch development. OPGL also generates signals for B and T cell development and will be discussed in that context in Chapters 7 and 9.

Signaling via TNFRI can lead to the activation of the NFκB and Jun N-terminal Kinase (JNK) pathways and also can contribute to apoptosis via caspase 8 activation (Figure 4.46). TNFRI signaling can mediate cellular activation, proliferation, death, and protection from death in different cells and in different situations. In general, TNFRI signaling does not mediate death unless protein synthesis is inhibited. The activation of NFκB and AP-1 by the receptor often contributes to the suppression of apoptosis.

All receptors of the TNFR superfamily contain variable numbers of extracellular cysteine-rich domains (CRDs) that are each about 40 amino acids long and contain six cysteine residues in conserved positions. There are four CRDs in TNFRI. The cytosolic portions of TNFRI (and of Fas and a few other death receptors) share a 70-amino acid motif known as a death domain. The death domain is merely a specialized protein–protein interaction domain that connects these receptors to cytosolic proteins that contain similar domains, some of which may be involved in the induction of death. Deletion of the death domain prevents the induction of cell death. Crosslinking of TNFRI leads to the recruitment of TRADD (TNFR1-associated death domain protein) and RIP (receptor-interacting protein) to the receptor complex. TRADD contains a death domain and functions as an adaptor linking the receptor to two other death domain-containing molecules, RIP and FADD (Fas-associated death domain protein). RIP is a serine-threonine kinase that indirectly connects TNFRI to TRAF2 and

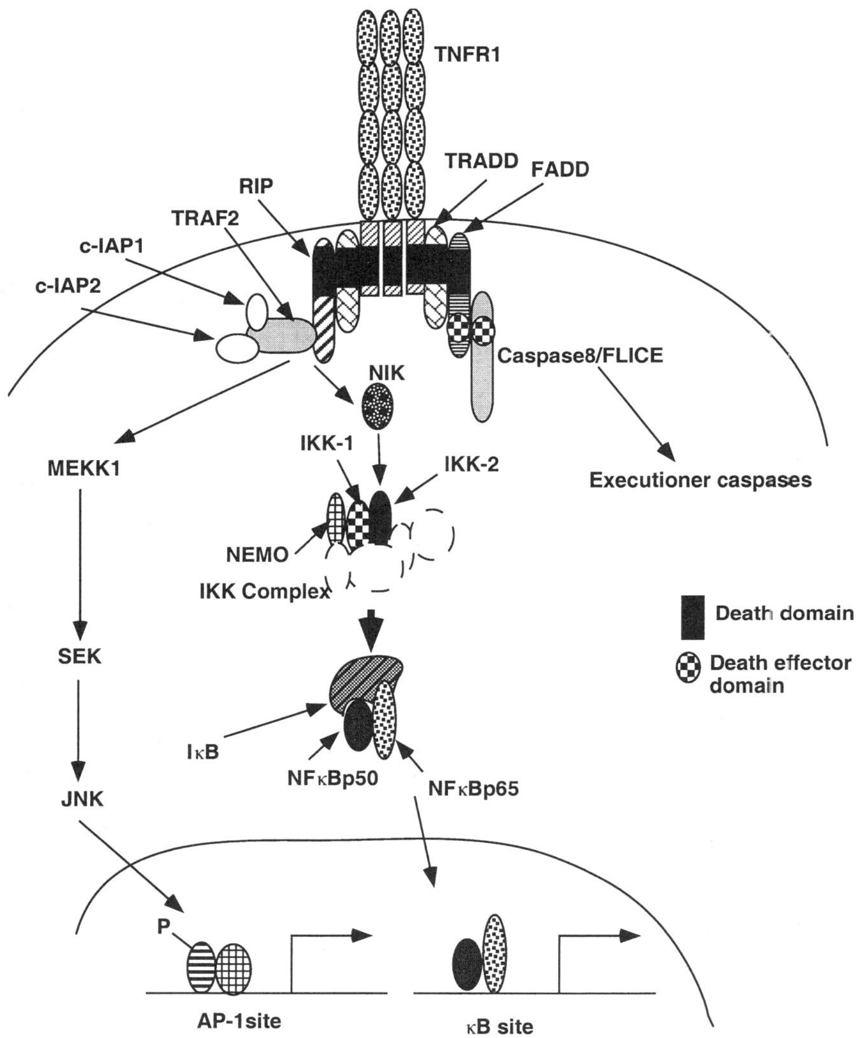

FIGURE 4.46. An overview of signaling via the Type I TNF receptor (TNFR1, or p55). The receptor trimerizes after it sees its ligand and recruits the adaptor TRADD via a death domain–death domain interaction. TRADD recruits TRAFs such as TRAF2, possibly using the death domain-containing serine-threonine kinase, RIP, as an intermediary. RIP is required for NFκB activation, while TRAF2 is absolutely required for JNK activation. These signals may protect cells from apoptosis mediated by caspase-8, which may also be recruited to the receptor via TRADD and FADD.

potentially other TRAF family members (discussed in connection with CD40 signaling). FADD is discussed in Chapter 5 in connection with Fas signaling. It contains an additional protein–protein interaction motif called a death effector domain, which can recruit caspase-8, which contains a similar domain.

TRAF2 may be involved both in the activation of the JNK pathway as well as the activation of NFκB. The exact mechanism by which TRAF2 activates JNK is not known, though this may possibly be achieved via the activation of MEKK1. MEKK1 does not physically associate with TRAF2, and it is unclear whether Rho family GTPases are involved in an intermediate step. TRAF2 or some other TRAF molecule may interact with and activate NFκB-inducing Kináse (NIK) (discussed in the earlier section on NFκB) and thus contribute to the activation of NFκB. Examination of cells from *Traf-2*$^{-/-}$ mice suggests that TRAF2 is not required for NFκB activation but is required for the activation of JNK. In contrast *Rip*$^{-/-}$ mice have revealed that RIP is absolutely required for NFκB activation but is dispensable for the activation of the JNK pathway.

In addition to its role in the activation of JNK and possibly in NFκB activation, TRAF-2 also binds to cIAP1 and cIAP2, which are cellular inhibitors of apoptosis.

Type IV Cytokine Receptors: IL-1 and IL-18 Signaling

IL-1α and IL-1β are highly related proinflammatory cytokines synthesized in the cytosol of activated macrophages as 33-kd precursors. They are cleaved by interleukin-1-converting enzyme (ICE, caspase-1) or related proteases in the cytosol into 17-kd mature forms and are "secreted" by a poorly understood signal peptide-independent mechansim. Both forms of IL-1 bind to the same IL-1 receptor complex discussed below and mediate identical functions. IL-1 can induce the production of acute phase proteins and also of other inflammatory cytokines such as TNF and IL-6. It can also function as a costimulant for T cell proliferation.

The IL-1 receptor consists of two chains: IL-1R and the IL-1R-accessory protein (IL-1R-AcP). IL-1R is a member of the Ig superfamily, but its cytoplasmic tail resembles the cytoplasmic tail of the *Drosophila* Toll receptor. Toll is involved in the generation of dorsal-ventral polarity in *Drosophila* and also in innate immunity. The cytoplasmic tail of IL-1R associates with an adaptor protein called Myd88. Myd88 is a cytoplasmic protein that contains an N-terminal death domain and a C-terminal domain that is homologous to the cytoplasmic tails of IL-1R and Toll. Ligation of the IL-1R leads to the Myd 88-dependent recruitment of a serine-threonine kinase called the IL-1 receptor-associated kinase (IRAK). IRAK dissociates from the receptor complex, associates with TRAF6, and in a poorly understood way induces the degradation of IκB and the activation of NFκB. IL-1 can also activate the JNK pathway in a Myd88-dependent manner.

IL-18 and IL-12 both contribute to the induction of IFN-γ expression in T and NK cells. IL-18 potentiates IL-12–induced Th1 cell development. IL-18 and IL-12 are not related structurally. IL-18 is homologous to IL-1 and IL-18R, the major receptor chain for IL-18 (also known as the IL-1 receptor-related protein) is structurally related to IL-1R and Toll. The IL-18 receptor complex, like the IL-1 receptor complex, transduces signals via the Myd88 adaptor and activates NFκB and the JNK signaling pathway.

Summary

In this chapter, we have examined numerous signal transduction pathways that are relevant from the viewpoint of lymphocyte development. We have examined, in some detail, how signals are initiated from antigen receptors and have asked how coreceptors and costimulators facilitate or cooperate with antigen receptors. In the last section of this chapter, we have discussed cytokine receptors that are of relevance from the standpoint of lymphocyte development and have summarized some of the signaling events downstream of these receptors.

Selected Reviews

Artavanis-Tsakonas, S., Matsuno, K., and Fortini, M. E. (1995). Notch signaling. Science *268*, 225–232.

Bauerle, P. A., and Baltimore, D. (1988). IκB: a specific inhibitor of the NFκB transcription factor. Science *242*, 540–546.

Bauerle, P. A., and Baltimore, D. (1996). NFκB: ten years after. Cell *87*, 13–20.

Bray, S. (1998). A notch affair. Cell *93*, 499–503.

Cadigan, K. M., and Nusse, R. (1997). Wnt signaling: a common theme in animal development. Genes Dev. *11*, 3286–3305.

Cambier, J. C., Bedzyk, W. K. C., Chien, N., Friedrich, J., Harwood, A., Jensen, W., Pleiman, C., and Clark, M. C. (1993). The B cell antigen receptor: structure and function of primary, secondary, tertiary and quaternary components. Immunol. Rev. *132*, 85–106.

Candeias, S., Muegge, K., and Durum, S. K. (1997). IL-7 receptor and VDJ recombination: trophic versus mechanistic actions. Immunity *6*, 501–508.

DeFranco, A. L. (1997). The complexity of signaling pathways activated by the BCR. Curr. Opin. Immunol. *9*, 296–308.

Dyson, N. (1998). The regulation of E2F by pRb-family proteins. Genes Dev. *12*, 2245–2262.

Germain, R. N. (1997). T cell signaling: The importance of receptor clustering. Curr. Biol. *7*, R640–R644.

Ghosh, S., May, M. J., and Kopp, E. B. (1998). NFκB and Rel proteins: evolutionarily conserved mediators of immune responses. Annu. Rev. Immunol. *16*, 225–260.

Hall, A. (1998). Rho GTPases and the actin cytoskeleton. Science *279*, 509–514.

Ihle, J. N., Wittuhn, B. A., Quelle, F. W., Yamamota, K., and Sivennoinen, O. (1995). Signaling through the hematopoietic cytokine receptors. Annu. Rev. Immunol. *13*, 369–398.

Kehrl, J. H. (1998). Heterotrimeric G protein signaling: Roles in immune function and fine tuning by RGS proteins. Immunity *8*, 1–10.

Lanzavecchia, A., Iezzi, G., and Viola, A. (1999). From TCR engagement to T cell activation: a kinetic view of T cell behavior. Cell *96*, 1–4.

Leonard, W. J., and O'Shea, J. J. (1998). JAKs and STATs: Biological implications. Annu. Rev. Immunol. *16*, 293–322.

Liu, K. D., Gaffen, S. L., and Goldsmith, M. A. (1998). JAK/STAT signaling by cytokine receptors. Curr. Opin. in Immunol. *10*, 271–278.

Lowell, C. A., and Soriano, P. (1996). Knockout of Src-family kinases: Stiff bones, wimpy T cells, and bad memories. Genes Dev. *10*, 1845–1857.

Newton, A. C. (1997). Regulation of protein kinase C. Curr. Opin. Cell Biol. *9*, 161–167.

Osborne, B., and Miele, L. (1999). Notch and the immune system. Immunity *11*, 653–663.

Penninger, J. M., and Crabtree, G. R. (1999). The actin cytoskeleton and lymphocyte activation. Cell *96*, 9–12.

Rao, A., Luo, C., and Hogan, P. G. (1997). Transcription factors of the NFAT family: regulation and function. Annu. Rev. Immunol. *15*, 707–747.

Robey, E. (1997). Notch in vertebrates. Curr. Opin. Genet. Dev. *7*, 551–557.

Rudd, C. E. (1999). Adaptors and molecular scaffolds in immune cell signaling. Cell *96*, 5–8.

Scharenberg, A. M., and Kinet, J.-P. (1998). PtdIns-3,4,5-P3: A regulatory nexus between tyrosine kinases and sustained calcium signals. Cell *94*, 5–9.

Stancovski, I., and Batimore, D. (1997). NFκB activation: the IκB kinase revealed? Cell *91*, 299–302.

Sha, W. C. (1998). Regulation of immune responses by NFκB/Rel transcription factors. J. Exp. Med. *187*, 143–146.

Taniguchi, T. (1995). Cytokine signaling through nonreceptor tyrosine kinases. Science *268*, 251–256.

Tapon, N., and Hall, A. (1997). Rho, Rac, and Cdc 42 GTPases regulate the organization of the actin cytoskeleton. Curr. Opin. Cell Biol. *9*, 86–92.

Wange, R. L., and Samelson, L. E. (1996). Complex complexes: signaling at the TCR. Immunity *5*, 197–205.

Weiss, A., and Littman, D. (1994). Signal transduction by lymphocyte antigen receptors. Cell *76*, 263–274.

Zamoyska, R. (1998). CD4 and CD8: modulators of T cell receptor recognition of antigen and of immune responses? Curr. Opin. Immunol. *10*, 82–87.

Some Relevant References

Adachi, O., Kawai, T., Takeda, K., Matsumoto, M., Tsutsui, H., Sakagami, M., Nakanishi, K., and Akira, S. (1998). Targeted disruption of the Myd88 gene results in loss of IL-1 and IL-18 mediated function. Immunity *9*, 143–150.

Akasaka, T., Tsuji, K.-I., Kawahira, H., Kanno, M., Harigaya, K.-I., Hu, L., Ebihara, Y., Nakahata, T., Tetsu, O., Taniguchi, M., and Koseki, H. (1997). The role of mel-18, a mammalian polycomb group gene, during IL-7 dependent proliferation of lymphocyte precursors. Immunity *7*, 135–146.

Alberola-Ila, J., Forbush, K. A., Seger, R., Krebs, E. G., and Perlmutter, R. M. (1995). Selective requirement for MAP kinase activation in thymocyte differentiation. Nature *373*, 620–623.

Alberola-Ila, J., Hogquist, K. A., Swan, K. A., Bevan, M. J., and Perlmutter, R. M. (1996). Positive and negative selection invoke distinct signaling pathways. J. Exp. Med. *184*, 9–18.

Alexander, W. S., Starr, R., Fenner, J. E., Scott, C. L., Handman, E., Sprigg, N. S., Corbin, J. E., Cornish, A. L., Darwiche, R., Owczarek, C. M., Kay, T. W. H., Nicola, N. A., Hertzog, P. J., Metcalf, D., and Hilton, D. J. (1999). SOCS1 is a critical inhibitor of interferon gamma signaling and prevents the potentially fatal neonatal actions of this cytokine. Cell *98*, 597–608.

Amakawa, R., Hakem, A., Kundig, T. M., Matsuyama, T., and Simard, J. J. et al. (1996). Impaired negative selection of T cells in Hodgkin's disease antigen CD30-deficient mice. Cell *84*, 551–562.

Appleby, M. W., Gross, J. A., Cooke, M. P., Levin, S. D., Qian, X., and Perlmutter, R. M. (1992). Defective T cell receptor signaling in mice lacking the thymic isoform of p59fyn. Cell *750*, 751–763.

Appleby, M. W., Gross, J. A., Cooke, M. P., Levin, S. D., Qian, X., and Perlmutter, R. M. (1995). Involvement of p59FynT in interleukin-5 receptor signaling. J. Exp. Med. *182*, 811–820.

Bachmaier, K., Krawczyk, C., Kozieradzki, I., Kong, Y.-Y., Sasaki, T., et al. (2000). Negative regulation of lymphocyte activation and autommunity by the molecular adaptor Cbl-b. Nature *403*, 211–216.

Barnner, M., Mohrs, M., Brombacher, F., and Kopf, M. (1998). Differences between IL-4 Rα deficient and IL-4 deficient mice reveal a role for IL-13 in the regulation of Th2 responses. Curr. Biol. *8*, 669-672.

Beals, C. R., Sheridan, C. M., Turck, C. W., Gardner, P., and Crabtree, G. R. (1997). Nuclear export of NF-ATc enhanced by glycogen synthase kinase-3. Science *275*, 1930–1934.

Beg, A. A., Sha, W. C., Bronson, R. T., and Baltimore, D. (1995a). Constitutive NFκB activation, enhanced granulopoiesis, and neonatal lethality in IκBα deficient mice. Genes Dev. *9*, 2736–2746.

Beg, A. A., Sha, W. C. Bronson, R. T., Ghosh, S., and Baltimore, D. (1995b). Embryonic lethality and liver degeneration in mice lacking the RelA component of NFκB. Nature *376*, 167–170.

Bernad, A., Kopf, M., Kulbacki, R., Weich, N., Kohler, G., and Gutirrez-Ramos, J. C. (1994). Interleukin-6 is required *in vivo* for the regulation of stem cells and committed progenitors of the immune system. Immunity *1*, 725–731.

Bondeva, T., Pirola, L., Bularelli-Leva, G., Rubio, I., Wetzker, R., and Wymann, M. P. (1998). Bifurcation of lipid and protein kinase signals of PI3Kgamma to the protein kinases PKB and MAPK. Science *282*, 293–296.

Boothby, M. R., Mora, A. L., Scherer, D. C., Brockman, J. A., and Ballard, D. W. (1997). Perturbation of the T lymphocyte lineage in transgenic mice expressing a constitutive repressor of NFκB. J. Exp. Med. *185*, 1897–1907.

Boniface, J. J., Rabinowitz, J. D., Wulfing, C., Hampl, J., Reich, Z., Altman, J. D., Kantor, R. M., Beeson, C., McConnell, H. M., and Davis, M. M. (1998). Initiation of signal transduction through the T cell receptor requires the multivalent engagement of peptide/MHC ligands. Immunity *9*, 459–466.

Boriack-Sjodin, P. A., Margarit, S. M., Bar-Sagi, D., and Kuriyan, J. (1998). The structural basis of the activation of Ras by Sos. Nature *394*, 337–344.

Burkly, L., Hession, C., Ogata, L., Reilly, C., Marconi, L. A., Olson, D., Tizard, R., Cate, R., and Lo, D. (1995). Expression of RelB is required for the development of thymic medulla and dendritic cells. Nature *373*, 531–536.

Byth, K. F., Conroy, L. A., Howlett, S., Smith, A. J., May, J., Alexander, D. R., and Holmes, N. (1996). CD45 null transgenic mice reveal a positive regulatory role for CD45 in early thymocyte development, in the selection of $CD4^+CD8^+$ thymocytes and B cell maturation. J. Exp. Med. *183*, 1707–1718.

Caamano, J. H., Rizzo, C. A., Durham, S. K., Barton, D. S., Raventos-Suarez, C., and Bravo, R. (1998). NFκB2 (p100/p52) is required for normal splenic microarchitecture and B cell mediated immune responses. J. Exp. Med. *187*, 185–196.

Cai, Y. C., Cefai, D., Schneider, H., Raab, M., Nabavi, N., and Rudd, C. E. (1995). Selective CD28 pYMNM mutations implicate phosphatidylinositol 3-kinase in CD86-CD28 mediated costimulation. Immunity *3*, 417–426.

Calvo, C. R., Amsen, D., and Kruisbeek, A. M. (1997). Cytotoxic T lymphocyte antigen 4 (CTLA-4) interferes with extracellular signal regulated kinase (ERK) and Jun NH2-terminal kinase (JNK) activation but does not affect phosphorylation of T cell receptor zeta and ZAP70. J. Exp. Med. *186*, 1645–1653.

Cao, X., Shores, E. W., Hu-Li, J., Anver, M. R., Kelsall, B. L., Russell, S. M., Drago, J., Noguchi, M., Grinberg, A., and Bloom, E. T. (1995). Defective lymphoid development in mice lacking expression of the common cytokine receptor gamma chain. Immunity *2*, 223–238.

Carrasco, D., Cheng, J., Lewin, A., Warr, G., Yang, H., Rizzo, C., Rosas, F., Snapper, C., and Bravo, R. (1998). Multiple hemopoietic defects and lymphoid hyperplasia in mice lacking the transcriptional activation domain of the c-Rel protein. J. Exp. Med. *187*, 973–984.

Chan, V. W., Meng, F., Soriano, P., DeFranco. A. L., and Lowell, C. A. (1997). Characterization of the B lymphocyte populations in Lyn deficient mice and the role of Lyn in signal initiation and down-regulation. Immunity *7*, 69–81.

Cheng, A. M., Negishi, I., Anderson, S. J., Chan, A. C., Bolen, J., Loh, D. Y., and Pawson, T. (1997). The Syk and ZaP-70 SH2 containing tyrosine kinases are implicated in pre-T cell receptor signaling. Proc. Natl. Acad. Sci. USA *94*, 9797–9801.

Cheng, A. M., Rowley, B., Pao, W., Hayday, A., Bolen, J. B., and Pawson, T. (1995). Syk tyrosine kinase required for mouse viability and B cell development. Nature *378*, 303–307.

Chiang, Y. J., Kole, H. K., Brown, K., Naramura, M., Fukuhara, S., et al. (2000). Cbl-b regulates the CD28 dependence of T-cell activation. Nature *403*, 216–220.

Chow, C. W., Rincon, M., Cavanagh, J., Dickens, M., and Davis, R. J. (1997). Nuclear accumulation of NFAT4 opposed by the JNK signal transduction pathway. Science *278*, 1638–1641.

Clements, J. L., Yang, B., Ross-Barta, S. E., Eliason, S. L., Hrstka, R. F., Williamson, R. A., and Koretzyk, G. A. (1998). Requirement for the leukocyte specific adapter protein SLP-76 for normal T cell development. Science *281*, 416–419.

Crespo, P., Schuebel, K. E., Ostrom, A. A., Gutkind, J. S., and Bustelo, X. R. (1997). Phospho-tyrosine dependent activation of Rac-1 GDP/GTP exchange by the vav proto-oncogene product. Nature *385*, 169–172.

Delon, J., Gregoire, C., Malissen, B., Darche, S., Lemaitre, F., Kourilsky, P., Abastado, J. P., and Trautmann, A. (1998). CD8 expression allows T cell signaling by monomeric peptide-MHC complexes. Immunity *9*, 467–474.

DiSanto, J. P., Muller, W., Guy-Grand, D., Fishcher, A., and Rajewsky, K. (1995). Lymphoid development in mice with a targeted deletion of the interleukin-2 receptor gamma chain. Proc. Natl. Acad. Sci. USA *92*, 377–381.

Doi, T. S., Takahashi, T., Taguchi, O., Takachika, A., and Obata, Y. (1997). NFκB RelA deficient lymphocytes: normal development of T cells and B cells, impaired production of IgA and IgG1 and reduced proliferative responses. J. Exp. Med. *185*, 953–961.

Dong, C., Yang, D. D., Wysk, M., Whitmarsh, A. J., Davis, R. J., and Flavell, R. A. (1998). Defective T cell differentiation in the absence of Jnk1. Science *282*, 2092–2095.

Druey, K. M., Blumer, K. J., Kang, V. H., and Kehrl, J. M. (1996). Inhibition of G protein mediated MAP kinase activation by a new mammalian gene family. Nature *379*, 742–746.

Durbin, J. E., Hackenmiller, R., Simon, M. C., and Levy, D. E. (1996). Targeted disruption of the mouse Stat 1 gene results in compromised innate immunity to viral disease. Cell *84*, 443–450.

Dustin, M. L., Olszowy, M. W., Holdorf, M. D., Li, J., Bromley, S., Desai, N., Widder, P., Rosenberger, F., van der Merwe, P. A., Allen, P. M., and Shaw, A. S. (1998). A novel adaptor protein orchestrates receptor patterning and cytoskeletal polarity in T cell contacts. Cell *94*, 667–677.

Ellisen, L. W., Bird, J., West, D. C., Soreng, A. L., Reynolds, T. C., Smith, S. D., and Sklar, J. (1991). TAN-1, the human homolog of the Drosophila notch gene, is broken by chromosomal translocations in T lymphoblastic neoplasms. Cell *66*, 649–661.

Esslinger, C. W., Wilson, A., Sardat, B., Beermann, F., and Jongeneel, C. V. (1997). Abnormal T lymphocyte development induced by targeted overexpression of IκBα. J. Immunol. *158*, 5075–5078.

Fischer, K.-D., Zmuidzinas, A., Gardner, S., Barbacid, M., Bernstein, A., and Guidos, C. (1995). Defective T cell receptor signalling and positive selection of Vav deficient $CD4^+CD8^+$ thymocytes. Nature *374*, 474–477.

Franzoso, G., Carlson, L., Poljak, L., Shores, E. W., Epstein, S., Leonardi, A., Grinberg, A., Tran, T., Scharton-Kesten, T., Anver, M., Love, P., Brown, K., and Siebenlist, U. (1998). Mice deficient in NFκB/p52 present with defects in humoral responses, germinal center rections, and splenic microarchitecture. J. Exp. Med. *187*, 147–159.

Franzoso, G., Carlson, L., Scharton-Kersten, T., Shores, E. W., Epstein, S., Grinberg, A., Tran, T., Shacter, E., Leonardi, A., Anver, M., Love. P., Sher, A., and Siebenlist, U. (1997a). Critical roles for the Bcl-3 oncoprotein in T cell mediated immunity, splenic microarchitecture and germinal center reactions. Immunity *6*, 479–490.

Franzoso, G., Carlson, L., Xing, L., Poljak, L., Shores, E. W., Brown, K. D., Leonardi, A., Tran, T., Boyce, B. F., and Siebenlist, U. (1997b). Requirement for NFκB in osteoclast and B cell development. Genes Dev. *11*, 3482–3496.

Fu, C., Turck, C. W., Kurosaki, T., and Chan, A. C. (1998). BLNK: a central linker protein in B cell activation. Immunity *9*, 93–103.

Galandrini, R., Henning, S. W., and Cantrell, D. A. (1997). Different functions of the GTPase Rho in pro-thymocytes and late pre-T cells. Immunity *7*, 163–174.

Grakoui, A., Bromley, S. K., Sumen, C., Davis, M. M., Shaw, A. S., Allen, P. M., and Dustin, M. L. (1999). The immunological synapse: a molecular mechine controlling T cell activation. Science *285*, 221–227.

Groves, T., Smiley, P., Cooke, M. P., Forbush, K., Perlmutter, R. M., and Guidos, C. J. (1996). Fyn can partially substitute for Lck in T lymphocyte development. Immunity *5*, 417–428.

Grumont, R. J., and Gerondakis, S. (1994). The subunit composition of NFκB complexes changes during B cell development. Cell Growth Differ. *5*, 1321–1331.

Grumont, R. J., Rourke, I. J., O'Reilly, L. A., Strasser, A., Miyake, K., Sha, W., and Gerondakis, S. (1998). B lymphocytes differentially use the Rel and NF-κB1 transcription factors to regulate cell cycle progression and apoptosis in quiescent and mitogen-activated cells. J. Exp. Med. *187*, 663–674.

Han, J., Luby-Phelps, K., Das, B., Shu, X., Xia, Y., Mosteller, R. D., Krishna, U. M., Falck, J. R., White, M. A., and Broek, D. (1998). Role of substrates and products of PI-3 kinase in regulating activation of Rac-related guanosine triphosphatases by Vav. Science *279*, 558–560.

Hendriks, R. W., deBruijn, M. F. T. R., Maas, A., Dingjan, G. M., Karis, A., and Grosveld, F. (1996). Inactivation of Btk by insertion of lacZ reveals defects in B cell development only past the pre-B stage. EMBO J. *15*, 4862–4872.

Henning, S., Galandrini, R., Hall, A., and Cantrell, D. A. (1997). The GTPase Rho has a critical regulatory role in thymocyte development. EMBO J *16*, 2397–2407.

Hibbs, M. L., Tarlinton, D. M., Armes, J., Graill, D., Hodgson, G., Maglitto, R., Stacker, S. A., and Dunn, A. R. (1995). Multiple defects in the immune system of Lyn deficient mice, culminating in autoimmune disease. Cell *83*, 301–311.

Hill, C. S., Wynne, J., and Treisman, R. (1995). The Rho family GTPases RhoA, Rac1, and CDC42hs regulate transcriptional activation by SRF. Cell *81*, 1159–1170.

Holsinger, L. J., Graef, I. A., Swat, W., Chi, T., Bautista, D. M., Davidson, L., Lewis, R. S., Alt, F. W., and Crabtree, G. R. (1998). Defects in actin-cap formation in Vav deficient mice implicate an actin requirement for lymphocyte signal transduction. Curr. Biol. *8*, 563–572.

Horwitz, B. H., Scott, M. L., Cherry, S. R., Bronson, R. T., and Baltimore, D. (1997). Failure of lymphopoiesis after adoptive transfer of NFκB deficient fetal liver cells. Immunity *6*, 765–772.

Hutloff, A., Dittrich, A. M., Beier, K. C., Eljaschewitsch, B., Kraft, R., Anagnostopoulos, I., and Kroczek, R. A. (1999). ICOS is an inducible T cell costimulator structurally and functionally related to CD28. Nature *397*, 263–266.

Imada, K., Bloom, E. T., Nakajima, H., Horvath-Arcidiacono, J. A., Udy, G. B., Davey, H. W., and Leonard, W. J. (1998). Stat5b is essential for natural killer cell-mediated proliferation and cytolytic activity. J. Exp. Med. *188*, 2067–2074.

Imamoto, A., and Soriano, P. (1993). Disruption of the csk gene, encoding a negative regulator of Src-family tyrosine kinases, leads to neural tube defects and embryonic lethality in mice. Cell *73*, 1117–1124.

Iritani, B. M., Forbush, K. A., Farrar, M. A., and Perlmutter, R. M. (1997). Control of B cell development by Ras-mediated activation of Raf. EMBO J. *16*, 7019–7031.

Ishikawa, H., Carrasco, D., Claudio, E., Ryseck, R.-P., and Bravo, R. (1997). Gastric hyperplasia and increased proliferative responses of lymphocytes in mice lacking the COOH-terminal ankyrin domain of NFκB2. J. Exp. Med. *186*, 999–1014.

Ishikawa, H., Claudio, E., Dambach, D., Raventos-Suarez, C., Ryan, C., and Bravo, R. (1998). Chronic inflammation and susceptibility to bacterial infections in mice lacking the p105 precursor but expressing p50. J. Exp. Med. *187*, 985–996.

Jacinto, E., Werlen, G., and Karin, M. (1998). Cooperation between Syk and Rac1 leads to synergistic JNK activation in T lymphocytes. Immunity *8*, 31–42.

Jain, J., McCaffrey, P. G., Miner, Z., Kerppola, T. K., Lambert, J. N., Verdine, G. L., Curran, T., and Rao, A. (1993). The T cell transcription factor NFATp is a substrate for calcineurin and interacts with Fos and Jun. Nature *365*, 352–355.

Joazeiro, C. A. P., Wing, S. S., Huang, H.-K., Leverson, J. D., Hunter, T., and Liu, Y. -C. (1999). The tyrosine kinase negative regulator c-Cbl as a RING-type E2 dependent ubiquitin protein ligase. Science *286*, 309–312.

Kanazawa, S., Ilic, D., Hashiyama, T., Nomura, T., Yamamoto, T., Suda, T., and Aizawa, S. (1996). p125 Fak cooperation in development of $CD4^+$ $CD8^+$ thymocytes. Blood *87*, 865–870.

Kaplan, M. H., Schindler, U., Smiley, S. T., and Grusby, M. J. (1996a). Stat6 is required for mediating responses to IL-4 and for the development of Th2 cells. Immunity *4*, 313–319.

Kaplan, M., Sun, Y., Hoey, T., and Grusby, M. (1996b). Impaired IL-12 responses and impaired development of Th2 cells in Stat 4 deficient mice. Nature *382*, 174–177.

Kao, H.-Y., Ordentlich, P., Koyano-Nakagawa, N., Tang, Z., Downes, M., Kintner, C. R., Evans, R. M., and Kadesch, T. (1998). A histone deacetylase corepressor complex regulates the Notch signal transduction pathway. Genes Dev. *12*, 2269–2277.

Kashiwada, M., Shirakata, Y., Inoue, J.-I., Nakano, H., Okazaki, K., Okumura, K., Yamamoto, T., Nagaoka, H., and Takemori, T. (1998). TRAF6 stimulates ERK activity in CD40 signaling along a Ras independent pathway. J. Exp. Med. *187*, 237–244.

Kawasaki, K., Gao, Y. H., Yokose, S., Kaji, Y., Nakamura, T., Suda, T., Yoshida, K., Taga, T., Kishimoto, T., and Kataoka, H. et al. (1997). Osteoclasts are present in gp130 deficient mice. Endocrinology *138*, 4959–4965.

Kerner, J. D., Appleby, M. W., Mohr, R. N., Chien, S., D. J., R., Maliszewski, C. R., Witte, O. N., and Perlmutter, R. M. (1995). Impaired expansion of mouse B cell progenitors lacking Btk. Immunity *3*, 301–312.

Khan, W. N., Alt, F. W., Gerstein, R. M., Malynn, B. A., Larsson, I., Rathbun, G., Davidson, L., Muller, S., Kantor, A. B., Herzenberg, L. A., Rosen, F. S., and Sideras, P. (1995). Defective B cell development and function in Btk deficient mice. Immunity *3*, 283–299.

Kishihara, K., Penninger, J., Wallace, V. A., Kundig, T. M., Kawai, K., Wakeham, A., Timms, E., Pfeffer, K., Ohashi, P. S., and Thomas, M. L. (1993). Normal B lymphocyte development but impaired T cell maturation in CD45 exon 6 protein tyrosine phosphatase deficient mice. Cell *74*, 143–156.

Klement, J. F., Rice, N. R., Car, B. D., Abbondanzo, S. J., Powers, G. D., Bhatt, G., Chen, C. H., Rosen, C. A., and Stewart, C. L. (1996). IκBα deficiency results in a sustained NFκB response and severe widespread dermatitis in mice. Mol. Cell Biol. *16*, 2341–2349.

Kondo, M., Akashi, K., Domen, J., Sugamura, K., and Weissman, I. L. (1997). Bcl-2 rescues T lymphopoiesis but not B or NK cell development in common γ chain-deficient mice. Immunity *7*, 155–162.

Kontgen, F., Grumont, R. J., Strasser, A., Metcalf, D., Li, R., Tarlinton, D., and Gerondakis, S. (1995). Mice lacking the c-Rel protooncogene exhibit defects in lymphocyte proliferation, humoral immunity and interleukin-2 expression. Genes Dev. *9*, 1965–1977.

Kopf, M., Baumann, H., Freer, G., Freudenberg, M., Lamers, M., Kishimoto, T., Zinkernagel, R., Buethmann, H., and Kohler, G. (1994). Impaired immune and acute-phase responses in interleukin-6 deficient mice. Nature *368*, 39–342.

Kopf, M., Brombacher, F., Hodgkin, P. D., Ramsay, A. J., Milbourne, E. A., Dai, W. J., Ovington, K. S., Behm, C. A., Kohler, G., Young, I. G., and Matthaei, K. I. (1996). IL-5 deficient mice have a developmental defect in $CD5^+$ B-1 cells and lack esoinophilia but have normal antibody and cytotoxic T cell responses. Immunity *4*, 15–24.

Kopf, M., Herren, S., Wiles, M. V., Pepys, M. B., and Kosco-Vilbois, M. H. (1998). Interleukin 6 influences germinal center development and antibody production via a contribution of C3 complement component. J. Exp. Med. *188*, 1895–1906.

Kopf, M., LeGros, G., Bachmann, M., Lamers, M. C., Bluethmann, H., and Kohler, G. (1993). Disruption of the murine IL-4 gene blocks Th2 cytokine responses. Nature *362*, 245–248.

Kuhn, R., Lohler, J., Rennick, D., Rajewsky, K., and Muller, W. (1993). Interleukin-10 deficient mice develop chronic enterocolitis. Cell *75*, 263–274.

Kuhn, R., Rajewsky, K., and Muller, W. (1991). Generation and analysis of interleukin-4 deficient mice. Science *254*, 707–710.

Kundig, T. M., Schorle, H., Bachmann, M. F., Hengartner, H., Zinkernagel, R., and Horak, I. (1993). Immune responses in interleukin-2 deficient mice. Science *262*, 1059–1061.

Kurts, C., Carbone, F. R., Krummel, M. F., Koch, K. M., Miller, J. F., and Heath, W. R. (1999). Signaling through CD30 protects against autoimmune diabetes mediated by CD8 T cells. Nature *398*, 341–344.

Lee, F. S., Hagler, J., Chen, Z. J., and Maniatis, T. (1997). Activation of the IκBα kinase complex by MEKK1, a kinase of the JNK pathway. Cell *88*, 213–222.

Lee, S. Y., Reichlin, A., Santana, A., Sokol, K. A., Nussenzweig, M. C., and Choi, Y. (1997). TRAF2 is essential for JNK but not NFκB activation and regulates lymphocyte proliferation and survival. Immunity *7*, 703–713.

Liao, X. C., and Littman, D. R. (1995). Altered T cell receptor signaling and disrupted T cell development in mice lacking Itk. Immunity *3*, 757–769.

Liao, X. C., Littman, D. R., and Weiss, A. (1997). Itk and Fyn make independent contributions to T cell activation. J. Exp. Med. *186*, 2069–2073.

Liu, K. Q., Bunnell, S. C., Gurniak, C. B., and Berg, L. J. (1998). T cell receptor-initiated calcium release is uncoupled from capacitative calcium entry in Itk deficient T cells. J. Exp. Med. *187*, 1721–1727.

Lodolce, J. P., Boone, D. L., Chai, S., Swain, R. E., Dassopoulos, T., Trettin, S., and Ma, A. (1998). IL-15 receptor maintains lymphoid homeostasis by supporting lymphocyte homing and proliferation. Immunity *9*, 669–676.

Magram, J., Connaughton, S. E., Warrier, R. R., D. M., C., Wu, C. Y., Ferrante, J., Stewart, C., Sarmiento, U., Faherty, D. A., and Gately, M. K. (1996). IL-12 deficient mice are defective in IFN gamma production and type 1 cytokine responses. Immunity *4*, 471–481.

Marengere, L. E. M., Watrehouse, P., Duncan, G. S., Mattrucker, H.-W., Feng, G.-S., and Mak, T. W. (1996). Regulation of T cell receptor signaling by tyrosine phosphatase SYP association with CTLA-4. Science *272*, 1170–1174.

Marine, J.-C., Topham, D. J., McKay, C., Wang, D., Parganas, E., Stravopodis, D., Yoshimura, A., and Ihle, J. N. (1999). SOCS1 deficiency causes a lymphocyte-dependent perinatal lethality. Cell *98*, 609–616.

McKenzie, G. J., Emson, C. L., Bell, S. E., Anderson, S., Fallon, P., Zurawski, G., Murray, R., Grencis, R., and McKenzie, A. N. J. (1998). Impaired development of Th2 cells in IL-13 deficient mice. Immunity *9*, 423–432.

Minegishi, Y., Rohrer, J., Coustan-Smith, E., Lederman, H. M., Pappu, R., Campana, D., Chan, A. C., and Conley, M. E. (1999). An essential role for BLNK in human B cell development. Science *286*, 1954–1957.

Meraz, M. A., White, J. M., Sheehan, K. C., Bach, E. A., Rodig, S. J., Dighe, A. S., Kaplan, D. H., Riley, J. K., Greenlund, A. C., Campbell, D., Carver-Moore, K., DuBois, R. N., Clark, R., Aguet, M., and Schreiber, R. D. (1996). Targeted disruption of the Stat1 gene in mice reveals unexpected physiologic specificity in the JAK-STAT signaling pathway. Cell *84*, 431–432.

Molina, T. J., Kishihara, K., Siderovski, D. P. et al. (1992). Profound block in thymocyte development in mice lacking p56lck. Nature *357*, 161–164.

Monks, C. R. F., Freiberg, B. A., Kupfer, H., Sciaky, N., and Kupfer, A. (1998). Three dimensional segregation of supramolecular activation clusters in T cells Nature *395*, 82–86.

Monks, C. R. F., Kupfer, H., Tamir, I., Barlow, A., and Kupfer, A. (1997). Selective modulation of protein kinase C-theta during T cell activation. Nature *385*, 83–86.

Moore, P. A., Belvedere, O., Orr, A. et al. (1999). BLyS: member of the tumor necrosis family and B lymphocyte stimulator. Science *285*, 260–263.

Moriggl, R., Topham, T. J., Teglund, S., Sexl, V., McKay, C., Wang, D., Hoffmeyer, A., van Deursen, J., Sangster, M. Y., Bunting, K. D., Grosveld, G. C., and Ihle, J. N. (1999). Stat 5 is required for IL-2 induced cell cycle progression of peripheral T cells. Immunity *10*, 249–259.

Nada, S., Yagi, T., Takeda, H., Tokunaga, T., Nakagawa, H., Ikawa, Y., Okada, M., and Aizawa, S. (1993). Constitutive activation of Src-family kinases in mouse embryos that lack Csk. Cell *73*, 1125–1135.

Nakajima, H., Liu, X. W., Wynshaw-Boris, A., Rosenthal, L. A., Imada, K., Finbloom, D. S., Henninghausen, L., and Leonard, W. J. (1997). An indirect effect of Stat5a in IL-2 induced proliferation: a critical role for Stat5a in IL-2 mediated IL-2 receptor α induction. Immunity *7*, 691–701.

Negishi, I., Motoyama, N., Nakayama, K., Nakayama, K., Senju, S., Hatakeyama, S., Zhang, Q., Chan, A. C., and Loh, D. Y. (1995). Essential role for Zap-70 in both positive and negative selection of thymocytes. Nature *376*, 435–438.

Nimnual, A. S., Yatsula, B. A., and Bar-Sagi, D. (1998). Coupling of Ras and Rac guanosine triphosphatases through the Ras exchanger Sos. Science *279*, 560–563.

Nishizumi, H., Taniuchi, I., Yamanashi, Y., Kitamura, D., Ilic, D., Mori, S., Watanabe, T., and Yamamoto, T. (1995). Impaired proliferation of peripheral B cells and indication of autoimmune disease in lyn-deficient mice. Immunity *3*, 549–560.

Noben-Trauth, N., Schulz, L. D., Brombacher, F., Urban, J. F., Gu, H., and Paul, W. E. (1997). An IL-4 independent pathway for $CD4^+$ T cell IL-4 production is revealed in interleukin-4 receptor deficient mice. Proc. Natl. Acad. Sci. USA *94*, 10838–10843.

Nosaka, T., van Deursen, J. M. A., Tripp, R. A., Thierfelder, W. E., Wittuhn, B. A., McMickle, A. P., Doherty, P. C., Grosveld, G. C., and Ihle, J. N. (1995). Defective lymphoid development in mice lacking Jak3. Science *270*, 800–802.

Nguyen, L. T., Duncan, G. S., Mirtsos, C., Ng, M., Speiser, D. E., Shahinian, A., Marino, M. W., Mak, T. W., Ohashi, P. S., and Yeh, W.-C. (1999). TRAF2 deficiency results in hyperactivity of certain TNFR1 signals and impairment of CD40 mediated responses. Immunity *11*, 379–389.

Ordentlich, P., Lin, A., Shen, C. P., Blaumueller, C., Matsuno, M., Artavanis-Tsakonas, S., and Kadesch, T. (1998). Notch inhibition of E47 supports the existence of a novel signaling pathway. Mol. Cell. Biol. *18*, 2230–2239.

Ohteki, T., Ho, S., Suzuki, H., Mak, T. W., and Ohashi, P. S. (1997). Role for IL-15/IL-15 receptor β chain in natural killer 1.1^+ T cell receptor αβ cell development. J. Immunol. *159*, 5931–5935.

Otipoby, K. L., Andersson, K. B., Draves, K. E., Klaus, S. J., Farr, A. G., Kerner, J. D., Perlmutter, R. M., Law, C. L., and Clark, E. A. (1996). CD22 regulates thymus independent responses and the lifespan of B cells. Nature *384*, 634–637.

Pappu, R., Cheng, A. M., Li, B., Gong, Q., Chiu, C., Griffin, N., White, M., Sleckman, B. P., and Chan, A. C. (1999). Requirement for B cell linker protein (BLNK) in B cell development. Science *286*, 1949–1954.

Parganas, E., Wang, D., Stravopodis, D., Topham, D. J., Marine, J. C., Teglund, S., Vanin, E. F., Bodner, S., Colamonici, O. R., van Deursen, J. M., Grosveld, G., and Ihle, J. N. (1998). Jak 2 is essential for signaling through a variety of cytokine receptors. Cell *93*, 385–395.

Park, S. Y., Saijo, K., Takahashi, T., Osawa, M., Arase, H., Hirayama, N., Miyakie, K., Nakauchi, H., Shirasawa, T., and Saito, T. (1995). Developmental defects of lymphoid cells in JAK3 kinase deficient mice. Immunity *3*, 771–782.

Pear, W. S., Aster, J. C., Scott, M. L., Hasserjian, R. P., Soffer, B., Sklar, J., and Baltimore, D. (1996). Exclusive development of T cell neoplasms in mice transplanted with bone marrow expressing activated notch alleles. J. Exp. Med. *183*, 2283–2291.

Peschon, J. J., Morrissey, P. J., Grabstein, K. H., Ramsdell, F. J., Maraskovsky, E., Gliniak, B. C., Park, L. S., Ziegler, S. F., Williams, D. E., Ware, C. B. et al. (1994). Early lymphocyte expansion is severely impaired in interleukin-7 deficient mice. J. Exp. Med. *180*, 1955–1960.

Pivniouk, V., Tsitsikov, E., Swinton, P., Rathbun, G., Alt, F. W., and Geha, R. S. (1998). Impaired viability and profound block in thymocyte development in mice lacking the adaptor protein SLP-76. Cell *94*, 229–238.

Pui, J. C., Allman, D., Xu, L., DeRocco, S., Karnell, F. G., Bakkour, S., Lee, J. Y., Kadesch, T., Hardy, R. R., Aster, J. C., and Pear, W. S. (1999). Notch 1 expression in early lymphopoiesis influences B versus T lineage determination. Immunity *11*, 299–308.

Radtke, F., Wilson, A., Stark, G., Bauer, M., van Meerwijk, J., MacDonald, H. R., and Aguet, M. (1999). Deficient T cell fate specification in mice with an induced inactivation of notch 1. Immunity *10*, 547–558.

Ranger, A. M., Hodge, M. R., Gravallese, E. M., Oukka, M., Davidson, L., Alt, F. W., de la Brousse, F. C., Hoey, T., Grusby, M., and Glimcher, L. H. (1998). Delayed lymphoid repopulation with defects in IL-4 driven responses produced by inactivation of NF-ATc. Immunity *8*, 125–134.

Reich, Z., Boniface, J. J., Lyons, D. S., Borochov, N., Wachtel, E. J., and Davis, M. M. (1997). Ligand specific oligomerization of T cell receptor molecules. Nature *387*, 617–620.

Rodig, S. J., Meraz, M. A., White, J. M., Lampe, P. A., Riley, J. K., Arthur, C. D., King, K. L., Sheehan, K. C., Yin, L., Pennica, D., Johnson, E. M. J., and Schreiber, R. D.

(1998). Disruption of the Jak 1 gene demonstrates obligatory and nonredundant roles of the Jaks in cytokine-induced biologic responses. Cell *93*, 373–383.

Romani, L., Mencacci, A., Cenci, E., Spaccapelo, R., Toniatti, C., Puccetti, P., Bistoni, F., and Poli, V. (1996). Impaired neutrophil response and T helper cell 1 development in interleukin-6 deficient mice infected with *Candida albicans*. J. Exp. Med. *183*, 1345–1355.

Rothe, M., Sarma, V., Dixit, V. M., and Goeddel, D. V. (1995). TRAF2 mediated activation of NFκB by TNF Receptor 2 and CD40. Science *269*, 1424–1427.

Sadlack, B., Merz, H., Schorle, H., Schimpl, A., Feller, A. C., and Horak, I. (1993). Ulcerative colitis like disease in mice with a disrupted interleukin-2 gene. Cell *75*, 253–261.

Sanchez, I., Hughes, R. T., Mayer, B. J., Yee, K., Woodgett, J. R., Avruch, J., Kyriakis, J. M., and Zon, L. I. (1994). Role of SAPK/ERK kinase-1 in the stress activated pathway regulating transcription factor c-Jun. Nature *372*, 794–798.

Schaeffer, E. M., Debnath, J., Yap, G., McVicar, D., Liao, X. C., Littman, D. R., Sher, A., Varmus, H. E., Lenardo, M. J., and Schwartzberg, P. L. (1999). Requirement for Tec kinases Rlk and Itk in T cell receptor signaling and immunity. Science *284*, 638–641.

Scher, I., Ahmed, A., Strong, D. M., Steinberg, A. D., and Paul, W. E. (1975). X linked B lymphocyte defect in CBA/N mice. I. Studies of the function and composition of spleen cells. J. Exp. Med. *141*, 788–803.

Schneider, P., Mackay, F., and Steiner, V. et al. (1999). BAFF, a new ligand of the tumor necrosis factor family stimulates B cell growth. J. Exp. Med. *189*, 1747–1756.

Schwarz, E. M., Krimpenfort, P., Berns, A., and Verma, I. M. (1997). Immunological defects in mice with a targeted disruption in Bcl-3. Genes Dev. *11*, 187–197.

Schorle, H., Holtschke, T., Hunig, T., Schimpl, A., Feller, A. C., and Horak, I. (1991). Development and function of T cells in mice rendered interleukin-2 deficient by gene targeting. Nature *352*, 621–624.

Sen, R., and Baltimore, D. (1986). Multiple nuclear factors interact with the immunoglobulin enhancer sequences. Cell *46*, 705–716.

Sgroi, D., and Stamenkovic, I. (1994). C22, a B cell Ig superfamily receptor with sialic acid-binding lectin activity. Immunologist *2*, 161–166.

Sha, W. C., Hsiou-Chi, L., Toumanen, E. I., and Baltimore, D. (1995). Targeted disruption of the p50 subunit of NFκB leads to multifocal defects in immune responses. Cell *80*, 321–330.

Shimoda, K., van Deursen, J., Sangster, M. Y., Sarawar, S. R., Carson, R. T., Tripp, R. A., Chu, C., Vignali, D. A. A., Doherty, P. C., Grosveld, G. C., and Ihle, J. N. (1996). Lack of IL-4 induced Th2 response and IgE class switching in mice with disrupted Stat6 gene. Nature *380*, 630–633.

Snapper, C. M., Rosas, F. R., Zelazowski, P., Moorman, M. A., Kehry, M. R., Bravo, R., and Weih, F. (1996a). B cells lacking RelB are defective in proliferative responses, but undergo normal B cell maturation to Ig secretion and Ig class switching. J. Exp. Med. *184*, 1537–1541.

Snapper, C. M., Zelazowski, P., Rosas, F. R., Kehry, M. R., Tian, M., Baltimore, D., and Sha, W. C. (1996b). B cells from p50/NFκB knockout mice have selective defects in proliferation, differentiation, germline CH transcription, and Ig class switching. J. Immunol. *156*, 183–191.

Stein, P. L., Lee, H. M., Rich, S., and Soriano, P. (1992). pp59Fyn mutant mice display differential signaling in thymocytes and peripheral T cells. Cell *70*, 741–750.

Suzuki, H., Duncan, G. S., Takimoto, H., and Mak, T. W. (1997a). Abnormal development of intestinal intraepithelial lymphocytes and peripheral natural killer cells in mice lacking the IL-2 receptor β chain. J. Exp. Med. *185*, 499–505.

Suzuki, H., Kundig, T. M., Furlonger, C., Wakeham, A., Timms, E., Matsuyama, T., Schmits, R., Simard, J. J., Ohashi, P. S., Griesser, H. et al. (1995). Deregulated T cell activation and autoimmunity in mice lacking interleukin-2 receptor beta. Science *268*, 1472–1476.

Suzuki, Y., Rani, S., Liesenfeld, O., Kojima T., Lim, S., Nguyen, T. A., Dalrymple, S. A., Murray, R., and Remington, J. S. (1997b). Impaired resistance to the development of toxoplasmic encephalitis in interleukin-6 deficient mice. Infect. Immun. *65*, 2339–2345.

Swallow, M. M., Wallin, J. J., and Sha, W. C. (1999). B7h, a novel costimulatory homolog of B7.1 and B7.2 is induced by TNFα. Immunity *11*, 423–432.

Swan, K. A., Alberola-Ila, J., Gross, J. A., Appleby, M. W., Forbush, K. A., Thomas, J. F., and Perlmutter, R. M. (1995). Involvement of p21ras distinguishes positive and negative selection in thymocytes. EMBO J. *14*, 276–285.

Swat, W., Fujikawa, K., Ganaiatsas, S., Yang, D., Xavier, R. J., Harris, N. L., Davidson, N. L., Ferrini, R., Davis, R. J., Labow, M. A., Flavell, R. A., Zon, L. I., and Alt, F. W. (1998). SEK1/MKK4 is required for maintenance of a normal peripheral lymphoid compartment but not for lymphocyte development. Immunity *8*, 625–634.

Swat, W., Shinkai, Y., Cheng, H. L., Davidson, L., and Alt, F. W. (1996). Activated Ras signals differentiation and expansion of $CD4^+ 8^+$ thymocytes. Proc. Natl. Acad. Sci. USA *93*, 4683–4687.

Takata, M., and Kurosaki, T. (1996). A role for Bruton's tyrosine kinase in B cell antigen receptor mediated activation of phospholipase C γ-2. J. Exp. Med. *184*, 31–40.

Takeda, K., Tanaka, T., Shi, W., Matsumoto, M., Minami, M., Kashiwamura, S., Nakanishi, K., Yoshida, N., Kishimoto, T., and Akira, S. (1996). Essential role of Stat6 in IL-4 signaling. Nature *380*, 627–630.

Takeda, K., Noguchi, K., Shi, W., Tanaka, T., Matsumoto, M., Yoshiad, N., Kishimoto, T., and Akira, S. (1997). Targeted disruption of the mouse Stat3 gene leads to early embryonic lethality. Proc. Natl. Acad. Sci. USA *94*, 3801–3804.

Takeda, K., Tsutsui, H., Yoshimoto, T., Adachi, O., Yoshida, N., Kishimoto, T., Okamura, H., Nakanishi, K., and Akira, S. (1998). Defective NK cell activity and Th1 response in IL-18 deficient mice. Immunity *8*, 383–390.

Tarakhovsky, A., Turner, M., Schaal, S., Mee, P. J., Duddy, L. P., Rajewsky, K., and Tybulewicz, V. L. J. (1995). Defective antigen receptor mediated proliferation of B and T cells in the absence of Vav. Nature *374*, 467–470.

Teglund, S., McKay, C., Schuetz, E., van Deursen, J. M., Stravopodis, D., Wang, D., Brown, M., Bodner, S., Grosveld, G., and Ihle, J. N. (1998). Stat5a and Stat5b proteins have essential and nonessential, or redundant, roles in cytokine responses. Cell *93*, 841–850.

Thierfelder, W. E., van Deursen, J. M., Yamamoto, K., Tripp, R. A., Sarawar, S. R., Carson, R. T., Sangster, M. Y., Vignali, D. A., Doherty, P. C., Grosveld, G. C., and Ihle, J. N. (1996). Requirement for Stat 4 in interleukin-12 mediated responses of natural killer and T cells. Nature *382*, 171–174.

Thomis, D. C., Gurniak, C. B., Tivol, E., Sharpe, A. H., and Berg, L. J. (1995). Defects in B lymphocyte maturation and T lymphocyte activation in mice lacking Jak3. Science *270*, 794–797.

Tivol, E. A., Borriello, F., Schweitzer, A. N., Lunch, W. P., Bluestone, J. A., and Sharpe, A. H. (1995). Loss of CTLA-4 leads to massive lymphophroliferation and fatal multiorgan tissue destruction, revealing a critical negative regulatory role of CTLA-4. Immunity *3*, 541–547.

Tomita, K., Hattori, M., Nakamura, E., Nakanishi, S., Minato, N., and Kageyama, R. (1999). The bHLH gene *hes1* is essential for expansion or early T cell precursors. Genes Dev. *13*, 1203–1210.

Tooze, R. M., Doody, G. M., and Fearon, R. T. (1997). Counterregulation by the coreceptors CD19 and CD22 of MAP kinase activation by membrane immunoglobulin. Immunity *7*, 59–67.

Turner, M., Mee, P. J., Costello, P. S., Williams, O., Price, A. A., Duddy, L. P., Furlong, M. T., Geahlen, R. L., and Tybulewicz, V. (1995). Perinatal lethality and blocked B cell development in mice lacking the tyrosine kinase Syk. Nature *378*, 298–303.

Turner, M., Mee, P. J., Walters, A. E., Quinn, M. E., Mellor, A. L., Zamoyska, R., and Tybulewicz, V. L. (1997). A requirement for the Rho family GTP exchange factor Vav in positive and negative selection of thymocytes. Immunity *4*, 451–460.

Van Oers, N. S. C., Lowin-Kropf, B., Finlay, D., Connolly, K., and Weiss, A. (1996). $\alpha\beta$ T cell development is abolished in mice lacking both Lck and Fyn protein tyrosine kinases. Immunity *5*, 429–436.

Viola, A., Schroeder, S., Sakakibara, Y., and Lanzavecchia, A. (1999). T lymphocyte costimulation mediated by reorganization of membrane microdomains. Science *283*, 680–682.

von Freeden-Jeffry, U., Solvason, N., Howard, M., and Murray, R. (1997). The earliest T-lineage committed cells depend on IL-7 for Bcl-2 expression and normal cell cycle progression. Immunity *7*, 147–154.

von Freeden-Jeffry, U., Vieira, P., Lucian, L. A., McNeil, T., Burdach, S. E., and Murray, R. (1995). Lymphopenia in IL-7 gene deleted mice identifies IL-7 as a nonredundant cytokine. J. Exp. Med. *181*, 1519–1526.

Wang, J., Koizumi, T., and Watanabe, T. (1996). Altered antigen receptor signaling and impaired Fas-mediated apoptosis of B cells in Lyn deficient mice. J. Exp. Med. *184*, 831–838.

Waterhouse, P., Penninger, J. M., Timms, E., Wakeham, A., Shahinian, A., Lee, K. P., Thompson, C. B., Riesser, H., and Mark, T. W. (1995). Lymphoproliferative disorders with early lethality in mice deficient in CTLA-4. Science *270*, 985–988.

Weih, F., Carrasco, D., Durham, S. K., Barton, D. S., Rizzo, C. A., Ryseck, R.-P., Lira, S. A., and Bravo, R. (1995). Multiorgan inflammation and hematopoietic abnormalities in mice with a targeted disruption of Rel B, a member of the NF-κB/Rel family. Cell *80*, 331–340.

Weih, F., Durham, S. K., Barton, D. S., Sha, W. S., Baltimore, D., and Bravo, R. (1997a) p50-NF-kB complexes partially compensate for the absence of RelB: severely increased pathology in p50-/-relB-/- double knockout mice. J. Exp. Med. *195*, 1359–1370.

Weih, F., Warr, G., Yang, H., and Bravo, R. (1997b). Multifocal defects in immune responses in relB deficient mice. J. Immunol. *158*, 5211–5218.

Wen, T., Zhang, L., Kung, S. K., Molina, T. J., Miller, R. G., and Mak, T. W. (1995). Allo-skin graft rejection, tumor rejection and natural killer cell activity in mice lacking p56lck. Eur. J. Immunol. *25*, 3155–3159.

Werlen, G., Jacinto, E., Xia, Y., and Karin, M. (1998). Calcineurin preferentially synergizes with PKC-theta to activate JNK and IL-2 promoter in T lymphocytes. EMBOJ *17*, 3101–3111.

Wiest, D. L., Ashe, J. M., Howcroft, T. K., Lee, H. M., Kemper, D. M., Negishi, I., Singer, D. S., Singer, A., and Abe, R. (1997). A spontaneously arising mutation in the DLAARN motif of murine Zap-70 abrogates kinase activity and arrests thymocyte development. Immunity *6*, 663–671.

Willerford, D. M., Chen, J., Ferry, J. A., Davidson, L., Ma, A., and Alt, F. W. (1995). Interleukin-2 receptor α chain regulates the size and content of the peripheral lymphoid compartment. Immunity *3*, 521–530.

Woronicz, J. D., Gao, X., Cao, Z., Rothe, M., and Goeddel, D. V. (1997). IκB kinase-β. NFκB activation and complex formation with IκB kinase-α and NIK. Science *278*, 866–869.

Wu, H., Kwong, P. D., and Hendrickson, W. A. (1997). Dimeric association and segmental variability in the structure of human CD4. Nature *387*, 527–530.

Wu, M., Lee, H., Bellas, R. E., Schauer, S. L., Arsura, M., Katz, D., FitzGerald, M. J., Rothstein, T. L., Sherr, D. H., and Sonenshein, G. E. (1996). Inhibition of NFκB/Rel induces apoptosis of B cells. EMBO J. *17*, 4682–4690.

Wulfing, C., and Davis, M. M. (1998). A receptor/cytoskeletal movement triggered by costimulation during T cell activation. Science *282*, 2266–2269.

Xanthoudakis, S., Viola, J. P. B., Shaw, K. T. Y., Luo, C., Wallace, J. D., Bozza, P. T., Curran, T., and Rao, A. (1996). An enhanced immune response in mice lacking the transcription factor NFAT1. Science *272*, 892–895.

Yablonski, D., Kuhne, M. R., Kadlecek, T., and Weiss, A. (1998). Uncoupling of non receptor tyrosine kinases from PLC-γ1 in an SLP-76 deficient T cell. Science *281*, 413–416.

Yang, D. D., Conze, A., Whitmarsh, A. J., Barrett, T., Davis, R. J., Rincon, M., and Flavell, R. A. (1998). Differentiation of $CD4^+$ T cells to Th1 cells requires MAP kinase JNK2. Immunity *9*, 575–585.

Yoshida, T., Ikuta, K., Sugaya, H., Maki, K., Takagi, M., Kanazawa, H., Sunaga, S., Kinashi, T., Yoshimura, K., Miyazaki, J.-I., Takaki, S., and Takatsu, K. (1996a). Defective B-1 cell development and impaired immunity against *Angiostrongylus cantonensis* in IL-5Rα deficient mice. Immunity *4*, 483–494.

Yoshida, H., Nishina, H., Takimoto, H., Marengere, L. E. M., Wakeham, A. C., Bouchard, D., Kong, Y.-Y., Ohteki, T., Shahinian, A., Nachmann, M., Ohashi, P. S., Penninger, J. F., Crabtree, G. R., and Mak, T. W. (1998). The transcription factor NFATc1 regulates lymphocyte proliferation and Th2 cytokine production. Immunity *8*, 115–124.

Yoshida, K., Taga, T., Saito, M., Suematsu, S., Kumanogoh, A., Tanaka, T., Fujiwara, H., Hirata, M., Yamagami, T., and Nakahata, T. et al. (1996b). Targeted disruption of gp130, a common signal transducer for the interleukin 6 family of cytokines, leads to myocardial and hematological disorders. Proc. Natl. Acad. Sci. USA *93*, 407–411.

Yoshinaga, S. K., Whoriskey, J. S., Khare, S. D., Sarmiento, U., and Guo, J. et al. (1999). T cell costimulation through B7RP-1 and ICOS. Nature *402*, 827–832.

Zandi, E., Rothwarf, D. M., Delhase, M., Hayakawa, M., and Karin, M. (1997). The IκB kinase complex (IKK) contains two kinase subunits, IKKα and IKKβ, necessary for IκB phosphorylation and NFκB activation. Cell *91*, 243–252.

Zhang, R., Alt, F. W., Davidson, L., Orkin, S. H., and Swat, W. (1995). Defective signaling through the T and B Cell antigen receptors in lymphoid cells lacking the vav proto-oncogene. Nature *374*, 470–473.

Zhang, Y., Yao, B., Delikat, S., Bayoumy, S., Lin, X.-H., Basu, S., McGinley, M., Chan-Hui, P.-Y., Lichenstein, H., and Kolesnick, R. (1997). Kinase suppressor of Ras is ceramide-activated protein kinase. Cell *89*, 63–72.

Zhang, W., Sloan-Lancaster, J., Kitchen, J., Trible, R. P., and Samelson, L. E (1998). LAT: The Zap-70 tyrosine kinase substrate that links T cell receptor to cellular activation. Cell *92*, 83–92.

Zhang, W., Sommers, C. L., Burshtyn, D. N., Stebbins, C. C., DeJarnette, J. B., Trible, R. P., Grinberg, A., Tsay, H. C., Jacobs, H. M., Kessler, C. M., Long. E. O., Love, P. E., and Samelson, L. E. (1999). Essential role of LAT in T cell development. Immunity *10*, 323–332.

Zhu, J., Shibasaki, F., Price, R., Guillemot, J.-C., Yano, T., Dotsch, V., Wagner, G., Ferrara, P., and McKeon, F. (1998). Intramolecular masking of nuclear import signal on NFAT4 by casein kinase I and MEKK1, Cell *93*, 851–862.

5
Molecular Mechanisms of Lymphocyte Apoptosis

Apoptosis refers to a form of death in which a cell initiates a suicide program and goes "quietly into the night." In contrast, necrotic cells erupt in explosive incandescence, singeing their neighbors and generally trashing the neighborhood. Apoptotic death is accompanied by characteristic morphological changes that include chromatin condensation, nucleolar disruption, cytoplasmic contraction, and membrane blebbing. In this form of death, DNA is generally cleaved into a "ladder" of oligonucleosome-length fragments. Apoptosing cells may be cleared as virtually whole cells by macrophages or go through a process of being "dismantled" to permit the endocytic clearance of stripped down packages. This controlled form of non-necrotic death is also referred to as *programmed cell death*, *physiological cell death*, or *active cell death*.

Apoptotic death was observed for decades by pathologists and developmental biologists, but its significance was poorly appreciated. In 1972, Wyllie and colleagues drew the attention of biologists to the importance and distinct nature of this form of physiological death in a number of cell types including developing thymocytes. In the mid-1980s, the field came to life with the discovery of genes that could regulate the process. In the nematode *Caenorhabditis elegans*, 131 of the organism's 1031 cells undergo programmed cell death during the process of development. Studies of this nematode revealed that two genes, *ced-3* and *ced-4* (*ced* stands for *C. elegans* death defective), were required for the loss of all 131 of the cells that undergo programmed cell death. A genetic pathway emerged from these studies (Figure 5.1) suggesting that *ced-4* was upstream of *ced-3* and that another gene, *ced-9*, encodes an antiapoptotic protein that can inhibit the death-promoting effects of *ced-4* and *ced-3*. As we shall see, many vertebrate homologs for these three genes have been identified and characterized.

Before all these details from the study of death in *C. elegans* actually emerged, the examination of a particular type of human B cell lymphoma provided another stunning insight into the process of apoptosis. Most human lymphomas happen to be B cell tumors, and the majority of these

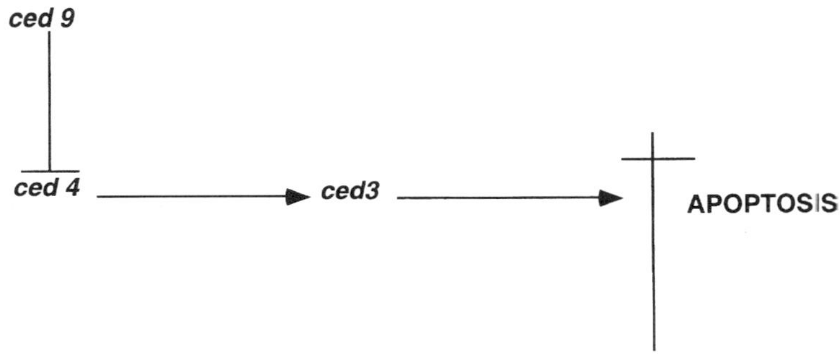

FIGURE 5.1. The *Caenorhabditis elegans* death-defective genes *ced-4* and *ced-3* are essential for the developmental death of 131 of the 1031 cells in this worm. The *ced-9* gene can prevent *ced-3* and *ced-4* from initiating the death process.

tumors originate from cells in the germinal center. Follicular lymphomas are characterized by a t(14;18) translocation that links regulatory sequences at the immunoglobulin (Ig) heavy chain locus on chromosome 14 to a gene on chromosome 18 known as *bcl-2*. (for B cell lymphoma-2). These translocations utilize V(D)J recombinational signal sequences. The totally unexpected result that emerged from the study of *bcl-2* was that it represented an oncogene that apparently transformed cells *not* because it was an inducer of proliferation but because it could prevent cells from dying. *Bcl-2* was identified as a "survival" gene or as an "antiapoptosis" gene. Interestingly, *bcl-2* has turned out to be a vertebrate homolog of the *C. elegans* gene *ced-9* that prevents *ced-3* and *ced-4* from initiating programmed cell death.

During the 1990s, an enormous amount of information has accumulated about the process of programmed cell death. A common death pathway has been identified, and it is clear that different cells and apoptogenic stimuli access this pathway by distinct routes. The focus in this chapter will be to initially examine the common pathway and then to examine the different ways by which this pathway may be accessed or blocked during lymphocyte development.

Programmed cell death is initiated many times during lymphoid ontogeny—during the process of negative selection and whenever lymphocytes fail to be positively selected. The termination or downregulation of immune responses is achieved, in part, by the induction of the apoptotic death of activated T cells using a death-inducing receptor known as Fas. Aberrantly activated B cells are also eliminated by the Fas pathway. Cytotoxic T lymphocytes as well as natural killer (NK) cells express specific molecules such as perforin and granzyme B that are designed to assist in the execution of targets by apoptosis. The Fas pathway may also be used by $CD4^+$ T cells to kill infected targets.

The Common Death Pathway and Caspases

Over the past few years, a molecular understanding of a common apoptotic death pathway has emerged from a range of studies in a number of organisms and cell types. This pathway is summarized in Figure 5.2.

The induction of death is generally linked to the activation of a set of cysteine proteases that cleave proteins immediately after aspartic acid residues. The first such enzyme to be identified was the mammalian homolog of the *C. elegans* cell death gene, *ced-3*. This protein was identical to an enzyme previously described as the interleukin-1β converting enzyme (ICE), which was known to be responsible for proteolytically converting pro-interleukin-1β to mature IL-1. It soon became apparent that the ICE protease was a member of a large family of related enzymes, most of which appear to be involved in cell death, and some of which may facilitate inflammatory responses. These enzymes contain a conserved QACRG motif, the cysteine being a critical residue in the active site. These proteases are are now referred to as caspases (cysteine-requiring aspartate proteases). A number of distinct caspases are employed as inducers or executioners during apoptosis in different cells and in different pathways.

Caspases may autoproteolytically activate themselves, especially when they are aggregated, or they may cleave and activate other caspases. Cleavage of caspases leads to the release of an N-terminal fragment, referred to as a prodomain, which generally serves a regulatory function; two active fragments are also generated, a large and a small proteolytic subunit. Crystal structures of two activated caspases have revealed that two large and two small subunits associate to form catalytically active tetramers. Each tetramer contains two identical but physically separated catalytic sites. A small connecting piece is also released that contains a number of cleavage sites that might be targeted by a number of noncaspase proteases.

"Inducer" caspases may initiate caspase cascades by cleaving downstream "executioner" caspases, which in turn cleave substrates that contribute to the death process. A number of caspase substrates have been identified. The relevance of many of these substrates to the initiation or progression of apoptotic death is unclear. It has been pointed out that many substrates of caspases are proteins involved in RNA splicing, DNA repair, and scaffold formation in the nucleus and cytoplasm. It has been suggested that when the decision has been made to initiate apoptosis, the cell seeks to prevent futile attempts at repair and essentially prepares to be dismantled and neatly packaged for endocytic clearance. It is assumed that many of the substrates targeted by caspases are essential for repair processes or are degraded to facilitate the "dismantling" and packaging of the dying cell. A particularly interesting substrate of caspase-3 is the DNA fragmentation

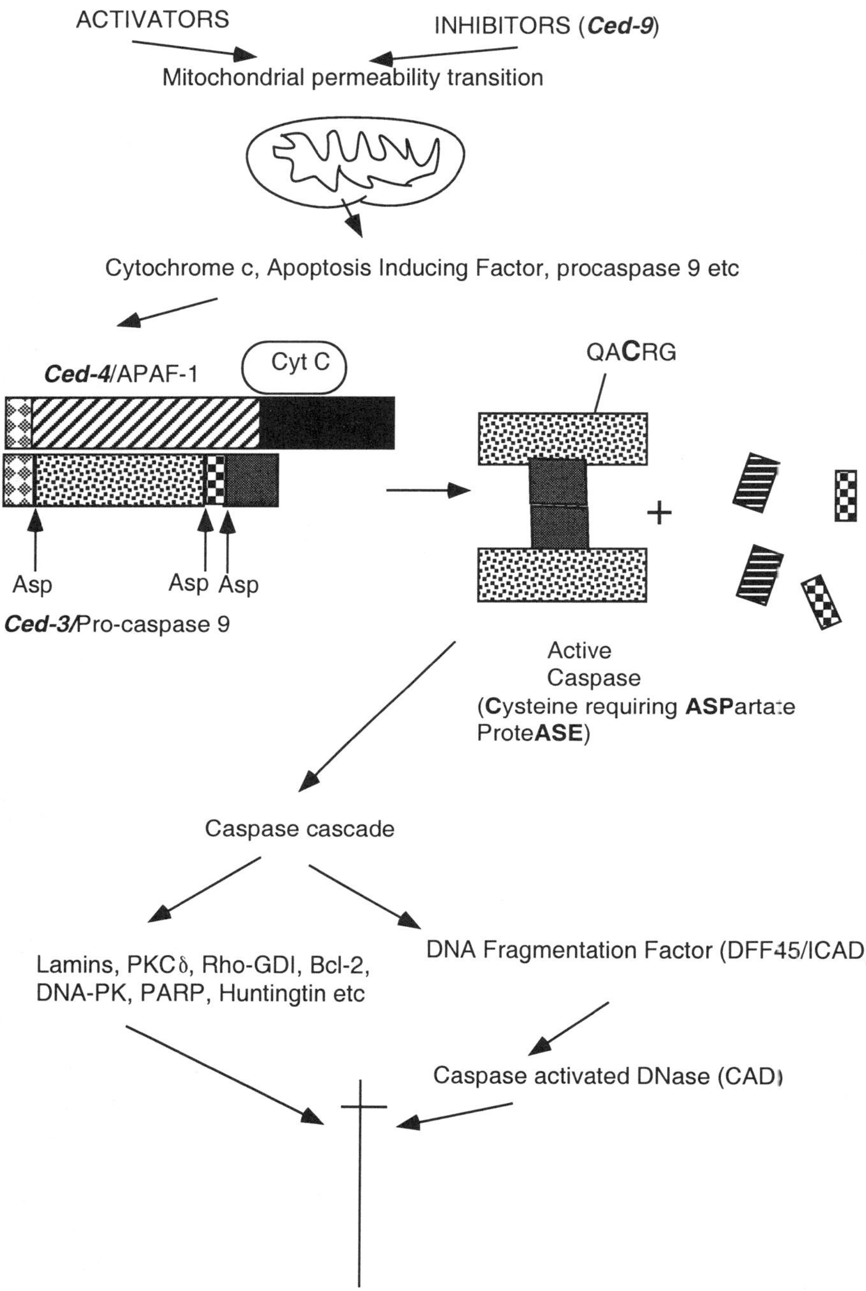

FIGURE 5.2. The caspase-9–mediated common death pathway. The mitochondrial permeability transition results in the release of factors including cytochrome c that are permissive for apoptosis. Cytochrome c can associate with the vertebrate Ced-4 homology, APAF-1, and thus contribute to the activation of the caspase cascade. This cascade contributes to the proteolytic cleavage of caspase targets including the DFF protein or ICAD, which sequesters CAD.

factor (DFF45), more commonly known as ICAD (inhibitor of caspase-activated deoxyribonuclease). ICAD forms a complex with a Ca^{++}/Mg^{++}–dependent endonuclease known as CAD (caspase-activated deoxyribonuclease), or DFF40, and thus prevents CAD from cleaving DNA. CAD cannot be translated in the absence of ICAD, which acts as a kind of chaperone for CAD synthesis. This requirement also ensures that CAD is not synthesized in the absence of ICAD and can therefore always be regulated. Activated executioner caspases cleave ICAD and thus allow CAD to cleave DNA. The ICAD/CAD complex thus broadly resembles the IκB/NFκB complex discussed in Chapter 4. However, it appears that this complex, unlike the cytosolic IκB/NFκB complex, may actually be nuclear. *ICAD/DFF45*–deficient mice develop normally, are unable to induce DNA fragmentation, but essentially have no defect in inducing apoptosis (see Table 5.4, later). DNA fragmentation is obviously not the most important consequence of caspase activation.

Caspases may cleave and inactivate Bcl-2 and Bcl-2 family members such as Bid (discussed below) and thus contribute to an increase in mitochondrial leakiness which sustains caspase activation. Cleavage by caspases of lamins leads to the destruction of the nuclear lamina and thus may contribute to chromosome condensation. Cleavage of cytoskeletal proteins such as gelsolin and focal-adhesion kinase contributes to the reorganization of the dying cell and the formation of apoptotic bodies. Caspase activation is also required for the exposure of phosphatidylserine on the outer leaflet of the plasma membrane and for the generation of other surface changes that contribute to the clearance of apoptotic cells and the packaged apoptotic bodies by local macrophages.

There are two broad ways in which caspase activation is initiated during apoptosis. A major route to the induction of apoptosis involves the mitochondrial permeability transition, discussed below. This transition targets the activation of caspase-9, the vertebrate Ced-3 homologue. A separate approach to the activation of the caspase cascade involves the triggering of death receptors such as Fas, and this involves the activation at the cell surface of another initiator caspase, caspase-8. These two routes of caspase activation are not mutually exclusive, as will be discussed below. Death receptors may initiate caspase activation without initially influencing mitochondrial permeability, but amplification of the death receptor-initiated signal might involve modulation of the mitochondrial permeability transition. Cytotoxic or NK cell-mediated killing (also discussed later in the chapter) involves the introduction of a caspase-8– and caspase-3–activating serine protease, granzyme B, into the target cell's cytosol.

Mitochondrial Permeability, Cytochrome c, and Cell Death

A general mechanism of caspase activation involves the induction of an increase in mitochondrial permeability during apoptosis and the release of activators of caspases, particularly cytochrome c, from the intermembrane spaces of mitochondria (between the outer and inner mitochondrial membranes) into the cytoplasm. Factors other than cytochrome c are also released from mitochondria to promote apoptosis. One such protein is apoptosis-inducing factor (AIF), a flavoprotein that normally resides in the mitochondrial intermembrane space, and that migrates to the nucleus, where it participates in nuclear disassembly. In certain cell types, procaspase 9, procaspase 2, and procaspase 3 are also stored in mitochondria and are released when apoptosis is initiated. There are at least three ways in which this release of caspases and caspase activators from the mitochondrion may occur.

The mitochondrial permeability transition (PT) involves the opening of large cyclosporin-inhibitable pores in the inner mitochondrial membrane, which may contribute to the release of mitochondrial Ca^{++} stores into the cytosol, dissipation of the voltage gradient ($\Delta\Psi$) across the inner mitochondrial membrane, and the extrusion of cytochrome c and other factors into the cytosol. It is believed that this "leakiness" after PT pore opening might result from the osmotic swelling of the mitochondrial matrix, which contains a high concentration of solutes. The swelling results in rupture of the outer membrane and the release of mitochondrial contents. The exact mechanisms by which apoptosis-inducing signals such as Ca^{2+}, reactive oxygen species, and caspases contribute to the opening of the PT pore is somewhat hazy. A second, more specific, mechanism for the egress of cytochrome c and other proteins could involve the opening of protein-transporting channels formed by Bax, a proapoptotic Bcl-2 family member see section below). The third potential way in which cytochrome c may be released could involve swelling and rupture of the outer mitochondrial membrane without the loss of $\Delta\Psi$ usually associated with the permeability transition.

It is possible that an increase in mitochondrial permeability and the release of cytochrome c is an essential event in all forms of apoptosis. When apoptosis is induced by direct activation of caspases such as caspase-8 and caspase-10, these inducing caspases may induce the release of cytochrome c by cleaving the proapoptosis Bcl-2 family member known as Bid, as discussed below. The mitochondrial pathway may serve as an amplification mechanism in apoptosis induced by death receptors.

Caspases may also be activated by Ced-4. In *C. elegans* the *ced-4* gene is believed to functionally upstream of *ced-3* (the worm equivalent of caspase-3). Ced-4 can associate with the cytoplasmic tail of Bcl-2 and be prevented

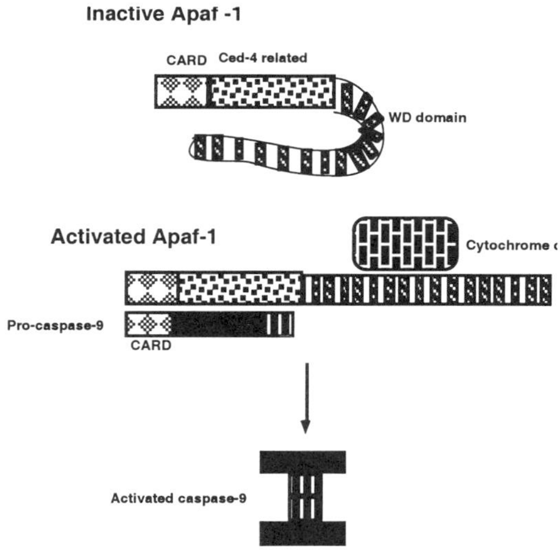

FIGURE 5.3. Cytochrome c activates the mammalian Ced-4 homologue, Apaf-1. Apaf-1 contains an N-terminal CARD domain, a Ced-4-like region, and a C-terminal stretch of WD repeats. Cytochrome c binds to the WD repeats and facilitates the interaction of Apaf-1 with pro-caspase-9.

from activating Ced-3 or caspases. These scenarios will be discussed below when discussing the mechanism by which Bcl-2 mediates its antiapoptotic function. In mammalian cells, the only known homologue of Ced-4 is a protein known as Apaf-1. Apaf-1 has a domain that is homologous to Ced-4 but also contains two additional domains, an N-terminal caspase recruitment or CARD domain that resembles the prodomain of certain caspases and a C-terminal domain that contains multiple WD repeats. Apaf-1 has the ability to interact directly with cytochrome c, which is released from mitochondria, and to associate directly with the head pieces of certain caspases, thereby activating the catalytic activity of the latter. Cytochrome c released from mitochondria presumably associates with the WD-rich C-terminus of Apaf-1, causing a conformational alteration in the latter, which then associates with caspases such as caspase-9. The association with caspase-9 is mediated via the N-terminal CARD domain as well as the Ced-4 homology region of Apaf-1. The activated caspase-9 or equivalent caspase then initiates the caspase cascade and apoptosis (Figure 5.3).

Caspase Specificity

Why are there so many caspases in vertebrates, and what do they do? Are all caspases involved in apoptosis? Vertebrate caspases may be categorized into three broad families based on their substrate specificity (Table 5.1).

TABLE 5.1. Mammalian caspase subfamilies

Subfamily	Other names	Consensus cleavage site	Function
Caspase-1 subfamily		(W/L) EHD	Inflammation
caspase-1	ICE		
caspase-4	ICE rel II, TX, ICH-2		
caspase-5	ICErel III, TY		
caspases-11, 13			
caspase-12			ER stress induced apoptosis
Caspase-8 subfamily		(I/V/L) EXD	Inducers of apoptosis
caspase-6	Mch2		? executioner
caspase-8	FLICE, Mach, Mch-5		Inducer
caspase-9	ICE-LAP6, Mch-6		Inducer
caspase-10	Mch-4, FLICE-2		Inducer
Caspase-3 subfamily		DEXD	Executioners of apoptosis
caspase-2	ICH-1		? inducer
caspase-3	CPP32, Yama, Apopain		
caspase-7	Mch3, ICE-LAP3, CMH-1		

Some caspases are involved in the maturation of pro-inflammatory cytokines, while others are key players in apoptosis. The primary function of caspase-1 (ICE), the prototypic member of the caspase family, is the cytosolic cleavage of precursor forms of IL-1β and of IL-18, permitting the generation of forms that can be exported from the cytosol by a signal peptide-independent mechanism. *Caspase-1* knockout mice have no developmental abnormalities, suggesting that this caspase is not critical for apoptosis. It is possible that the primary function of the caspase-1–related proteases—caspase-4 and caspase-5, and perhaps also caspases 11, and 13—is also to generate mature cytokines by cleavage after specific aspartate residues. Pro-caspase 12 is a member of this family which localizes to the endoplasmic reticulum (ER) membrane. This caspase may be required for ER stress induced apoptosis. The consensus tetrapeptide cleavage site for this group of caspases is (W/L)EHD.

The remaining members of the caspase family in vertebrates are believed to be involved in apoptosis, with certain members being of particular importance in specific cell types or in selected pathways. In the Fas pathway (discussed below), for instance, caspase-8 serves as an "initiator" of death signals, whereas caspase-3 is activated by caspase-8 and serves as an "executioner" protease. The caspase family members that are important for apoptosis can therefore be divided into caspase-8–like members, which cleave after the aspartate in proteins that contain a (I/V/L)EXD consensus tetrapeptide sequence, and caspase-3–like enzymes, which recognize a

DEXD consensus tetrapeptide sequence in target proteins. Caspase-8–like family members include caspase-10, caspase-9, and caspase-6. Caspase-3 may well be the prototypic "executioner" protease. Caspase-7 and caspase-2 are the other members of the caspase-3 subfamily of caspases. Caspase-2 has a large prodomain and may actually be an inducer. In general, caspases with large prodomains tend to be inducers of apoptosis, whereas those with small N-terminal prodomains tend to be executioners.

Most apoptotic signals are delivered along one of two broad pathways: one dependent on the activation of caspase-8 downstream of death receptors, and the other downstream of noxious stimuli that induce mitochondrial changes and caspase-9 activation (Figure 5.4). Caspase-8–deficient mice are resistant to the induction of apoptosis by TNF (tumor necrosis factor) or by triggering of Fas or DR3 (death receptor-3). However, fibroblasts from these mice can still be induced to undergo apoptosis by apoptogenic triggers such as UV-radiation, staurosporine, or etoposide. On the other hand, fibroblasts from caspase-9–deficient mice are sensitive to Fas but cannot be killed by dexamethasone or γ-radiation. There appear to be two

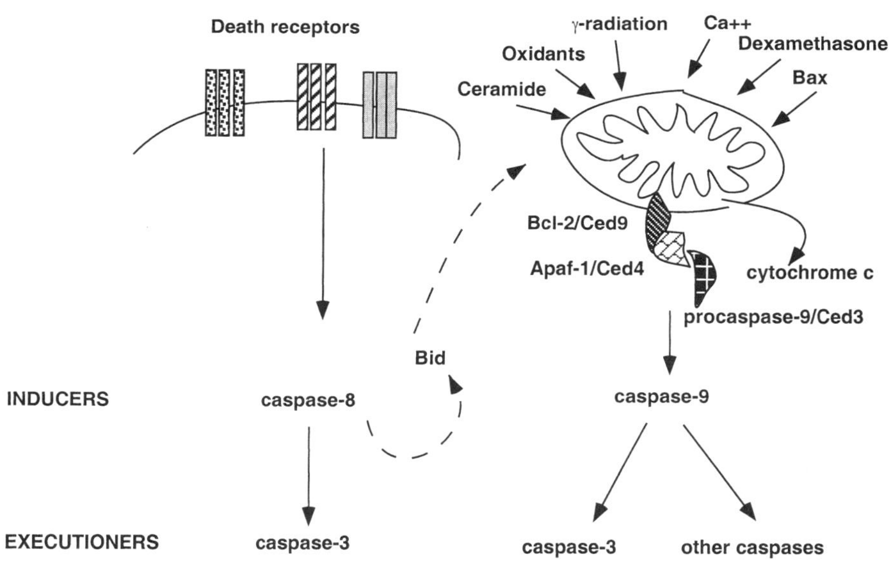

FIGURE 5.4. Two broad pathways for the induction of apoptosis. Death receptors such as the Type I TNF receptor, Fas, and death receptor 3 (DR3) trigger the activation of caspase-8, which is recruited to the receptors, as described in Figure 5.9. A number of other signals for the induction of apoptosis including exposure of cells to dexamethasone or γ-irradiation contribute to the activation of caspase-9. While caspase-8–dependent apoptosis is mediated largely via caspase-3, caspase-9 is upstream of caspase-3 in neuronal cells and of other executioner caspases in thymocytes.

types of caspase-9–dependent apoptotic pathways. In ES cells and in the central nervous system, caspase-9–mediated death requires caspase-3. However, in thymocytes, caspase-9 mediates apoptosis in a caspase-3–independent manner, presumably by triggering some other executioner caspase. Some of the phenotypes of caspase knockout mice are listed in Table 5.4, at the end of the chapter.

Inhibitors of Caspases

As will be discussed below, the major inhibitors of apoptosis are certain members of the Bcl-2 family that prevent caspase activation either by sequestering activators of caspases or by inhibiting the mitochondrial permeability transition. These proteins do not directly inhibit caspase activity. One of the strategies used by viruses to prevent apoptotic death of their hosts or to inhibit inflammatory responses is the inhibition of caspase activity. Some viruses have evolved proteins that can bind to caspases and inhibit their activity. Examples include the baculovirus protein p35 and the poxvirus protein CrmA. CrmA is a serpin that binds most efficiently to caspase-1 and caspase-8. In cowpox infections, its major role may be to prevent IL-1β and IL-18 production in infected cells and possibly to inhibit the Fas pathway or the initiation of apoptosis by other stimuli that activate initiator caspases such as caspase-8 or caspase-10. CrmA is not a particularly good inhibitor of the "executioner" caspases such as caspase-3, caspase-6, and caspase-7. p35 can inhibit the activity of caspase-8 and is also a potent inhibitor of caspase-3, caspase-6, and caspase-7.

The only endogenous caspase inhibitors in mammals are members of the IAP or Inhibitor of Apoptosis Protein family. One such protein is the human X-linked inhibitor of apoptosis (X-IAP), which can inhibit caspase-3 and caspase-7. The exact biological function of X-IAP remains unknown. It could be a modulator of caspase activity in apoptotic cells, or it could potentially protect certain cells such as cytotoxic T cells and possibly some professional antigen-presenting cells from apoptosis. Other IAPs, such as c-IAP-1 and c-IAP2, have been identified that bind to TRAF2. Since the Type I TNF receptor may recruit and activate caspase-8 to induce apoptosis, the recruitment of IAPs to the vicinity might contribute to the modulation of activated caspase activity. Both c-IAP1 and c-IAP2 have been shown to be induced by NFκB, and this may explain why this transcription factor is anti-apoptotic in certain cells.

An interesting competitive inhibitor of caspase-8 recruitment is known as c-FLIP (flice inhibitor protein) and has also been called Casper, I-FLICE, FLAME, or CASH). FLIP contains a death-effector domain and structurally resembles caspase-8 or FLICE. It will be discussed below in the section on Fas-induced apoptosis.

Regulation of the Common Death Pathway by Members of the Bcl-2 Family

An important inhibitor of apoptotic death in lymphocytes was first identified in the course of the molecular characterization of a chromosomal translocation between chromosomes 14 and 18 seen in patients with follicular lymphoma. This translocation brings the Ig heavy chain locus on chromosome 14 in apposition with a gene on chromosome 18. The gene activated at this breakpoint, and that is thus inappropriately expressed, plays a role in the process of oncogenic transformation, and is known as Bcl-2. The Bcl-2 protein is a 26-kd molecule that is found in the outer mitochondrial membrane as well as in the ER membrane. Bcl-2 inhibits apoptosis in a number of cell types, and this gene is the mammalian homolog of the *C. elegans ced-9* gene (which antagonizes death induced by *ced-4* and *ced-3*). Genetic analyses in the worm, as mentioned above, have revealed that *ced-4* is an upstream activator of *ced-3* (the worm caspase equivalent) and that *ced-9* is upstream of *ced-4* and can inhibit the activity of the latter.

A large family of Bcl-2–like proteins has been characterized in mammalian cells. Some members of the family are, like Bcl-2, antagonists of cell death, while others may induce apoptosis (Table 5.2). In terms of lymphocyte development, Bcl-2 and Bcl-X_L (the long form of Bcl-X) are particularly important antiapoptotic members of the Bcl-2 family. Bcl-X is a more readily induced member of this family and can be alternatively spliced to yield a long form (Bcl-X_L), which is antiapoptotic, and a short form (Bcl-X_S), which is proapoptotic. Bcl-2 may be viewed as a maintenance factor for survival, while Bcl-X_L is apparently induced by external signals when cells need to be instructed to survive. Accordingly, the levels of Bcl-2 appear to be downregulated during periods of lymphocyte development when cells need to be negatively or positively selected. In the B lineage, for instance, Bcl-2 levels are high in the earliest pro-B stages, but late in pro-B development, Bcl-2 levels drop and stay low through the pre-B and immature B

TABLE 5.2. Some proapoptotic and antiapoptotic members of the Bcl-2 family

Proapoptotic	Antiapoptotic
Bax	Bcl-2
Bak	Bcl-X_L
Bok	Bcl-w
Bad	NR-13
Bcl-X_S	Bfl-1
Bik	Brag-1
Hrk	Mcl-1
Bid	A1
EGL-1	CED-9

stages. This drop presumably is permissive for the elimination/selection of lymphocytes at late pro-B, pre-B, and immature B cell stages. Bcl-2 is expressed at relatively high levels in long-lived IgM^+/IgD^+ follicular B cells. In the germinal center, Bcl-2 levels drop once again in dark zone centroblasts and basal light zone centrocytes, presumably to permit selection processes to occur in the immediate aftermath of somatic mutation (Figure 5.5). Similarly, in the T lineage, Bcl-2 is expressed at relatively high levels in double-negative ($CD4^-/CD8^-$) T cells, levels drop in the double-positive stage ($CD4^+/CD8^+$) when positive and negative selection events occur, and high levels of this protein are expressed in peripheral $CD4^+$ and $CD8^+$ stages.

Bcl-X_L is structurally very similar to Bcl-2 (Figure 5.6) and is induced when B and T cells are activated. Bcl-X_L is expressed in pre-B cells at a time when signals are delivered via the pre-B receptor, and it is probably induced as part of the positive selection process. Other members of the Bcl-2 family that are antagonists of apoptosis include Bcl-w, A1, and Mcl-1.

Members of the Bcl-2 family have very similar structures in certain regions that have been described as the BH1, BH2, BH3, and BH4 regions. Bcl-2 and Bcl-X_L can form homodimers or heterodimerize with antagonists such as Bax. The BH1, BH2, and BH3 regions form a hydrophobic groove into which the BH3 domain of a partner might slide during the formation of dimers. Both Bcl-2 and Bcl-X_L have C-terminal transmembrane anchor domains. A relative excess of antiapoptotic Bcl-2 members in a cell is assumed to promote survival, while an excess of proapoptotic proteins such as Bax presumably leads to apoptosis. This "dueling dimer" model is still

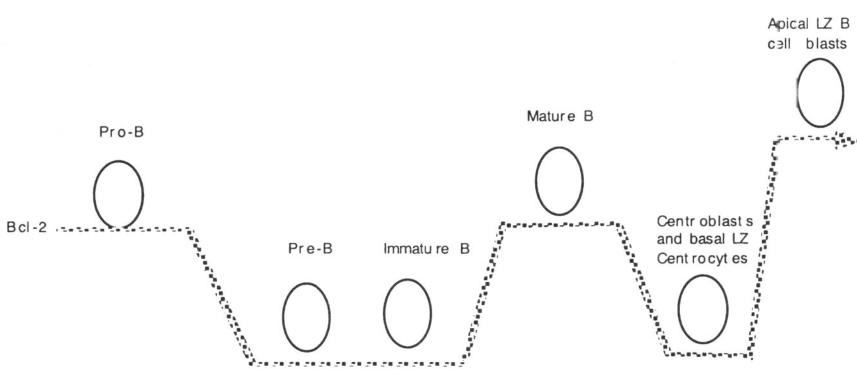

FIGURE 5.5. The levels of Bcl-2 are regulated during B cell differentiation. High levels of Bcl-2 are expressed in pro-B cells, at the IgM^{hi}/IgD^{hi} and IgM^{lo}/IgD^{hi} follicular B cell stages and in B cell blasts in the apical light zones of germinal centers. When cells need to go through a selection event, Bcl-2 levels may be reduced, facilitating a "default" death state.

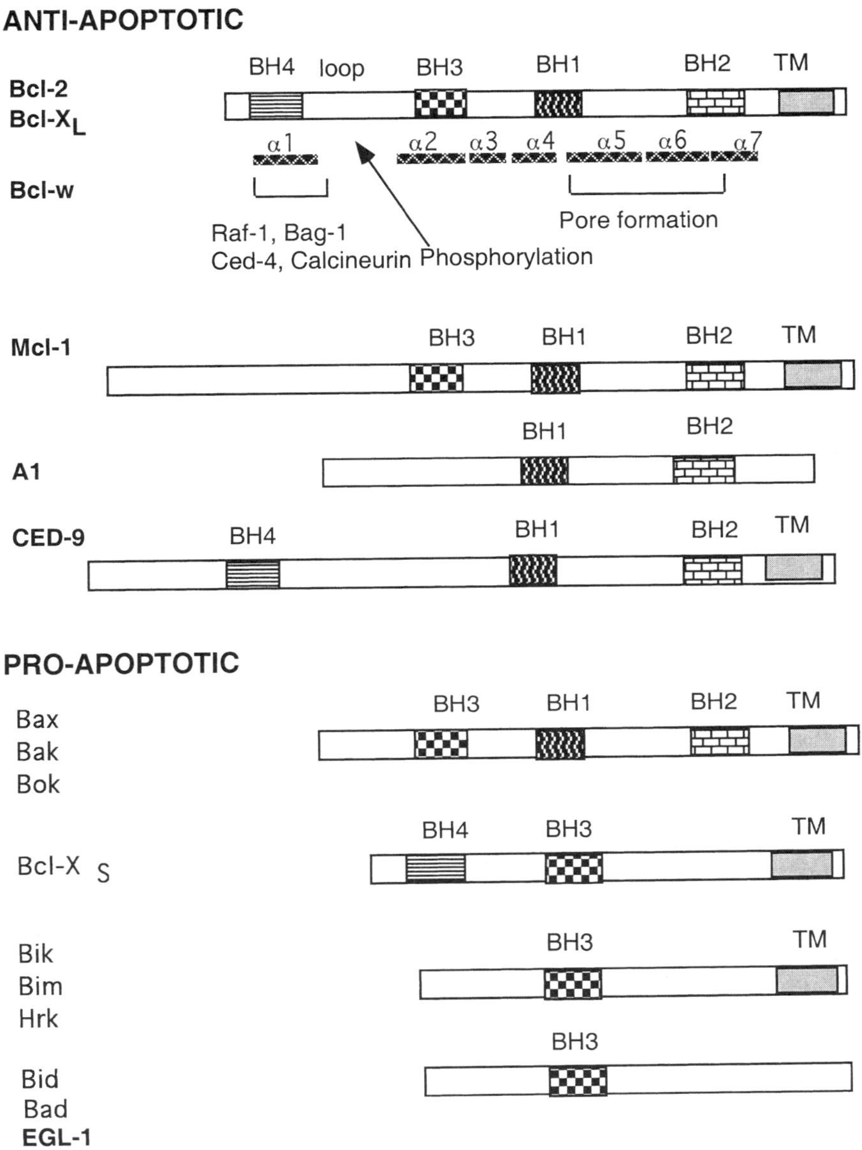

FIGURE 5.6. Structural features of Bcl-2 and family members. The BH1, BH2, and BH3 regions form a hydrophobic groove into which the BH3 region of a partner may slide. C-terminal hydrophobic anchors are present in most family members. The α5 and α6 helices are probably essential for the channel-forming activities of Bcl-2, Bcl-X_L, and Bax. The flexible loop region between the first two helices is a site for phosphorylation that contributes to negative regulation of function. The BH4 region may be essential for the association with Raf-1, Bag-1, Ced-4 and calcineurin. CED-9 is the *C. elegans* Bcl-2 homolog, and EGL-1 is a proapoptotic protein in nematodes.

likely to be generally valid, although there are now two distinct models for the topology and function of the Bcl-2 family of proteins.

A number of Bcl-2 family members function as death agonists. These include Bax, Bad, Bak, Bok, Bik, Bid, and Bcl-X_S. Bcl-X_S is an alternatively spliced version of Bcl-XL, that lacks the BH1 and BH2 regions, which play an important role in Bcl-2 and Bcl-XL. In general, death agonists can be divided into two classes. Proteins such as Bax, Bak, and Bok contain BH3, BH1, and BH2 domains. Most other death agonists contain a BH3 domain but no other areas of homology with Bcl-2. Some of these BH3 domain-containing proteins, such as Bik, Bim, and Hrk contain a C-terminal membrane anchor, whereas others such as Bad and Bid do not.

It is unclear whether Bcl-2 family members function primarily within intracellular membranes. Growing evidence suggests that both Bax and Bcl-2 can function as channels. Bax may function in the outer mitochondrial membrane to contribute to the mitochondrial permeability transition that accompanies apoptosis. Cyclosporin-inhibitable megapores formed during apoptosis are located in the inner mitochondrial membrane, but Bax channels in the outer mitochondrial membrane may play a role in the efflux of cytochrome c and AIF into the cytosol.

Bcl-X_L has also been demonstrated to form channels *in vitro*. The $\alpha 4$ and $\alpha 5$ helices are very hydrophobic and may potentially span the membrane, allowing the other five amphipathic helices to pen out in an umbrella-like fashion at the interface with the intermembranous space. While Bax channels might contribute to apoptosis, Bcl-2 and Bcl-X_L might form counter-channels that functionally antagonize pores formed by Bax, transporting cytochrome c back into the mitochondrion, for instance. They may contribute to occluding Bax channels when they form heterodimers (Figure 5.7). Recent evidence suggests that Bcl-X_L functions as an ion channel in the outer membrane of mitochondria, preventing osmotic swelling of this organelle. These channels may be cation selective and may transport potassium out of the mitochondrion to prevent swelling and subsequent rupture or leakiness of the outer mitochondrial membrane.

All channel-based models assume that Bcl-2 and at least one in every pair of Bcl-2–related proteins must be membrane anchored. Although most members of this family are Type II membrane proteins, in some experiments, removal of the transmembrane anchor at the C-terminus did not significantly compromise Bcl-2 function. One of the alternative models for Bcl-2 function suggests that membrane-anchored Bcl-2 may sequester Ced-4 associated with caspase-9/Ced-3 and prevent caspase-mediated cleavage of death targets. When Bcl-2 associates with a death-inducing partner such as Bax, it can no longer hold back the Ced-4–caspase, complex and death ensues. Ced-4 has been shown to interact with both Ced-9 and mammalian Bcl-X_L and to form a physical bridge between Ced-9 and Ced-3. In *C. elegans*, a BH3 domain-containing protein called EGL-1 may promote apoptosis by displacing a complex of Ced-4 and Ced-3 from

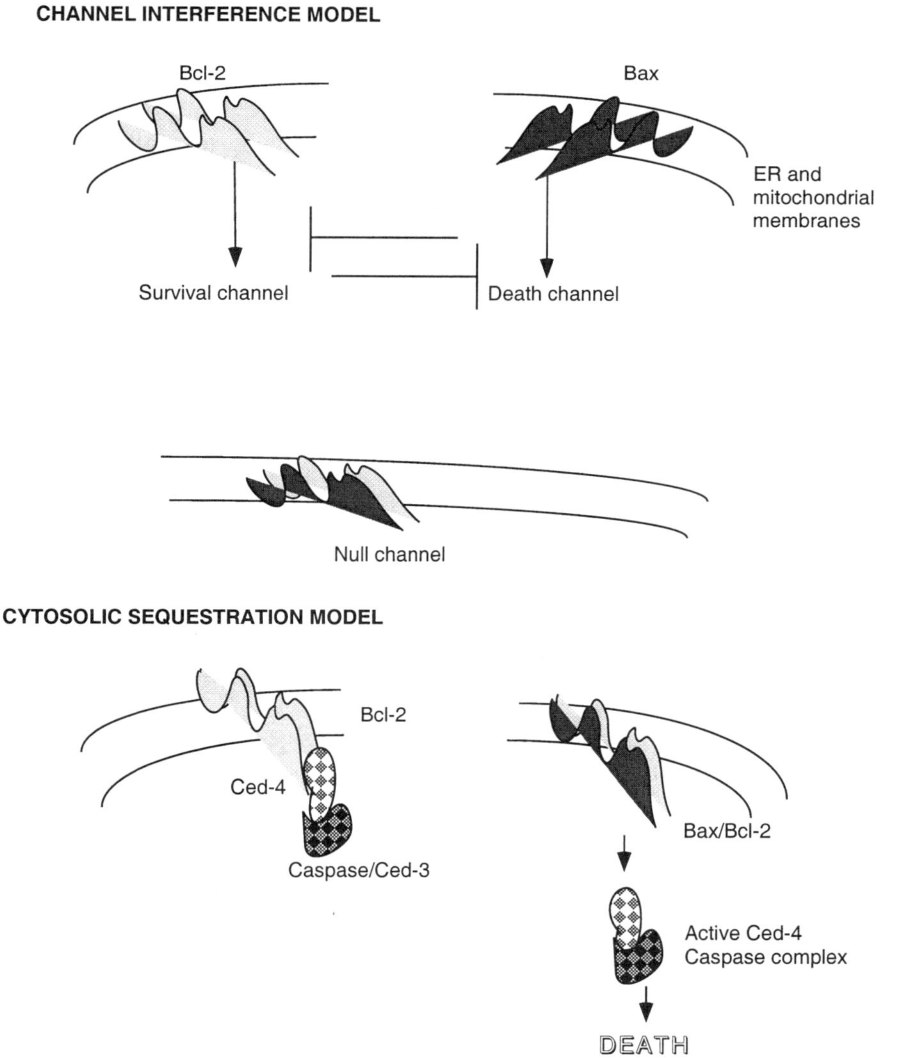

FIGURE 5.7. Mechanisms of anti- and proapoptotic activity of Bcl-2 family members. Regulation by phosphorylation is not considered in this figure. Bax and other death agonists may form channels permitting the efflux of apoptogenic proteins into the cytosol. Bcl-2 may form channels that antagonize Bax channels or may heterodimerize with Bax to neutralize Bax channel activity. Alternatively, Bcl-2 might, in a channel-independent manner, sequester Ced-4–caspase complexes, preventing death.

membrane-attached Ced-9, thus permitting the activation of Ced-3 in the cytosol.

How do signals from the cell surface influence the activity or function of Bcl-2 family members? The expression of Bcl-2 can be induced in germinal center B cells by signals that promote the selection and survival of B cells. The molecular basis for the upregulation of Bcl-2 expression—presumably by antigen receptor signaling collaborating with CD40—is poorly understood. A Bcl-2 homolog, A1, is expressed in myeloid cells that have been induced by granulocyte-macrophage colony–stimulating factor. A1 is also expressed in mature peripheral B cells. Another Bcl-2 homologue that is expressed in myeloid cells is Mcl-1. The *mcl-1* gene can be induced in these cells by phorbol esters. In the case of apoptosis that is induced in a p53-dependent manner (in response to DNA damage or γ-irradiation), an important event appears to be the transcriptional induction of *bax*. The *bax* promoter can be activated by p53 and contains a p53-binding site.

Bcl-2 itself can be inactivated by serine phosphorylation of residues in the loop region. How exactly this event is regulated by extracellular signals remains to be established. Growth factor receptors may sustain cell survival by inducing the serine kinase-dependent phosphorylation of proapoptotic members of the Bcl-2 family such as Bad. Bad lacks a C-terminal hydrophobic domain. In its active nonphosphorylated form, it can associate with Bcl-2 or Bcl-X_L and neutralize their antiapoptotic activity. However, when Bad is phosphorylated on a critical serine residue in the flexible loop region between the first two α-helices, it preferentially associates with 14-3-3 proteins in the cytosol and can no longer efficiently inhibit Bcl-2 or Bcl-XL. The role of PI3 kinase and Akt in the serine phosphorylation and inactivation of Bad is discussed later in this chapter.

The N-terminal BH4 domain of Bcl-2 and Bcl-X_L is essential for the association of these survival factors with c-Raf, Bag-1, calcineurin, and Ced-4. This BH4 structure is found in antiapoptotic members of the Bcl-2 family and not in proapoptotic relatives. The specific association of death-antagonizing Bcl-2 family members with Ced-4 and the consequent inhibition of caspase activation may best explain why BH4 domains are seen only in antiapoptotic members of the Bcl-2 family. It has also been argued that the recruitment of c-Raf and calcineurin may help Bcl-2 and Bcl-X_L mediate survival functions. c-Raf is a serine/threonine kinase and may phosphorylate Bad, thus inhibiting the ability of the latter to induce apoptosis. How calcineurin recruitment to Bcl-2 or a Bcl-2-like protein might facilitate survival is less clear. It has been suggested that cytosolic calcineurin, a Ca^{++}-regulated phosphatase, may promote apoptosis by dephosphorylating Bad, suggesting that sequestration by Bcl-2 in intracellular membranes might counter this.

Bcl-2 was originally discovered as an oncogene, mediating cell survival rather than proliferation, and was constitutively expressed in a B lymphoma because its expression was driven by the Ig heavy chain enhancer. It is of interest to note that during evolution a number of viruses have co-opted Bcl-2–like proteins, and these proteins permit the transformation of mammalian cells by these viruses. Notable examples of Bcl-2–like proteins include the adenovirus E1B 19K protein and the Epstein-Barr virus BHRF-1 protein. Both of these proteins functionally mimic Bcl-2. The African swine fever virus also has a Bcl-2 homologue, LMMW5-HL, which has not been extensively studied.

Chromatin Condensation and Nuclear Fragmentation

One of the major hallmarks of apoptosis, as noted at the beginning of this chapter, is the altered shrunken morphology of the nucleus in which chromatin is condensed and clumped. This process is called pyknosis, and it is often accompanied by fragmentation of the nucleus or karyorrhexis.

Chromatin condensation and DNA fragmentation are generally caspase-dependent events, but they may also be initiated without activating caspases. We have previously discussed the cleavage by executioner caspases of ICAD. This cleavage event releases CAD, which enters the nucleus (if it is not already in the nucleus), can facilitate the condensation of chromatin, and cleaves DNA into nucleosome size (approximately 200-bp) fragments.

Another protein, called Acinus (apoptotic chromatin condensation inducer in the nucleus) is also activated in apoptotic cells by caspase-dependent and independent cleavages. Caspase-3 in combination with unidentified proteases cleaves Acinus into a small fragment that migrates to the nucleus and induces chromosomal condensation.

A protein that we have discussed earlier in this chapter, AIF, may contribute to the peripheral condensation of chromatin and the cleavage of DNA into large, approximately 50-kb, fragments. AIF is released from the intermembrane spaces of mitochondria during apopotosis.

Dipsosal of the Corpse

Apoptotic bodies as well as apoptotic cells that have not been fully packaged into apoptotic bodies are phagocytosed by macrophages. Macrophages recognize surface changes on apoptotic cells, including the increased levels of phosphatidylserine. They ingest apoptotic cells and apoptotic bodies but are not activated in the process and therefore do not induce inflammation. Macrophages may use a number of receptors, including the vitronectin receptor or a specific receptor for phosphatidylserine, to ingest apoptotic cells. Other cell surface proteins that have been implicated in this event include the CD36 adhesion molecule, the *Drosophila* croquemort protein, class A scavenger receptors, the ABC-1 transporter (which is an ATP-binding cassette protein structurally homologous to TAP-1 and TAP-2, the transporters associated with antigen presentation, in the major histocompatibility complex class I pathway), and macrophage lectins.

A number of *C. elegans* genes have been identified whose products are involved in the disposal of the apoptotic corpse. The worm proteins involved in this disposal process include CED-1, CED-2, CED-5, CED-6, CED-7, and CED-10. CED-7 has been cloned and is a nematode ABC transporter. It may be the *C. elegans* homologue of ABC-1. It is likely that CED-7 and ABC-1 mediate some transport function in macrophages that is required during the engulfment process. Cytoplasmic proteins involved in the process of phagocytosis are likely to be involved in the engulfment process. Rho family proteins and regulators are obvious candidates. CED-5 is homologous to human Dock 180, and this large adaptor protein participates in the activation of Rac. It may be involved in regulating cytoskeletal rearrangement during the engulfment process. CED-6 is an adaptor protein that contains an N-terminal PTB domain and a C-terminal proline/serine-rich region. It is likely that tyrosine kinase activity is important during the engulfment process. Genetic analysis in *C. elegans* suggests that *ced-1* and *ced-7* are upstream of *ced-6*.

Signals That Lead to the Common Death Pathway

A vast number of different kinds of signals, or the lack of signals, may lead to the initiation of apoptosis in different cell types. There are three broad categories of mechanisms by which apoptosis may be regulated in lymphocytes and in their cellular targets, and these are summarized in Table 5.3. In one category, apoptosis may be "directly" triggered by some signals from the cell surface or the cytosol without the need for any new gene expression. In some circumstances, which constitute the second category of

TABLE 5.3. Signals that regulate the common death pathway during lymphocyte development

1. "Direct" activation of death
   a. Death mediated by Fas, TNFRI, and other death domain-containing receptors (Activation-induced cell death, bystander death, T cell-mediated homicide)
   b. Death mediated by granzyme B (cytotoxic T cells, NK cells)
2. Death secondary to the induction of gene expression
   a. Ionizing radiation, UV, and the p53 pathway
   b. The death pathway in immature lymphocytes—Myc, E2F1, and Nur 77 (negative selection)
   c. Glucocorticoid receptor-mediated death (? Death by neglect)
3. The inhibition of default death
   a. Growth factors, PI3 kinase, Akt, and Bad phosphorylation
   b. NFκB and cell survival
   c. The inhibition of default death by preantigen and antigen receptors (positive selection I and II)

mechanisms that we will consider, transcriptional activation is a necessary prelude to apoptosis. The final set of regulatory mechanisms that we will discuss reflect signals that overcome pathways for "default" death.

The classification used below is somewhat arbitrary. For instance, while the Fas ligand (FasL)–Fas pathway appears to represent a "direct" pathway for apoptosis, it is worth keeping in mind that the induction of the FasL on activated T cells requires transcriptional activation. As we shall see, the activity and availability of FasL may be modulated by transcriptional activators such as c-myc and by Jun N-terminal kinase (JNK)-mediated signals. Similarly, the induction of death by transcription factors such as p53 may depend on their ability to modulate direct pathways of death activation such as the Fas pathway or the related TRAIL/DR5 pathway (considered below).

"Direct" Activation of the Common Death Pathway

The two best examples for the "direct" activation of the death pathway are death induced by Granzyme B and death induced by Fas. Cytotoxic T cells and NK cells induce apoptosis by injecting an activator of the caspase cascade into the cytosol of target cells (Figure 5.8). Perforin secreted by these killer lymphocytes inserts itself into target membranes and generates

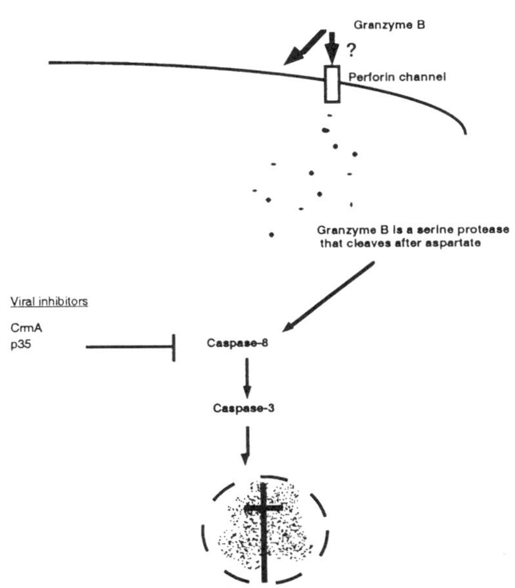

FIGURE 5.8. Apoptotic death induced by Granzyme B. Cytotoxic T cells and NK cells introduce Granzyme B into target cells perhaps via perforin channels. Granzyme B is a serine protease that cleaves after aspartate residues and activates the caspase cascade.

channels that can cause osmotic lysis of the target cells. Another protein extruded in the granules of killer lymphocytes in Granzyme B, which is a serine protease with the unique ability (among noncaspase proteases) to cleave targets after aspartate residues. The activation of the caspase cascade by this protease leads to the death of target cells.

Death Mediated by Fas

Another example of the "direct" activation of caspase activity is via the Fas pathway. The FasL is a trimeric type II membrane protein expressed on the surface of activated T cells (but may also be expressed by some other cells). Fas is a member of the TNF receptor family discussed in Chapter 4 and is a membrane protein expressed by most cells in vertebrates. In B and T cells, the levels of Fas increase following antigen-mediated activation. FasL induces the crosslinking of Fas to induce signal transduction. The cytoplasmic tail of Fas contains a protein–protein interaction motif, which is known as a *death domain*. This domain facilitates the interaction of Fas with another death domain found in the C-terminus of a protein known as FADD (Fas-associated death domain), also known as MORT1. The Fas–FADD interaction occurs via a homotypic death domain-mediated interaction. FADD serves to recruit and activate caspase-8, also known as FLICE (Fas-linked interleukin-1 cleaving enzyme) or MACH Mort-1 associated ced3 homolog, which initiates caspase-3 activation and subsequent apoptosis (Figure 5.9). The interaction between FADD and caspase-8 involves the N-terminal domain of FADD and the N-terminal propiece of caspase-8. TNFR1 can also interact with FADD via an adaptor protein called TRADD.

FasL–Fas interaction may not only activate death via the direct activation of caspase-8 but may also separately activate the JNK/SAPK pathway via a molecule known as Daxx. Daxx itself lacks a death domain, but the C-terminal portion of Daxx associates with the death domain of Fas. Another segment of Daxx binds and activates a JNKKK (a JNK kinase kinase) called ASK1 (apoptosis signal-regulating kinase 1) and thus contributes to JNK/SAPK activation and apoptosis. While FADD-mediated apoptosis cannot be blocked by Bcl-2, apoptosis mediated via Daxx and JNK may be inhibited by this survival factor. Inhibition of caspases does not block JNK activation but does abrogate Daxx-induced death, suggesting that caspase activation may occur downstream of JNK.

CTLs, NK cells, and the occasional CD4 cell that assumes a killer phenotype all use the Fas pathway for disposing of targets. Fas is critical for the phenomenon of activation-induced cell death (AICD), which modulates T cell responses and which will be discussed in some detail in Chapter 10. It is also of importance in eliminating nonspecifically triggered "Bystander" B cells, a phenomenon that will be discussed in Chapter 8. The inhibitor

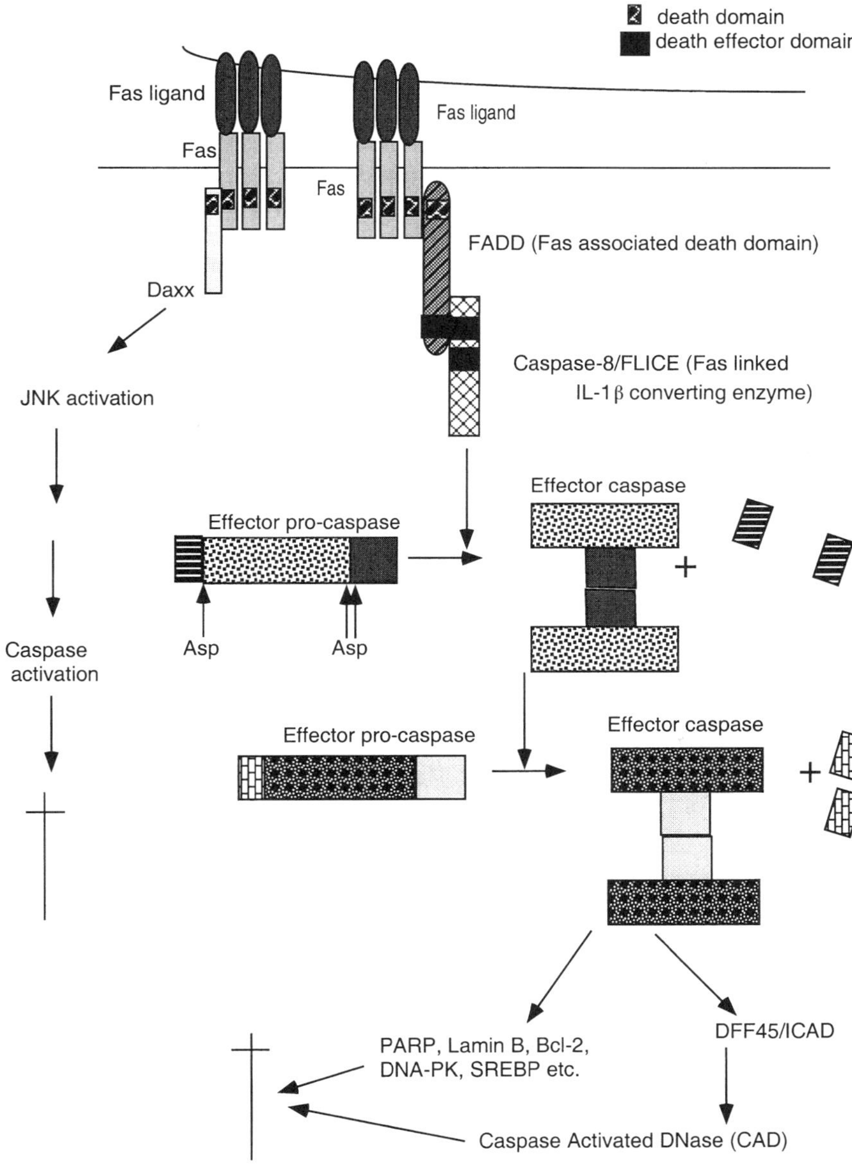

FIGURE 5.9. Fas activates caspases separately via FADD and Daxx. FADD interacts with caspase-8, while Daxx interacts with a JNK kinase kinase called ASK1 and activates the JNK pathway. In certain cell types, JNK activation can result in the induction of apoptosis.

protein, FLIP, structurally resembles caspase-8 but is not a protease. It functions as a competitive inhibitor of Fas signaling. It can bind to FADD and inhibit the recruitment of caspase-8. FLIP may be induced by IL-2 receptor signaling and thus potentiate AICD. It may be induced by the Ras/Raf/ERK pathway downstream of the BCR and thus protect antigen-specific B cells from being killed when activated T cells expressing FasL interact with B lymphocytes.

Two naturally occurring mouse mutations that have been studied in great detail are the *lpr* mouse (which carries a recessive mutatation in the *fas* gene) and the *gld* mouse (which has a mutant *fasL* gene). Homozygous mutation in these genes lead to hypergammaglobulinemia, autoantibody production, vasculitis, arthritis, glomerulonephritis, and lymphoproliferation. The proliferating cells are $CD3^{+}CD4^{-}CD8^{-}$ cells, which are derived from single-positive peripheral T cells, which were not eliminated by AICD. Exactly why they downregulate coreceptors is unclear; perhaps signals similar to those that occur during lineage commitment and coreceptor downregulation are responsible for this phenotype. The autoimmune manifestations in these mice presumably result from the failure to eliminate self-reactive T cells by AICD and also a failure to eliminate some bystander B cells that might be autoreactive. Other mice with a similar phenotype include mice that lack IL-2 or the IL-2Rα chain, discussed in Chapter 4. A rare human disease involving fas mutations with many features reminiscent of the *lpr* mouse has been described and will be discussed in Chapter 12. Interestingly, mice engineered to lack FADD differ from the *lpr* and *gld* mice in many ways (see Table 5.4 at the end of the chapter). FADD is required during embryonic development and $FADD^{-/-}$ mice day at embryonic day 9. In $Rag\text{-}1^{-/-}$ reconstituted mice, FADD is absolutely required for Fas-mediated apoptosis. However, lymphoproliferation is not observed, apparently because FADD is somehow also required for T cell proliferation.

Fas does not appear to be critical for negative selection of T cells and B cells, though this issue remains to be resolved. Although FasL appears to be expressed mainly by activated T cells and perhaps in some "immunologically privileged" tissue sites in which it is believed to serve a "veto" role eliminating autoimmune T cells, a more general but yet to be substantiated role for Fas–FasL interactions has been recently suggested. In a number of cell types, expression of high levels of c-myc leads to apoptosis (as will be discussed below). It has been suggested that c-myc is somehow responsible for inducing FasL levels in nonlymphoid cells and thus contributes to the induction of suicide partly via a Fas–FasL mechanism. It may well be that Fas plays a much wider role in apoptosis than it is currently given credit for.

A few other receptors appear to use pathways similar to those used by Fas to induce apoptosis. One of these is TNFRI, which activates caspase-8 via TRADD and FADD mentioned above. Although TNF plays an important role in lymphocyte development, it is unclear as to when apoptosis mediated by TNF is actually important *in vivo*. A few other TNFR family death domain receptors have been identified. These include DR2, DR3, DR4 and DR5 (for death related). The ligand for DR4 and DR5 is a protein called TRAIL (TNF related apoptosis inducing ligand). TRAIL may be important in p53-mediated apoptosis.

Apoptosis Linked to Gene Activation

In some forms of apoptosis, transcriptional activation of certain genes is a necessary prelude to the activation of the common death pathway. We will consider a few selected examples to illustrate potential apoptotic pathways that depend on the induction of gene expression.

p53-Mediated Apoptosis

An intensively studied apoptotic pathway that is relevant to most vertebrate cells is the pathway downstream of p53. p53 is a tumor-suppressor gene that is mutated in a wide variety of human tumors. This gene encodes a transcription factor whose primary role is to respond to genotoxic agents and induce either cell cycle arrest or apoptosis. There are a number of reasons to connect p53 pathways with lymphocyte development. DNA-damaging agents have been noted to stimulate V(D)J recombination in thymocytes (but not in B cell precursors). It has been suggested that p53 might play a role in B cell development from studies in cell lines. p53-mediated cell cycle arrest of pre-B cells has been suggested as a mechanism for further differentiation in the B lineage (discussed in Chapter 7). However, there is no obvious defect in B or T cell development in the p53 knockout mouse. Ataxia telangiectasia is a disease in which immunodeficiency and lymphoid tumors are frequently seen (see Chapter 12 for details). The ATM (ataxia telangiectasia mutated) gene product is, in some stress situations, critical for the activation of p53. Understanding p53-dependent pathways (Figure 5.10) is of importance when considering any cell type, but it is also of importance from the standpoint of lymphocyte development.

Genotoxic agents—either by activating the ATM protein (a protein serine/threonine kinase as well as possibly a lipid kinase of the PI-3 kinase family) or by other unknown means—may mediate the stabilization of the p53 protein. The increase in p53 levels can lead to either cell cycle arrest or to apoptosis. p53 probably mediates cell cycle arrest by transcriptional activation of the p21 gene, which encodes a CDK inhibitor. p53 may also mediate apoptosis by a number of different possible means. In some cells, apoptosis may be mediated by the induction of Bax. This protein may generate pores in the outer mitochondrial membrane or may induce caspase activation by functionally inactivating Bcl-2 or related antiapop-

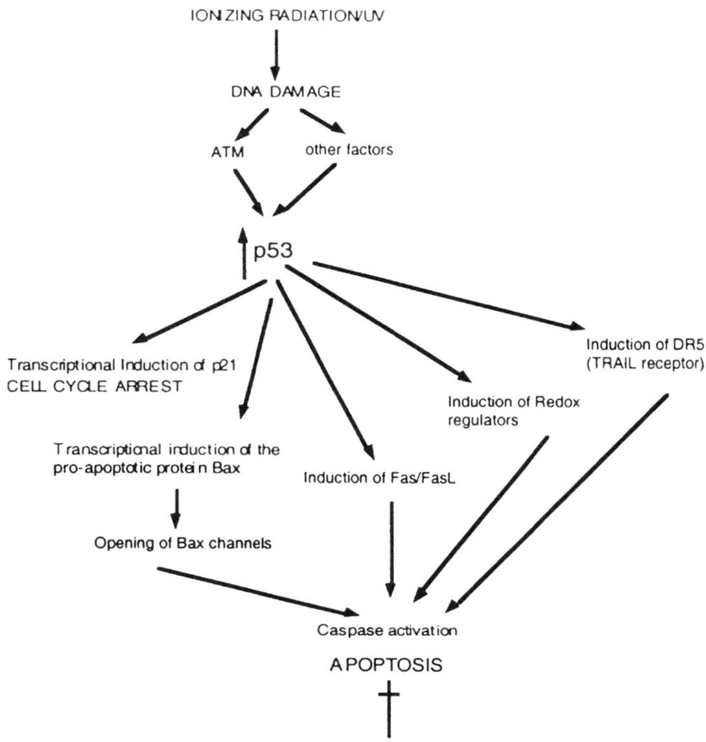

FIGURE 5.10. p53 is induced following DNA damage and transcriptionally activates targets that help mediate cell cycle arrest and apoptosis.

totic proteins. Alternatively, p53 has been suggested to induce a number of genes that might contribute to the generation of reactive oxygen species that could induce the mitochondrial permeability transition and apoptosis. p53 has also been implicated in the induction of DR5 (also referred to as KILLER), which is a receptor for the death-inducing TRAIL ligand (a member of the TNF superfamily). Another suggested mechanism involves the potential activation by p53 of the FasL–Fas pathway, perhaps via the induction of FasL.

Transcription Factors, Cell Cycle Progression, and Apoptosis

Transcriptional regulators have been implicated in the apoptotic death of immature lymphocytes that are in the process of being positively or negatively selected during ontogeny. These processes will be considered in some detail in Chapters 7 and 9. Transcription factors that may be involved in negative selection are discussed here briefly because of the connections that have sometimes been made between cell cycle progression and apoptosis.

Myc is a transcription factor that mediates both proliferative and apoptotic functions. On the one hand, Myc is believed to be necessary for G1 progression as a result of its ability to induce Cdc25A, the phosphatase responsible for the dephosphorylation of the negative regulatory tyrosine residues (Thr14, Tyr15) in cyclin-dependent kinases. In keeping with such a role, the inability to induce c-myc in an immature B cell line, WEHI 231, has been linked to the proclivity of this cell line to undergo programmed cell death following antigen-receptor ligation. The existence of an NFκB site in the c-myc promoter has led to the demonstration that NFκB may be a critical regulator of c-myc, and that because they fail to activate NFκB, WEHI 231 cells are particularly prone to die in response to antigen-receptor ligation. While these findings point to an anti-apoptotic role for NFκB and c-myc, the overexpression of c-myc in a number of cell types has been correlated with the induction of apoptosis. The pro-apoptotic function of c-myc has been linked to the induction of FasL and the induction of Fas–FasL mediated apoptosis.

During G1 progression, the activation of Cdk4 and Cdk6 leads to the hyperphosphorylation of Rb, as discussed in Chapter 4. Hyperphosphorylated Rb releases the E2F transcription factors, E2F-1, E2F-2, E2F-3, and E2F-4. While these transcription factors in general promote S phase progression, a number of studies have established that E2F-1 actually functions to promote apoptosis. Negative selection in the thymus might be impaired in E2F-1 null mice, and death might be induced by this transcription factor only in late G1 or during S phase. It is therefore conceivable that in some circumstances apoptosis may be promoted by factors (such as c-myc) that actually promote G1 progression. It is unclear as to what the targets of E2F-1 are *in vivo* that promote apoptosis and how these unknown targets might connect with the common death pathway. Some of these issues are depicted in Figure 5.11.

Nur-77 and Thymocyte Apoptosis

Another transcription factor that has been implicated in apoptosis in T cells is a member of the steroid-receptor family called Nur-77. Nur-77 was originally identified as an immediate early gene, which is transiently induced by serum, growth factors, and nerve growth factor (NGF). Nur-77 and a related protein, Nor-1 are activated to high levels during T cell-receptor (TCR)–mediated apoptosis. A dominant negative version of Nur-77 can block negative selection, and a Nur-77 transgene greatly enhances apoptosis of thymocytes. Transcription initiated from the nur-77 promoter can be regulated by a factor known as MEF-2 (myocyte enhancer factor-2). TCR signaling leads to an increase in intracellular calcium and the calcium-calmodulin dependent release of MEF-2 from a nuclear attenuator of signaling called Cabin 1. At the same time, calcineurin is activated by calcium-calmodulin complexes and this leads to the dephosphorylation and activation of NFAT proteins. NFAT and MEF-2 proteins collaborate to

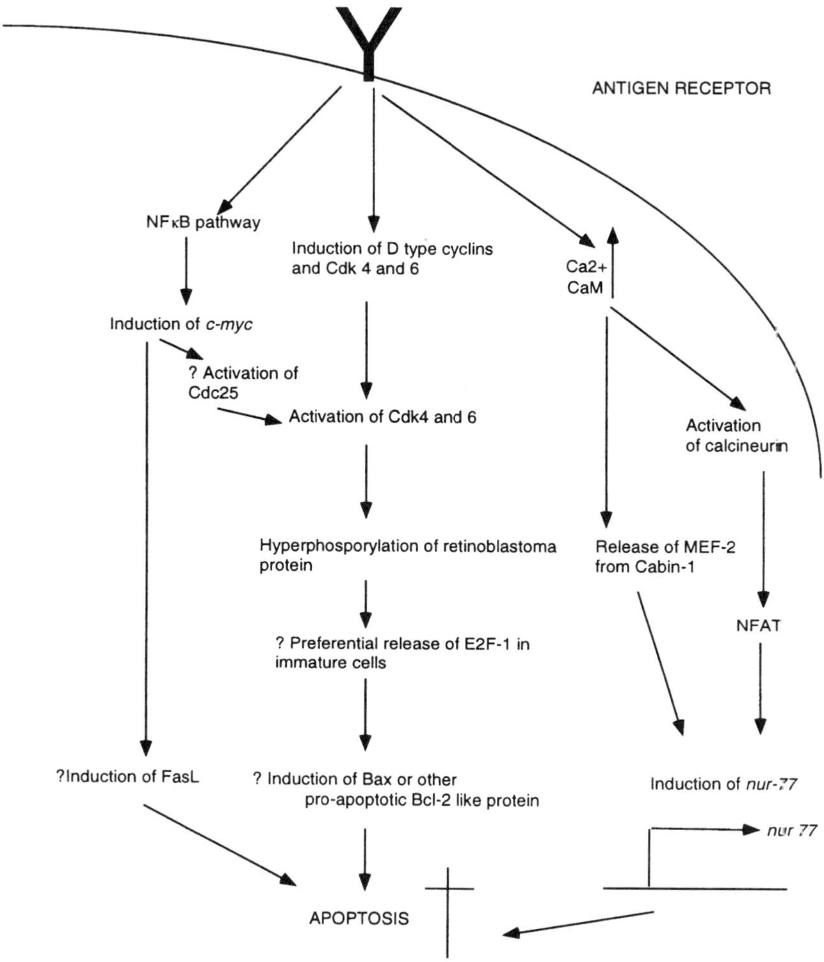

FIGURE 5.11. Transcription factors may induce cell cycle progression as well as apoptosis. Progression through the cell cycle may sometimes facilitate the induction of apoptosis. Calcium signaling may induce the activation of NFAT as well as of MEF-2, which cooperate to induce transcription of nur 77. While in some cells c-Myc might induce FasL expression, the exact targets c-Myc, E2F-1, or Nur-77 that contribute to apoptosis remain to be identified.

induce high levels of Nur-77. The targets of Nur-77 that are responsible for apoptosis remain to be identified.

Glucocorticoids and Thymocyte Apoptosis

It has long been observed that immature double-positive thymocytes are exquisitely susceptible to the induction of apoptosis by high doses of exogenous glucorticoids. At times of systemic stress, high levels of endogenous steroids may induce thymic atrophy. However, lower levels of steroid may serve an antiapoptotic function. Endogenously produced steroids in the thymus may, at their physiological low levels, antagonize TCR signaling in double-positive thymocytes induced by moderate-to-low–avidity ligands and thus facilitate positive selection. TCR signaling induced by high-avidity ligands may be too strong to be abrogated by glucocorticoids, and it has been suggested that endogenous glucocorticoids may thus help set thresholds for TCR signaling to mediate deletion versus positive selection. It has been suggested that "death by neglect" in the thymus of T cells that cannot recognize self-MHC with any affinity might depend on T cells being triggered by endogenous steroids.

Signals That Abrogate Default Apoptosis

A number of pathways are believed to be antiapoptotic. Many of these pathways culminate in the activation of an antiapoptotic member of the Bcl-2 family or the inactivation of a proapoptotic member of the same family. Early in T cell development, activation of the IL-7 receptor provides survival signals, possibly by influencing the levels of Bcl-2. Antigen and preantigen receptor signals can presumably contribute to the upregulation of Bcl-X_L. As noted in the section above, NFκB activation has been linked to the prevention of apoptosis. It is possible that NFκB contributes to the induction of antiapoptotic Bcl-2 family members, but this remains to be formally established. NFκB could help abrogate apoptosis via the induction of c-myc, as discussed above, a phenomenon that has so far only been shown to be relevant in a B cell line. It has been suggested that in some cells NFκB activated downstream of TNFRI (TNF receptor type I) might inhibit apoptosis by the transcriptional activation of c-IAP1 and c-IAP2. A few growth factor receptors including the IL-3 receptor have been shown to prevent the induction of apoptosis by negatively regulating the function of proapoptotic proteins such as Bad and caspase-9 via a PI3K-dependent survival pathway, which may also contribute to the activation of NFκB and forkhead transcription factors, as outlined in Figure 5.12.

Growth factors such as IL-3 activate the PI3 kinase pathway. Activation of PI3 kinase may occur by a number of means including recruitment of the regulatory p85 subunit to phosphotyrosinylated proteins or via Ras

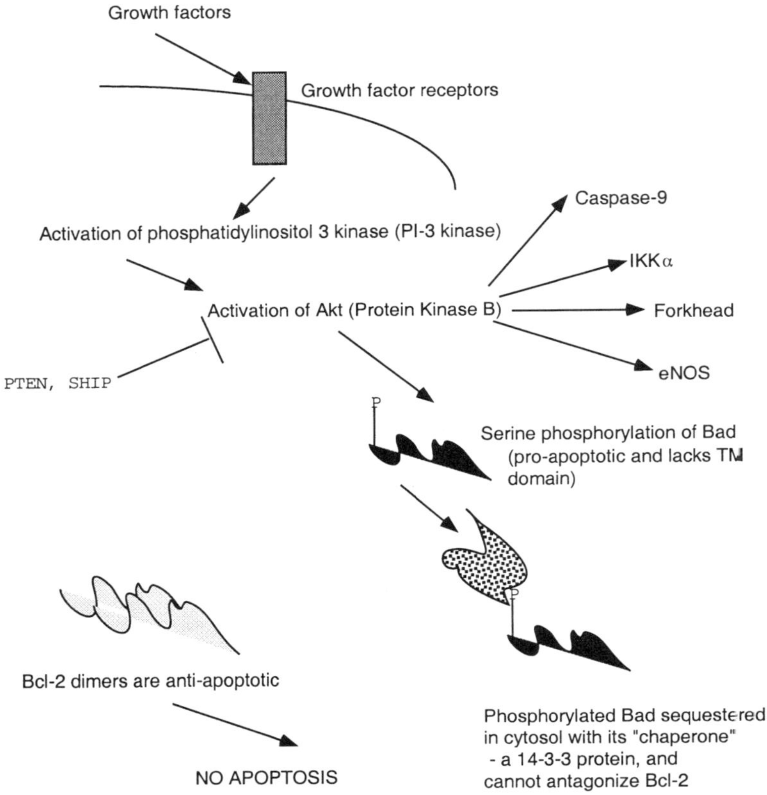

FIGURE 5.12. Abrogating default death by interfering with dueling dimers and other apoptotic mechanisms. Growth factor receptors may activate PI3 kinase. PIP3 generation may be opposed by SHIP and by protein tyrosine phosphatase with homology to tensin (PTEN), as described in Chapter 4. Activation of Akt may contribute to the abrogation of death signals in a number of ways. Potential consequences of Akt activation include the inactivation of Bad, the activation of NFκB, the inactivation of procaspase 9, the reduced synthesis of FasL (as a consequence of forkhead phosphorylation), and the activation of eNOS or endothelial NO synthase.

activation. These events have been discussed in Chapter 4. A target of PI-3 kinase is the serine threonine kinase known as Akt (also known as protein kinase B or RACα). Akt can phosphorylate a proapoptotic Bcl-2 family member known as BAD on a critical serine residue. BAD lacks a transmembrane domain and can interfere with the function of Bcl-2 or Bcl-X_L. When BAD is phosphorylated on serine 112 or serine-136 by Akt or by other serine-kinases, it is inactivated. Inactivation involves the sequestering of BAD by a protein known as 14-3-3, which prevents BAD from associating with and inhibiting the function of Bcl-2.

Another way in which Akt contributes to cell survival is by the induction of pro-caspase-9 phosphorylation. Phosphorylation of a critical serine

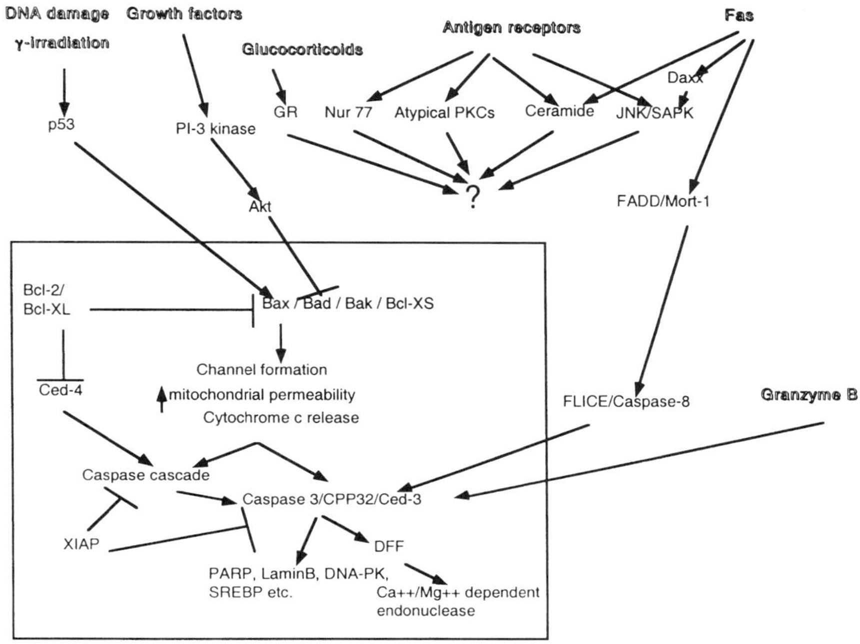

FIGURE 5.13. An overview of the signals that participate in the induction or prevention of apoptotic death.

residue (serine-196), on pro-caspase-9, prevents the activation of this procaspase in a cytochrome c-dependent manner. Akt may also contribute to the induction of antiapoptotic transcriptional regulators, particularly members of the winged-helix/forkhead family. Phosphorylated forkhead proteins may be retained in the cytosol and thus the transcription of specific proapoptotic genes including the gene encoding the Fas ligand may be inactivated. Akt can also phosphorylate and activate IKKα (see Chapter 4), thus contributing to the activation of NFκB. A less well understood mechanism of protecting cells from apoptosis may depend on the Atk-dependent activation of endothelial nitric oxide synthase, and the increased production of endothelial NO.

Summary

Apoptosis is a critical regulatory event during lymphocyte development. Numerous specialized pathways impinge upon a common death pathway that mediates programmed cell death. This common death pathway involves the activation of executioner caspases, generally by signals that promote the release of cytochrome c from mitochondria or by signals that activate inducer caspases. Caspase activation is regulated largely at the level of the

TABLE 5.4. Mouse mutations linked to apoptosis

Mutation	Phenotype	Reference
caspase-3$^{-/-}$	Perinatal death. Reduced apoptosis of neuroepithelial cells. Aberrant brain development.	Kuida et al., 1996
caspase-8$^{-/-}$	Death Ed13. Fas-, TNFR1-, and DR3-mediated apoptosis is abrogated.	Varfolomeev et al., 1998
caspase-9$^{-/-}$	Reduced apoptosis and caspase activation via the mitochondrial pathway. Fas sensitivity retained. Aberrant brain development.	Kuida et al., 1998 Hakem et al., 1998
caspase-12$^{-/-}$	Resistance to ER stress induced apoptosis	Nakagawa et al., 2000.
apaf-1$^{-/-}$	Reduced apoptosis and caspase activation via the mitochondrial pathway. Fas sensitivity retained. Aberrant brain development.	Cecconi et al., 1998 Yoshida et al., 1998
bcl-2$^{-/-}$	Normal early lymphocyte development. Increasing lymphoid apoptosis, beginning at 3–4 weeks of age. Enhanced apoptosis of neurons melanocytes and enterocytes. Renal disease.	Veis et al., 1993 Nakayama et al., 1993
bcl-X$^{-/-}$	Death at Ed13. Enhanced apoptosis in brain, spinal cord, and hematopoietic cells. Aberrant brain development. Chimera reveal defects in B but not T cell development.	Motoyama et al., 1995 Ma et al., 1995
bax$^{-/-}$	Increased cellularity of lymphocytes, granulosa cells, selected neurons, and immature germ cells.	Knudson et al., 1995
bcl-2$^{-/-}$/*bax*$^{-/-}$	Protection of lymphocytes and neurons from apoptotic loss.	Knudson and Korsmeyer, 1997
bid$^{-/-}$	Resistance to Fas-induced hepatocellular apoptosis.	Yin et al., 1999
lpr/lpr	Fas mutation. Autoimmunity, lymphoproliferation with accumulation of $CD3^{+}CD4^{-}CD8^{-}$ T cells. Defect in AICD.	Watanabe-Fukunaga et al., 1992
gld/gld	FasL mutation. Autoimmunity, lymphoproliferation with accumulation of $CD3^{+}CD4^{-}CD8^{-}$ T cells. Defect in AICD.	Takahashi et al., 1994 Lynch et al., 1994
fadd$^{-/-}$	Death at Ed9. Reconstitution in *Rag-1*$^{-/-}$ mice reveals defects in Fas-mediated apoptosis and a defect in activation-induced T cell proliferation.	Zhang et al., 1998a Yeh et al., 1998
icad/dff45$^{-/-}$	Marked reduction in DNA fragmentation. No developmental abnormalities.	Zhang et al., 1998b

outer mitochondrial membrane by dueling dimers of the Bcl-2 family. The common death pathway may be regulated by direct caspase activation, transcriptionally or at the post-translational level. An overview of these events is provided in Figure 5.13.

A list of mouse germline mutations of genes involved in apoptosis is provided in Table 5.4.

Selected Reviews

Adams, J. M., and Cory, S. (1998). The Bcl-2 protein family: arbiters of cell survival. Science *281*, 1322–1326.

Ashkenazi, A., and Dixit, V. M. (1998). Death receptors: signaling and modulation. Science *281*, 1305–1308.

Crispe, I. N. (1994). Fatal interactions: Fas induced apoptosis of mature T cells. Immunity *1*, 347–349.

Ellis, R. E., Yuan, J., and Horvitz, H. R. (1991). Mechanisms and functions of cell death. Annu. Rev. Cell. Biol. *7*, 663–698.

Evan, G., and Littlewood, T. (1998). A matter of life and cell death. Science *281*, 1317–1322.

Green D. R. (1998). Apoptotic pathways: the roads to ruin. Cell *94*, 695–698.

Green D. R., and Reed, J. C. (1998). Mitochondria and apoptosis. Science *281*, 1309–1312.

Hawley, S. R., and Friend, S. (1996). Strange bedfellows in even stranger places: the role of ATM in meiotic cells, lymphocytes, tumors, and in functional links to p53. Genes Dev. *10*, 2383–2388.

Henkart, P. A. (1996). ICE family proteases: mediators of all apoptotic cell death. Immunity *4*, 195–201.

Kastan, M. (1997). On the TRAIL from p53 to apoptosis. Nature Genet *17*, 130–131.

Kerr, J. F., Wyllie, A. H., and Currie, A. R. (1972). Apoptosis: a basic biological phenomenon with wide ranging implications in tissue kinetics. Br. J. Cancer *26*, 239–257.

Khwaja, A. (1999). Akt is more than just a Bad kinase. Nature *401*, 33–34.

Nagata, S. (1997). Apoptosis by death factor. Cell *88*, 355–365.

Nagata, S., and Golstein, P. (1995). The Fas death factor. Science *267*, 1449–1456.

Oltvai, Z. N., and Korsmeyer, S. J. (1994). Checkpoints of dueling dimers foil death wishes. Cell *79*, 189–192.

Reed, J. C. (1997). Cytochrome c: Can't live with it—Can't live without it. Cell *91*, 559–562.

Salvesen, G. S., and Dixit, V. M. (1997). Caspases: intracellular signaling by proteases. Cell *91*, 443–446.

Theophilopoulos, A. N., and Dixon, F. J. (1968). Murine models of systemic lupus erythematosus. Adv. Immunol. *37*, 269–305.

Thornberry, N. A., and Labeznik, Y. (1998). Caspases: enemies within. Science *281*, 1312–1316.

Vaux, D. L., and Strasser, A. (1996). The molecular biology of apoptosis. Proc. Natl. Acad. Sci. USA *93*, 2239–2244.

White, E. (1996). Life, death, and the pursuit of apoptosis. Genes Dev. *10*, 1–15.

Yang, E., and Korsmeyer, S. J. (1996). Molecular thanaptosis: a discourse on the Bcl-2 family and cell death. Blood *88*, 386–401.

Wyllie, A. H., Kerr, J. F. R., and Currie, A. R. (1980). Cell death: the significance of apoptosis. Int. Rev. Cytol. *68*, 251–306.

Yuan, J. (1997). Transducing signals of life and death. Curr. Opin. Cell Biol. *9*, 247–251.

Some Relevant References

Bennett, M., Macdonald, K., Chan, S.-W., Luzio, J. P., Simari, R., and Weissberg, P. (1998). Cell surface trafficking of Fas: a rapid mechanism of p53 mediated apoptosis. Science *282*, 290–293.

Broome, H. E., Dargan, C. M., Krajewski, S., and Reed, J. C. (1995). Expression of Bcl-2, Bcl-x and Bax after T cell activation and IL-2 withdrawal. J. Immunol. *155*, 2311–2317.

Brunet, A., Bonni, A., Zigmond, M. J., Lin, M. Z., Juo, P., Hu, L. S., Anderson, M. J., Arden, K. C., Blenis, J., and Greenberg, M. E. (1999). Akt promotes cell survival by phosphorylating and inhibiting a forkhead transcription factor. Cell *96*, 857–868.

Bump, N. J., et al. (1995). Inhibition of ICE family proteases by baculovirus antiapoptotic protein p35. Science *269*, 1885–1887.

Cardone, M. H., Roy, N., Stennicke, H. R., Salvesen, G. S., Franke, T. F., Stanbridge, E., Frisch, S., and Reed, J. C. (1998). Regulation of cell death protease caspase-9 by phosphorylation. Science *282*, 1318–1321.

Cecconi, F., Alvarez-Bolado, G., Meyer, B. I., Roth, K. A., and Gruss, P. (1998). Apaf-1 (CED-4 homolog) regulates programmed cell death in mammalian development. Cell *94*, 727–738.

Chang, H. Y., Nishitoh, H., Yang, X., Ichijo, H., and Baltimore, D. (1998). Activation of apoptosis signal regulating kinase 1 by the adapter protein Daxx. Science *281*, 1860–1863.

Cheng, E. H., Kirsch, D. G., Clem, R. J., Ravi, R., Kastan, M. B., Bedi, A., Ueno, K., and Hardwick, J. M. (1997). Conversion of Bcl-2 to a Bax-like death effector by caspases. Science *278*, 1966–1968.

Cheng, E. H.-Y., Levine, B., Boise, L. H., Thompson, C. B., and Hardwick, J. M. (1996). Bax-independent inhibition of apoptosis by Bcl-XL. Nature *379*, 557–560.

Chinnaiyan, A. M., O'Rourke, K., Lane, B. R., and Dixit, V. M. (1997). Interaction of CED-4 with CED-3 and CED-9: a molecular framework for cell death. Science *275*, 1122–1126.

Chou, J. J., Matsuo, H., Duan, H., and Wagner, G. (1998). Solution structure of the RAIDD CARD and model for CARD/CARD interaction in Caspase-2 and Caspase-9 recruitment. Cell *94*, 171–180.

Clement, M.-V., and Stamenkovic, I. (1994). Fas and tumor necrosis factor receptor mediated cell death: similarities and distinctions. J. Exp. Med. *180*, 557–567.

Clement, M.-V., and Stamenkovic, I. (1996). Superoxide anion is a natural inhibitor of fas-mediated cell death. EMBO J. *15*, 216–225.

Conradt, B., and Horvitz, H. R. (1998). The *C. elegans* protein EGL-1 is required for programmed cell death and interacts with the Bcl-2 like protein CED-9. Cell *93*, 519–529.

Datta, S. R., Dudek, H., Tao, X., Masters, S., Fu, H., Gotoh, Y., and Greenberg, M. E. (1997). Akt phosphorylation of Bad couples survival signals to the cell-intrinsic death machinery. Cell *91*, 231–241.

Enari, M., Sakahira, H., Yokoyama, H., Okawa, K., Iwamatsu, A., and Nagata, S. (1998). A caspase-activated DNase that degrades DNA during apoptosis and its inhibitor ICAD. Nature *391*, 43–50.

Hakem, R., Hakem, A., Duncan, G. S., Henderson, J. T., Woo, M., Soengas, M. S., Elia, A., de la Pompa, J. L., Kagi, D., Khoo, W., Potter, J., Yoshida, R., Kaufman, S. A., Lowe, S. W., Penninger, J. M., and Mak, T. W. (1998). Differential requirement for caspase 9 in apoptotic pathways *in vivo*. Cell *94*, 339–352.

Hay, B. A., Wasserman, D. A., and Rubin, G. M. (1995). Drosophila homologs of baculovirus inhibitor of apoptosis proteins function to block cell death. Cell *83*, 1253–1262.

Hsu, H., Shu, H.-B., Pan, M.-G., and Goeddel, D. V. (1996). TRADD-TRAF2 and TRADD-FADD interactions define two distinct TNF receptor 1 signal pathways. Cell *84*, 299–308.

Hueber, A.-O., Zornig, M., Lyon, D., Suda, T., Nagata, S., and Evan, G. I. (1997). Requirement for the CD95 receptor ligand pathway in c-Myc induced apoptosis. Science *278*, 1305–1309.

Imai, Y., Kimura, T., Murakami, A., Yajima, N., Sakamaki, K., and Yonehara, S. (1999). The CED-4 homologous protein FLASH is involved in Fas mediated activation of caspase-8 during apoptosis. Nature *398*, 777–785.

Israeli, D., Tessler, E., Haupt, Y., Elkeles, A., Wilder, S., Amson, R., Telerman, A., and Oren, M. (1997). A novel p53 inducible gene, PAG608, encodes a nuclear zinc finger protein whose overexpression promotes apoptosis. EMBO J. *14*, 4384–4392.

Kluck, R. M., Bossy-Wetzel, E., Green, D. R., and Newmeyer, D. D. (1997). The release of cytochrome c from mitochondria: a primary site for Bcl-2 regulation of apoptosis. Science *275*, 1132–1136.

Knudson, C. M., and Korsmeyer, S. J. (1997). Bcl-2 and Bax function independently to regulate cell death. Nat. Genet. *16*, 358–363.

Knudson, C. M., Tung, K. S., Tourtellotte, W. G., Brown, G. A., and Korsmeyer, S. J. (1995). Bax deficient mice with lymphoid hyperplasia and lymphoid cell death. Science *270*, 96–99.

Kothakota, S., Azuma, T., Reinhard, C., Klippel, A., Tang, J., Chur, K., McGarry, T. J., Kirschner, M. W., Koths, K., Kwiatowski, D. J., and Williams, L. T. (1997). Caspase-3 generated fragment of gelsolin: effector of morphological change in apoptosis. Science *278*, 294–298.

Kuida, K., Haydar, T. F., Kuan, C.-Y., Gu, Y., Taya, C., Karasuyama, H., Su, M. S.-S., Rakic, P., and Flavell, R. A. (1998). Reduced apoptosis and cytochrome c-mediated caspase activation in mice lacking caspase 9. Cell *94*, 325–337.

Kuida, K., Zheng, T. S., Na, S., Kuan, C., Yang, D., Karasuyama, H., Rakic, P., and Flavell, R. A. (1996). Decreased apoptosis in the brain and premature lethality in CPP32-deficient mice. Nature *384*, 368–372.

Lenardo, M. J. (1991). Interleukin-2 programs mouse αβ T lymphocytes for apoptosis. Nature *353*, 858–862.

Li, H., Zhu, H., Xu, C.-J., and Yuan, J. (1998). Cleavage of BID by Caspase 8 mediates the mitochondrial damage in the Fas pathway of apoptosis. Cell *94*, 491–501.
Li, P., Nijhawan, D., Brudihardjo, I., Srinivasula, S. M., Ahmad, M., Alnemri, E. S., and Wang, X. (1997). Cytochrome c and dATP dependent formation of Apaf-1/caspase 9 complex initaites an apoptotic protease cascade. Cell *91*, 479–489.
Lissy, N. A., Van Dyk, L. F., Becker-Hapak, M., Vocero-Akbani, A., Mendler, J. H., and Dowdy, S. F. (1998). TCR antigen-induced cell death occurs from a late G1 phase cell cycle check-point. Immunity *8*, 57–66.
Liston, P., Roy, N., Tamai, K., Lefebvre, C., Baird, S., Cherton-Horvat, C., Farahani, R., McLean, M., Ikeda, J.-E., MacKenzie, A., and Korneluk, R. G. (1996). Suppression of apoptosis in mammalian cells by NAIP and a related family of IAP genes. Nature *379*, 349–353.
Liu, Q. A., and Hengartner, M. O. (1998). Candidate adaptor protein CED-6 promotes the engulfment of apoptotic cells in *C. elegans*. Cell *93*, 961–972.
Liu, X., Zou, H., Slaughter, C., and Wang, X. (1997). DFF, a hetrodimeric protein that functions downstream of caspase-3 to trigger DNA fragmentation during apoptosis. Cell *89*, 175–184.
Luo, X., Budihardjo, I., Zou, H., Slaughter, C., and Wang, X. (1998). Bid, a Bcl-2 intercating protein, mediates cytochrome c release from mitochondria in response to activation of cell surface death receptors. Cell *94*, 481–490.
Lynch, D. H., Watson, M. L., Alderson, M. R., Baum, P. R., Miller, R. E., Tough, T., Gibson, M., Davis-Smith, T., Smith, C. A., Hunter, K., Bhat, D., Din, W., Goodwin, R. G., and Seldin, M. F. (1994). The mouse fas ligand gene is mutated in *gld* mice and is part of a TNF family gene cluster. Immunity *1*, 131–136.
Ma, A., Pena, J. C., Chang, B., Margosian, E., Davidson, L., Alt, F. W., and Thompson, C. B. (1995). BclX regulates the survival of double positive thymocytes. Proc. Natl. Acad, Sci. USA *92*, 4763–4767.
Minn, A. J., Velez, P., S. L., S., Liang, H., Muchmore, S. W., Fesik, S. W., Fill, M., and Thompson, C. W. (1997). Bcl-XL forms an ion channel in synthetic lipid membranes. Nature *385*, 353–357.
Motoyama, N., Wang, F., Roth, K. A., Sawa, H., Nakayama, K., Negishi, I., Senju, S., Zhang, Q., Fujii, S., and Loh, D. Y. (1995). Massive cell death of immature hematopoietic cells and neurons in Bcl-x deficient mice. Science *267*, 1506–1510.
Mueller, D. L., Seiffert, S., Fang, W., and Behrens, T. (1996). Differential regulation of bcl-2 and bcl-x by CD3, CD28 and the IL-2 receptor incloned CD4$^+$ helper T cells. J. Immunol. *156*, 1764–1771.
Nakagawa, T., Zhu, H., Morishima, N., Xu, J., Yankner, B. A., and Yuan, J. (2000). Caspase-12 mediates endoplasmic-reticulum-specific apoptosis and cytotoxicity by amyloid beta. Nature *403*, 98–103.
Nakayama, K., Nakayama, K., Negishi, I., Kuida, K., Shinkai, Y., Louie, M. C., Fields, L. E., Lucas, P. J., Stewart, V., and Alt, F. W., et al. (1993). Disappearance of the lymphoid system in Bcl-2 homozygous mutant chimeric mice. Science *261*, 1584–1588.
Nicholson, D. W., Ali, A., Thornberry, N. A., Vaillancourt, J. P., Ding, C. K., Gallant, M., Gareau, Y., Griffin, P. R., Labelle, M., Labeznik, Y. A., Munday, N. A., Raju, S. M., Smulson, M. E., Yamin, T. L., Yu, V. L., and Miller, D. K. (1995). Identification and inhibition of the ICE/CED-3 protease necessary for mammalian apoptosis. Nature *376*, 37–44.

Ozes, O. N., Mayo, L. D., Gustin, J. A., Pfeffer, S. R., Pfeffer, L. M., and Donner, D. B. (1999). NFκB activation by tumor necrosis factor requires the Akt serine-threonine kinase. Nature *401*, 82–86.

Romashkova, J. A., and Makarov, S. S. (1999). NFκB is a target of Akt in anti-apoptotic PDGF signaling. Nature *401*, 86–90.

Rothe, M., Pan, M.-G., Henzel, W. J., Ayres, T. M., and Goeddel, D. V. (1995). The TNFR2-TRAF signaling complex contains two novel proteins related to baculoviral inhibitor of apoptosis proteins. Cell *83*, 1243–1252.

Sahara, S., Aoto, M., Eguchi, Y., Imamoto, N., Yoneda, Y., and Tsujimoto, Y. (1999). Acinus is a caspase-3 activated protein required for apoptotic chromatin condensation. Nature *401*, 168–173.

Schneider, T. L., Grillot, D., Foote, L. C., Nunez, G. E., and Rothstein, T. L. (1997). Bcl-X protects primary B cells against Fas-mediated apoptosis. J. Immunol. *159*, 4834–4839.

Singer, G. G., and Abbas, A. K. (1994). The Fas antigen is involved in peripheral but not thymic deletion of T lymphocytes in T cell receptor transgenic mice. Immunity *1*, 365–371.

Susin, S. A., Lorenzo, H. K., Zamzami, N., Marzo, I., Snow, B. E., Brothers, G. M., Mangion, J., Jacotot, E., Constantini, P., Loeffler, M., Larochette, N., Goodlett, D. R., Aebersols, R., Siderovski, D. P., Penninger, J. M., and Kroemer, G. (1999). Molecular characterization of mitochondrial apoptosis-inducing factor. Nature *397*, 441–446.

Takahashi, T., Tanaka, M., Brannan, C. I., Jenkins, N. A., Copeland, N. G., Suda, T., and Nagata, S. (1994). Generalized lymphoproliferative disease in mice caused by a point mutation in the Fas ligand. Cell *76*, 969–976.

Tamura, T., Ishihara, M., Lamphier, M. S., Tanaka, N., Oishi, I., Aizawa, S., Matsuyama, T., Mak, T. W., Taki, S., and Taniguchi, T. (1995). An IRF-1 independent pathway of DNA damage induced apoptosis in mitogen activated T lymphocytes. Nature *376*, 596–599.

Vander Heiden, M. G., Chandel, N. S., Williamson, E. K., Schumacker, P. T., and Thompson, C. B. (1997). Bcl-XL regulates the membrane potential and volume homeostasis of mitochondria. Cell *91*, 627–638.

Varfolomeev, E. E., et al. (1998). Targeted disruption of the mouse caspase 8 gene ablates cell death induction by the TNF receptors, Fas/Apol, and DR3 and is lethal perinatally. Immunity *9*, 267–276.

Veis, D. J., Sorenson, C. M., Shutter, J. R., and Korsmeyer, S. J. (1993). Bcl-2 deficient mice demonstrate fulminant lymphoid apoptosis, polycystic kidneys, and hypopigmented hair. Cell *75*, 229–240.

Wang, C.-Y., Mayo, M. W., Korneluk, R. G., Goeddel, D. V., and Baldwin, A. S. (1998). NFκB antiapoptosis: induction of TRAF1 and TRAF-2 and c-IAP1 and c-IAP2 to suppress caspase 8 activation. Science *281*, 1680–1683.

Wang, Y., and Prives, C. (1995). Increased and altered DNA binding of human p53 by S and G2/M but not G1 cyclin dependent kinases. Nature *376*, 88–91.

Watanabe-Fukunaga, R., Brannan, C. I., Copeland, N. G., Jenkins, N. A., and Nagata, S. (1992). Lymphoproliferation disorder in mice explained by defects in Fas antigen that mediates apoptosis. Nature *356*, 314–318.

Wu, D., Wallen, H. D., and Nunez, G. (1997). Interaction and regulation of subcellular localization of CED-4 by CED-9. Science *275*, 1126–1129.

Wu, G. S. et al. (1997). KILLER/DR5 is a DNA damage-inducible p53-regulated death receptor gene. Nature Genet. *17*, 141–143.
Wu, Y. C., and Horvitz, H. R. (1998a). The *C. elegans* cell corpse engulfment gene ced-7 encodes a protein similar to ABC transporters. Cell *93*, 951–960.
Wu, Y. C., and Horvitz, H. R. (1998b). *C. elegans* phagocytosis and cell-migration protein CED-5 is similar to human DOCK 180. Nature *392*, 501–504.
Xue, D., and Horvitz, H. R. (1997). *Caenorhabditis elegans* CED-9 protein is a bifunctional cell-death inhibitor. Nature *390*, 305–308.
Yang, J., Liu, X., Bhalla, K., Kim, C. N., Ibrado, A. M., Cai, J., Peng, T.-I., Jones, D. P., and Wang, X. (1997). Prevention of apoptosis by Bcl-2: release of cytochrome c from mitochondria blocked. Science *275*, 1129–1133.
Yang, X., Khosravi-Far, R., Chang, H. Y., and Baltimore, D. (1997). Daxx, a novel Fas binding protein that activates JNK and apoptosis. Cell *89*, 1067–1076.
Yeh, W. C., Pompa, J. L., McCurrach, M. E., Shu, H. B., Elia, A. J., Shahinian, A., Ng, M., Wakeham, A., Khoo, W., Mitchell, K., El-Deiry, Lowe, S. W., Goeddel, D. V., and Mak, T. W. (1998). Fadd: essential for embryo development and signaling from some, but not all, inducers of apoptosis. Science *279*, 1954–1958.
Yin, X.-M., Wang, K., Gross, A., Zhao, Y., Zinkel, S., Klocke, B., Roth, K. A., and Korsmeyer, S. J. (1999). Bid-deficient mice are resistant to Fas-induced hepatocellular apoptosis. Nature *400*, 886–891.
Yoshida, R., Kong, Y.-Y., Yoshida, R., Elia, A. J., Hakem, A., Hakem, R., Penninger, J. M., and Mak, T. W. (1998). Apaf1 is required for mitochondrial pathwauys of apoptosis and brain development. Cell *94*, 739–750.
Youn, H.-D., Sun, L., Prywes, R., and Liu, J. O. (1999). Apoptosis of T cells mediated by Ca^{2+} induced release of the transcription factor MEF2. Science *286*, 790–794.
Zha, J., Harada, H., Yang, E., Jockel, J., and Korsmeyer, S. J. (1996). Serine phosphorylation of death agonist BAD in response to survival factor results in binding to 14-3-3 not Bcl-XL. Cell *87*, 619–628.
Zhang, J., Cado, D., Chen, A., Kabra, N. H., and Winoto, A. (1998a). Fas mediated apoptosis and activation induced T cell proliferation are defective in mice lacking FADD/Mort1. Nature *392*, 296–300.
Zhang, J., Liu, X., Scherer, D. C., van Kaer, L., Wang, X., and Xu, M. (1998b). Resistance to DNA fragmentation and chromatin condensation in mice lacking the DNA fragmentation factor 45. Proc. Natl. Acad. Sci. USA *95*, 12480–12485.
Zou, H., Henzel, W. J., Liu, X., Lutschg, A., and Wang, X. (1997). Apaf-1, a human protein homologous to *C. elegans* Ced-4, participates in cytochrome c dependent activation of caspase-3. Cell *90*, 405–413.

6

Lymphocyte Homing and the Generation of Secondary Lymphoid Organs

At every stage of lymphocyte development, cells receive instructions that direct them to specific sites. Hematopoietic stem cells migrate from the AGM (aorta gonad mesonophros) region and the yolk sac to the fetal liver and later to the bone marrow. T cell progenitors migrate in waves from the fetal liver and after birth in a steady daily flow from the bone marrow to the thymus. Homing mechanisms facilitate the entry of these precursors into the thymus. Naive T and B lymphocytes migrate from the thymus and bone marrow, respectively, to peripheral lymphoid tissue and recirculate between peripheral lymphoid organs. Specific retention mechanisms ensure that only mature lymphocytes leave the thymus and bone marrow, and homing mechanisms facilitate the entry of these naive cells into peripheral or secondary lymphoid organs such as the spleen and lymph nodes. Following an encounter with antigen, activated B cells and T cells may migrate within secondary lymphoid organs to interact with one another. Activated lymphocytes exit peripheral lymphoid organs and are attracted to the tissue site containing the source of the antigen. Mucosal sites and the skin use specific molecules to direct the flow of activated lymphocytes into these tissues.

There is a very clear and elegant functional logic to all the migration events that guide a lymphocyte from one site to another during its lifetime. Lymphocytes must be retained in central organs until receptor rearrangement and selection events are completed. When lymphocytes are mature, they must become independent of stromal cells and acquire the ability to leave these central organs. Naive lymphocytes need to be directed to peripheral lymphoid organs where they have the opportunity to be in proximity to antigen, professional antigen-presenting cells as well as other lymphocytes. Naive cells lack the ability to migrate to peripheral tissue sites. The molecules that guide cells to tissue sites are turned on in activated or memory lymphocytes.

In this chapter, the broad mechanisms involved in lymphocyte migration will be initially considered. The role of "adhesion" molecules and chemokines in lymphocyte trafficking as well as in lymphopoiesis and lymphocyte activation will be discussed. The molecular mechanisms postulated to occur in individual migration events will be examined in detail. We will also discuss our current knowledge of the cells, ligands, and receptors that are involved in the generation and architectural organization of secondary lymphoid organs.

Some General Features of Lymphocyte Migration

Two specific events must be considered in understanding the process of lymphocyte migration via the blood stream: emigration from one site and entry into another. A cell that leaves an anatomical site at a particular developmental stage probably does so because it has lost the ability to be retained in this location. For example, the presumed loss of the ability of surface immunoglobulin M (IgM)-expressing B cells to adhere to the bone marrow stroma permits their egress from this site. The functional loss of an adhesion molecule or molecules will most likely characterize the process of emigration.

Once a lymphocyte enters the circulation, it can be guided into a target lymphoid or nonlymphoid tissue site if it is recognized by the endothelial cells at the site. The lymphocyte is slowed down and eventually brought to a halt when it is identified by the endothelium. A multistep paradigm for the transendothelial trafficking of leukocytes has been developed (Figure 6.1). Initial recognition generally occurs by an interaction between carbohydrate ligands and adhesion receptors known as *selectins*. This interaction leads to a slowing down of leukocytes on the target endothelium, a process known as "tethering." Lymphocytes may "roll" on the endothelium,

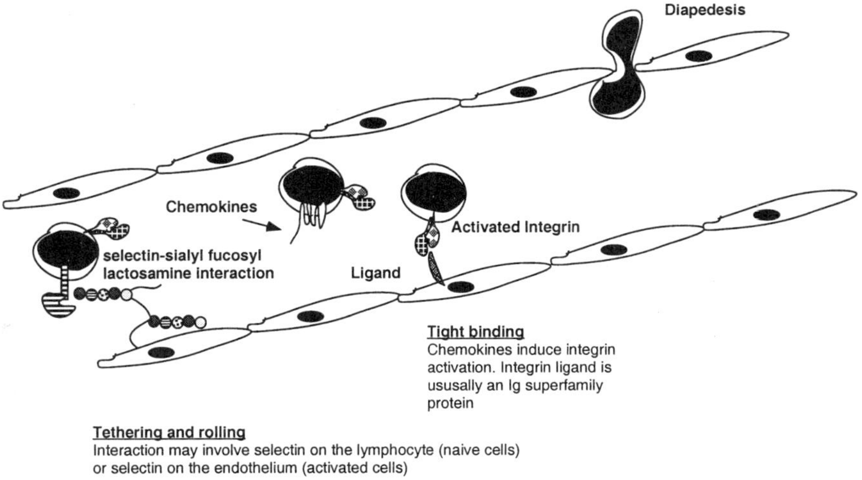

FIGURE 6.1. A multistep paradigm for lymphocyte migration across and endothelial barrier. The initial recognition event involves selectins. Chemokines activate integrins that contribute to tight binding. This is followed by diapedesis.

and in inflamed sites, lymphocytes may roll on other lymphocytes or on other white cells. An interaction between chemoattractant cytokines (also known as *chemokines*) available locally and their receptors leads to the activation of the lymphocyte and the post-translational modification of the cytosolic domains of cell surface receptors of the *integrin* family. These integrins are now activated to bind tightly to endothelial ligands that halt the lymphocyte's rolling behavior and facilitate entry of the lymphocyte into the target tissue. The process of tissue entry, also known as diapedesis, involves specific recognition events that permit the lymphocyte to pass between adjacent endothelial cells. The release by the migrating white cell of enzymes that digest extracellular matrix components facilitates the crossing of the underlying basement membrane. Given the general role played by selectins, chemokines, and integrins in the processes of lymphocyte homing as well as in lymphopoiesis, these molecules will be discussed in some detail before examining specific migration events. An additional protein outside these above categories that is important for lymphocyte trafficking is CD44, a cell surface protein that can bind hyaluronic acid and facilitates lymphocyte–extracellular matrix interactions.

Selectins

As their name suggests, selectins are lectins (proteins that bind to carbohydrate ligands). Three different selectin molecules have been described. L-selectin is found primarily on white cells, P-selectin is expressed on platelets and endothelial cells, and E-selectin is primarily expressed on the endothelium. All three selectins have broadly similar structures. They are Type I membrane glycoproteins and possess an N-terminal calcium-dependent C-type lectin domain, an epidermal growth factor-like domain, a number of short consensus repeat (SCR) motifs, a transmembrane domain, and a cytosolic tail. All the selectins bind carbohydrate ligands via the N-terminal lectin domain. They recognize the sialylated fucosylated lactosamines of which sialyl-Lewis^x and its isomer sialyl-Lewis^a represent the simplest examples. In addition to being adhesion molecules, selectins can also serve as signaling receptors, although this aspect of selectin function remains poorly studied.

L-selectin is expressed on white blood cells including T and B lymphocytes. Ligands for L-selectin are expressed in lymph nodes on the surface of the specialized plump endothelia cells in high-endothelial venules (HEVs). HEVs are specialized postcapillary venules, and naive lymphocytes carried by the bloodstream bind to these venules and migrate through them into the lymph node. L-selectin ligands are specific sulfated oligosaccharides. These ligands, along with other endothelial ligands for homing receptors, are referred to as *vascular addressins*. L-selectin ligands (which share the appropriate sulfated O-linked oligosaccharide side chains) on adult peripheral lymph node HEVs are collectively referred to as peripheral lymph node addressins, or *PNAds*. Different adhesion molecules are critical for lymph node entry in adult and fetal life, as will be discussed below. The sulfated sugar side chains that constitute PNAds are expressed on the luminal surfaces of lymph node HEVs beginning only a day or two after birth. Potential addressins that are decorated with the PNAd epitope include GlyCAM-1, MAdCAM-1, podocalyxin, and CD34.

GlyCAM-1 is a small mucin-type glycoprotein, which, in its correctly glycosylated and sulfated form, can be recognized with relatively high affinity by L-selectin. It is expressed, as would be predicted for an L-selectin ligand, in high endothelial venules and is also induced at sites of inflammation. However, GlyCAM 1 is unlikely to be an "adhesion" ligand for L-selectin, although it now appears to be a "signaling" ligand. It is a secreted protein that lacks a transmembrane domain but has been shown to transduce signals via L-selectin for the activation of integrins. GlyCAM-1 has so far been identified only in rodents, and a human homologue remains to be identified.

CD34 is a transmembrane mucin-type glycoprotein expressed on all endothelial cells but which is appropriately glycosylated and sulfated in HEVs of lymph nodes and can thus be recognized by L-selectin. This glycoprotein is likely to function as functionally important ligand for L-selectin. However, mice engineered to lack CD34 have no obvious defect in lymphocyte trafficking to lymph nodes, suggesting that CD34 may either be redundant or irrelevant. Two other potential ligands for L-selectin are of interest. There is a 200-kd glycoprotein, gp200, that has been identified in HEVs but which has as yet not been cloned. Another very interesting molecule has been found in the HEVs of Peyer's patches (organized peripheral lymphoid tissue, much like that seen in lymph nodes, and located in the small intestine) and of mesenteric lymph nodes. The carbohydrate side chains of this molecule, known as MadCAM-1, can function as a ligand for L-selectin, while the protein backbone serves as a ligand for the $\alpha_4\beta_7$ integrin.

A potential ligand for L-selectin that is expressed on white cells but not on endothelium is the glycoprotein PSGL-1 (P-selectin glycoprotein ligand-1). The PSGL-1-L-selectin interaction might mediate the phenomenon of lymphocytes rolling on other white cells at sites of inflammation. PSGL-1 is probably the major ligand on neutrophils for P-selectin. PSGL-1 needs to be correctly glycosylated and tyrosine sulfated to be recognized by selectins. All of the above L-selectin ligands are sialomucins; they present O-linked sugar side chains at a high density. Completely different carbohydrate ligands may also bind to L-selectins. These include heparan sulfate glycosaminoglycans, which are found on endothelial surfaces, and also sulfatides, which are negatively charged glycolipids. These ligands lack sialic acid or fucose but are also bound by the N-terminal lectin domain in a calcium-dependent manner.

From the viewpoint of the binding of lymphocytes to endothelial surfaces, P- and E-selectins on the endothelial cell may contact ligands on lymphocytes. The heat-stable antigen (HSA), also known as CD24, is found on recently generated lymphocytes and may bind to P-selectin. It is a somewhat unusual selectin ligand, since most of its sugars are N-linked rather than O-linked saccharide chains. An interesting E-selectin ligand, a 250-kd glycoprotein, has been discovered on bovine γδ T cells. Memory T lymphocytes, which are destined to home to the skin, express a specific E-selectin ligand on their surfaces, which facilitates their homing back to this organ. This glycoprotein is known as CLA-1.

Chemokines and Chemokine Receptors

Chemokines are chemoattractant cytokines. Chemokines have numerous functions; they play a critical role in lymphocyte migration by triggering the increased adhesiveness of integrins, a process known as "inside-out" signaling. Because specific lymphocyte subsets express distinct receptors, specificity for the homing of specific lymphocytes to defined anatomical sites may depend on the local production of a specific chemokine. Chemokines have also been shown to be of importance in lymphopoieisis. The receptors for chemokines are 7-transmembrane/serpentine receptors that are linked to heterotrimeric G proteins. Chemokine receptors have attracted a considerable amount of attention because some of them are used by the human immunodeficiency virus (HIV) to gain entry into T cells and monocytes.

Most chemokines are about 8 to 12 kd in size, and they contain four conserved cysteine residues. These molecules contain an N-terminal domain that contains two cysteine residues, followed by at least three β-pleated sheets and a C-terminal α-helix. These mediators are generally basic proteins, which, apart from binding specific signaling receptors on leukocytes, also bind to heparin and heparan sulfate proteoglycans. Binding to matrix proteoglycans may help set up chemokine gradients that direct migratory events. To date, about 40 chemokines have been characterized, and this number probably represents the tip of the iceberg.

There are four families of chemokines, but most fall into one of two broad categories based on whether their two N-terminal cysteines are adjacent to each other or are separated by a spacer residue. Members of one family have two adjacent cysteine residues near the N-terminus, and these chemokines are known as C-C chemokines. C-X-C chemokines have a spacer amino acid between the two N-terminal cysteines. Most C-C or β chemokines in humans are located at 17q11.2–12, and most human C-X-C or α chemokines map to 4q13. These genes appear to have arisen by a process of duplication. Although the data to support this view are quite limited, it has been speculated that chemokines have evolved fairly rapidly; some chemokines in humans (such as IL-8) lack murine counterparts. In addition to C-C and C-X-C chemokines, two chemokines are characterized as members of distinct groups. Lymphotactin is the only member of the γ or C-chemokine family and contains a single N-terminal cysteine. Fractalkine is the only CX3C chemokine (with three amino acids between the two N-terminal cysteines) and is unusual in that it is the only membrane-bound chemokine. Figure 6.2 lists selected chemokines and their receptors that are of importance from a lymphocyte migration standpoint.

Clearly, only a fraction of all chemokine receptors have been identified. At present there are five known human C-X-C receptors (CXCR1 through CXCR5) and nine C-C receptors (CCR1 through CCR9). As seen in Figure 6.2, multiple chemokines may signal via a given receptor. However C-C chemokines only signal via C-C receptors, and C-X-C chemokines activate only C-X-C receptors. In general, C-X-C chemokines have more restricted specificities than C-C chemokines. Chemokine receptors are linked to heterotrimeric G-proteins containing $G_{i\alpha}$ and can be inhibited by pertussis toxin. These G proteins typically contain a DRYLAIV motif in their second intracellular domains and are linked to phospholipase C isoforms that activate calcium signaling and protein kinase C (PKC). The activation of PKC might contribute to inside-out signaling and integrin activation. Chemokine receptor activation also leads to the activation of small guanosine triphosphatases (GTPases) of the Rho family, which, as discussed in Chapter 4, can influence cell motility by regulating actin-dependent processes such as membrane ruffling and the formation of actin stress fibers. Many other signaling events, including the activation of the Ras pathway, may be turned on in selected target cells by specific chemokines.

α or CXC chemokines

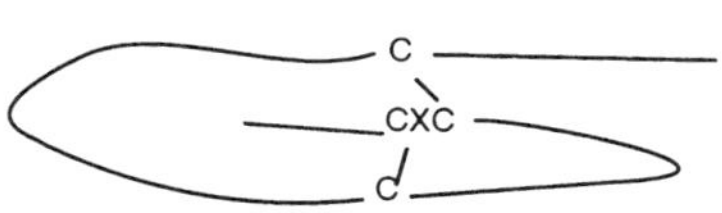

Receptor	Chemokine	Lymphoid/APC target
CXCR1	IL-8	NK, ?T
CXCR2	IL-8	NK, ?T
CXCR3	IP-10, Mig	NK Th1
CXCR4	SDF-1	Naive B and T, DC
CXCR5 (BLR-1)	BCA-1	Naive B

γ or C chemokines

XCR1	Lymphotactin	T cells

δ or CX3C chemokines

CX3CR1	Fractalkine	Act/Memory T

β or CC chemokines

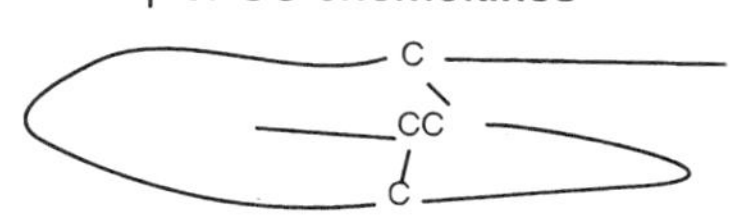

Receptor	Chemokine	Lymphoid/APC target
CCR1	RANTES, MIP-1α, MCP-3, MIP-5	Act/Mem T NK, DC
CCR2	MCP-1, MCP-2, MCP-3, MCP-4, MCP-5	Act/Mem T DC
CCR3	Eotaxin, MIP-5, MCP-2, MCP-3 MCP-4	Th2 DC (eosinophils)
CCR4	TARC	Th2
CCR5	MIP-1α, MIP-1β RANTES	Th1
CCR6	MIP-3α	Act/MemT
CCR7	MIP-3β, SLC	NaiveT, Act B "Central" Memory T
CCR8	I-309, TARC	Th2

FIGURE 6.2. A list of chemokines and chemokine receptors. Table 6.1 provides explanations of acronyms and alternative names for chemokines. Act = activated, Mem = memory, DC = dendritic cells.

Chemokines and Early B Cell Development

While chemokines generally influence lymphoid development because of their importance in migration events, at least one chemokine has also been shown to be of direct importance in development. The C-X-C chemokine, SDF-1/PBSF (see Table 6.1 for an explanation of acronyms) is secreted by bone marrow and other stromal cells and is a potent stimulator of pro-B cell proliferation. Its receptor is CXCR4 (previously known as Fusin and Lestr), which also assists in HIV entry. SDF-1 can prevent HIV from gaining access to cells. SDF-1, apart from its role in pro-B lymphocyte proliferation, also may play a role in the homing of both naive T and B cells to lymph nodes. In the absence of SDF-1, mice have defects in B cell generation, with a very marked reduction of pro-B cells (and of more differentiated cells that make up the B lineage). The defect appears to be at the level of commitment to the B lineage, since T cell development is unaffected. SDF-1 null mice also have a defect in bone marrow derived myelopoiesis (there is no such defect at the fetal liver stage) as well as a defect in ventricular septum generation in the heart. SDF-1 may also contribute to the migration of hematopoietic stem cells.

Chemokines and Naive T and B Lymphocytes

A few chemokines have been implicated in the entry of naive T lymphocytes into lymph nodes. These include two β chemokines, SLC/6C-kine/Exodus-2

TABLE 6.1. Acronyms and alternative names for selected chemokines

Acronym	Expansion	Other names
α/CXC		
IP-10	IFN-γ-inducible protein	
Mig	Monokine induced by IFN-γ	
SDF-1	Stromal cell-derived factor-1	PBSF
PBSF	Pre-B cell growth-stimulating factor	SDF-1
BCA-1	B cell-attracting chemokine-1	BLC
BLC	B lymphocyte chemoattractant	BCA-1
β/CC		
RANTES	Regulation on activation, normal T expressed and secreted	
MCP-1	Monocyte chemotactic protein-1	
MIP-5	Macrophage inflammatory protein-5	Leukotactin-1
MIP-3α	Macrophage inflammatory Protein-3α	LARC, Exodus
LARC	Liver and activation regulated chemokine	MIF-3α, Exodus
MIP-3β	Macrophage Inflammatory protein-3β	ELC, Exodus-3
ELC	EBI-1 (Ebstein-Barr induced-1) ligand chemokine	MIP-3β, Exodus-3
TARC	Thymus and activation regulated chemokine	
MDC	Macrophage derived chemokine	ABCD-1
ABCD-1	Activated B cell and dendritic cell CC chemokine	MDC
SLC	Secondary lymphoid-tissue chemokine	Exodus-2, 6Ckine
TCA-3	T cell activation gene-3 (murine)	I-309 (human)
TECK	Thymus expressed chemokine	

and ELC/MIP-3β/Exodus-3, and an α chemokine, SDF-1. All three molecules are expressed at high levels in lymph nodes (see Table 6.1 for definitions of acronyms). ELC and SLC are ligands for CCR7. SLC is expressed in high endothelial venules (HEVs) and is probably the major player in attracting navie T cells into lymph nodes. In mice with the naturally occurring *plt* (paucity of lymph node T cells) mutation, the absence of lymph node T cells is linked to a defect in the *SLC* gene. ELC is expressed constitutively in the T cell areas of secondary lymphoid organs. It has been implicated not only in the migration of naive T cells into these T cell areas but also in the migration of activated B cells, which also express CCR7, into the T cell areas in order to promote T-B interactions (see below). SDF-1 is synthesized by dendritic cells in lymph nodes and the receptor for this cytokine, expressed on naive T and B lymphocytes, is CXCR4.

The entry of naive B cells into the follicular areas of the spleen and lymph nodes may be mediated by two different chemokines. CXCR4 (the receptor for SDF-1) is expressed on naive B cells. CXCR5/BLR-1 is expressed on mature B cells as well as on memory T cells with a $CD45RO^+$ phenotype. CXCR-5 is required for the entry of B cells into follicles in the spleen and Peyer's patches, as well as for the migration of activated B cells into germinal center (see below). In the absence of CXCR-5, naive mature B cells remains in the T cell zones and fail to enter follicles. The ligand for BLR-1 is called BLC or BCA-1 and is made in follicles, presumably by follicular dendritic cells. An appreciation of the role played by this chemokine has led to the understanding that chemokines are involved in "basal" lymphocyte trafficking.

Chemokines Bring B and T Cells Together

Chemokines contribute in very sophisticated ways to bringing lymphocytes to specific tissue sites, to keeping them in specific compartments in lymphoid organs, and to bringing rare antigen specific B and T cells together in secondary lymphoid organs. As mentioned above, the triggering of CCR7 by SLC helps bring naive T and possibly B cells into lymph nodes across HEVs. The activation of CXCR5 by BLC contributes to the entry of B cells into follicles in some peripheral lymphoid organs. Once SLC has drawn naive and some "central" memory T cells into lymph nodes, both SLC and ELC expressed in a range of cell types contribute to maintaining CCR7 expressing T cells in the T cell areas of these organs. Dendritic cells also express CCR7 and are drawn into the T cell areas of lymph nodes by SLC and ELC. Dendritic cells themselves may secrete ELC and other naive T cell attracting molecules and thus bring naive T cells and dendritic cells into an ongoing dance in the T cell areas of secondary lymphoid organs.

Antigen specific activation of T helper cells, particular of Th2 cells, results in the shut-off of CCR7 expression and therefore a decreased ability of activated T cells to respond to SLC and ELC. As a result activated T cells may migrate away from the main source of SLC and ELC in the T cell areas and move toward the interface between the T cell areas and follicles. Some T cells which have received strong costimulatory signals via CD28 and OX40 may express high levels of CXCR5 and migrate towards the T cell area-follicular interface in response to signals from BLC. In addition, activated B cells may secrete chemokines such as MIP-1α, MIP-1β, and MDC/ABCD-1, that trigger receptors such as CCR1, CCR4, and CCR5 on activated T cells, and thus contribute to bringing antigen activated B and T cells together.

Chemokines on Activated and Memory T Cells

A large and confusing number of chemokine receptors are displayed on activated or memory T cells but not on naive lymphocytes (see Figure 6.2). MIP-3α/LARC/Exodus is a β-chemokine that is expressed in tissue sites and is upregulated during inflammatory responses. This chemokine can bind to CCR6 and thus activate β2 intergrins on memory T cells. This is likely to be a major signal for the entry of activated T cells into tissues.

The CCR1, CCR4, and CCR5 receptors are induced on activated and $CD45RO^+$ memory phenotype T cells. These receptors may also contribute to bringing activated B and T cells together as mentioned above. Another chemokine receptor that is expressed only on some (mainly Th1) activated T cells and on NK cells is the CXCR3 molecule that responds to the chemokines IP-10 and Mig. These chemokines are induced by interferon (IFN)-γ and are believed to be critical for the activation of T cells that are primed by IFN-γ to migrate into inflamed tissue sites. Delayed-type hypersensitivity (DTH) responses are dependent on the induction of IP-10.

Human memory T cells can be subdivided into two subsets based primarily on the presence or absence of CCR7. "Central" memory cells (and Th1 cells) express CCR7 and L-selectin and home to the T cell areas of secondary lymphoid organs. They are $CD45RA^-$ and can be distinguished from naive T cells which are $CD45RA^+$. "Effector" memory T cells home to tissue sites. They are $CD45RA^-$ $CCR7^-$ cells. Cells of this subset that are destined for the skin express the E-selectin ligand CLA (discussed below), the α4β1 integrin, and CCR4. CCR4 is the receptor for TARC, a ligand that is found largely in the skin. "Effector" memory T cells that are destined for mucosal sites express the α4β7 integrin and CCR5, but do not express CCR4.

Not surprisingly, chemokine receptors can be used to distinguish between Th1 and Th2 cells, but the correlations between chemokine and T helper cytokine expressions are not as tight as sometimes assumed. Some of the differences in chemokine receptor expression in polarized T helper cells are presumably induced as a consequence of the autocrine triggering of specific cytokine receptors; other receptors may be required specifically for the biological functions of Th1 or Th2 cells. As depicted in Figure 6.3, Th1 cells preferentially express CXCR3, CCR5, and CCR7. CXCR3 is the receptor for IP-10 and Mig, CCR5 can be triggered by MIP-1α, MIP-1β, and RANTES, and CCR7 is the receptor for SLC and ELC which is also

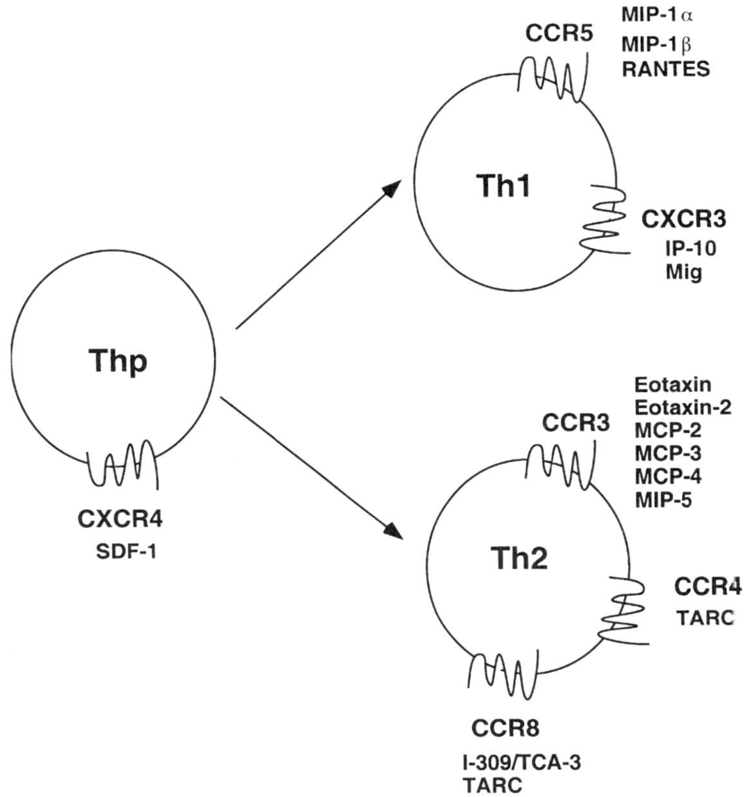

FIGURE 6.3. Distinct chemokine receptors are expressed on Th1 and Th2 cells. Some receptors such as CCR1 and CCR2 are expressed on both Th1 and Th2 cells.

expressed on naive T and B cells. CCR7 presumably facilitates the homing of Th1 cells to T cell areas in lymphoid organs while the lower levels of CCR7 on Th2 cells facilitates T-B collaboration as noted above. The possible role of CXCR3 in DTH responses has been considered above.

Th2 cells may preferentially express CCR3, and CXCR4. CCR3 can be triggered by eotaxin, a chemokine which may contribute to the migration both of Th2 cells and of eosinophils during allergic responses.

Integrins

Integrins are integral membrane heterodimers, each made up of an α chain and a β chain that are noncovalently associated. They play important roles in cell–cell interactions, cell-substrate interactions, cell migration, and hematopoiesis (Table 6.2). There are 8 different integrin β chains and 16 α chains, which combine to give rise to 22 identified heterodimers. These inte-

TABLE 6.2. Selected integrins that participate in lymphocyte development

Family	Function	Partners
β1 integrins		
α?β1 (α chain undetermined but not α4; there are 8 other α chains that associate with) β1)	Migration of hematopoietic stem cells to fetal liver.	cell–ECM (extracellular matrix) interactions (collagens, laminin, fibronectin, vitronectin)
α4β1 (VLA-4)	Not required for fetal hematopoiesis, other than for B1 cells. Required for adult B and T cell generation.	cell–cell (VCAM1), cell–ECM (fibronectin)
β2 integrins		
LFA-1 (αLβ2; CD11aCD18)	Lymphocyte migration Costimulation/T-APC synapse formation	ICAM-1, 2, and 3
Mac-1 (CR3; CD11bCD18)	Leukocyte migration, phagocytosis; cell-matrix association	iC3b, fibrinogen, ICAM-1,
p150, 95 (CR4; CD11cCD18)	Adhesion, phagocytosis	iC3b, fibrinogen
β7 integrins		
α4β7 (LPAM-1)	Lymphocyte migration to peripheral nodes in fetal life to mucosal sites in adult. Migration of T cells to Peyer's patches.	MadCAM-1
αEβ7	Lymphocyte adhesion to gut epithelium (for intraepithelial lymphocytes).	E-cadherin

grins are divided into 8 families, each of which share a given β chain. Integrins can be activated by signaling via other cell surface receptors, primarily receptors for chemokines, which can modulate the affinity of integrins for their ligands (inside-out signaling). Integrins can themselves induce signal transduction. Most integrins bind to extracellular matrix proteins, while some, particularly the two α4 integrins, and the three β2 integrins, mediate cell–cell interactions. The cell surface protein ligands for integrins involved in cell-cell interactions tend to be members of the Ig superfamily such as intercellular-adhesion molecule-1 (ICAM-1), ICAM-2, ICAM-3, vascular cell-adhesion molecule-1 (VCAM-1), and MadCAM-1. One cell surface integrin ligand, E-Cadherin, which binds to the αEβ7 integrin, does not belong to the Ig superfamily.

In the absence of β1 integrins, hematopoietic stem cells (HSCs) fail to migrate to and colonize the fetal liver, although they can give rise to all hematopoietic lineages. There are nine β1 integrins (six of which were once categorized as very late antigens—VLAs), and the specific member of this family that is required for the HSC migration event has not been identified. It is unlikely to be the α4β1 integrin, since the absence of α4 integrins results in a distinct phenotype, which is discussed below. The majority of the β1 integrins bind to extracellular matrix proteins such as collagens, fibronectin, laminin, and vitronectin. The α4β1 integrin can bind to both fibronectin as well as a cell-surface Ig superfamily protein known as vascular cell-adhesion molecule-1 (VCAM-1).

The absence of α4 integrins severely compromises B and T cell development (but not NK cell development) after birth. Development of T cells from fetal liver precursors is unaffected. Although B1 cells are not generated, B cell development at the fetal liver stage is almost normal. Only two α4 integrins are known, α4β1 and α4β7. Since the targeted disruption of the β7 integrin gene did not lead to a defect in lymphopoiesis, it is reasonable to assume that the α4β1 molecule is required for the development of B and T cells from bone marrow-derived stem cells. In the absence of α4 integrins, development of monocytes and NK cells is unaffected. It is therefore assumed that the requirement for the α4β1 integrin is not at the level of the HSC or even at the level of the common lymphoid progenitor. It is possible that this integrin is required in the bone marrow by a precursor that gives rise to only B and T cells but not to NK cells, although no precursor of this sort has yet been identified. It is likely therefore that α4 integrins are required separately for B and T cell development. In the absence of α4 integrins, B cell development is blocked prior to the pro-B stage, and the α4β1-VCAM-1 interaction may be an essential component of the stromal signals that are critical for commitment to the B lineage. However, this particular stromal requirement applies only to B lineage commitment in the bone marrow and not in the fetal liver. Exactly why the α4β1 molecule is required in the bone marrow for B cell development is unclear, though the most likely scenario is that signal transduction initiated via this receptor provides essential signals for commitment. We must assume therefore that signaling via some other unidentified receptor on lymphoid progenitor cells may be required for B lineage commitment in the fetal liver.

β2 and β7 integrins are important players in the process of lymphocyte trafficking. β2 integrins are required by all white cells for transendothelial migration, but one member of this family, leukocyte function antigen-1 (LFA-1), may have other functions in lymphocyte activation, which are discussed below. When lymphocytes begin to roll on an endothelial surface to which they are attracted, they remain long enough in the local vicinity of chemokines, which induce signals that can lead to the phosphorylation of the cytoplasmic tail of β2 integrins such as LFA-1. This cytoplasmic modification leads to an increase in integrin affinity for its extracellular ligands, a process widely referred to as "inside-out" signaling. The ligands for LFA-1 include ICAM-1 and ICAM-2 on the endothelial surface.

In addition to the role of LFA-1 in the adhesion of lymphocytes to endothelium, this integrin has an important role to play in the interactions of T cells with B cells, with other antigen-presenting cells (APCs) as well as with target cells. Interactions between LFA-1 on T cells and ICAM-1 on APCs may serve not only to bring T cells in close apposition with APCs but may also serve to provide co-stimulatory signals. Although it is accepted that the major receptor that triggers T cell costimulation is CD28, nevertheless, it is conceivable that integrin signaling also participates in the process of T cell activation. The importance of LFA-1 in the formation of immunological synapses has been discussed in Chapter 4.

Memory T cells express the α4β7 integrin, which directs these cells to mucosal sites. The ligand for this integrin is MadCAM-1, which is expressed on high endothelial venules in Peyer's patches and mesenteric lymph nodes. Intraepithelial lymphocytes (IELs) in the gut associate tightly with the basolateral surfaces of adjacent epithelial cells. This interaction depends on the interaction of αEβ7 molecules on IELs with E-cadherin on the epithelial surface.

Adhesion Molecules and Early Lymphopoiesis

Lymphocyte development begins in the adult in specialized niches in the bone marrow. This process may be influenced by adhesion molecules and by extracellular matrix components for a number of reasons. The migration of progenitor cells may require these molecules. Extracellular matrix components may hold back developing precursors in the bone marrow or thymus. Signals may be generated from "adhesion" receptors that are critical for survival, proliferation, or further development. Adhesion molecules may place lymphoid progenitors in close proximity to stromal cells, which produce the "goodies" required for further differentiation.

The early phase of B cell development is stromal dependent. Stromal cells produce signaling ligands such as SDF-1, FL (the Flk2/Flt3 ligand), IL-7 and IL-11. They also provide ligands for the α4β1 integrin, which has been implicated in the process of early B cell differentiation in bone marrow-derived but not fetal liver-derived B cells. In the absence of α4 integrins, B cell development is arrested prior to the pro-B stage, as discussed above. The interaction of the α4β1 integrin with its ligand VCAM-1 might provide an essential survival signal during early B cell development. Alternatively, integrin signaling in the B cell might be responsible for turning on some other event that is critical for B cell development. Another possibility is that a signaling event in the stromal cell may be generated that contributes to the release from the latter of trophic factors (such as SDF-1, FL, IL-7, or IL-11) that contribute to early B cell development. However, it is clear that neither IL-7 nor IL-11 signaling is essential for commitment to the B lineage, so if VCAM-1 contributes to stromal cell activation, the latter might be induced to secrete some other soluble factors that contribute to molding a progenitor cell into a committed pro-B lymphocyte.

Homing of Progenitor T Cells to the Thymus

T cell progenitors migrate from fetal liver and the post natal bone marrow to the thymus. They must be recognized by the thymic vascular endothelium and cross into the perivascular space separating the endothelium from a perivascular epithelial layer. They must then cross this perivascular layer and enter into the thymic parenchyma. Early thymic immigrants express L-selectin and CD44 (the receptor for hyaluronate). CD44 may be involved in the process of homing to the thymus, while L-selectin does not appear to be essential, since this process appears to be unaffected in L-selectin–deficient mice.

It has been suggested that α6 integrin expressed on the apical surfaces of thymic endothelial cells may be critical for the entry of thymic progenitors in the thymus. However, this integrin is expressed at a number of sites, and it is unclear how this molecule provides specificity. A late adhesion event involving a thymic stromal molecule. Vanin-1, may contribute to the specificity of this thymic entry process. Antibodies to Vanin-1 can block thymic colonization and regeneration but do not affect T cell maturation. Vainin-1 is a glycosylphosphatidyl inosital-(GPI)-anchored protein is structurally related to human biotinidase. Biotinidases are enzymes secreted by hepatocytes that can release biotin from biocytin. Whether Vanin-1 actually possesses biotinidase activity and if so, whether this activity is required for homing to the thymus, remains to be established.

The TNF Receptor Family, Chemokines, and the Biogenesis of Secondary Lymphoid Organs

The generation of lymph nodes begins in fetal life and depends on instructions being delivered to stromal cells initially by a specific nonlymphoid progenitor and later by migrating B cells. A few days before birth, similar inductive processes occur in the intestinal submucosa, setting up the formation of Peyer's patches. A $CD4^+CD3^-IL\text{-}7R\alpha^+$ cell that expresses LTβ is believed to be involved in providing the inductive signal for lymph nodes and Peyer's patches during embryogenesis, and this cell is probably a progenitor of natural killer (NK) cells, some unidentified follicular cells involved in lymph node biogenesis, and APCs. The development of this cell is compromised in the absence of the Id2 bHLH inhibitor. Presumably, Id2 blocks the bHLH-dependent induction of may lineages, permitting the generation of the $CD4^+CD3^-IL\text{-}7R\alpha^+$ precursor. In the absence of Id-2, peripheral lymph nodes and Peyer's patches do not develop, splenic architecture is normal, and

NK cells are not generated. This $CD4^+CD3^-IL\text{-}7R\alpha^+$ precursor cell is apparently the major source of LTβ in developing lymph nodes and in the gut. Some other cell presumably provides LTβ in the spleen during fetal development, since in the absence of LTβ, splenic microarchitecture is also distorted (see below). Postnatally, B cells are the major source of LTβ required for the maintenance of the architecture of secondary lymphoid organs.

"Stromal" cells receiving signals from the inducing $CD4^+CD3^-IL\text{-}7R\alpha^+$ cells belong to many lineages and include fibroblasts, macrophages, reticular cells, and endothelial cells. Once the basic architecture of the primordial lymph node is created, lymphocytes begin to migrate into the anlage and may continue to provide signals to the stromal components for maintenance of the structure of the secondary lymphoid organ. The architectural organization of secondary lymphoid organs depends on inductive signals delivered by members of the TNF (tumor necrosis factor) family of Type III cytokines, and the regulated migration of naive lymphocytes, orchestrated in part by specific chemokines and their receptors. While we have touched on the TNF family of ligands and their receptors in Chapter 4, we will discuss some members of this family in a little more detail here in the context of the architecture of lymph nodes and the generation of secondary lymphoid organs.

There are currently five receptors and six ligands in the LT/TNF family (Figure 6.4) that are required for the biogenesis of secondary lymphoid organs. The lymphotoxin β receptor (LTβR) is critical for the generation of all lymph nodes and Peyer's patches. Both LTβR and TNFR1 are required for the generation of lymphoid follicles and follicular dendritic cell (FDC) networks and therefore also for the architectural changes that characterize the generation of germinal enters during secondary immune responses. TNF itself, as discussed in Chapter 4, is generated by TNF α converting enzyme TACE-mediated cleavage of the extracellular domains of a trimeric type II membrane protein (which may itself be functional). TNF can bind to the TNFRI (p55) and TNFRII (p75). The lymphotoxin α chain (LTα) can form a secreted homotrimer (LTα3), which is the only ligand of the TNF family that is not synthesized as a membrane protein. It is structurally homologous to TNF and binds to and activates TNFRI, TNFRII, and an additional receptor of the TNFR family called herpes simplex virus entry mediator (HVEM), which is expressed on activated T cells. It does not bind to the lymphotoxin β receptor, LTβR. The lymphotoxin α chain can form membrane-bound heterotrimers with the Type II integral membrane lymphotoxin β chain (LTβ). LT $\alpha 2\beta 1$ heterotrimers trigger TNFR1 and TNFR2 and probably do not trigger LTβR. $LT\alpha 1\beta 2$ heterotrimers trigger LTβR. While the major ligand for LTβR is $LT\alpha 1\beta 2$, an additional Type II integral membrane homotrimer called LIGHT might also trigger this receptor. LIGHT also binds to HVEM.

Another TNF family member, the osteoprotegerin ligand (OPGL), is required for the development of all lymph nodes and exists in both membrane-bound and cleaved soluble forms. OPGL is also known as osteo-

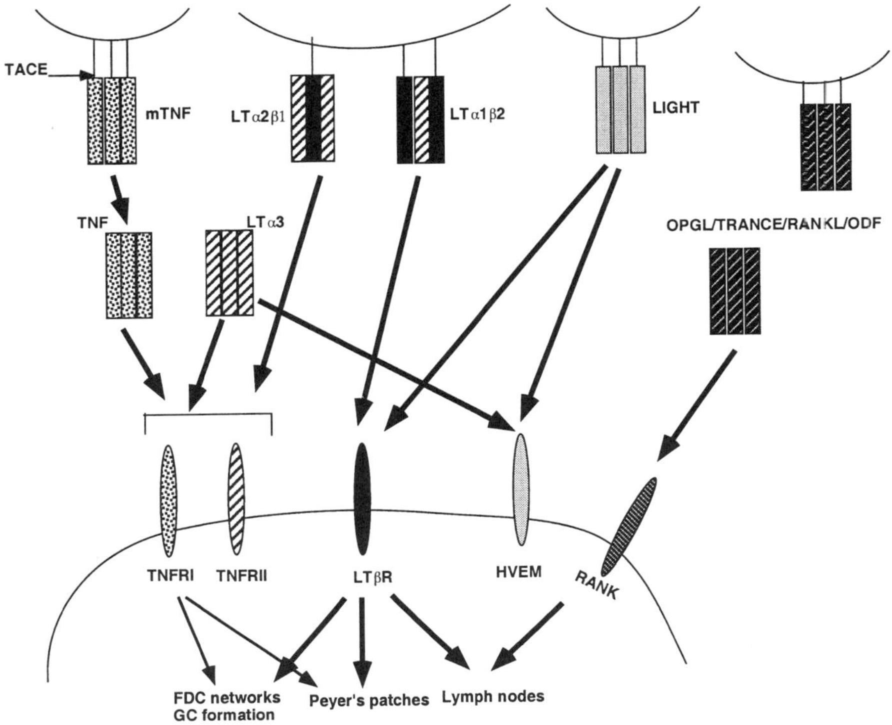

FIGURE 6.4. Ligands and receptors of the TNF family. TNF may function in a cleaved soluble trimeric form or as a membrane-bound trimer. LTα3 is a ligand for TNFRI and TNFRII but not for LTβR. LTα1β2 functions only as a membrane-bound ligand and is apparently not susceptible to cleavage by TACE.

clast differentiation factor (ODF), TRANCE (TNF-related activation induced cytokine), and as the RANK ligand, or RANKL. RANK (receptor activator of NFκB) is the TNFR family member that is the receptor for OPGL. Although OPGL is required for the generation of all lymph nodes, splenic architecture is generally normal in the absence of this cytokine, and *OPGL* null mice as well as *RANK*$^{-/-}$ mice have small but normal-looking Peyer's patches. The requirement for OPGL during lymph node biogenesis reflects a functional role of this ligand expressed in nonlymphoid cells involved in lymph node organogenesis. Normal bone marrow failed to rescue the lymph node generation defect in *OPGL*$^{-/-}$ mice. While OPGL expressed in nonlymphoid cells is critical for the generation of lymph nodes, OPGL expressed in cells of bone marrow origin influences the development of pre-B and pre-T cells. These defects will be considered in Chapters 7 and 9.

LTβR, the receptor for LTα1β2, is required for the development of all lymph nodes. LTβR is not expressed on lymphoid cells but is expressed on stromal cells. The LTα1β2 ligand for LTβR is initially made available during embryogenesis (around day 17 in the mouse) on the surface of a $CD4^+CD3^-IL\text{-}7R\alpha^+$ cell, which can give rise to antigen-presenting cells, NK cells, and some poorly defined follicular cells. In *LTβR*$^{-/-}$ mice, no lymph node development occurs, and no Peyer's patches are observed. The optimal development of FDC networks requires LTα3 and presumably LTα1β2 on B cells that migrate into secondary lymphoid tissues, presumably in a CXCR5 (BLR-1)- and CXCR4-dependent manner, to trigger both the LTβR as well as TNFRI on FDC precursors in these organs. In

TABLE 6.3. Type III cytokine/cytokine receptor knockouts

Knockout	Phenotype	References
LTα$^{-/-}$	No lymph nodes other than a few mesenteric lymph nodes. No Peyer's patches. Splenic B and T areas and marginal zones disorganized. FDC networks absent, and germinal center formation impaired. Some impairment of affinity maturation. Increased $\alpha_E\beta_7^{hi}$ cells.	DeTogni et al., 1994 Banks et al., 1995 Matsumoto et al., 1996
LTβ$^{-/-}$	No lymph nodes other than mesenteric, cervical, and some paraaortic/sacral lymph nodes. No Peyer's patches. Splenic B and T area organization largely intact. Defects in FDC networks and GC formation. Minimal impairment in affinity maturation. Increased $\alpha_E\beta_7^{hi}$ cells.	Alimzhanov et al., 1997 Koni et al., 1997
LTβR$^{-/-}$	No lymph nodes. No Peyer's patches. Splenic B and T areas disorganized, and marginal zones absent. FDC networks absent, and GC formation impaired. Some impairment of affinity maturation. Increased $\alpha_E\beta_7^{hi}$ cells.	Futterer et al., 1998
TNF$^{-/-}$	No splenic primary follicles.	Pasparakis et al., 1996
TNFRI$^{-/-}$ (*p55*$^{-/-}$)	No Peyer's patches, but lymph nodes develop normally. Impaired FDC network and GC formation. Minimal impairment in affinity maturation.	Pfeffer et al., 1993 Rothe et al., 1993 Le Hir et al., 1996 Neumann et al., 1996 Tkachuk et al., 1998
TNFRII$^{-/-}$ (*p75*$^{-/-}$)	No developmental defect.	Erickson et al., 1994
OPGL$^{-/-}$	Osteopetrosis and an absence of osteoclasts. Defects in development of T and B cells. Complete absence of lymph nodes.	Kong et al., 1999
RANK$^{-/-}$	Osteopetrosis and a block in osteoclast differentiation. Absence of peripheral lymph nodes. Deficiency in splenic B cells.	Dougall et al., 1999

the absence of LTβR signaling, the B and T cell areas in the spleen do not segregate properly, and the marginal zone cannot be distinguished. Since both the LTβR and TNFRI are required for FDC generation/maintenance, it is not surprising that germinal center formation and affinity maturation are also compromised in $LT\beta R^{-/-}$, $TNFRI^{-/-}$, $LT\alpha^{-/-}$, and $LT\beta^{-/-}$ mice.

TNF-receptor family members that are involved in the generation of secondary lymphoid organs probably communicate this information to the nucleus in part by activating NFκB signals. As discussed in Chapter 4 and as outlined in Table 4.5, NFκB family members, in particular NFκB2 (p52) and Bcl-3 appear to be of critical importance in communicating this information to the nucleus.

A naturally occurring homozygous mouse mutation of the *alymphoplasia* gene leads to the absence of lymph nodes and Peyer's patches. The *aly/aly* mouse also presents with distorted thymic architecture and reduced numbers of functionally compromised T cells in the blood and spleen. In the spleen, the white pulp is markedly atrophic, and follicles are not clearly demarcated. The *aly* gene has been cloned and is identical to NIK (NFκB-inducing kinase). As discussed in Chapter 4, NIK is activated by its interaction with TRAF2, which in turn participates in signaling downstream of the TNFR, CD40, and other related receptors. NIK can induce the catalytic activation of the IκB kinase complex and thus contribute to the activation of NFκB.

While inductive and maintenance signals for lymph node and Peyer's patch formation are delivered via TNF family members as outlined above, it is clear that chemokines must also be involved in these processes, perhaps downstream of LTβR and OPGL signaling. A chemokine that is produced in a local secondary lymphoid environment will attract specific lymphoid cells to that location. The entry of B cells into follicular areas, possibly as a consequence of BLR-1/CXCR5 and CXCR4 signaling, may be required for lymph node genesis, since B cells provide the ligands for LTβR and TNFR1-mediated generation of FDC networks and germinal centers during secondary immune responses. Mice lacking CXCR5 lack inguinal lymph nodes, and B cells are unable to form follicular structures in the spleen. CCR7 is a chemokine receptor expressed on B cells, T cells, and activated mature dendritic cells (DCs). In the absence of CCR7, naive T cells and DCs are not found in lymph nodes, and T cells accumulate in the blood, bone marrow, and the splenic red pulp. Adoptive transfer experiments have revealed that CCR7 expressed on B cells is required to retain these lymphocytes for sufficient time in T cell areas to promote T–B interactions. In the absence of CCR7, the microarchitecture of secondary lymphoid organs is distorted and primary immune responses are markedly delayed. Selected knockouts of chemokines and chemokine receptors that impact upon lymphocyte development and lymph node biogenesis are listed in Table 6.4.

TABLE 6.4. Selected chemokine and chemokine receptor knockouts

Mutant	Phenotype	Reference
SDF-1$^{-/-}$	Perinatal death. Defective B lymphopoiesis and bone-marrow myelopoiesis. Ventricular septal defect. Defective cerebellar neuron migration.	Nagasawa et al., 1996
CXCR4$^{-/-}$	Perinatal death. Defective B lymphopoiesis and bone-marrow myelopoiesis. Ventricular septal defect. Defective cerebellar neuron migration.	Ma et al., 1998 Tachibana et al., 1998 Zou et al., 1998
CXCR5$^{-/-}$ (*BLR-1*$^{-/-}$)	Defective or absent Peyer's patches. Severe impairment of B cell migration to follicles in spleen and Peyer's patches. No inguinal lymph nodes.	Forster et al., 1996
CCR7$^{-/-}$	Impaired migration of naive T cells and mature dendritic cells into lymph nodes. Disordered microarchitcture of secondary lymphoid organs. Inability of B cells to be retained in outer PALS (periarteriolar lymphoid sheath) Inability to mount rapid primary immune response.	Forster et al., 1999
plt/plt (*SLC* mutation)	Reduced T cell numbers in lymph nodes, Peyer's patches, and spleen.	Vassileva et al., 1999

Emigration of Mature Lymphocytes From Central Lymphoid Organs

In the B lineage, either immature (surface IgM positive) or mature (IgM and IgD positive) B cells must emigrate from the bone marrow, while less mature pro- and pre-B stage cells should be held back. Similarly, single-positive T cells should be capable of emigrating from the thymus, while double-negative and double-positive T cells should remain in the thymus. It has been suggested that less mature B lineage cells express fibronectin receptors and that more mature cells do not. This has led to the view that immature B lineage cells are held back by stromal fibronectin receptors in the bone marrow. The evidence for such a model remains quite limited.

Emigration from the bone marrow is also regulated by signals from the B cell receptor. These signals, which will be discussed in some depth in Chapter 7, contribute to the ability of B cells that have succesfully rearranged both heavy and light chain genes to leave the bone marrow and migrate to the spleen. In mice lacking most of the cytoplasmic tail of Igα (an important component of the B cell receptor), B cells do develop in the bone marrow but very few are seen in the periphery, possibly because of a defect in emigration. Very little is understood about this emigration event; it is apparently independent of Btk (a tyrosine kinase that contributes to B cell signaling). How exactly signals from the antigen receptor contribute to emigration from the bone marrow has not ben established.

Entry of Naive Lymphocytes Into Peripheral Lymph Nodes and of B Cells Into Lymphoid Follicles

Naive B and T cells express L-selectin and are therefore capable of binding to PNAds expressed on peripheral lymph node HEVs. As discussed above, PNAds are believed to be mucin-like glycoproteins embedded in the microvillar plasma membrane of these specialized endothelial cells. The carbohydrate side chains of these proteins include the appropriate sialylated fucosylated lactosamines that bind with high affinity to L-selectin. No single protein has been identified as representing the only crucial selectin ligand in HEVs. In the majority of peripheral lymph nodes, PNAds are probably represented by CD34 (which is appropriately glycosylated only in HEVs) and by a protein known as gp200, which remains to be cloned. In mesenteric lymph nodes and in Peyer's patches, a protein known as MadCAM (which is also the ligand for the α4β7 integrin) is appropriately decorated with the carbohydrates that make it an L-selectin ligand. In these sites, naive lymphocytes may be identified by an L-selectin–MadCAM-1 interaction.

Once a naive lymphocyte binds to the HEV via this selectin–PNAd interaction, it is basically slowed down and begins to roll in the HEV lumen. It thus has the attention of the locally available chemokines, which can now begin to instruct this cell that has been "chosen" to pause in the neighborhood. The chemokines of importance may include SLC and SDF-1, which are expressed at high levels in peripheral lymph nodes. SLC is expressed at high levels in HEVs. Apart from signals from these chemokines, which are delivered through CCR7 and CXCR4, chemokine receptors expressed on naive lymphocytes, signals may also be delivered via L-selectin itself. How L-selectin functions as a signaling receptor is poorly understood, but the ligand for this signaling event is GlyCAM-1, a soluble glycoprotein expressed at high levels in peripheral lymph node HEVs. Chemokine and GlyCAM-1 both contribute to "inside-out" signaling, which leads to the activation of β2 intergrins in these naive lymphocytes. These activated integrins bind with high affinity to ICAM-1 and ICAM-2 on the endothelial surface and bring the naive lymphocyte to a halt in the lumen of the HEV. The naive lymphocyte that has been suitably activated via chemokine signaling then begins the process of migrating between endothelial cells into the tissue. This process, also known as diapedesis, is poorly understood in molecular terms. After squeezing between endothelial cells, the migrating lymphocyte must secrete enzymes that will help it break down the underlying basement membrane. Migration in the underlying tissue is probably driven by a chemokine gradient, the chemokine being held in place by heparan sulfate proteoglycans.

An overview of the mechanisms involved in the binding of naive lymphocytes to the high endothelial venules in peripheral lymph nodes is provided in Figure 6.5.

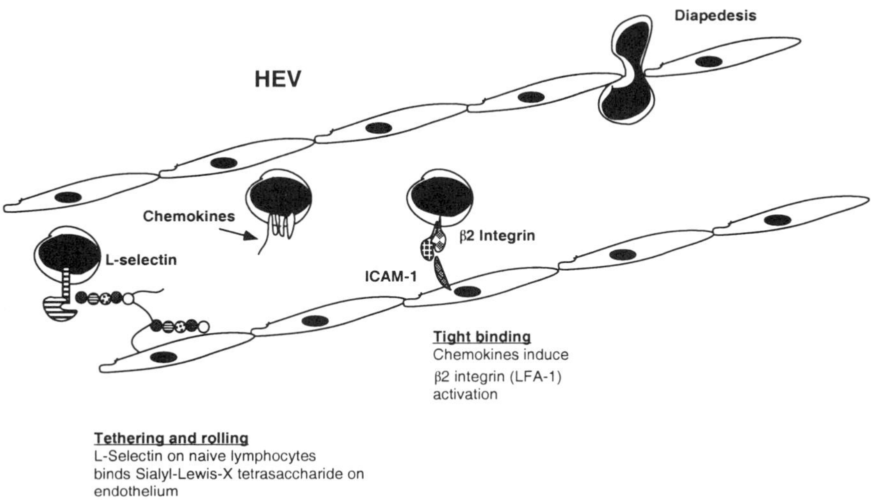

FIGURE 6.5. Multistep interaction of naive lymphocytes with high endothelial venule cells in peripheral lymph nodes.

Naive B cells that enter lymph nodes are organized in the cortex into specialized collections known as follicles. In the spleen, lymphocytes from the bloodstream enter via the marginal vein and the marginal sinus and migrate from the extrafollicular T cell area into follicles. In lymph nodes, B cells enter from HEVs into the periarteriolar T cell area and then migrate into follicles. This follicular migration event may depend on signals provided by BLC, which is the ligand for CXCR5.

Naive lymphocytes recirculate, moving from one secondary lymphoid organ to the next. While entry into these lymphoid organs is understood at least in outline, it is unclear what induces them to "move on" when they have tarried for a while in a particular site. Just as the entry of naive B cells into follicles is a regulated process, it is likely that signals are provided for naive B cells to leave follicles and for naive B and T cells to reenter the circulation when their time has come.

Entry of Activated Lymphocytes Into Target Tissues

Lymphocytes are generally activated in secondary lymphoid organs. It is important for activated T and B cells to emigrate from lymph nodes and to home to the tissue site or sites where the initiating antigen is located. However, some memory cells, known as "central" memory cells, must return to lymph nodes, much like naive T cells, to be able to respond to a new inoculum of antigen. It is useful for activated lymphocytes in general (and T cells in particular) to home to the site from which the antigen originated and to attempt to get rid of the source of the antigen. Cells with a memory/activated phenotype are $CD44^{hi}$ and $L\text{-selectin}^{lo}$. The low levels of L-selectin ensure that these cells are less efficient than naive lymphocytes in homing to lymph nodes. The high levels of CD44 presumably permit the cells to crawl more rapidly on tissue hyaluronate. Some activated/memory cells express the α4β7 integrin, and these memory cells are destined primarily for mucosal tissue and mucosa-associated lymphoid organs including Peyer's patches and mesenteric lymph nodes. Other memory cells do not express this integrin and home preferentially to peripheral tissues and nonmucosal peripheral lymph nodes. In human T cell populations, naive T cells usually have a $CD45RO^{-}$, $CD45RA^{hi}$ phenotype, while memory/activated T cells are $CD45RO^{+}$, $CD45RA^{lo}$. It is unclear why these tyrosine phosphatase isoforms have different extracellular domains and whether these serve any specific recognition function relevant to the migration of these T cell subsets. Table 6.5 lists some of the differences in adhesion molecules and chemokine receptors between naive and memory/activated T cells.

Endothelial cells that have been activated by exposure to IL-1 and TNF-α express P-selectin and E-selectin. Activated T cells lack L-selectin and therefore do not migrate into lymph nodes, but they do express selectin ligands. One such ligand expressed on at least some subsets of activated T cells is PSGL-1 (P-Selectin Glycoprotein Ligand-1), which in its appropriately glycosylated and tyrosine sulfated form, binds avidly to selectins. E-selectins on the endothelium may also bind to sialyl $Lewis^{x}$ moieties on activated lymphocytes. These moieties might be presented on CD24/heat-stable antigen expressed at high levels on activated lymphocytes. The initial selectin–carbohydrate interaction presumably causes activated T cells to roll on inflamed endothelial surfaces. Chemokines secreted in the local tissue site—the levels of which reflect an ongoing local response to a pathogen or some other source of antigen—may then induce the activation of integrins in the activated or memory T cell. A large number of chemokines have been identified as ligands for receptors expressed on activated T cells, and some of these are listed in Figure 6.2. IL-2 signaling can induce the expression of a number of CC chemokines, and IFN-γ induces IP-10 and Mig. IP-10 has been implicated in the generation of delayed type hypersensitivity responses.

Table 6.5. A comparison of adhesion molecules between naive and memory/activated T cells

	Naive	Effector/Memory
CD2	Low	High
LFA-3	–	+
L-selectin*	High	Low
α4 integrins	Low	High
CCR7*	+	–

* "Central" memory cells express CCR7 and high levels of L-selectin.

Chemokine exposure leads to the activation of integrins—α4β7 in the case of mucosal-type memory cells and the a4β1 integrin in the case of memory lymphocytes destined for the skin. The α4β7 integrin binds tightly to MadCAM-1, and after being thus brought to a halt the memory cell crosses the endothelium by diapedesis. Memory cells destined for the skin express high levels of a specialized E-selectin ligand called CLA-1. Once CLA-1 has interacted with E-selectin, the entry of activated T cells into inflamed cutaneous sites may be regulated by the specific local secretion of TARC, a chemokine that activates CCR4 which *in vivo* appears to be present on memory CD4 cells that may be destined to acquire either a Th1 or a Th2 phenotype. Table 6.6 lists some of the known differences in the surface phenotypes of memory T cells destined for mucosal sites and the skin. A view of the mechanisms involved in the extravasation of activated lymphocytes into inflamed tissue sites is provided in Figure 6.6.

The regulated wave-like migration of γδ T cells into the skin and mucosal sites during the fetal phasee of lymphocyte development probably depends on very specific homing mechanisms, which remain poorly understood. The migration of epidermal Langerhans cells from the skin into draining lymph nodes probably requires α6 integrins. Nonactivated dendritic cells express the CCR6 chemokine receptor, and when activated, these cells express CCR7, which is required for the migration of mature DC into lymph nodes. In general, memory T cells destined for mucosal sites and memory T cells destined for the skin have very different patterns of cell surface molecules.

Summary

Developing lymphocytes migrate between central and peripheral lymphoid organs and between lymphoid organs and tissues in a highly regulated manner. This regulation is dependent on specific adhesion molecules, chemokines, and chemokine receptors being expressed at the appropriate time and in the appropriate places. Cytokines of the lymphotoxin/TNF family and their receptors are critical for the generation and maintenance of secondary lymphoid organs.

TABLE 6.6. Expression patterns of adhesion molecules and chemokine receptors on T cells destined for mucosa and the skin

	Mucosa	Skin
αEβ7	+	–
α4β7	High	Low
α4β1	Low	High
α6	Low	High
CLA-1	–	+
CCR4	–	+
CCR5	+	–

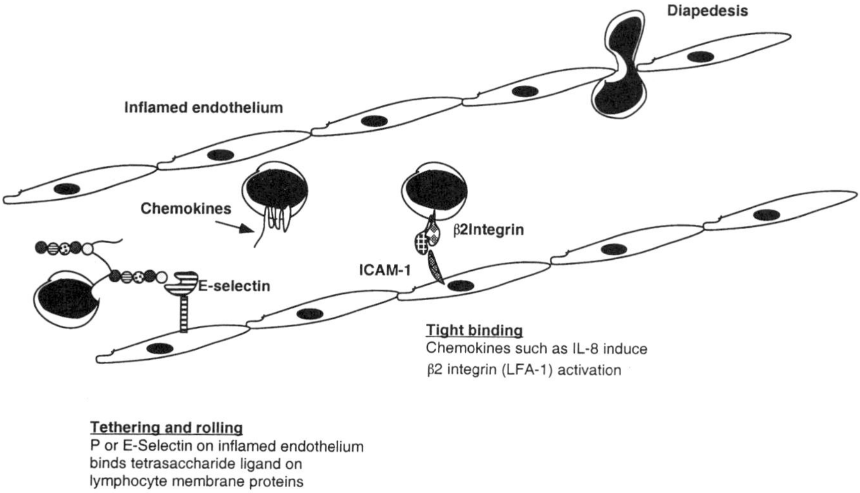

FIGURE 6.6. Migration of activated lymphocytes across the endothelium at sites of inflammation.

Selected Reviews

Baggiolini, M. (1998). Chemokines and leukocyte traffic. Nature *392*, 565–569.

Baggiolini, M., Dewald, B., and Moser, B. (1997). Human chemokines: an update. Annu. Rev. Immunol. *15*, 675–705.

Butcher, E. C. (1991). Leukocyte-endothelial cell recognition: three (or more) steps to specificity and diversity. Cell *67*, 1033–1036.

Butcher, E. C., and Picker, L. J. (1996). Lymphocyte homing and homeostasis. Science *272*, 60–66.

Chaplin, D. D., and Fu, Y.-X. (1998). Cytokine regulation of secondary lymphoid organ development. Curr. Opin. Immunol. *10*, 289–297.

Cyster, J. G. (1999). Chemokines and cell migration in secondary lymphoid organs. Science *286*, 2098–2102.

Luster, A. D. (1998). Chemokines-chemotactic cytokines that mediate inflammation. N. Engl. J. Med. *338*, 436–445.

Melchers, F., Rolink, A. G., and Schaniel, C. (1999). The role of chemokines in regulating cell migration during humoral immune responses. Cell *99*, 351–354.

Rollins, B. J. (1997). Chemokines. Blood *90*, 909–928.

Springer, T. A. (1994). Traffic signals for lymphocyte recirculation and leukocyte emigration: the multi-step paradigm. Cell *76*, 301–314.

Some Relevant References

Aiuti, A., Webb, I. J., Bleul, C., Springer, T., and Gutirrez-Ramos, J. C. (1997). The chemokine SDF-1 is a chemoattractant for human CD34$^+$ hematopoietic progenitor cells and provides a new mechanism to explain the mobilization of CD34$^+$ progenitors to peripheral blood. J. Exp. Med. *185*, 111–120.

Alimzhanov, M. B., Kuprash, D. V., Kosco-Vilbois, M. H., Luz, A., Turestskaya, R. L., Tarakhovsky, A., Rajewsky, K., Nedospasov, S. A., and Pfeffer, K. (1997). Abnormal development of secondary lymphoid organs in lymphotoxin β deficient mice. Proc. Natl. Acad. Sci. USA *94*, 9302–9307.

Arroyo, A. G., Yang, J. T., Rayburn, H., and Hynes, R. O. (1996). Differential requirements for α4 integrins during fetal and adult hematopoiesis. Cell *85*, 997–1008.

Arroyo, A. G., Yang, J. T., Rayburn, H., and Hynes, R. O. (1999). α4 integrins regulate the proliferation/differentiation balance of multilineage hematopoietic progenitors *in vivo*. Immunity *11*, 555–566.

Aurrand-Lions, M., Galland, F., Bazin, H., Zakhareyev, V. M., Imhof, B. A., and Naquet, P. (1996). Vanin-1, a novel GPI-linked perivascular molecule involved in thymus homing. Immunity *5*, 391–405.

Banks, T. A., Rouse, B. T., Kerley, M. K., Blair, P. J., Godfrey, V. L., Kuklin, N. A., Bouley, D. M., Thomas, J., Kanangat, S., and Mucenski, M. L. (1995). Lymphotoxin α deficient mice. Effects on secondary lymphoid organ formation and humoral immune responsiveness. J. Immunol. *155*, 1685–1693.

Bonecchi, R., Bianchi, G., Bordignon, P. P., D'Ambrosio, D., Lang, R., Borsatti, A., Sozzani, S., Allavena, P., Gray, P. A., Montovani, A., and Sinigaglia, F. (1998). Differential expression of chemokine receptors and chemotactic responsiveness of Type 1 T helper cells (Th1s) and Th2s. J. Exp. Med. *187*, 129–134.

Browning, J. L., Ngam-ek, A., Lawton, P., Demarinis, J., Tizard, R., Chow, E. P., Hession, C., O'Brien-Greco, B., Foley, S. F., and C. F., W. (1993). Lymphotoxin β, a novel member of the TNF family that forms a heteromeric complex with lymphotoxin on the cell surface. Cell *72*, 847–856.

Caamano, J. H., Rizzo, C. A., Durham, S. K., Barton, D. S., Raventos-Suarez, C., Snapper, C. M., and Bravo, R. (1998). NFκB2 (p100/p52) is required for normal splenic microarchitecture and B cell mediated immune responses. J. Exp. Med. *187*, 185–196.

Campbell, J. J., Haraldsen, G., Pan, J., Rottman, J., Qin, S., Ponath, P., Andrew, D. P., Warnke, R., Ruffing, N., Kassam, N., Wu, L., and Butcher, E. C. (1999). The chemokine receptor CCR4 in vascular recognition by cutaneous but not intestinal memory T cells. Nature *400*, 776–780.

Campbell, J. J., Hedrick, J., Zlotnik, A., Siani, M. A., Thompson, D. A., and Butcher, E. C. (1998). Chemokines and the arrest of lymphocytes rolling undeer flow conditions. Science *279*, 381–384.

Cepek, K. L., Shaw, S. K., Parker, C. M., Russel, G. J., Morrow, J. S., Rimm, D. L., and Brenner, M. B. (1994). Adhesion between epithelial cells and T lymphocytes mediated by E-cadherin and the αEβ7 integrin. Nature *372*, 190–193.

De Togni, P., et al. (1994). Abnormal development of peripheral lymphoid organs in mice deficient in lymphotoxin. Science *264*, 703–707.

Dougall, W. C., Glaccum, M., Charrier, K., et al. (1999). RANK is essential for osteoclast and lymph node development. Genes Dev. *13*, 2412–2424.

Ellies, L. G., Tsuboi, S., Petryniak, B., Lowe, J. B., Fukuda, M., and Marth, J. D. (1998). Core 2 oligosaccharide biosynthesis distinguishes between selectin ligands essential for leukocyte homing and inflammation. Immunity *9*, 881–890.

Erickson, S. L., deSauvage, F. J., Kikly, K., Carver-Moore, K., Pitts-Meek, S., Gillett, N., Sheehan, K. C. F., Schreiber, R. D., Goeddel, D. V., and Moore, M. W. (1994). Decreased sensitivity to tumor necrosis factor but normal T cell development in TNF receptor-2 deficient mice. Nature *372*, 560–563.

Forster, R., Mattis, A. E., Kremmer, E., Wolf, E., Brem, G., and Lipp, M. (1996). A putative chemokine receptor, BLR1, directs B cell migration to defined lymphoid organs and specific anatomic compartments of the spleen. Cell *87*, 1037–1047.

Forster, R., Schubel, A., Breitfeld, D., Kremmer, E., Renner-Muller, I., Wolf, E., and Lipp, M. (1999). CCR7 coordinates the primary immune response by establishing functional microenvironments in secondary lymphoid organs. Cell *99*, 23–34.

Futterer, A., Mink, K., Luz, A., Kosco-Vilbois, M. H., and Pfeffer, K. (1998). The lymphotoxin β receptor controls organogenesis and affinity maturation in peripheral lymphoid tissues. Immunity *9*, 59–70.

Gerber, B. O., Zanni, M. P., Uguccioni, M., Loetscher, M., Mackay, C. R., Pichler, W. J., Yawalkar, N., Baggiolini, M., and Moser, B. (1997). Functional expression of the eotaxin receptor, CCR3, in T lymphocytes co-localizing with eosinophils. Curr. Biol. *7*, 836–843.

Gunn, M. D., Ngo, V. N., Ansel, K. M., Ekland, E. H., Cyster, J. G., and Williams, L. T. (1998). A B cell homing chemokine made in lymphoid follicles activates Burkitt's lymhoma receptor-1. Nature *391*, 799–803.

Higgins, J. M. G., Mandlebrot, D. A., Shaw, S. K., Russell, G. J., Murphy, E. A., Chen, Y. T., Nelson, W. J., Parker, C. M., and Brenner, M. B. (1998). Direct and regulated interaction of integrin αEβ7 with E-cadherin. J. Cell. Biol. *140*, 197–210.

Hirsch, E., Iglesias, A., Potocnik, A. J., Hartmann, U., and Fassler, R. (1996). Impaired migration but not differentiation of haematopoietic stem cells in the absence of β1 integrins. Nature *380*, 171–175.

Kong, Y.-Y., Yoshida, H., Sarosi, I., Tan, H.-L., Timms, E., Capparelli, C., Morony, S., Oliviera-dos-Santos, A. J., Van, G., Itie, A., Khoo, W., Wakeham, A., Dunstan, C. R., Lacey, D. L., Mak, T. W., Boyle, W. J., and Penninger, J. M. (1999). OPGL is a key regulator of osteoclastogenesis, lymphocyte development, and lymph-node organogenesis. Nature *397*, 315–323.

Koni, P. A., Sacca, R., Lawton, P., Browning, J. L., Ruddle, N. H., and Flavell, R. A. (1997). Distinct roles in lymphoid organogenesis for lymphotoxins α and β revealed in lymphotoxin β deficient mice. Immunity *6*, 491–500.

Le Hir, M., Bluethmann, H., Kosco-Vilbois, M. H., Muller, M., di Pardova, F., Moore, M., Ryffel, B., and Eugster, H. P. (1996). Differentiation of follicular dendritic cells and full antibody responses require tumor necrosis factor receptor-1 signaling. J. Exp. Med. *183*, 2367–2372.

Legler, D. F., Loetscher, M., Roos, R. S., Clark-Lewis, I., Baggiolini, M., and Moser, B. (1998). B cell attracting chemokine 1, a human CXC chemokine expressed in lymphoid tissues, selectively attracts B lymphcytes via BLR1/CXCR5. J. Exp. Med. *187*, 655–660.

Loetsher, P., Uguccioni, Mo, Bordoli, L., Baggiolini, M., Moser, B., Chizzolini, C., and Dayer, J.-M. (1998). CCR5 is characteristic of Th1 lymphocytes. Nature *391*, 344–345.

Ma, O., Jones, D., Borghesani, P. R., Segal, R. A., Nagasawa, T., Kishimoto, T., Bronson, R. T., and Springer, T. A. (1998). Impaired B-lymphopoiesis, myelopoiesis and derailed cerebellar neuron migration in CXCR-4 and SDF-1 deficient mice. Proc. Natl. Acad. Sci. USA *95*, 9448–9453.

Matsumoto, M., Maraiathasan, S., Nahm, M. H., Baranyay, F., Peschon, J. J., and Chaplin, D. D. (1996). Role of lymphotoxin and the type I TNF receptor in the formation of germinal centers. Science *271*, 1289–1291.

McEvoy, L. M., Sun, H., Frelinger, J. G., and Butcher, E. C. (1997). Anti-CD43 inhibition of T cell homing. J. Exp. Med. *185*, 1493–1498.

Mebius, R. E., Rennert, P., and Weissman, I. L. (1997). Developing lymph nodes collect $CD4^+CD3^-LT\beta^+$ cells that can differentiate to APC, NK cells, and follicular cells but not T or B cells. Immunity *7*, 493–504.

Mebius, R. E., Streeter, P. R., Michie, S., Butcher, E. C., and Weissman, I. L. (1996). A developmental switch in lymphocyte homing receptor and endothelial vascular address in expression regulates lymphocyte homing and permits $CD4^+CD3^-$ cells to colonize lymph nodes. Proc. Natl. Acad. Sci. USA *93*, 11019–11924.

Miyawaki, S., Nakamura, Y., Suzuka, H., Koba, M., Yasumizu, R., Ikehara, S., and Shibata, Y. (1994). A new mutation, aly, that induces a generalized lack of lymph nodes accompanied by immunodeficiency in mice. Eur. J. Immunol. *24*, 429–434.

Nagasawa, T., Hirota, S., Tachibana, K., Takakura, N., Nishikawa, S., Kitamura, Y., Yoshida, N., Kikutani, H., and Kishimoto, T. (1996). Defects of B-cell lymphopoiesis and bone-marrow myelopoiesis in mice lacking the CXC chemokine PBSF/SDF-1. Nature *382*, 635–638.

Neumann, B., Luz, A., Pfeffer, K., and Holzmann, B. (1996). Defective Peyer's patch organogenesis in mice lacking the 55kD receptor for tumor necrosis factor. J. Exp. Med. *184*, 259–264.

Ngo, V. N., Korner, H., Gunn, M. D., Schmidt, K. N., Riminton, D. S., Cooper, M. D., Browning, J. L., Srdgwick, J. D., and Cyster, J. G. (1999). Lymhotoxin alpha/beta and tumor necrosis factor are required for stromal cell expression of homing chemokines in B and T cell areas of the spleen. J. Exp. Med. *189*, 403–412.

Ngo, V. N., Tang, H. L., and Cyster, J. G. (1998). Epstein Barr virus induced molecule 1 ligand chemokine is expressed by dendritic cells in lymphoid tissues and strongly attracts naive T cells and activated B cells. J. Exp. Med. *188*, 181–191.

Oritani, K., Kanakura, Y., Aoyama, K., Yokota, T., Copeland, N. G., Gilbert, D. J., Jenkins, N. A., Tomiyama, Y., Matsuzawa, Y., and Kincade, P. W. (1997). Matrix glycoprotein SC1/ECM2 augments B lymphopoiesis. Blood *90*, 3404–3413.

Pasparakis, M., Alexopoulou, L., Episkopou, V., and Kollias, G. (1996). Immune and inflammatory responses in TNF alpha deficient mice: a critical requirement for TNF α in the formation of primary B cell follicles, follicular dendritic cell networks and germinal centers, and in the maturation of the humoral immune response. J. Exp. Med. *184*, 1397–1411.

Pfeffer, K., Matsuyama, T., Kundig, T. M., Wakeham, A., Kishihara, K., Shahinian, A., Wiegmann, K., Ohashi, P. S., Kronke, K., and Mak, T. W. (1993). Mice deficient for the 55kd tumor necrosis factor receptor are resistant to endotoxic shock, yet succumb to *L. monocytogenes* infection. Cell *73*, 457–467.

Price, A. A., Cumberbatch, M., Kimber, I., and Ager, A. (1997). $\alpha 6$ integrins are required for Langerhans cell migration from the epidermis. J. Exp. Med. *186*, 1725–1735.

Randolph, D. A., Huang, G., Carruthers, C. J. L., Bromley, L. E., and Chaplin, D. D. (1999). The role of CCR7 in Th1 and Th2 cell localization and delivery of T cell help. Science *286*, 2159–2162.

Rennert, P. D., Browning, J. L., Mebius, R., Mackay, F., and Hochman, P. S. (1996). Surface lymphotoxin α/β complex is required for the development of peripheral lymphoid organs. J. Exp. Med. *184*, 1999–2006.

Rothe, J., Lesslauer, W., Lotscher, H., Lang, Y., Koebel, P., Kontgen, F., Althage, A., Zinkernagel, R., Steinmetz, M., and Bluethmann, H. (1993). Mice lacking the tumor necrosis factor receptor 1 are resistant to TNF mediated toxicity but highly susceptible to infection by *Listeria monocytogenes*. Nature *364*, 798–802.

Sallusto, F., Lenig, D., Forster, R., Lipp, M., and Lanzavecchia, A. (1999). Two subsets of memory T lymphocytes with distinct homing potentials and effector functions. Nature *401*, 708–712.

Sallusto, F., Lenig, D., Mackay, C. R., and Lanzavecchia, A. (1998). Flexible programs of chemokine receptor expression on human polarized T helper 1 and 2 lymphocytes. J. Exp. Med. *16*, 875–883.

Sallusto, F., Mackay, C. R., and Lanzavecchia, A. (1997). Selective expression of the eotaxin receptor CCR3 by human T helper 2 cells. Science *227*, 2005–2007.

Shinkura, R., Kitada, K., Matsuda, F., Tashiro, K., Ikuta, K., Suzuki, M., Kogishi, K., Serikawa, T., and Honjo, T. (1999). Alymphoplasia is caused by a point mutation in the mouse gene encoding NFκB inducing kinase. Nature Genet. *22*, 74–77.

Stein, J. V., Rot, A., Luo, Y., Narasimhaswamy, M., Nakano, H., Gunn, M. D., Matsuzawa, A., Quackenbush, E. J., Dorf, M. E., and von Andrian, U. H. (2000). The CC chemokine thymus derived chemotactic agent 4 (TCA-4, secondary lymphoid tissue chemokine, 6Ckine, exodus-2) triggers lymphocyte function associated antigen 1-mediated arrest of rolling T lymphocytes in peripheral lymph node high endothelial venules. J. Exp. Med. *191*, 61–76.

Tachibana, K., Hirota, S., Iizasa, H., Yoshida, H., Kawabata, K., Kataoka, Y., Kitamura, Y., Matsushima, K., Yoshida, N., Nishikawa, S.-I., Kishimota, T., and Nagasawa, T. (1998). The chemokine receptor CXCR4 is essential for vascularization of the gastrointestinal tract. Nature *393*, 591–594.

Tkachuk, M., Bolliger, S., Ryffel, B., Pluschke, G., Banks, T. A., Herren, S., Gisler, R. H., and Kosco-Vilbois, M. H. (1998). Crucial role of tumor necrosis factor receptor 1 expression on nonhematopoietic cells for B cell localization within the splenic white pulp. J. Exp. Med. *187*, 469–477.

Vassileva, G., Soto, H., Zlotnik, A., Nakano, H., Kakiuchi, T., Hedrick, J. A., and Lira, S. A. (1999). The reduced expression of 6Ckine in the plt mouse results from the deletion of one of two 6Ckine genes. J. Exp. Med. *190*, 1183–1188.

Wagner, N., Lohler, J., Kunkel, E. J., Ley, K., Leung, E., Krissansen, G., Rajewsky, K., and Muller, W. (1996). Critical role for β7 integrins in formation of the gut associated lymphoid tissue. Nature *382*, 366–370.

Warnock, R. A., Askari, S., Butcher, E. C., and von Andrian, U. H. (1998). Molecular mechanisms of lymphocyte homing to peripheral lymph nodes. J. Exp. Med. *187*, 205–216.

Williams, M. B., and Butcher, E. C. (1997). Homing of naive and memory T lymphocyte subsets to Peyer's patches, lymph nodes, and spleen. J. Immunol. *159*, 1746–1752.

Yokota, Y., Mansouri, A., Mori, S., Sugawara, S., Adachi, S., Nishikawa, S.-I., and Gruss, P. (1999). Development of peripheral lymphoid organs and natural killer cells depends on the helix-loop-helix inhibitor Id2. Nature *397*, 702–706.

Zou, Y.-R., Kottmann, A. H., Kuroda, M., Taniuchi, I., and Littman, D. R. (1998). Function of the chemokine receptor CXCR4 in haematopoiesis and cerebellar development. Nature *393*, 595–599.

7

The Generation of Naive B Lymphocytes

B cells that have completed their developmental program in the postnatal bone marrow leave this site to mature further in the periphery and eventually inhabit the B cell areas or follicles in peripheral lymphoid organs. These cells make up the conventional B cell pool and are called *naive B cells*, *mature B cells*, or *follicular* or *recirculating B cells*. They express both surface immunoglobulin M (IgM) and IgD, high levels of CD23, intermediate levels of CD21/CR2 (the type 2 complement receptor), and low levels of the heat-stable antigen (HSA/CD24). These cells do not express the pB130–140 marker (also called the 493 antigen) found on newly formed B cells. Newly formed "virgin" or "immature" B cells represent cells on their way to becoming naive B cells and are found in the bone marrow and in the extrafollicular regions of the spleen. They express high levels of surface IgM, CD24/HSA, pB130–140, and little or no IgD. They do not express CD23 and can be distinguished from follicular cells as well as from marginal zone B cells (see below) by their low to absent levels of surface CD21. While the development of naive B cells has generally been thought to be antigen independent, it is likely that both self-antigens as well as antigens derived from endogenous flora influence the differentiation of these cells, both positively as well as negatively. B cell development is initiated in the fetal liver prior to birth and in the bone marrow, in the postnatal period. While hematopoietic progenitors are first seen in fetal liver around day 11 of fetal life, mature B cells are first seen on embryonic day 17. In this chapter, we will concern ourselves with the generation, in central lymphoid organs, and subsequently in the periphery, of mature follicular B cells. A subset of B cells, referred to as B-1 cells, or the B-1 lineage, originates primarily from fetal liver-derived stem cells. Since cells of this lineage resemble activated B cell in some ways, they will be discussed in Chapter 8. Conventional B cells originate from bone marrow-derived HSCs and are sometimes referred to as B-2 cells. While it is unclear whether B-1 cells truly represent naive B cells or activated B cells, a third B cell population, marginal zone (MZ) B cells, will also be discussed in the next chapter. These cells are probably derived from conventional follicular B cells and are distinct from B-1 cells. While, like B-1 cells, they share some of the features of activated B cells, they may actually represent an intriguing and distinct population of naive B cells.

While the initial steps of B lymphoid commitment and development depend, in part, upon signals from stromal cells, unique B lineage-specific selection events drive the developmental program after Ig heavy chain gene rearrangement is initiated. Some of these events are based on the nature of the rearrangements that have occurred at the heavy chain locus and will be described in detail in this chapter. In some pro-B cells, following D to J_H rearrangement, a particular D segment reading frame can be utilized to generate a truncated heavy chain protein. Cells that express this truncated protein may be selected against. This phenomenon is poorly understood, and it is unclear whether this pro-B selection event occurs in species other than the mouse. Approximately one-third of developing pre-B cells in the bone marrow make in-frame V to DJ_H rearrangements, and these cells are positively selected (positive selection I). There may be differences in the way cells are selected at the pre-B stage in the bone marrow as compared with the fetal liver. These potential differences will be discussed later in this chapter. In the bone marrow, pre-B receptor signaling mediates both positive selection I as well as allelic exclusion at the IgH locus. Once both heavy and light chain genes are rearranged and expressed, surface IgM^+ immature B cells may also be positively selected in response to antigen receptor-mediated signals (positive selection II). If these cells express self-reactive antigen receptors, they may be subject to receptor editing, or they may be clonally eliminated or anergized. An overview of the different stages of B cell development, following the nomenclature developed by Hardy and colleagues (1991), is provided in Figure 7.1. Broadly similar schemes proposed by Rolink and Melchers (1991) and Osmond (1993) will be described later.

When Is a Cell Irreversibly Committed to the B Lineage?

Cells that express specific cell surface markers and μ0 transcripts are "committed" to the B lineage. These heavy chain locus germline transcripts are initiated upstream of the J_{H1} gene segment (see Chapter 2 for details) and are spliced, incorporating RNA segments corresponding to Cμ exons. It is important to reiterate that commitment is most likely a gradual stepwise event and is not always completed in a single stroke. When is a cell committed to the B lineage in the sense that there is no going back? In the absence of Pax-5, pro-B cells that have completed D to J_H rearrangement and appear to have progressed well into the B lineage have been demonstrated to be multipotent with the ability to give rise to many other

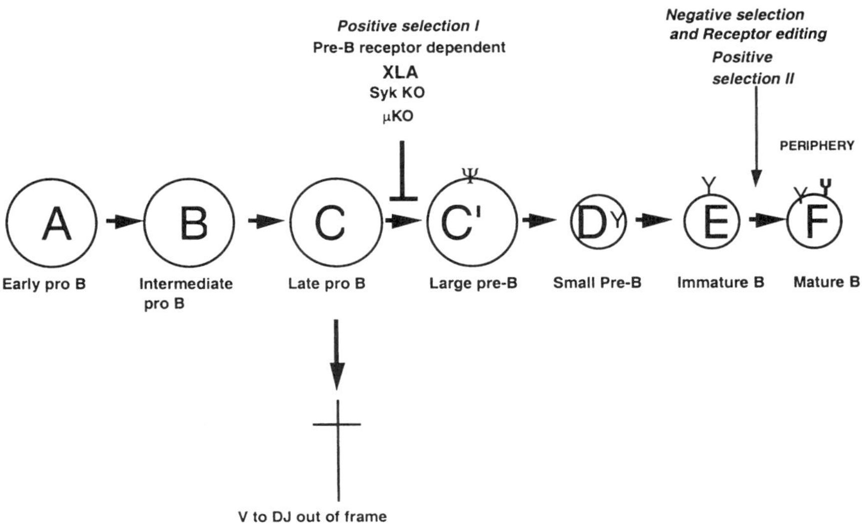

FIGURE 7.1. A simplified overview of B cell development. The A-F nomenclature is that of Hardy and colleagues (1991), while the names of stages follow the simplified nomenclature discussed in Figure 7.3. Details regarding substages of stages A and F, the surface characteristics of each stage, and alternative nomenclatures, are discussed later in this chapter. Stage A may include uncommitted progenitors as well as the earliest cells in the bone marrow, which have entered the B lineage but have yet to rearrange their Ig heavy chain genes. D to J_H rearrangement peaks during stage B. V to DJ_H rearrangement is initiated during stage C. Cells that make productive heavy chain gene rearrangements receive signals via the pre-B receptor and expand into a population of proliferating cells represented by stage C′. In humans (but not in mice), this transition requires Btk-derived signals. The gene encoding this kinase is mutated in boys with X-linked agammaglobulinemia (XLA).

hematopoietic lineages. In the absence of Pax-5, irreversible commitment to the B lineage apparently cannot occur. On the other hand, studies with wild-type B lineage cells have demonstrated that commitment to the B lineage occurs prior to D to J_H rearrangement (in Fractions A1 and A2, described later in this chapter).

Commitment probably begins when cells begin to express certain critical regulators of gene expression in B cells. These include *E2a*-derived proteins such as E12 and E47, EBF, and Pax-5. These factors have been briefly considered in Chapter 2 and will be discussed below. Commitment in an irreversible sense probably requires that a number of distinct molecular events must be completed; it now appears that these events require Pax-5, but are completed even before the onset of heavy chain gene rearrangement.

Precursors for the B Lineage

Most B cells presumably originate from the common lymphoid progenitor discussed in Chapter 2. This is a cell that has no myeloid potential but can give rise to B, T, natural killer (NK), and lymphoid dendritic cells. A second type of B-myeloid progenitor has been described that apparently can give rise to either B cells or macrophages. Fetal liver cells that are $AA4.1^+$, $B220^-$, $Mac\text{-}1^-$, and $Ly\text{-}6A^+$ may have the potential to clonally give rise to B cells and macrophages. It is unclear at this time as to what proportion of B cells are derived from a common lymphoid progenitor and what proportion arise from a specialized B-myeloid progenitor. It is also conceivable, though there is no evidence to support this possibility, that the B-myeloid progenitor is a precursor of the common lymphoid progenitor and is capable of giving rise to macrophages, T, B, NK, and LD cells.

A unique link between B cell and macrophage development, however, has been suggested from other unrelated studies. $CD5^+$ B cell lines have been described that can acquire a macrophage-like phenotype, retaining their clonal Ig gene rearrangements while they express macrophage-specific genes and repress B lineage-type characteristics. Whether a separate B-myeloid type of progenitor for B cells actually exists either in fetal liver or in the postnatal bone marrow, remains to be confirmed.

Growing evidence suggests that a cell fate decision may drive a lymphoid progenitor into the T lineage instead of the B lineage or vice versa. Apparently, when a common lymphoid progenitor receives signals via Notch1, it is driven into the T lineage at the expense of assuming a B cell fate. Mice with a conditional mutation in Notch1 lack T cells but present with the exaggerated development of thymic T cells. Conversely, overexpression of an activated form of Notch1 leads to the preferential development of T cells and a marked inhibition of B cell development.

Stromal Factors and the Development of B Cell Progenitors From Hematopoietic Stem Cells

As we have discussed in Chapter 2, HSCs presumably respond to external stimuli that commit them to distinct differentiation programs. These external stimuli may initiate signaling pathways that induce specific nuclear factors to drive the process of differentiation. A challenge in the field is to connect specific ligands, receptors, and signaling pathways, to transcription factors that drive the developmental process. A number of extracellular factors may contribute to the process of B cell commitment, and some of these are considered in Figure 7.2.

Cell surface receptors that may be required for B cell commitment include the $\alpha4\beta1$ integrin, the Flt3 receptor tyrosine kinase, the c-Kit receptor tyrosine kinase, the interleukin-7 receptor (IL-7R), the chemokine

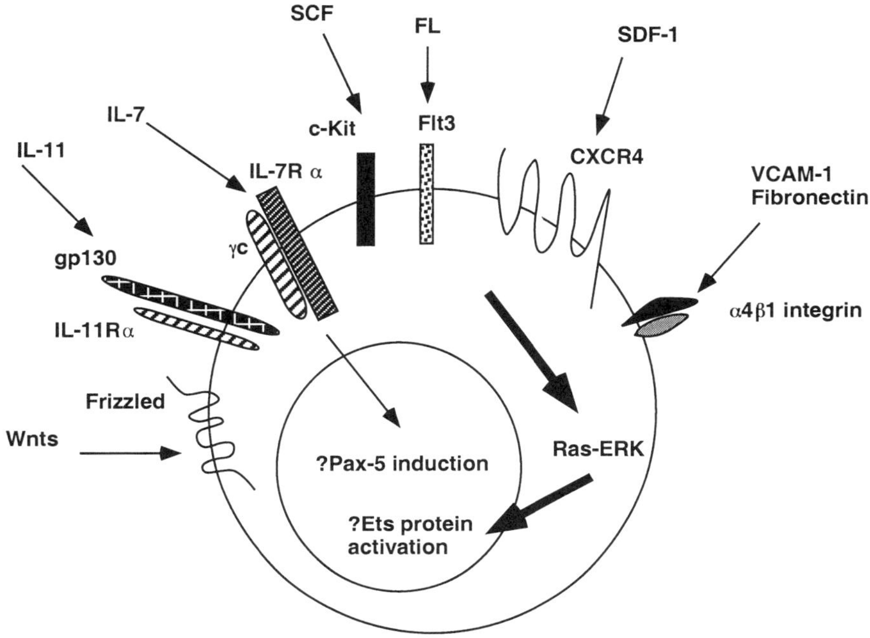

FIGURE 7.2. Some of the factors that may contribute to B lineage commitment and differentiation. Cell surface receptors that are critical include the Flk2/Flt3 receptor tyrosine kinase, the c-Kit receptor tyrosine kinase, the α4β1 integrin, the interleukin-7 (IL-7) receptor, the IL-11 receptor, and the CXCR4 chemokine receptor. The Pax-5 transcription factor might be induced downstream of the IL-7R, and the Ras/ERK pathway might contribute to the activity of Ets family transcription factors that drive development. Wnts secreted by stromal cells may signal through specific Frizzled receptors on pro-B cells.

receptor CXCR4, and the IL-11 receptor. Flt3/Flk2 is a receptor tyrosine kinase that is activated by a soluble protein secreted by stromal cells known as the Flt3 ligand (FL). In *flk2* null mice, the generation of the earliest B cell progenitors is compromised. The lack of pro-B cells is particularly striking in *c-kit/flk2* double-mutant mice. c-Kit is also a receptor tyrosine kinase, which like Flk2 is expressed on hemopoietic stem cells (HSCs) capable of long-term reconstitution of hematopoietic cells. The ligand for c-Kit is known as stem cell factor (SCF), or Steel, and may be secreted or exist in a membrane-bound form on stromal cells. While B cell development is not significantly compromised in *c-Kit* mutant mice (such as the viable *W/Wv* mouse), flk2$^{-/-}$ homozygotes on a *W/Wv* background have a profound defect in hematopoiesis, and make very few B and T cell progenitors. These receptor tyrosine kinases probably have overlapping roles in the generation, maintenance, or expansion of pro-B cells or their progenitors.

The α4β1 integrin may be required for B lineage commitment in the bone marrow but not in the fetal liver. IL-7 and the IL-7R may not be absolutely required for entry into the B lineage but may be necessary for progression through the B lineage (in mice, but not in humans). As described in Chapter 2, the IL-7R is first expressed on common lymphoid progenitor cells. In the absence of IL-7Rα, B cell generation is impaired, though not completely abrogated, and the rearrangement of heavy chain genes and the progression of apparently committed cells through the process of differentiation is compromised. The defect, albeit incomplete, in development seen in the absence of IL-7Rα (or of downstream signaling components such as γc or Jak-3) may not reflect the absence of IL-7 signaling alone. Apart from responding to IL-7, the IL-7R may also be triggered by an additional ligand or ligands such as TSLP (thymic stromal lymphopoietin, see Chapter 4). This might explain why the phenotype of IL-7 null mice is less pronounced than that of IL-7R null mice.

While IL-11 might contribute to early B cell development as evidenced by its ability to influence B lymphopoiesis in *in vitro* studies, knockout experiments have revealed that IL-11 is not essential for the development of this lineage. In the absence of the chemokine stromal cell-derived factor-1 (SDF-1), commitment to the B lineage is apparently not compromised but progenitor B cells (in stages A and B) fail to proliferate.

Ligands of the Wnt family secreted by stromal cells may trigger receptors of the Frizzled group on B lineage progenitors. The significance of Wnt–Frizzled signaling in B lineage commitment and developmental progression remain unclear at this time.

Characterization of B Cell Development in the Bone Marrow and Periphery Using Surface Markers

There are a number of useful ways in which to categorize the stages of B cell development in the bone marrow. A scheme developed by Hardy (Hardy et al., 1991), and another by Rolink and Melchers (1991), both using a range of surface markers, are widely used in the field. Another classification that relies in part on intracellular staining for TdT and cytoplasmic μ chains was developed by Osmond (see review by Osmond, 1993). In the scheme of Hardy and colleagues, the earliest B lineage cells (stage A) express the B220 cell surface marker (the long isofom of the CD45 receptor tyrosine phosphatase). They also express high levels of CD43 or leukosialin but do not express CD24/HSA. A "revised" and more detailed categorization of Stage A cells reveals the existence of $CD45^-$ and $CD45^+$ cells as described below. Stage A "pre-pro B" cells mature into $B220^+$ $CD43^+$ cells which also express high levels of CD24/HSA. These latter pro-B cells (stage B in the Hardy classification), express the CD19 marker, initiate D_H to J_H rearrangement, and progress to stage C when they express the BP-1 cell surface marker as well. Stage C represents cells in which D to J_H rearrangement is completed. V_H to DJ_H rearrangement initiates soon after, peaking in stage C′. Once a heavy chain protein is made, the pre-B receptor drives stage C′ cells to enter the cell cycle and expand. These cells express higher levels of CD24, synthesize μ heavy chains, which may associate with surrogate light chains but are retained in the endoplasmic reticulum (so-called cytoplasmic μ), and may express low levels of the pre-B receptor on the cell surface. Cells at stage C/C′ that fail to productively rearrange their heavy chain genes drop out of differentiation and die. All subsequent stages in the maturation process (stages D-F) are generally considered to involve no further cell division. This view has been recently questioned and it seems likely that both naive B and T cells may actually divide occasionally, perhaps as part of a homeostatic process.

Fraction D consists of small pre-B cells that have downregulated surrogate light chain gene expression but that still express cytoplasmic μ heavy chains. These cells have stopped dividing, do not express CD43, and progress into the surface IgM-expressing immature B cell stage (Fraction E). Both positive and negative selection events occur at the immature B stage, and these will be considered later in this chapter. Mature B cells express both IgM and IgD (Fraction F) and emerge in the periphery initially as surface $IgM^{hi}IgD^{lo}$ cells. The further maturation of B cells in the periphery will be discussed toward the end of this chapter.

By definition a B cell precursor that does not make a heavy chain protein is a pro-B cell, whereas a precursor that synthesized μ is a pre-B cell. An attempt has been made to reconcile the schemes for the stages of B cell development according to Hardy (1991), Osmond (see review, 1993), and Rolink and Melchers (see review, 1991), in Figure 7.3. While these schemes are broadly similar, it is sometimes difficult to match stages in a totally consistent manner. In general, we will refer to subpopulations using the Hardy Fractions A-F nomenclature, since the majority of gene knockouts that affect the B lineage are characterized using this approach (an increasing number of studies do use the Rolink and Melchers classification). B lineage cells that have not yet completed V_H to DJ_H rearrangements, and therefore do not express the μ protein, are referred to as pro-B cells. For the sake of simplicity, Fractions A, B, and C are called early, intermediate, and late pro-B cells respectively (combining the simpler terminology from the Osmond nomenclature with the Fractions characterized by Hardy). Cells expressing cytoplasmic μ are called pre-B cells. Fraction C′ cells represent large pre-B cells and Fraction D is made up of small pre-B cells.

The Earliest Identifiable B Lineage Cells

Fraction A cells were originally classified as $B220^+$, $CD43^+$, HSA^- cells. They are now more conveniently categorized as $AA4.1^+$, $CD43^+$, HSA^- cells which include both B cell progenitors as well as "committed" B lineage cells.

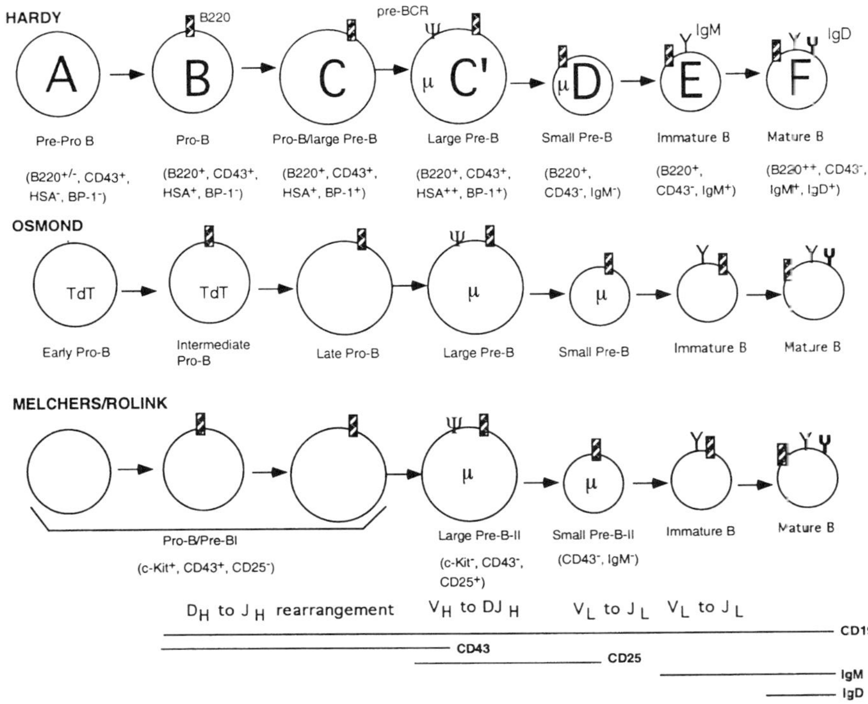

FIGURE 7.3. Characterizing the stages of early B cell development. Schemes suggested by Hardy (Hardy et al., 1991), Osmond (see reviews, 1986, and Osmond et al., 1998), and Rolink and Melchers (see reviews 1991, and Osmond et al., 1998) are presented in outline. Although the names used by the various authors are mentioned in this figure, for the sake of simplicity we will refer to pro-B cells in Fractions A, B, and C as early, intermediate, and late pro-B cells respectively.

Fraction A0 includes cells that do not express B220 but express low levels of CD4 (Figure 7.4). These cells may include the common lymphoid progenitor discussed in Chapter 3. Fraction A1 cells are the earliest B cells that express B200; they also express low levels of CD4. Fraction A2 cells are derived from Fraction A1 cells and do not express CD4.

Regulatory Events in Pro- and Pre-B Cell Differentiation

The primary goal of a B cell progenitor is to successfully rearrange Ig heavy and light chain genes and to thus generate a B cell with a functional antigen receptor. A number of selection events are critical during this process of

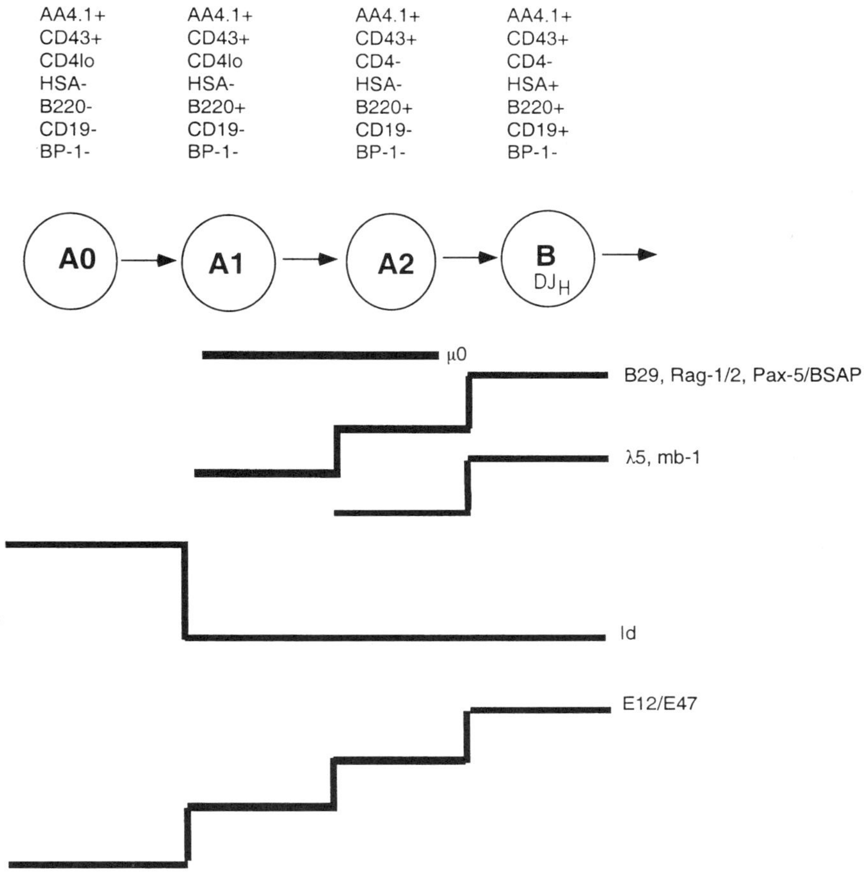

FIGURE 7.4. A more detailed view of the early stages of B cell development. In uncommitted A0 cells, high levels of Id, an inhibitory nonactivating E box-binding protein, are expressed that drop when cells enter the B lineage. Some of the genes turned on in A1 stage cells include the B29 gene (which encodes the Igβ protein), the *Rag1* and *Rag2* genes, and the *Pax-5/BSAP* gene. μ0 transcripts are also seen soon after commitment.

differentiation. Many of these cell fate decisions are driven by signals from receptors that contain the membrane form of Ig heavy chains. Progression involves the successful negotiation of checkpoints by the "fortunate" cells that have made the right kind of rearrangement product or products. It makes sense therefore that most B lineage-specific genes that are expressed during early development should either encode proteins that are essential for setting up the rearrangement machinery or that contribute to the generation of signals that are critical at these developmental checkpoints. The mechanisms involved in generating functionally relevant Ig heavy chain gene rearrangements have been considered in some depth in Chapter 3.

As discussed above, stromal signals are responsible for the commitment and proliferation of the earliest B cell progenitors. These signals influence the expression or function of specific DNA-binding proteins that are required for V(D)J rearrangement or that cooperate to induce the expression of proteins required at early checkpoints during differentiation. While the Ikaros and PU.1 transcriptional regulators are essential for some unknown aspects of commitment, extracellular ligands such as FL and the $\alpha4\beta1$ integrin, may contribute to the B lineage commitment process. Signals from SDF-1 are essential for proliferation of early progenitors but might also contribute to the differentiation process. Rearrangement of the heavy chain locus first involves a D_H–J_H joining event followed by the joining of a V_H segment to the rearranged DJ_H locus. For rearrangement to proceed at the heavy chain locus, the locus must be "open" and accessible in terms of chromatin structure. The lymphoid-specific *Rag-1*, *Rag-2*, and *TdT* (*terminal deoxynucleotidyl transferase* genes) genes need to be expressed, and genes that are involved in signaling events that influence the selection of pro-B and pre-B cells also need to be turned on. IL-7 and IL-7R signaling may also contribute to the efficiency of the D_H–J_H rearrangement process and may be essential for the subsequent rearrangement of distal V_H segments. It has been postulated that IL-7R signaling might induce the expression of the BSAP/Pax-5 transcription factor. Since Pax-5 appears to be essential for progenitor lymphoid cells to commit to the B lineage, IL-7 may actually be an important factor for commitment to the B lineage.

A dominant negative form of the Ras protein (N-17 Ras) that was transgenically targeted to the B lineage-compromised B cell development at the earliest detectable stage, possibly at the commitment step itself (Fraction A was drastically reduced). This inhibition could be overcome by overexpression of an activated (membrane-attached) form of Raf in the B lineage. This suggests that the Ras/ERK MAP kinase pathway is critical at a very early stage of B cell development. In Drosophila eye development, signaling via the Sevenless receptor tyrosine kinase results eventually in the extracellular signal-regulated kinase (ERK)-dependent phosphorylation of the Ets family transcription factors Yan and Pointed. Phosphorylation of these Ets family proteins is essential during development. Similarly, in B cell development, Ets family transcription factors play a critical role. It is possible that signaling via the Ras pathway is required for the transcriptional activity of Ets family members (such as PU.1, which is probably required very early) that drive B cell development. In the dominant negative Ras experiment, the transgene was expressed under the regulation of the Lck promoter and the IgH enhancer. Given that Fraction A was virtually missing, it is conceivable that the transgene was expressed prior to B cell commitment and that it interfered with Ras signaling at or before the commitment stage. It is possible that Ras might be important downstream of FL signaling and that this pathway targets an Ets protein, possibly PU.1 itself (which may be required at earlier stages of hematopoietic development as well as at the B lineage commitment point) or some other critical Ets family protein.

Transcription Factors and Early B Cell Development

The early stages of B cell development depend on the expression of transcription factors, some of which were considered in Chapter 2. These include Ikaros and PU.1, which are required prior to B lineage commitment; early B cell factor (EBF) and E2A, which are necessary soon after commitment or during the commitment process; and Sox-4 and Pax-5/BSAP, which were generally thought to be required during progression through early B cell development. It is now clear that Pax-5 is essential for "fixing" the commitment process. The requirement for EBF, E2A proteins, Pax-5, and Sox-4 will be considered below. Apart from the potential role of IL-7 in activating BSAP/Pax-5, the external signals that drive the expression of these transcription factors remain to be identified.

Before B cell commitment, the common lymphoid progenitor expresses relatively high levels of Id (an E box-binding protein that lacks a transactivation domain) and thus E box-binding E2A proteins may be functionally repressed. Some combination of stromal signals may contribute to a signaling process that leads to a drop in the levels of Id and the induction of the EBF transcription factor. The drop in Id levels might permit the activation of E2A proteins such as E12 and E47. E2A proteins and EBF collaborate and probably contribute jointly to some of the earliest events in the progression of B cell development (if not to the commitment process itself). These (and other) transcription factors probably contribute to making the μ locus accessible for transcription, and the generation of $\mu 0$ transcripts may reflect the initiation of this process. It must be stressed that the proteins that actually contribute to the opening of the IgH locus *in vivo* remain to be be identified. E2A, EBF, and other transcriptional regulators may induce the expression of the *B29* gene (which encodes the Igβ protein), the *λ5* and *vpreB* surrogate light chain genes, and the *mb-1* gene (which encodes the Igα protein). These proteins are essential for membrane Ig-derived checkpoint signals during development and must be synthesized by the cell prior to the generation of rearrangement products. E2A may also be required for the transcriptional induction of Pax-5. Some combination of the above transcriptional regulators also induces the *Rag-1* and *Rag-2* genes in stage A1, and *TdT* is also induced around this time. In the absence of EBF or E2A, B lymphocyte development does not progress beyond stage A2 (see Figures 7.4 and 7.9). In the absence of the Pax-5/BSAP transcription factor or of the Sox-4 protein, B cell development is blocked at stages

B and C, and, as mentioned earlier and discussed below, cells fail to be irreversibly committed to the B lineage.

Pax-5/BSAP (B lineage-specific activator protein) is a member of the Pax family of proteins, which contain a paired domain DNA-binding motif. Structurally, Pax-5 contains an N-terminal paired box domain, an octapeptide motif that is found in Pax family proteins, and a partial homeodomain. The protein is post-translationally glycosylated (the sugar added is O-linked *N*-acetylglucosamine—part of a cytosolic/nuclear form of glycosylation whose functional relevance for transcription is unclear). Four differentially spliced versions of Pax-5 exist, and one or another isoform is expressed at every stage of B cell differentiation, including the plasma cell stage. B cell-specific genes that might be regulated by Pax-5/BSAP include *CD19, vpreB1*, and *λ5*. In mice in which the *pax-5* gene has been targeted by homologous recombination, bone marrow-derived B cell development progresses until the D-J_H rearrangement stage, but V to DJ_H recombination occurs inefficiently (the block is between the B and C stages). It is possible that this phenotype reflects, in part, a defect in IL-7R signaling. Strikingly, the pro-B cells that accumulate in the absence of Pax-5 remain uncommitted to the B lineage and can be induced to give rise to a range of other hematopoietic cell types.

In the absence of Pax-5, no B cell development occurs in the fetal liver. When *pax-5*$^{-/-}$ fetal liver stem cells are transferred to adult recipients, B cell development progresses to Fraction B, suggesting that in the fetal liver Pax-5 is induced in B lineage cells by some external factor that is required in the fetal environment. Pax-5 might have a role to play in later stages of B cell development, and these aspects will be discussed in Chapter 8.

The Sox-4 protein contains an N-terminal high-mobility group (HMG) box, which contributes to the binding of specific DNA sequences, and a serine-rich C-terminus, which serves as an activation domain. Sox-4 is a homologue of the Y-chromosomal gene *Sry*, which is involved in mammalian sex determination. The sequence recognized by Sox-4 is an AACAAAG motif, which has been found in the enhancers a number of genes including *CD4* and *CD3ε*. The HMG box binds to the AACAAAG motif in a somewhat unusual manner in the minor groove and may contribute to inducing a strong bend in the DNA. Many HMG box-containing proteins are believed to influence gene expression in part by the induction of DNA bending. Although *Sox-4* is expressed at several sites during embryogenesis, in the adult mouse, it is expressed primarily in the early stages of B and T cell development. In the absence of Sox-4, mice have defects in cardiac septation and die of circulatory failure at day E14. However, Sox-4-deficient fetal liver stem cells have been introduced into lethally irradiated mice to examine hemopoiesis. B cell development is blocked around stage C, mimicking the defect seen in Pax-5/BSAP-deficient mice. A Sox-4 motif has been noted in a putative *λ5* regulatory region. The exact transcriptional targets of Sox-4 in the B lineage that are critical for this block remain to be ascertained.

Membrane Ig Based Signals in Pro-B and Pre-B Cells

A rearranged Ig heavy chain gene encodes two forms of the heavy chain protein, a membrane form and a secreted form. mRNAs specific for these two forms are generated as a result of selective polyadenylation and splicing. It is the membrane form of the Ig heavy chain that is involved in signal transduction, as part of the pro-B receptor in pro-B cells, the pre-B receptor in pre-B cells, and the antigen receptor in B cells.

Broad differences in membrane and secretory Ig heavy chains are highlighted in Figure 7.5.

The membrane and secretory forms of Ig heavy chains of all isotypes are generated by alternative RNA processing. As seen in Figure 7.6, cleavage directed by an upstream polyadenylation site in the μ locus yields the secretory, or μs RNA, while the use of a distal polyadenylation site results in the generation of μm, or the membrane form of the μ heavy chain. The process is actually quite complex. When the distal polyadenylation site is used, alternative splicing results in the exclusion of a portion of the μs-coding sequence (that encodes the tailpiece of secretory IgM) and also excludes the proximal μs-specific polyadenylation site. Through most stages of B cell differentiation, RNA processing generates both forms of heavy chain

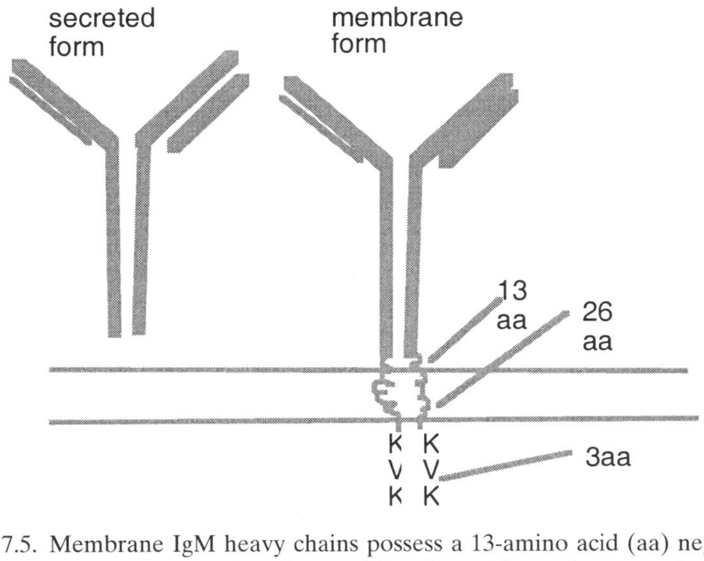

FIGURE 7.5. Membrane IgM heavy chains possess a 13-amino acid (aa) negatively charged extracellular connecting piece, a 26-amino acid membrane anchor, and a short cytoplasmic tail (KVK).

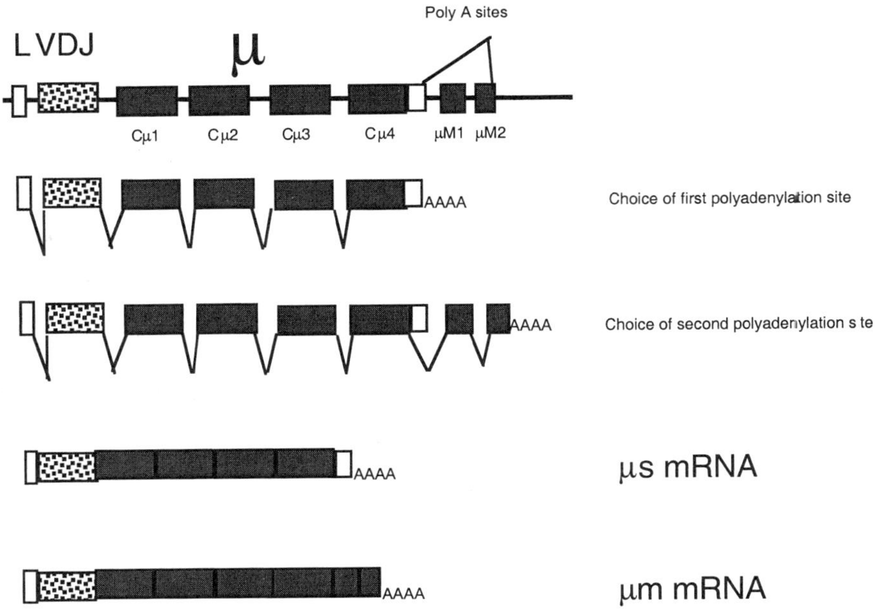

FIGURE 7.6. Membrane and secretory forms of the μ heavy chain are generated by alternative splicing and selective polyadenylation. The relative levels of one general polyadenyaltion factor, CstF-64, may influence the ratios of the membrane and secretory forms that are generated.

message, with more μm being produced in pre-B and B cells. However, in plasma cells, the secretory form is far more efficiently generated. It appears that an increase in the efficiency of 3′ end cleavage and polyadenylation is sufficient for plasma cells to make the switch from μm to μs RNA. Different levels of splicing and polyadenylation factors during B cell development modulate the relative levels of μm and μs during development. One potential regulator of the relative levels of μm and μs RNA has been identified. This factor is part of the basal polyadenylation machinery (Figure 7.7).

A large number of proteins are involved in pre-mRNA polyadenylation. The polyadenylation signal in all pre-mRNAs is an AAUAA sequence. Downstream of this signal, a GU-rich region influences the efficiency of polyadenylation. The cleavage-polyadenylation specificity factor (CPSF) contains 3 to 4 subunits. A 160-kd subunit is involved in contacting the AAUAA motif. The cleavage stimulation factor (CstF) binds to the GU-rich region. It is made up of three subunits CstF-50, CstF-64, and CstF-77. CstF-77 forms a bridge between the GU-rich region-binding CstF-64 subunit and CstF-50 and also physically contacts CPSF. Formation of a complex between CPSF and CstF leads to the recruitment of other factors involved in cleavage and polyadenylation such as cleavage factor I and

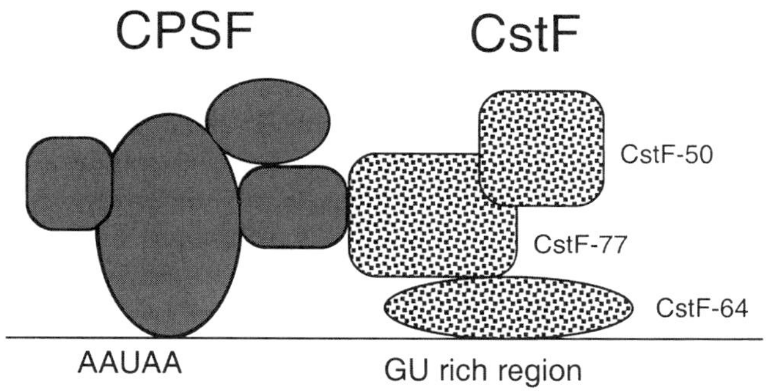

FIGURE 7.7. CstF-64 is a component of the basal polyadenylation machinery. Levels of this protein as well as other factors that influence selective polyadenylation and splicing of the μ gene are regulated during B cell development.

cleavage factor II, as well as polyA polymerase. CstF-64 levels are induced by LPS activation, and this factor probably contributes in a major way to the switch toward μs in activated B and plasma cells.

The Pro-B Receptor

Two selection events occur during the pro-B and pre-B stages of B cell differentiation (see Figure 7.10, later, for an overview of checkpoints and signals during the early stages of B cell differentiation). The first selection event occurs in a subset of pro-B cells soon after DJ rearrangement has occurred. Pro-B cells—in which a particular reading frame (RF2) of the D segment is in frame with the J segment and the constant regions of the immunoglobulin heavy chain—can synthesize a protein known as the Dμ protein. The membrane form of this protein associates with the *ω* and *ι* surrogate light chains encoded by the *λ5* and *vpreB genes.*

Surrogate light chains were first described in association with the μ heavy chain in pre-B cells. The term *surrogate light chains* was originally coined to refer to both of the proteins mentioned above; it is now sometimes used in the singular form to refer to a single λ5-vPreB unit. The λ5 chain contains a C-terminal Ig domain made up of seven β strands, much like a conventional C-domain. A cysteine residue in the penultimate position associates covalently with the corresponding cysteine in the C-terminal portion of the CH1 domain. The N-terminal 50 amino acids of the mature λ5 protein (a relatively large signal peptide is cleaved in the ER) is not homologous to any known proteins and constitutes a unique domain. This N-terminal unique domain functions as an intramolecular chaperone, which prevents the λ5 protein from fully folding and exiting the ER in the absence of vPreB. In between the N-terminal unique domain and the 7 β-strand

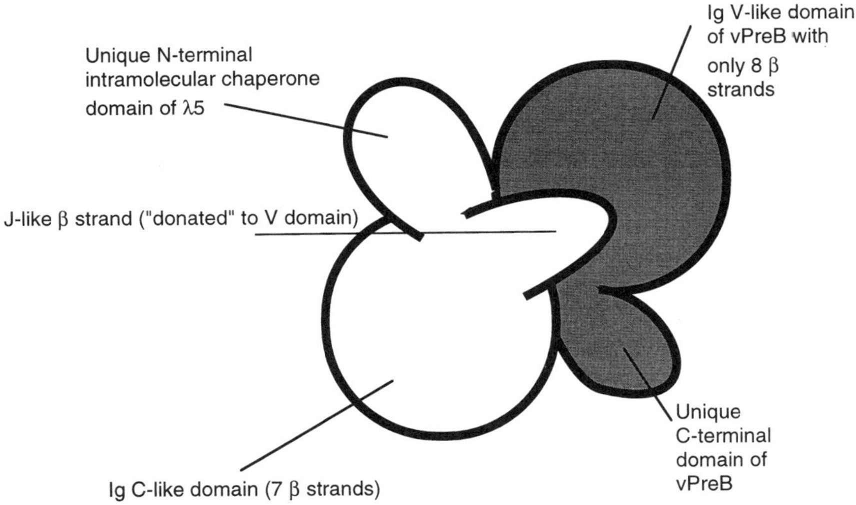

FIGURE 7.8. The vpreB and λ5 surrogate light chains associate with each other to generate a two-domain light chain-like unit. The complete unit is itself sometimes referred to as a *surrogate light chain*.

containing C-like domain is an additional stretch containing a single β strand with homology to a J region. This β strand is "donated" to vPreB when λ5 and vPreB assemble together to form a complete light chain-like unit containing two Ig domains. VpreB/ι non-covalently associates with the pro-B and pre-B receptor complexes. The N-terminal portion of vPreB contains an Ig domain made up of only eight β strands, which therefore looks like an incomplete V-domain. A typical V domain contains 9 β strands. The association of vPreB with λ5 depends on the intercalation of the J-like β strand in λ5 with the eight β strands of vPreB to form a complete V-like Ig domain (Figure 7.8). The function of a 24-amino acid unique region in the C-terminus of vPreB remains unclear. There are two vPreB genes in the mouse; they encode almost identical proteins, which presumably serve the same function.

The membrane form of Dμ also associates with the Igα/β heterodimer, and this entire complex is known as the pro-B receptor (Figure 7.9). Cells that express the pro-B receptor may be selected "against" during-B cell development. The Igα/β heterodimer plays a critical role in signal transduction mediated by Src-family kinases and Syk, as described in Chapter 4. Interestingly, the mRNAs encoding Dμ heavy chains initiate translation from a conserved ATG upstream of the recombinational signal sequence (RSS) that flanks the D segment. While RSSs do not in general get transcribed or translated, in this case, the RSS nucleotides are downstream of translation initiation site (containing an ATG) and provide a sequence encoding an in-frame stretch of hydrophobic amino acids that can consti-

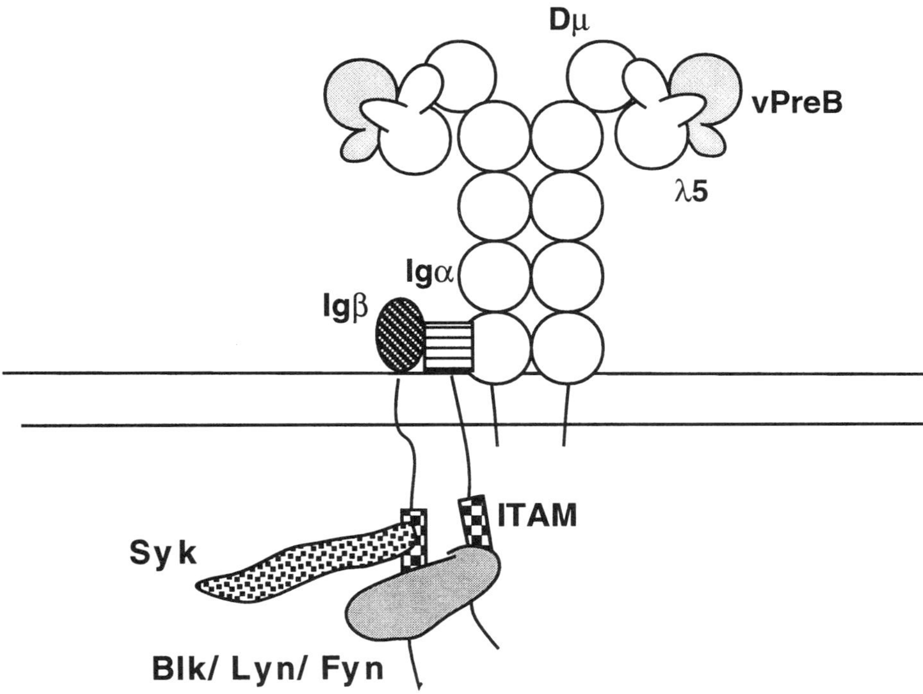

FIGURE 7.9. Structure of the pro-B receptor. The λ5 surrogate light chain, once referred to as the ω Ig light chain, associated with the CH1 domain of the μ heavy chain. This association is covalent and occurs via a disulfide bridge similar to the association of H-chain with conventional light chains. λ5 and vPreB form a noncovalently assembled heterodimeric unit. Igα and Igβ form a disulfide-linked transmembrane heterodimer that links tyrosine kinases to the μ heavy chain.

tute a signal or leader peptide. This peptide permits the co-translational entry of the heavy chain protein into the endoplasmic reticulum via the signal recognition particle machinery and the Sec61 channel. Pro-B receptor signaling apparently leads either to the elimination of pro-B cells that use the "wrong" RF2 reading frame, or as has been suggested from studies involving expression of a Dμ transgene, leads to a block in V to DJ_H rearrangement and consequently a failure in further development of clones using RF2. Signals delivered by the pro-B receptor may therefore mimic the signal for allelic exclusion delivered by the pre-B receptor in more differentiated cells. Mice engineered to lack the transmembrane domains of μ, or to lack the λ5 gene or to lack the Syk gene, all fail to eliminate, or inhibit the differentiation of, pro-B cells in which RF2 has been used during heavy chain rearrangement. In the mouse, therefore, RF2 is selected against, RF3 cannot be used because of the presence of numerous stop codons in the D segment when this reading frame is employed, and only RF1 cells survive. It is difficult to explain why this happens. Why is reading frame 1 considered particularly useful? While it could be argued that perhaps there are structural reasons for wanting to use RF1, this argument is not supported by any evidence, and indeed, the reading frame through the D segment could well be altered following V to DJ_H rearrangement. Whether Dμ signaling represents a freak accident of nature or a brilliant part of nature's grand design that we are much too dense to fully understand remains an unanswered issue.

A Summary of Signals at the Pro-B Stage

Some of the critical signals involved in the development and further differentiation of pro-B cells are summarized in Figure 7.10. While IL-7, FL, SCF (the c-Kit ligand), SDF-1, and the α4β1 integrin contribute to the survival, proliferation, and differentiation of pro-B cells, IL-3, estrogens, and IL-1 can negatively influence B cell growth. Not shown in this figure are Wnt proteins, which also presumably contribute to pro-B cell survival or development. The decline in lymphocyte numbers during pregnancy may reflect the inhibitory effects of estrogens, which presumably impair the ability of stromal cells to support lymphopoiesis. As discussed above, in some cells, signals delivered via Dμ may also negatively influence survival.

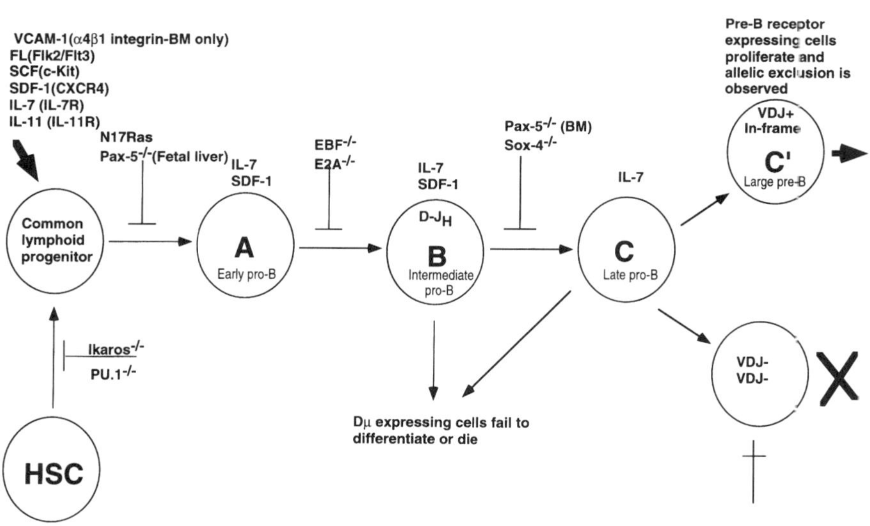

FIGURE 7.10. A summary of regulatory factors and events during the early stages of B cell differentiation.

Other B lineage-specific receptors might influence development at the pro-B stage. Complexes of surrogate light chains with other proteins (sometimes described as surrogate heavy chains, though this is probably a rather loosely used term) have been described in pro-B cells. No evidence suggests that these complexes are actually signaling receptors, and a functional role for surrogate light chains prior to the pre-B receptor driven pro-B to pre-B transition is not supported by data from the *λ5* knockout mouse (Table 7.1). Since a knockout of the *B29* gene that encodes the Igβ protein also appears to block development at an early pro-B stage (Table 7.1), it has been speculated that some receptor containing Igβ might contribute to B cell development at the pro-B stage. No such receptor has been convincingly characterized. As discussed above, a number of transcriptional

TABLE 7.1. Mutations that impair development at the pro-B and pre-B stages

Knockout	Phenotype	References
Integrins, cytokines, chemokines		
α4β1 (VLA-4)$^{-/-}$	Not required for fetal hematopoiesis, other than for B1 cells. Required for adult B and T cell generation.	Arroyo et al., 1996
Flk2$^{-/-}$	Impaired pro-B cell generation.	Mackarehtschian et al., 1995
Flk2$^{-/-}$*/W/W*^{v} (*Flk2/c-Kit* double mutant)	Marked impairment in B and T cell development.	Mackarehtschian et al., 1995
IL-7$^{-/-}$	Marked impairment of B and T cell development.	von Freeden-Jeffry et al., 1995
IL-7Rα$^{-/-}$	Marked impairment of B and T cell development.	Peschon et al., 1994
IL-11Rα$^{-/-}$	Normal hematopoiesis. Female sterility.	Nandurkar et al., 1997
γc$^{-/-}$	Marked impairment of B, T, and NK cell development.	Cao et al., 1995 DiSanto et al., 1995
Jak3$^{-/-}$	Severe combined immunodeficiency (SCID) resembling X-linked SCID. Marked impairment of B, T, and NK cell development.	Thomis et al., 1995 Nosaka et al., 1995 Park et al., 1995
SDF-1$^{-/-}$	Perinatal death; defective B lymphopoiesis and bone marrow myelopoiesis; ventricular septal defect.	Nagasawa et al., 1996
OPGL$^{-/-}$	Defect in pro-B to pre-B transition. Defects in T cell development and osteoclast generation. No lymph nodes.	Kong et al., 1999
RANK$^{-/-}$	Marked B cell deficiency in spleen. Defects in osteoclast differentiation. No lymph nodes.	Dougall et al., 1999

TABLE 7.1. (*Cont.*)

Knockout	Phenotype	References
Transcription factors		
Ebf$^{-/-}$	Block in pro-B cell development prior to Fraction B.	Lin and Grosschedl, 1995
E2a$^{-/-}$	Block in pro-B cell development prior to Fraction B.	Bain et al., 1994 Zhuang et al., 1994
Pax-5/BSAP$^{-/-}$	Complete block of fetal liver-derived B cell precursors. In bone marrow-derived B cells, block in pro-B cell development at Fraction C. Pax-5–deficient pro-B cells are multipotent.	Urbanek et al., 1994 Nutt et al., 1997 Nutt et al., 1999 Rolink et al., 1999
Sox-4$^{-/-}$	Death at embryonic day 14; defect in cardiac outflow tract formation. Partial block in B cell development at Fraction B and C/C′ noted after reconstitution in lethally irradiated recipients.	Schilham et al., 1996
Pre-B receptor signaling		
μTM segment$^{-/-}$	Block in B cell development at the pro-B to pre-B transition; RF2 cells persist.	Kitamura et al., 1991 Kitamura and Rajewsky, 1992; Gu et al., 1991a Ehlich et al., 1993
JH$^{-/-}$	Block in B cell development at the pro-B to pre-B transition.	Chen et al., 1993 Ehlich et al., 1993
λ5$^{-/-}$	Block in B cell development at the pro-B to pre-B transition; RF2 cells persist.	Kitamura et al., 1992
B29/Igβ$^{-/-}$	Block in B cell development at the pro-B stage.	Gong and Nussenzweig 1996
Syk$^{-/-}$	Block in B cell development at the pro-B to pre-B transition; RF2 cells persist.	Cheng et al., 1995 Turner et al., 1995
Blnk$^{-/-}$	Partial defect in pro-B to pre-B transition; peripheral B cells respond poorly to mitogens; No B-1 cells.	Jumaa et al., 1999 Pappu et al., 1999
Bcl-X$^{-/-}$	Embryonic lethal at dE13. Chimera reveals defects in pre-B development.	Motoyama et al., 1995 Ma et al., 1995
Human		
Btk	X-linked agammaglobulinemia (XLA). Block in development at pre-B stage.	Tsukada et al., 1993 Vetrie et al., 1993
μ	Phenotype resembling XLA.	Yel et al., 1996
λ5/14.1	Phenotype resembling XLA.	Minegishi et al., 1998
Blnk	Phenotype resembling XLA.	Minegishi et al., 1999b

regulators are required during early B cell development, and these are also summarized in Table 7.1.

Cell Selection and Allelic Exclusion Mediated by the Pre-B Receptor

Pre-B cells that have made in-frame rearrangements at the heavy chain locus after V to DJ_H rearrangement and that can therefore synthesize a μ protein are positively selected. This selection event depends on signals from the pre-B receptor Figure 7.11. Pre-B receptor-mediated survival signals require the presence of the membrane Ig heavy chain, surrogate light chains, the Igα/β heterodimer, Syk, and Btk (the latter is relevant in humans, but not critically so in mice). This latter kinase is defective in X-linked agammaglobulinemia, an inherited disease in which the pre-B to B cell transition is blocked (primarily at the level of pre-B survival). Formal evidence indicating that Src-family kinases are involved in the generation of pre-B receptor-mediated survival signals is lacking, presumably because of the

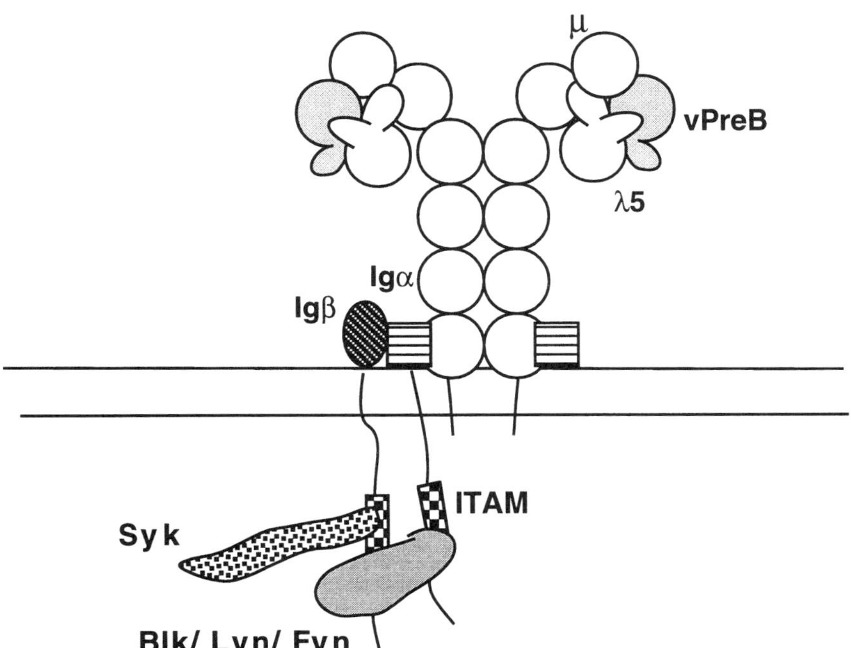

FIGURE 7.11. The pre-B receptor is made up of the membrane form of the heavy chain, surrogate light chains, and the Igα/β heterodimer. Src-family kinases and Syk are associated with the receptor.

large number of Src-family kinases that can associate with the pre-B and B cell receptors. However, since both Syk and Btk are activated in part by Src-family kinases, it is extremely likely that members of this family of kinases that are associated with the pre-B receptor play a role in mediating pre-B cell survival and allelic exclusion. The pre-B receptor is largely intracellular, although it can be detected with some difficulty on the surface of a small fraction of pre-B cells. No ligand has been identified for this receptor, and it is likely that it signals constitutively, being triggered by the act of preantigen-receptor assembly.

It is formally possible that a yet to be identified ligand exists for the pre-B receptor. However, truncated Ig heavy chains lacking ligand-binding regions can participate in the generation of pre-B survival and progression signals. Preantigen receptors, unlike most other signaling receptors, are generated, not to respond to external clues, but to monitor whether the "right" kind of rearrangement has taken place within a given cell. It is therefore not unreasonable to suspect that such a receptor does not require an external ligand for signal initiation. The fully assembled receptor, because of some structural feature of the surrogate light chains, may be constitutively "on" or may respond to a ligand made by the pre-B cell itself.

Allelic exclusion may be achieved by completely different mechanisms at Ig heavy and light chain loci. Allelic exclusion at the heavy chain locus is a regulated process dependent on a feedback signaling mechanism; however, for at least one light chain locus (κ), this process might occur by the intrinsic restriction of accessibility factors for gene expression to a single chromosome in each cell. The feedback regulation of allelic exclusion may depend upon both the reduced expression of Rag-2 mRNA and protein as well as the regulated alteration of accessibility at the heavy chain locus. Signals from the pre-B receptor may contribute to the degradation of the Rag-2 protein and possibly to a transient downregulation of *Rag-2* gene expression. Since the pre-B receptor delivers a proliferative signal, it probably influences G1 progression by inducing cyclin D2 and activating cyclin-dependent kinases such as cyclin-dependent kinase-4 (CDK4) and CDK6. Activation of CDKs presumably mediates (among other things) the phosphorylation of the C-terminus of Rag-2 at threonine 490 (adjacent to proline 491). This phosphorylation event is probably critical for the polyubiquitination and proteasomal degradation of Rag-2. Although evidence linking Rag-2 degradation to allelic exclusion remains to be obtained, the evidence to the contrary is weak, and it should still be considered formally possible that heavy chain allelic exclusion depends in part on the induction of Rag-2 degradation by CDKs.

Since light chain rearrangement is induced at the small pre-B stage (Fraction D), allelic exclusion of the heavy chain cannot be achieved solely by a transient reduction of Rag expression. A separate but poorly understood event that occurs as a consequence of pre-B receptor signaling is an alteration of accessibility for V(D)J recombination specifically at the IgH locus.

A Role of the Pre-B Receptor in Repertoire Modification in the B Lineage: Differences Between Fetal and Adult B Cell Receptor Repertoires

An interesting aspect of pre-B receptor signaling is the apparent role of this process in shaping the B cell repertoire. B cells generated in fetal and neonatal life preferentially express certain V_H genes, most (but not all) of which lie proximal to the D cluster. In adult life, a wider range of B cells are generated that include Ig heavy chains derived from rearrangements that involve distal V_H genes as well. This process has been referred to as repertoire maturation. While IL-7R signaling might play a role in generating accessibility to distal V_H genes during the development of adult bone marrow-derived B cells, an interesting mechanism involving the pre-B receptor has emerged that might contribute significantly to the shaping of the fetal and the adult immune repertoires.

It is pertinent to ask why there needs to be a different repertoire during fetal and adult life. In addition to V_H usage differences, fetal liver B cells do not express TdT, and junctions tend to be far more diverse in adult bone marrow-derived Ig heavy chains because of the introduction of N regions. The teleological explanations for why repertoire maturation occurs at all are somewhat speculative. It might be useful early on to specifically generate a battery of B cells that can readily respond to a range of common microbial antigens, particularly microbial polysaccharides. The proximal V_H genes tend to contribute to an anticarbohydrate repertoire and thus potentially produce an "intermediate" line of defense that is more diverse than that provided by agglutinins of the innate immune system. It has also been suggested that it might be useful to keep the fetal B cell repertoire restricted to decrease the likelihood of an immune reponse that might be directed against maternal structures. These hand-waving considerations notwithstanding, the differences in the fetal and adult repertoires are significant.

When examining the frequency of productive versus nonproductive rerrangements in bone marrow-derived small pre-B cells versus pro-B cells, it was noted that the frequency was much higher than expected without selection for V_H 36–60 but considerably lower than expected for V_H81X (this is the most proximal of V_H genes; see Figure 3.6). It has been demonstrated that not all μ heavy chain proteins readily associate with surrogate light chains. It appears that in the adult bone marrow a subset of rearranged heavy chain genes that tend to be derived from the rearrangement of distal V_H genes can more readily associate with surrogate light chains and therefore assemble a signaling competent pre-B receptor. This subset of heavy chain proteins, presumably because they can deliver the necessary selection signal, are then preferentially represented in the B cell repertoire. As we shall see, the repertoire may be further shaped by positive and negative selection events at the immature B cell stage.

Intriguingly, while pre-B receptor signaling may positively select pre-B cells that have productively rearranged the right (usually distal V_H containing) kind of heavy chain gene segments in bone marrow-derived B cells, this receptor apparently provides a negative signaling function in fetal liver-derived pre-B cells. As a result, the very V_H rearrangements that are selected for in adult pre-B cells are selected against during fetal development. Therefore, the majority of fetally derived pre-B cells that survive this signaling event may be cells that have productively rearranged proximal V_H genes. Obviously, this raises a number of questions. Are fetal liver-derived pre-B cells wired differently from bone-marrow derived pre-B cells, leading to apoptotic signaling in the fetal liver stage but survival in bone marrow-derived cells? It is assumed that at stage C, bone marrow-derived pro-B cells cells cease expressing a receptor for an essential stromal signal and become totally dependent on the pre-B receptor for survival. The default fate is therefore death unless productive rearrangements occur. In fetal and neonatal pre-B cells, survival signals might be delivered in a λ5 independent fashion by some other pre-B receptor complex, and strong association with the λ5 protein might lead to the generation of a death signal. It might also be the case that surrogate light chains are only required for positive selection in adult-type pre-B cells because these latter cells express TdT and therefore their heavy chains must be checked out for the structural ability to properly associate with conventional light chains. Surrogate light chain-mediated testing may be therefore relevant to positive selection only in adult bone marrow-derived B lineage cells.

Some Mechanistic Aspects of Pre-B-Receptor Signaling

Two basic questions are worth considering for preantigen receptors in general; we will discuss these here in the pre-B context;

1. Is there a ligand for the pre-B receptor, or is signaling constitutive and dependent primarily on assembly of the receptor?
2. What are the molecules and signaling pathways downstream of the pre-B receptor?

In spite of considerable effort over a decade, no ligand has so far been discovered for the pre-B or pre-T receptors, though one might well be found in the future. The model suggesting that signaling might be constitutive is supported by experiments in which a truncated μ chain has been found to drive cells through the pro-B to pre-B transition. Analagous experiments have been performed in the T lineage. Although in pre-B cells that are expressing surrogate light chains at high levels prior to light chain gene rearrangement the pre-B receptor has been described on the cell surface at low levels, no hard evidence indicates that it is actually expressed in significant amounts on the cell surface. Experiments in which the pre-T receptor has been held intracellularly have attempted to address this issue but might suffer from the problem that ER-retention motifs compromise association with signaling chains. However, even if surface expression is required—possibly for the receptor to be carried to caveoli like detergent-insoluble glycolipid (DIG) domains to be in proximity to signaling molecules—these receptors may well be activated in a ligand-independent fashion.

Signal transduction via the pre-B receptor probably requires the activation of Src-family kinases, Btk, and Syk. In the mouse, Btk is not required at the pro-B stage. In humans, mutations in Btk are responsible for X-linked agammaglobulinemia and most likely reflect the failure of pre-B receptor signaling and hence the failure to generate B cells in boys with this disease. Mutation of the λ5 homologue in humans leads to a B cell developmental phenotype identical to that observed in X-linked agammaglobulinemia.

While mutations in pre-B receptor components or in signal transducing molecules linked to this receptor lead to a block in B lineage development at the pro-B to pre-B transition, an unexpected block in this transition was observed in mice in which a mutation in the *OPGL* gene was engineered. OPGL is the osteoprotegerin ligand and is a member of the TNF family. It is also known as TRANCE, RANKL, or ODF, and this protein is required for the generation of lymph nodes. A more detailed consideration of OPGL and its functions may be found in Chapter 6. In the absence of OPGL, developing $B220^+$ $CD43^+$ $CD25^-$ pro-B cells fail to efficiently mature into $B220^+CD43^-CD25^+$ pre-B cells. As a result, there is a significant reduction in the number of peripheral B cells. When $OPGL^{-/-}$ mice were reconstituted with normal bone marrow, B cell development progressed unimpaired. For B cell development to occur normally, OPGL must be synthesized by a bone marrow-derived cell, either a B lineage progenitor itself, or a stromal cell of bone marrow origin. It is presumed that OPGL signaling in some way facilitates pre-B receptor signal transduction. A knockout of RANK (the TNFR family protein that is the receptor for OPGL) also leads to a profound block in B cell development. The $RANK^{-/-}$ mouse also presents with defects in osteoclast differentiation and the development of peripheral lymph nodes. The development of Peyer's patches in not compromised. OPGL/RANK signaling is apparently required for pre-B cell survival and further differentiation.

The Induction of Light Chain Gene Rearrangement

Pre-B-receptor signaling may be directly or indirectly responsible for a multitude of molecular events (Figure 7.12). As discussed above, this receptor may provide survival and proliferative signals and may also be required for allelic exclusion at the heavy chain locus. In addition, it may contribute to the efficient induction of κ light chain gene rearrangement and also provide signals for the shut-off of surrogate light chain gene expression. Clearly, in

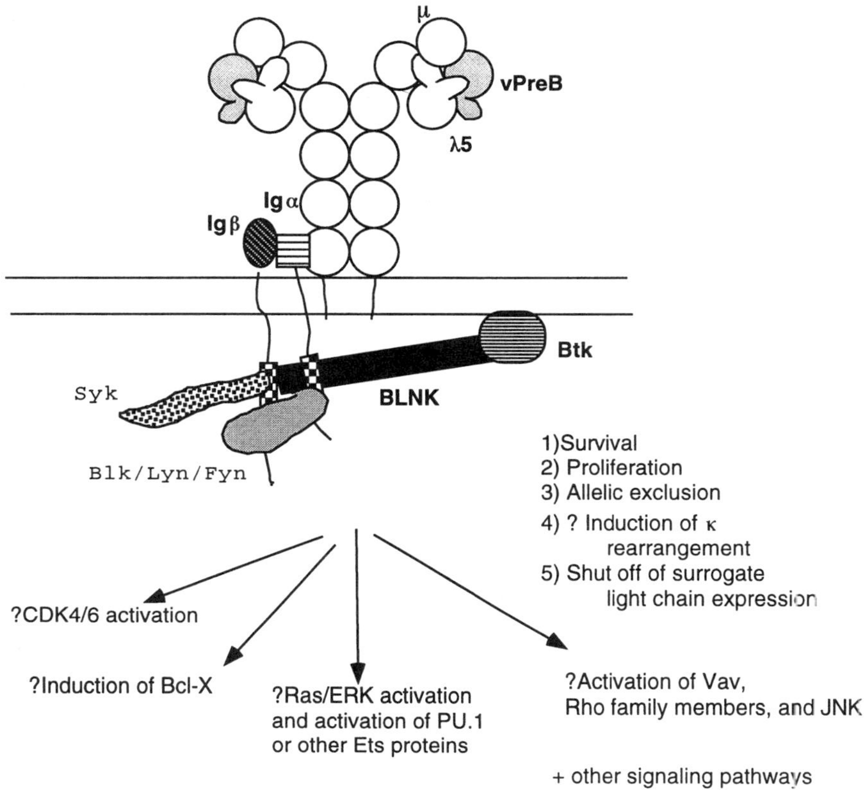

FIGURE 7.12. Signal transduction by the pre-B receptor and its consequences.

the absence of pre-B receptor signaling, the κ locus is at least minimally accessible for recombination. However, signals from the pre-B receptor contribute to the induction of nuclear factors that enhance accessibility at the κ locus and thereby contribute to more efficient rearrangement in small preB cells.

Signal transduction by the pre-B receptor directly or indirectly contributes to activation of the κ locus. *In vivo* footprinting studies have revealed that the NFκB site in the κ intronic enhancer (Figure 7.13) is occupied even at the pre-B stage, but as cells progress from the pro-B to the pre-B stage, differences were noted in the set of transcriptional regulators bound to the 3′κ enhancer (which is located about 10-kb downstream of the Cκ region). A Pax-5/BSAP site is occupied in proB cells, but the Pax-5 protein is apparently displaced in pre-B cells by proteins binding to the PU.1 site, the adjacent Pip site, and a poorly characterized CRE (a cAMP-response element) site. Pip is a lymphoid-specific interferon-regulated factor (IRF), which binds to DNA cooperatively with PU.1. Some *in vivo*

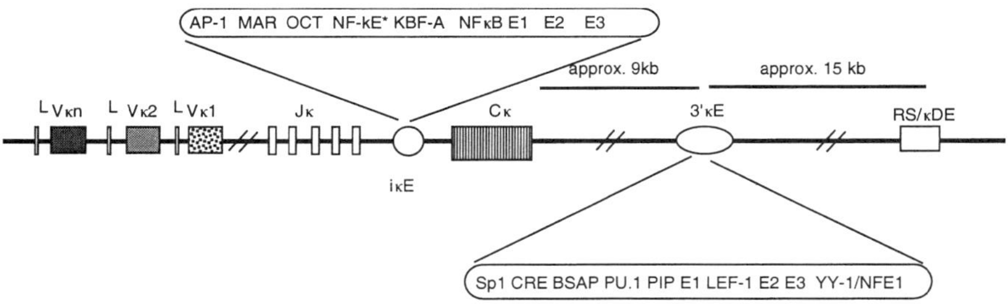

FIGURE 7.13. The murine κ light chain locus. A large number of V segments (a few hundred assorted into 20 families) lie a considerable distance upstream of four functional Jκ segments (Jκ3 has a mutant 3′ splice donor site and is nonfunctional). The Jκ-Cκ intron contains the intronic κ enhancer (iκE). A second enhancer, the 3′κ enhancer, is located approximately 9 kb downstream of the Cκ segment. Approximately 25 kb downstream of the Cκ region is the recombining segment (RS), which in humans is aptly described as the κ-deleting element (κDE). This element mediates deletion of the κ locus during receptor editing.

experiments have suggested a role for the PU.1 site in the 3′ enhancer in Vκ to Jκ rearrangement during B cell development. Deletion of the 3′κ enhancer alters the κ to λ ratio, as discussed below, but this enhancer is not absolutely required for κ rearrangement.

Mutations that impair development at the pre-B stage are listed in Table 7.1.

The κ Versus λ Choice

In mice, approximately 95% of B cells express κ light chains, while only about 5% express λ light chains. There are three expressed λ alleles, λ1, λ2, and λ3, and λ1 is the most abundant of the three. When Abelson virus-derived pre-B cell lines were analyzed in the early studies of B cell development, most were found to have rearranged their κ loci. The majority of κ-expressing cell lines and hybridomas contain λ light chain genes in germline configuration, while most λ-expressing cells appear to have rearranged and often deleted their κ genes. These observations led to the enunciation of a regulated model that suggested that κ rearrangement preceded λ rearrangement and that perhaps κ rearrangement was necessary to initiate λ rearrangement. This model is no longer considered valid. An alternative model is supported by a large body of evidence from a number of studies examining B cell development and light chain gene rearrangement in mice lacking the κ locus or the κ enhancer. These studies indicate that the selection of κ or λ is stochastic but that accessibility for rearrangement is more efficiently achieved at the κ than at the λ locus.

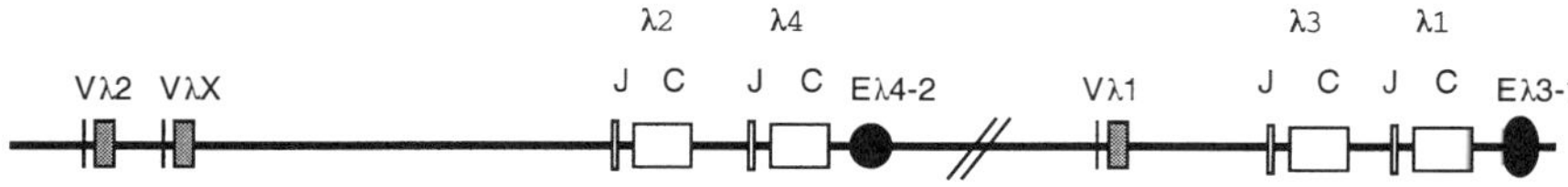

FIGURE 7.14. The murine λ light chain locus. A mutation in the Jλ4 spice donor site compromises the expression of this form of λ light chain. Eλ4-2 and Eλ3-1 refer to the two described enhancers in this locus.

The λ locus in the mouse (Figure 7.14), in contrast to the κ locus, has very limited diversity. Only three Vλ gene segments (only two of which may be actually used) are linked to four J-C units (one of which is nonfunctional). It has previously been argued that λ-bearing cells contribute to a limited antigen receptor repertoire and therefore may not be expanded as extensively during B cell maturation. However, the more limited repertoire of λ-containing antibodies cannot account for the observations made on cell lines, as described above.

The κ locus both in mice and humans contains a conserved element about 25 kb downstream of the Cκ exon (see Figure 7.13) known as the recombining segment (RS) in mice and the κ-deleting element (κDE) in humans. This element contains an RSS and in λ expressing cells is often found to have been involved in a V(D)J-type rearrangement event deleting much of the κ locus (when recombination involved an upstream Vκ segment) or to have rearranged to a conserved heptamer sequence in the Jκ-Cκ intron, which results in the deletion of Cκ and the 3′κ enhancer. Transcripts have been found spanning the RS region in cell lines actively rearranging their λ light chains and in the process of deleting their κ loci. These transcripts most likely reflect the generation of accessibility during the recombination process, and no evidence suggests that they are required for λ locus rearrangement, as was once postulated. The RS element may be particularly important in immature B cells undergoing receptor editing, and this will be discussed below.

Mice in which the κ locus has been deleted have fairly normal B cells and antibodies in which the light chains are exclusively λ. It appears likely that both κ and λ light chain genes are targeted for rearrangement around the same time in development. Apparently, κ rearrangement does not need to be completed for λ rearrangement to be initiated. The two λ enhancers lack NFκB sites, and this locus may be less efficiently activated in the mouse than the κ locus, thus accounting for a stochastically determined relative increase in accessibility of the κ gene in this species. Deletion of the 3′κ enhancer results in a higher ratio of Igλ versus Igκ B lymphocytes. The 3′κ enhancer influences the normal κ versus λ ratio but is not absolutely required for κ rearrangement. It is not required for κ expression in activated B cells.

Allelic Exclusion and Isotypic Exclusion at Light Chain Loci

There are major differences in the way allelic exclusion occurs at the Ig heavy and light chain loci. Allelic exclusion at the heavy chain locus is a feedback process that depends on signals delivered via the pre-B receptor. Although the maintenance of allelic exclusion at the κ locus may depend on the shutoff of *Rag* gene expression downstream of maturational/positive selection signals delivered by the B cell receptor, the induction of allelic exclusion at this locus appears *not* to be a signal-dependent feedback process. The initial establishment of allelic exclusion is a consequence of the differential availability of the two κ alleles for rearrangement. At the small pre-B stage, before rearrangement, a single κ allele is demethylated. This demethylation event requires the presence in cis of the intronic and 3′κ enhancers and probably involves the limiting generation of accessibility, and therefore of rearrangement, at a single-κ allele in a developing pre-B cell.

Exactly how isotypic exclusion (the inability of a λ producer to rearrange κ and vice versa) occurs is unclear. The stochastic choice of either κ or λ by a very limiting amount of an accessibility factor that favors one or the other locus is one explanation. A feedback process may also contribute to isotypic exclusion. The formation of a complete antibody molecule at the immature B cell stage contributes to the process of positive selection, which will be discussed in some detail below. While "strong" signals through the B cell receptor (BCR) either lead to deletion or receptor editing, less well-understood nonmitogenic signals delivered through the antigen receptor contribute to B cell maturation. One of the consequences of this maturation process is the feedback regulation of V(D)J rearrangement. It is likely that maturation signals, in a poorly understood manner, shut off *Rag* gene expression (and probably inactivate accessibility factors as well), and these events might contribute to the process of isotypic exclusion.

Clonal Deletion and Receptor Editing During B Cell Development in the Bone Marrow

A proportion of the B cells that are generated each day express antigen receptors that are directed against self-antigens. B cells bearing receptors that are cognate for multivalent self-antigens (polysaccharides, membrane proteins, plasma membrane glycolipids) need to be either eliminated or tolerized in some other manner. At the immature B cell stage, crosslinking of the B cell receptor by a multivalent ligand might induce signals that could lead to the induction of apoptotic death. Such a view has been supported by studies in which "monoclonal mice" have been generated, which express specific Ig heavy and light chain transgenes that form a receptor directed against a multivalent antigen (a membrane-anchored protein, for example). Initial studies of this sort suggested that a strong crosslinking stimulus could lead to the elimination of immature B cells, thus leading to the generation of tolerance. While this remains an important modality by which tolerance may be achieved, it has now become apparent that B cells that receive a strong crosslinking stimulus at the immature B cell stage may be given "another chance" or chances to reform themselves. They do this by undergoing further rearrangements primarily of light chain genes and also by presumably deleting their existing rearranged light chain genes. If a given B cell has already made a productive Vκ to Jκ rearrangement and is triggered to edit its receptor, the next Vκ to Jκ rearrangement that it makes will delete and replace the previously rearranged VκJκ unit. As a result, this cell may now make a receptor that no longer is triggered by multivalent self and will both evade elimination as well as have the potential to serve as a useful citizen in a future pathogen-directed immune response. While it is generally assumed that receptor editing evolved as a sophisticated tolerance mechanism, it is likely that this process contributes in a major way to repertoire diversification during B cell development.

Receptor editing may also involve rearrangement of a λ light chain in a cell that previously made a self-reactive B cell receptor that expresses κ light chains. In this case, the κ locus might be deleted as a result of RS recombination using the RS/κDE element discussed above. The apparent importance of deletion of the preexisting rearranged light chain VJ unit during receptor editing has been underscored by elegant studies in which the rearranged Ig heavy and light chain genes encoding an antibody directed against a specific major histocompatibility complex (MHC) class I protein were "knocked-in" to the mouse germline. The VDJ_H and VJ_κ units were therefore placed where they should be in the right chromosomal locations (in contrast to typical transgenes, which are expressed from random locations in the genome). When the appropriate MHC class I gene was bred into this Ig "knock-in" mouse, immature B cells were not deleted but underwent receptor editing. Because the rearranged transgenes were in the "right" place, it was possible to delete the self-reactive light chain gene in the process of generating a new L-chain rearrangement. If the original light chain was not deleted, the initial self-reactive B cell receptor would presumably still be around, delivering unwelcome death signals.

From a mechanistic standpoint, it is unclear whether a strong signal to a self-reactive B cell actually leads to higher levels of expression of *Rag* genes (which have not really been shut off at the immature B cell stage) or merely prevents the cell from moving on to a stage when *Rag* genes are shut-off. The end result is a cell that continues to rearrange its light chain genes. Indeed, multiple light chain loci might have been generated by gene duplication and preserved during evolution in part to facilitate the process of receptor editing. Receptor editing is primarily a light chain phenomenon.

V_H gene replacement is known to occur at the heavy chain locus where a new V_H segment might recombine with a "cryptic" heptamer in the coding region of a previously rearranged V_H gene.

Receptor Editing and Clonal Deletion May Occur in Different Cell Types

Both receptor editing and clonal deletion may play significant roles in achieving B cell tolerance against multivalent self but may operate at different stages of differentiation. Immature B cells in the bone marrow express low levels of IgM and low levels of intracellular Bcl-2. A proportion of these immature cells emigrate initially to the spleen (as will be discussed below), where they mature into extrafollicular newly formed/virgin B cells (Figure 7.15). These cells express high levels of IgM and higher levels

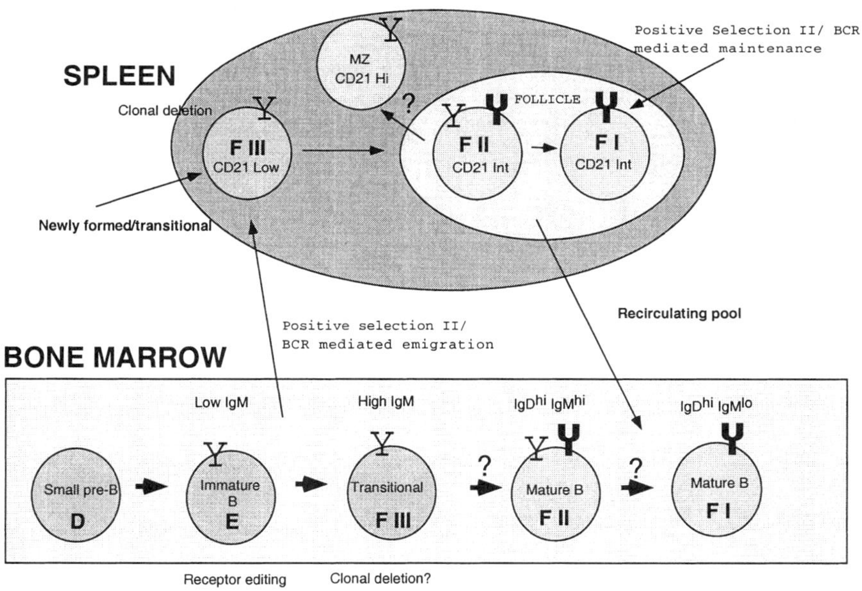

FIGURE 7.15. An overview of selection events at the immature B cell stage and the peripheral maturation of B cells. Self-reactive immature B cells in the bone marrow may undergo receptor editing. If self-reactivity cannot be dealt with by this means, newly formed/transitional cells may be clonally eliminated either in the periphery or in the bone marrow itself. The emigration of immature or newly formed B cells from the bone marrow to the spleen may depend on signals from the BCR. Maintenance in the periphery may also require signals from the antigen receptor. The emigration of newly formed B cells and maintenance in the periphery may represent positive selection events.

of Bcl-2 than immature B cells. They express low levels of IgD and CD23 and can be distinguished from MZ B cells on the basis of their lack of CD21 expression. Cells that are phenotypically identical to splenic virgin B cells develop in the bone marrow as well and have been occasionally called "transitional" B cells. Splenic newly formed/virgin B cells have also been more recently described as "transitional" B cells. Both immature and newly formed B cells fail to proliferate in response to anti-IgM crosslinking, while follicular B cells proliferate far more readily. Multivalent self-antigens that can access the bone marrow probably induce tolerance by provoking antigen-receptor signal-mediated receptor editing in immature bone marrow B cells. Some self-antigens may be recognized primarily by the variable domain of the heavy chain; a new light chain may not be able to alter such a B cell clone's self-reactivity. In such cases, deletion might represent the fallback option. It has been suggested that deletion primarily occurs in peripheral transitional B cells (another name for newly formed, or virgin, B cells). It is unclear whether clonal deletion occurs in the bone marrow as well, and if it does, it has not been established whether this event results only in the elimination of bone marrow transitional B cells and not of immature B cells. Some studies suggest that immature B cells in the bone marrow are protected from apoptotic death by the presence of a protective cell type in the bone marrow that is not found in the spleen. Whether this protection actually extends to bone marrow transitional B cells is unclear.

Other modalities of tolerance induction, such as anergy and follicular exclusion, may be relevant at least for a subset of soluble self-antigens, and these will be discussed in a later section in this chapter.

Positive Selection II and Peripheral Maturation

For most students of the immune system, the positive selection of T cells during thymic development seems to make intuitive sense when one considers the need to select MHC-restricted T cells. It has not been readily apparent why a positive selection phenomenon might be of relevance for recently generated B cells. Of the large number of B cells generated in the bone marrow every day (about 10^8 cells), only about 15% actually make it to the periphery. It is estimated that about 2×10^7 mature peripheral B cells are made every 24 hours. It is presumed that most newly generated B cells have a short life-span that can be measured in days, but a proportion of naive B cells become long-lived cells that may recirculate in the periphery for 6 to 8 weeks. The possibility has been raised in some studies that a proportion of newly generated B lymphocytes might in a nonrandom way (based on receptor specificity) be selected to develop in the periphery and to become long-lived B cells. Considerable cell death occurs during the pro-B and pre-B stages of development, but little direct evidence suggests that much cell death occurs in immature or in newly formed/transitional B cells.

Where most B cells are lost after they are generated in the bone marrow remains a mystery. Afficionados will recognize that even the BrdU-based life span measurements of peripheral B cells should be viewed with caution, and the accepted life span of 6 to 8 weeks for so-called long-lived B cells should not be considered to be written in stone.

It is now believed that immature B lymphocytes not only may be subject to receptor editing and clonal deletion but also need to be positively selected. In mice carrying an engineered mutation in the cytoplasmic tail of Igα, B cell development in the bone marrow was largely unaffected, but emigration to the periphery appeared to be severely compromised. In an antigen-specific "knock-in" mouse model in which B cell receptor genes were conditionally deleted in the periphery, the maintenance of these peripheral B cells was severely compromised. Signals from the B cell receptor are therefore required both for the emergence of B cells in the periphery as well as for maintenance or survival. Follicular entry might require a minimal signal from the BCR (while stronger signals from the BCR may impede this process, as discussed below). Downstream of the B cell receptor, Syk might play a role in the peripheral emigration/follicular entry step, whereas Btk may be required for the maintenance of long-lived peripheral B cells.

Immature B cells (stage E) emerge from the bone marrow between 2 and 3 days after they receive the pre-B receptor signal in stage C/C′ and can be observed in the spleen as an $IgM^{hi}IgD^{lo}$ population (Fraction FIII, according to the classification developed by Hardy and colleagues (1982); see Figure 7.16). These newly formed/virgin/transitional Fraction III cells are distributed initially in the T cell zone and in the red pulp of the spleen. Low-level BCR signals and signals received from chemokines might contribute to the entry of these cells into follicles. These newly formed cells express low levels of CD21, are HSA^{hi}, express pB130–140 antigen, and are $CD23^{-}$. Newly generated Fraction III cells mature in the periphery over the next 1 to 3 days sequentially into $IgM^{hi}IgD^{hi}$ (Fraction II) B cells and $IgD^{hi}IgM^{lo}$ (Fraction I) B cells, which represent the most mature peripheral B cell population. Most Fraction II and Fraction I cells express low levels of CD24/HSA, intermediate levels of CD21, are $CD23^{+}$, and acquire the ability to recirculate, migrating constantly between follicular areas in the spleen and other peripheral lymphoid organs. The bulk of these cells are apparently long lived and do not express the pB130–140 antigen. Indeed, in just a day after bone marrow cells enter the spleen (soon after they enter Fraction II in the follicles), they acquire the ability to recirculate and seed lymph nodes.

In secondary lymphoid organs, IgD^{hi} (Hardy Fractions FII and FI) B cells reside primarily in lymphoid follicles. The follicular environment may, in some unknown way, contribute to their extended life span. It is now clear that in the absence of Btk-derived signals, cells that enter Fraction I are unable to become long-lived B cells and rapidly die. The steady-state level of Fraction I cells is markedly diminished in the absence of Btk. This

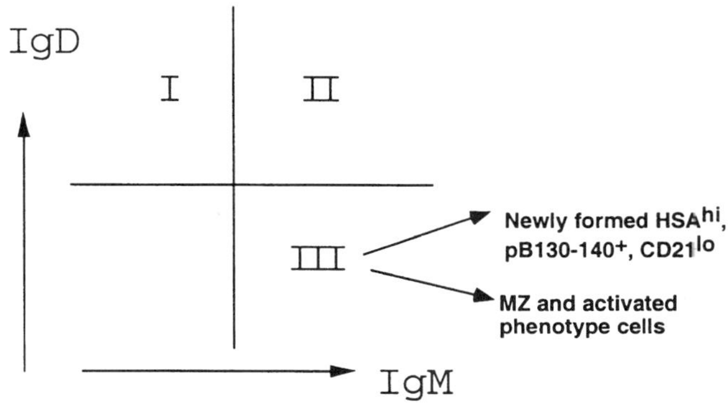

	IgM	IgD	CD21	CD22	CD23	CD24/HSA	CD1
F I	Low	High	Int	High	High	Low	Low
F II	High	High	Int	High	High	Low	Low
F III (NF)	High	-/Low	Low	Low	Low	High	Low
F III (MZ)	High	-/Low	High	High	-/Low	Int	High

FIGURE 7.16. Phenotypic characterization of peripheral B cell populations. Fraction III ($IgM^{hi}IgD^{lo}$) contains both newly formed (NF) as well as marginal zone (MZ) cells. Some MZ cells have switched isotypes and are found in the double-negative quadrant. MZ cells and NF cells are both $CD23^{lo}$, but MZ cells express high levels of CD1 and CD21. This categorization is based on Hardy et al. (1982), Waldschmidt et al. (1991), Allman et al. (1992), Oliver et al. (1997), and Roark et al. (1998). Newly formed cells also express the pB130–140 antigen (Rolink et al., 1998).

fraction is also decreased in the absence of CD45, which is required for Src-family kinase activation and therefore presumably for Btk activation as well. The proportion of Fraction I cells increases in the absence of CD22 (which functions as a negative regulator of the BCR). It is considered likely therefore that nonmitogenic, low-intensity signals are delivered by the BCR via the Btk pathway and that these signals are essential for the maintenance of long-lived naive B cells. Some of the checkpoints involved in the generation and maturation of peripheral B cells are summarized in Figure 7.15.

What drives positive selection? Does the B cell receptor signal constitutively, as has been suggested, or is it triggered in a nonmitogenic fashion by cross-reactive self antigens? Is a constitutive signal generated primarily to tell a B cell that its bone marrow developmental program is complete and that it should now shut off its V(D)J recombination machinery and move out to where the real action will be? If the signal is constitutive and ligand independent, it might be reasonable to suggest that this process is *not* positive "selection" but merely represents a maturational process. What is being selected might be an appropriate fit between heavy and light chain. No conclusive answer is available at present for any of these questions. However, it is conceivable that this process is driven by endogenous antigens and that this presumed selection event actually evolved for a defense-related reason. It is conceivable that positive selection is generally driven by non-crosslinking self-antigens and that in a viremic or otherwise infected host non-crosslinking foreign antigens may target emerging B cells and bias the peripheral B cell repertoire toward the pathogen that requires to be targeted. This could potentially result in more of the right kind of B cells in the periphery to drive T-dependent responses that are likely to be of protective value. Alternatively, positive selection might have evolved both in B and T cells to facilitate the maintenance of naive lymphocytes.

In summary, signals from the B cell receptor are required for the maturation of naive immature B cells into recirculating follicular long-lived lymphocytes. The most mature long-lived population is made up of $IgD^{hi}IgM^{lo}$ cells. This maturation process may be referred to as positive selection II (Figures 7.15 and 7.17). To distinguish it from positive selection I, which is mediated by the pre-B receptor. These signals, if they are indeed antigen induced, must represent nonmitogenic, trophic, "tickles" rather than the full-fledged stimulation associated with cross-linking antigens. While it appears likely that some B cell loss occurs somewhere between the bone marrow and the spleen, where cells are actually lost is unclear. No hard evidence suggests that only a proportion of the emerging B cell repertoire is selected positively or negatively at the immature B cell stage. Low-intensity BCR signals delivered via the Syk tyrosine kinase might assist in emigration of B cells from the bone marrow and therefore facilitate follicular entry (triggering of the BCR by cognate antigens actually contributes to the exclusion of activated B cells from follicles; this will be discussed later and in Chapter 8). Signals delivered via Btk may be required for peripheral B cell maintenance and therefore for the generation of long-lived B cells.

Peripheral B Cell Populations and Their Maintenance

Peripheral B cells first emerge in the spleen. As mentioned above, soon after entering primary follicles (follicles that lack germinal centers), they acquire the ability to recirculate and move from one follicular site to the other. This migratory behavior enables them to be available for any cognate antigen

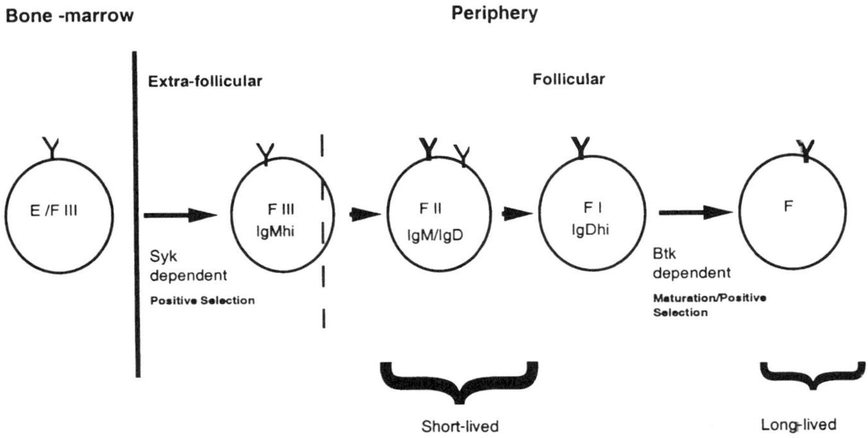

FIGURE 7.17. Positive selection II: Immature B cells may be positively selected to exit the bone marrow and enter lymphoid follicles by Syk-dependent mechanisms and to mature into long-lived B cells by a Btk-dependent pathway.

that might emerge at a given site and to be in proximity to recirculating T cells that might provide help. The following splenic B cell populations have been discussed above (see Figures 7.15 and 7.18) and are summarized again for convenience.

1. *Newly formed B cells* that are IgM^{hi}, $IgD^{lo/-}$, $CD23^{lo/-}$, and $CD21^{lo}$. These newly generated naive B cells cells, sometimes referred to as *virgin B cells*, and occasionally and more recently called *transitional* B cells, form a portion of Faction III, and they express high levels of CD24/HSA. They also express the pB 130–140 antigen. These cells may be found in the red pulp of the spleen and in the T cell areas (the periarteriolar lymphoid sheath in the spleen).

2. *Follicular B cells* are derived from the recently generated Fraction III cells. These cells colonize lymphoid follicles, have the ability to recirculate, and initially have an IgM^{hi} IgD^{hi} $CD23^{+}$ phenotype (Fraction II). They mature very rapidly into IgM^{lo} IgD^{hi} $CD23^{+}$ cells (Fraction I). In older mice, the majority of follicular cells express low levels of HSA/CD24 and do not express pB130–140. However, a small proportion of these cells at any given time (more so in very young mice) will express high levels of HSA/CD24 and are pB130–140 positive. They represent cells that have recently entered the follicular pool. Follicular cells can be quite tightly distinguished from newly formed cells by virtue of CD21 expression (intermediate levels in follicular cells). These naive B cells are variously referred to as *mature B cells*, *naive B cells*, *follicular B cells*, or *recirculating B cells*.

3. *Marginal zone B cells*. The MZ surrounds the follicular area and is populated by B cells that are IgM^{hi}, IgD^{-}, $CD23^{-}$, $CD21^{hi}$, and $CD1^{hi}$. These

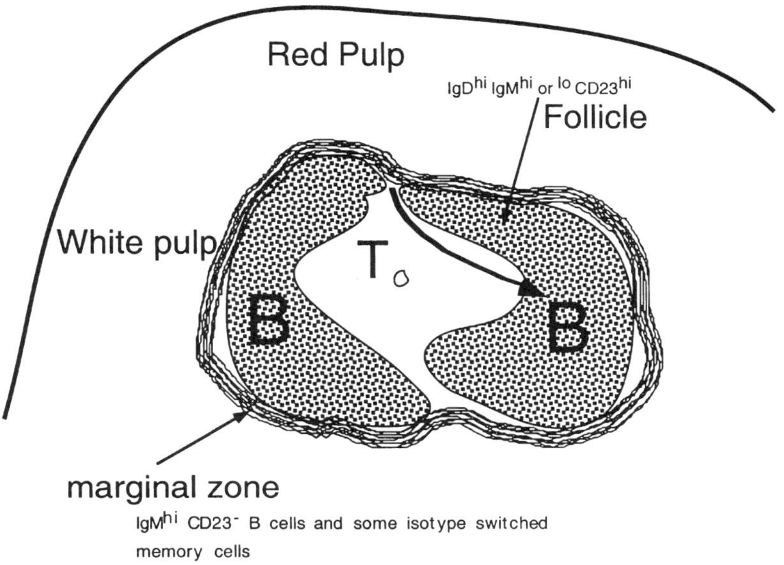

FIGURE 7.18. A schematic view of a segment of the spleen. The white pulp is made up of follicles that are B cell areas and the periarteriolar lymphoid sheath, which represents a T cell zone. Both BLR1/CXCR5 and CXCR4 (the receptor for the SDF-1 chemokine) participate in the entry of recirculating B cells into follicles. Follicular entry requires that B cells *not* be triggered by high-affinity cognate antigens.

cells can be distinguished from newly generated Fraction III cells by virtue of their higher levels of CD21 and also by the fact that they express high levels of CD1. The marginal zone also contains marginal zone macrophages, and a distinct population of marginal zone metallophilic macrophages. While some conventional memory cells and also inhabit the marginal zone, MZ B cells defined by the above antigens are nonrecirculating cells that have sometimes been considered to represent a memory population generated exclusively in the spleen by multivalent antigens such as bacterial polysachharides that are made available on the surface of the specialized macrophages in the area. MZ B cells are mainly found in the spleen and may account for the critical role played by the spleen (as opposed to other peripheral lymphoid sites) in responses to blood-borne carbohydrate antigens. These B cells have been considered by some investigators to represent a separate lineage. They might represent a very specialized population of naive B cells. MZ B cells will be discussed in the next chapter. Other B cell populations such as B-1 cells, germinal center B cells, isotype switched memory B cells, and plasma cells will also be considered in Chapter 8.

While recently generated peripheral B cells might require nonmitogenic signals derived from the Syk tyrosine kinase to emigrate from the bone marrow and possibly to enter follicles, the normal entry of naive B cells into follicles occurs when B cells have *not* been triggered by high-affinity cognate antigens. In mice that have a mutant form of Igα, B cells are generated and express surface IgM, but they are unable to seed the periphery, probably because of a defect in emigration. Similarly, the pre-B defect in $Syk^{-/-}$ mice is partially rescued by Ig heavy and light chain transgenes, but these B cells have a defect in the emigration of B cells to the periphery (which has been also interpreted as a defect in follicular entry). It is likely that "low-intensity" BCR signals are probably required for emigration from the bone marrow and therefore indirectly for follicular entry. "High-intensity" signals via the BCR in the periphery *prevent* follicular entry.

In Chapter 6, we discussed the role of B cells in the maintenance of follicular structure and the requirement for TNF family ligands and receptors in this process. Entry (and repeated re-entry) of B cells into follicles may also require signals delivered via follicular chemokines that serve as chemoattractants for B cells (also discussed in Chapter 6). One of these is a C-X-C chemokine known as B lymphocyte chemoattractant (BLC) or B cell-attracting chemokine-1 (BCA-1), which signals via a chemokine receptor that is present on mature B cells known as BLR-1 (for Burkitt's lymphoma receptor-1), now reclassified as CXCR5. BLC was first described in mouse tissues, and BCA-1 is the human homologue of BLC. Another B cell chemoattractant is SDF-1, which activates CXCR4 and is also essential for the expansion of pro-B cells in the bone marrow (discussed earlier in this chapter). CXCR5 may be more critical for migration into follicles in the spleen and Peyer's patches (since the CXCR5 knockout mouse fails to develop follicles at these locations), while CXCR4 may be crucial at other sites. Recirculating B cells emerge from the marginal sinus or from the central arteriole, migrate through the outer T cell zone, and if they are not challenged by antigen, enter the adjacent follicles.

Once B cells seed the periphery, maintenance or survival also requires signals via the BCR. Signals via Src-family kinases, CD45, and Btk are required for the generation of long-lived peripheral B cells. Downstream of BCR signaling, transcription factors such as NFκB may be required for peripheral B cell maintenance. Transcription factors such as Oct-2 and OCA-B may be required for B cell emigration as well. Some of these factors will be discussed when we examine B cell activation in Chapter 8.

Table 7.2 summarizes some of the information on mutants that compromise BCR-mediated emigration or survival during maturation or positive selection II.

There are at present no satisfactory explanations for why two mutant mice strains, mice that lack the invariant chain, and MHC class II-deficient mice present with defects in B cell maintenance and maturation. This is not

TABLE 7.2. Defects in BCR signaling that might influence emigration or maintenance

Mutation	Emigration defect	Maintenance/ maturation defect	Reference
BCR (conditional)	Not assayed	Yes	Lam et al., 1997
Igα tail deletion	Yes	Yes	Torres et al., 1996
$Lyn^{-/-}$	Not assayed	Yes	Hibbs et al., 1995 Nishizumi et al., 1995 Wang et al., 1996 Chan et al., 1997
$Btk^{-/-}$ and *Xid*	No Accelerated emigration	Yes	Khan et al., 1995 Kerner et al., 1995 Hendriks et al., 1996 Cariappa et al., 1999
$Syk^{-/-}$	Yes	Not assayed	Turner et al., 1997
$CD45^{-/-}$	No	Yes	Byth et al., 1996
$NF\kappa B1^{-/-}$	Not assayed	Yes	Sha et al., 1995 Snapper et al., 1996
$NF\kappa B1^{-/-}/NF\kappa B2^{-/-}$	Not assayed	Yes	Franzoso et al., 1997
$Oct\text{-}2^{-/-}$	Not assayed	Yes	Corcoran and Karvelas, 1994 Humbert and Corcoran, 1997
$Oca\text{-}B/OBF\text{-}1/Bob\text{-}1^{-/-}$	Yes*	Yes	Kim et al., 1996 Schubart et al., 1996 Nielsen et al., 1996
$Bcl\text{-}2^{-/-}$	Not assayed	Yes	Veis et al., 1993 Nakayama et al., 1993

related to concomitant defects in T cell development, since many mutant mice that lack T cells have perfectly normal B cells.

μ and δ Heavy Chains Are Generated From a Rearranged Ig Locus by RNA Processing

Immature B cells express no IgM and mature in follicles into cells expressing high levels of IgD. After activation, the cells revert to an IgM^{hi} phenotype. IgM and IgD are generated from long transcripts that initiate upstream of a rearranged V(D)J unit and extend across the μ and δ constant regions. The ratios of IgM and IgD at different stages of peripheral

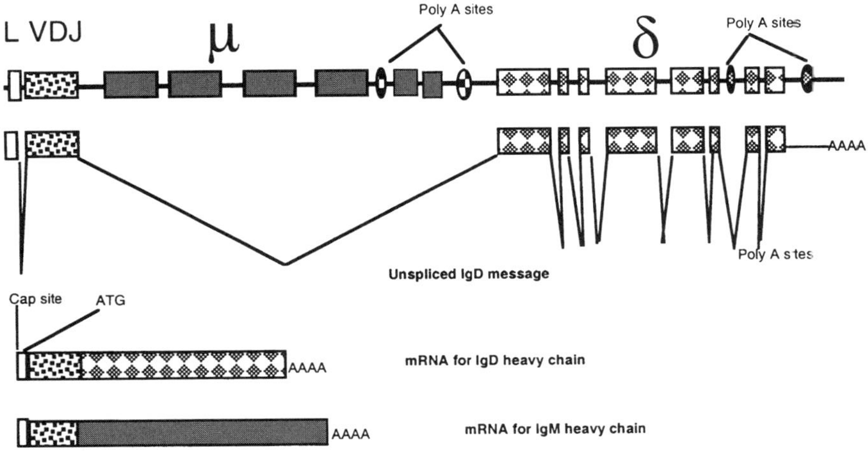

FIGURE 7.19. μ and δ heavy chains are generated by RNA processing of a long transcript.

differentiation are regulated at the level of RNA processing (Figure 7.19). While polyadenylation site selection most likely plays a critical role in this process, the molecular basis for the regulation of IgD during development is poorly understood.

Central and Peripheral Tolerance and the Putative Function of IgD

Cells expressing IgD represent mature peripheral B cells and are therefore not as prone to being tolerized as immature B cells in the bone marrow. The presence of IgD has therefore long been considered as a hallmark of maturity, but we remain in the dark as to whether the IgD molecules on these cells function in a distinct way from IgM. IgD knockout mice have been generated, but few clues have been obtained from these mice as to the functional significance of IgD. No firm evidence indicates that IgD induces signaling differently from IgM. The function of IgD remains an enigma.

We have already discussed two potential mechanisms for B cell tolerance (see above and Figure 7.20). While the induction of receptor editing and clonal deletion has been studied in the context of multivalent membrane-bound self-antigens, a model of tolerance induced by a transgenically expressed soluble antigen, hen egg lysozyme, has led to the view that soluble self-antigens might induce anergy in the periphery. Keep in mind that this model antigen not only has a high affinity for the antilysozyme antibody

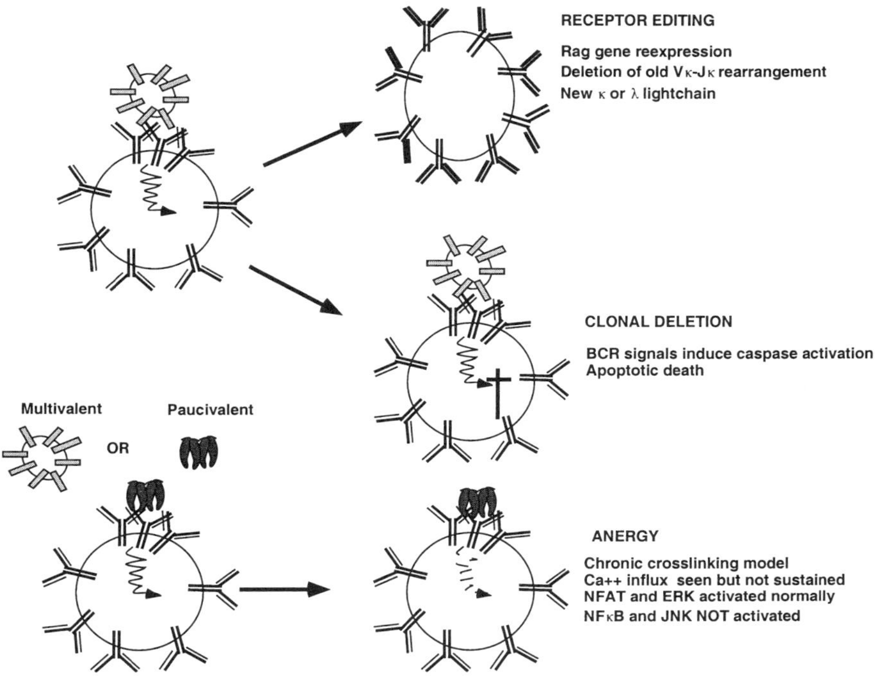

FIGURE 7.20. Models for B cell tolerance include receptor editing, deletion, and anergy.

expressed transgenically on all B cells in these double transgenic mouse models but that "soluble" lysozyme is a potent crosslinking antigen that can induce extracellular signal-regulated kinase (ERK) activation in cognate B cells as efficiently as anti-IgM or anti-IgD antibodies. The soluble lysozyme model therefore mimics the chronic exposure of cognate B cells to a high-affinity, paucivalent, crosslinking, self-antigen. Clearly, the efficiency of crosslinking may be less than that of a high-affinity membrane-bound antigen, but this is not a model that involves the kind of antigen that might be postulated to be responsible for "tickling" B cells for the induction of positive selection/maturation at the level of emigration or maintenance, as discussed above.

In such a model, the chronic activation of B cells by a potent, crosslinking, soluble, self-antigen leads to the induction of anergy. B cells from these mice can no longer be triggered to proliferate by exposure *in vitro* to the same antigen. When these anergic B cells are transferred to another recipient, they lack the ability to respond to the specific antigen *in vivo*. Apart from B cells directed against soluble lysozyme, B cells in some mice that express anti-DNA heavy and light chain transgenes also seem to undergo changes suggestive of anergy. In most other transgenic models for autoantigen-specific B cells, including other anti-DNA transgenic mice, developing B lymphocytes are deleted. In the soluble lysozyme anergic model, chronic stimulation through the antigen receptor leads to some distinct biochemical alterations that correlate with the anergic state. When these anergic B cells are triggered *in vitro* via the antigen receptor, they do flux calcium, but the increase in intracellular calcium is not sustained. These B cells can activate the ERK MAP kinase pathway in response to BCR triggering but cannot trigger NFκB or JNK. In this model, the development of the anergic state requires chronic triggering via the BCR.

Summary

Hematopoietic stem cells are triggered by a variety of signals to give rise to progenitors of the B lineage. Some of the ligands, receptors, and transcription factors that are involved in B lymphopoiesis have been identified, but most of the connections between cell surface signals and nuclear events remain to be made. B lymphopoiesis may be initiated from a common lymphoid progenitor or possibly from a common B-myeloid progenitor. Signals required for lymphopoiesis may be delivered by the Flk2 and c-Kit receptor tyrosine kinases, by the $\alpha4\beta1$ integrin, by the IL-7 receptor, and by CXCR4, the receptor for the SDF-1 chemokine. c-Kit and Flk2 are probably required before progenitor cells enter the B lineage and may provide survival or proliferation signals for progenitors. Both the Ikaros and PU.1 transcription factors are required for a hematopoietic progenitor to differentiate into a B lineage progenitor. Early in the commitment process, the EBF and E2A transcription factors are essential for progression through the pro-B stage and for irreversible commitment to the B lineage. Although Pax-5/BSAP and Sox-4 appear to be essential a little later in pro-B cell development, they may also participate in the stepwise commitment process. In the mouse, B cells that make a DJ_H rearrangement (using reading frame 2), which can give rise to a truncated μ heavy chain protein (Dμ), fail to differentiate further. Cells that make productive heavy chain gene V to DJ_H rearrangements express the pre-B receptor, which contains the membrane form of μ, the λ5 and vPreB surrogate light chains, and signaling components. This receptor mediates pre-B cell survival and proliferation, is required for allelic exclusion at the Ig heavy chain locus, and signals to induce κ light chain gene rearrangement. Immature B cells express surface IgM and may respond to cognate multivalent self-antigens by either editing their receptors or by being induced to undergo apoptosis. Other immature B cells may differentiate further but require antigen receptor-mediated signals to emigrate to the periphery and to survive as long-lived IgM- and IgD-expressing follicular B cells.

Selected Reviews

Burrows, P. D., and Cooper, M. D. (1997). B cell development and differentiation. Curr. Opin. Immunol. *9*, 239–244.

Cambier, J. C., Bedzyk, W. K. C., Chien, N., Friedrich, J., Harwood, A., Jensen, W., Pleiman, C., and Clark, M. C. (1993). The B cell antigen receptor: structure and function of primary, secondary, tertiary and quaternary components. Immunol. Rev. *132*, 85–106.

Carsetti, R. (2000). The development of B cells in the bone marrow is controlled by the balance between cell-autonomous mechanisms and signals from the microenvironment. J. Exp. Med. *191*, 5–8.

Cedar, H., and Bergman, Y. (1999). Developmental regulation of immune system gene rearrangement. Curr. Opin. Immunol. *11*, 64–69.

Coutinho, A. (1993). Lymphocyte survival and V-region repertoire selection. Immunol. Today *14*, 38–40.

DeFranco, A. L. (1997). The complexity of signaling pathways activated by the BCR. Curr. Opin. Immunol. *9*, 296–308.

Goodnow, C. C. (1996). Balancing immunity and tolerance: deleting and tuning lymphocyte repertoires. Proc. Natl. Acad. Sci. USA *93*, 2264–2271.

Gorman, J. R., and Alt, F. W. (1998). Regulation of immunoglobulin light chain isotype expression. Adv. Immunol. *8*, 743–750.

Glimcher, L. H., and Singh, H. (1999). Transcription factors in lymphocyte development—T and B cells get together. Cell *96*, 13–23.

Hertz, M., and Nemazee, D. (1998). Receptor editing and commitment in B lymphocytes. Curr. Opin. Immunol. *10*, 208–213.

Klinman, N. (1996). The clonal selection hypothesis and current concepts of B cell tolerance. Immunity *5*, 189–195.

Kraal, G. (1992). Cells in the marginal zone of the spleen. Int. Rev. Cytology *132*, 31–74.

MacLennan, I. C. M. (1998). B cell receptor regulation of peripheral B cells. Curr. Opin. Immunol. *10*, 220–225.

MacLennan, I. C. M., Gulbranson-Judge, A., Toellner, K.-M., Casamayor-Palleja, M., Chan, E., Sze, D. M.-Y., Luther, S. A., and Acha-Orbea, H. (1997). The changing preference of T and B cells for partners as T-dependent antibody responses develop. Immunol. Rev. *156*, 53–66.

Melchers, F., Karasuyama, H., Haasner, D., Bauer, S., Kudo, A., Sakaguchi, N., Jameson, B., and Rolink, A. (1993). The surrogate light chain in B cell development. Immunol. Today *14*, 60–68.

Neuberger, M. S. (1997). Antigen receptor signaling gives lymphocytes a long life. Cell *90*, 971–973.

Nussenzweig, M. C. (1998). Immune receptor editing: revise and select. Cell *95*, 875–878.

Osmond, D. G. (1986). Population dynamics of bone marrow B lymphocytes. Immunol. Rev. *93*, 103–124.

Osmond, D. G., Rolink, A., and Melchers, F. (1998). Murine lymphopoiesis: towards a unified model. Immunol. Today *19*, 65–68.

Papavisiliou, F., Jankovic, M., Gong, S., and Nussenzweig, M. C. (1997). Control of immunoglobulin gene rearrangements in developing B cells. Curr. Opin. Immunol. *9*, 233–238.

Pillai, S. (1991). Immunoglobulin transport in B cell development. Int. Rev. Cytology *130*, 1–34.
Pillai, S. (1999). The chosen few? Positive selection and the generation of naive B lymphocytes. Immunity *10*, 493–502.
Rajewsky, K. (1996). Clonal selection and learning in the antibody system. Nature *381*, 751–758.
Sprent, J. (1993). Lifespans of naive, memory, and effector lymphocytes. Curr. Opin. Immunol. *5*, 433–438.
Reth, M. (1992). Antigen receptors on B lymphocytes. Ann. Rev. Immunol. *10*, 97–121.
Reya, T., and Grosschedl, R. (1998). Transcriptional regulation of B cell differentiation. Curr. Opin. Immunol. *10*, 158–165.
Rolink, A., and Melchers, F. (1991). Molecular and cellular origins of B lymphocyte diversity. Cell *66*, 1081–1094.
Roth, P. E., and DeFranco, A. L. (1996). Receptor tails unlock developmental checkpoints for B lymphocytes. Science *272*, 1752–1754.

Some Relevant References

Allman, D. M., Ferguson, S. E., and Cancro, M. P. (1992). Peripheral B cell maturation I. Immature peripheral B cells in adults are heat stable antigen^{high} and exhibit unique signaling characteristics. J. Immunol. *149*, 2533–2540.
Allman, D. M., Ferguson, S. E., Lentz, V. M., and Cancro, M. P. (1993). Peripheral B cell maturation II. Heat stable antigen hi splenic B cells are an immature developmental intermediate in the production of long lived marrow derived B cells. J. Immunol. *151*, 4431–4444.
Allman, D., Li, J., and Hardy, R. R. (1999). Commitment to the B lymphoid lineage ocurs before DH-JH recombination. J. Exp. Med. *189*, 735–740.
Alt, F., Rosenberg, N., Lewis, S., Thomas, E., and Baltimore, D. (1981). Organization and reorganization of immunoglobulin genes in A-MuLV transformed ells: rearrangement of heavy but not light chain genes. Cell *27*, 381–390.
Aoki, Y., Isselbacher, K., and Pillai, S. (1994). Bruton's tyrosine kinase is tyrosine phosphorylated and activated in pre-B lymphocytes and receptor ligated B cells. Proc. Natl. Acad. Sci. USA *91*, 10606–10609.
Arroyo, A. G., Yang, J. T., Rayburn, H., and Hynes, R. O. (1996). Differential requirements for α4 integrins during fetal and adult hematopoiesis. Cell *85*, 997–1008.
Bain, G., Maandag, E. C. R., te Reile, H. P. J., Feeney, A. J., Sheehy, A., Schlissel, M., Shinton, S. A., Hardy, R. R., and Murre, C. (1997). Both E12 and E47 allow commitment to the B lineage. Immunity *6*, 145–154.
Bain, G., Robanus-Maandag, E., te Riele, H., Izon, D., Amsen, D., Kruisbeek, A., Krop, I., Schlissel, M., Feeny, A. J., van Roon, M., van der Valk, M., Berns, A., and Murre, C. (1994). E2A proteins are required for proper B cell development and initiation of immunoglobulin gene rearrangements. Cell *79*, 885–892.
Bikah, G., Carey, J., Ciallella, J. R., Tarakhovsky, A., and Bondada, S. (1996). CD5-mediated negative regulation of antigen receptor-induced growth signals in B-1 B cells. Science *274*, 1906–1909.
Brouns, G. S., de Vries, E., Neefjes, J. J., and Borst, J. (1996). Assembled pre-B receptor complexes are retained in the endoplasmic reticulum by a mechanism that is not selective for the pseudo light chain. J. Biol. Chem. *271*, 19272–19278.

Byth, K. F., Conroy, L. A., Howlett, S., Smith, A. J. H., May, J., Alexander, D. R., and Holmes, N. (1996). CD45-null transgenic mice reveal a positive regulatory role for CD45 in early thymocyte development, in the selection of $CD4^+$ $CD8^+$ thymocytes, and in B cell maturation. J. Exp. Med. *183*, 1707–1718.

Caamano, J. H., Rizzo, C. A., Durham, S. K., Barton, D. S., Raventos-Suarez, C., and Bravo, R. (1998). NF-κB2 (p100/p52) is required for normal splenic microarchitecture and B cell mediated immune responses. J. Exp. Med. *187*, 185–196.

Cao, X., Shores, E. W., Hu-Li, J., Anver, M. R., Kelsall, B. L., Russell, S. M., Drago, J., Noguchi, M., Grinberg, A., and Bloom, E. T. (1995). Defective lymphoid development in mice lacking expression of the common cytokine receptor gamma chain. Immunity *2*, 223–238.

Cariappa, A., Kim, T. J., and Pillai, S. (1999). Accelerated emigration of B lymphocytes in the Xid mouse. J. Immunol. *162*, 4417–4423.

Carsetti, R., Kohker, G., and Lamers, M. C. (1993). A role for immunoglobulin D: interference with tolerance induction. Eur. J. Immunol. *23*, 168–178.

Carsetti, R., Kohler, G., and Lamers, M. C. (1995). Transitional B cells are the target of negative selection in the B cell compartment. J. Exp. Med. *181*, 2129–2140.

Chan, V. W., Meng, F., Soriano, P., DeFranco, A. L., and Lowell, C. A. (1997). Characterization of the B lymphocyte populations in Lyn deficient mice and the role of Lyn in signal initiation and down-regulation. Immunity *7*, 69–81.

Chen, C., Nagy, Z., Prak, E. L., and Weigert, M. (1995). Immunoglobulin heavy chain gene replacement: a mechanism of receptor editing. Immunity *3*, 747–755.

Chen, C., Prak, E. L., and Weigert, M. (1997). Editing disease-associated autoantibodies. Immunity *6*, 97–105.

Chen, J., Trounstine, M., Alt, F. W., Young, F., Kurahara, C., Loring, J. F., and Huszar, D. (1993). Immunoglobulin gene rearrangement in B cell deficient mice generated by targeted deletion of the J_H locus. Int. Immunol. *5*, 647–656.

Cheng, A. M., Rowley, B., Pao, W., Hayday, A., Bolen, J. B., and Pawson, T. (1995). Syk tyrosine kinase required for mouse viability and B cell development. Nature *378*, 303–307.

Cherayil, B. J., and Pillai, S. (1991). The ω/λ5 surrogate immunoglobulin light chain is expressed on the surface of transitional B lymphocytes in the murine bone marrow. J. Exp. Med. *173*, 111–116.

Coleclough, C., Perry, R. P., Karjalainen, K., and Weigert, M. (1981). Aberrant rearrangements contribute significantly to the allelic exclusion of immunoglobulin gene expression. Nature *290*, 372–378.

Cook, M. C., Basten, A., and Fazekas de St. Groth, B. (1997). Outer periarteriolar lymphoid sheath arrest and subsequent differentiation of both naive and tolerant immunoglobulin transgenic B cells is determined by B cell receptor occupancy. J. Exp. Med. *186*, 631–643.

Corcoran, A. E., Riddell, A., Kroshoop, D., and Venkitaraman, A. R. (1998). Impaired immunoglobulin gene rearrangement in mice lacking the IL-7 receptor. Nature *391*, 904–907.

Corcoran, L. M., and Karvelas, M. (1994). Oct-2 is required early in T-independent B cell activation for G1 progression and for proliferation. Immunity *1*, 635–645.

Corcoran, L. M., Karvelas, M., Nossal, G. J. V., Ye, Z., Jacks, T., and Baltimore, D. (1993). Oct-2, although not required for early B cell development, is critical for later B cell maturation. Genes. Dev *7*, 570–581.

Corcos, D., Dunda, O., Butor, C., Cesbron, J. Y., Lores, P., Bucchini, D., and Jami, J. (1995). Pre-B cell development in the absence of λ5 in transgenic mice expressing a heavy chain disease protein. Curr. Biol. *5*, 1140–1148.

Cumano, A., Paige, C. J., Iscove, N. N., and Brady, G. (1992). Bipotential precursors of B cells and macrophages in murine fetal liver. Nature *356*, 612–615.

Cyster, J. G., and Goodnow, C. C. (1995). Protein tyrosine phosphatase 1C negatively regulates antigen receptor signaling in B lymphocytes and determines thresholds for negative selection. Immunity *2*, 13–24.

Cyster, J. G., Hartley, S. B., and Goodnow, C. C. (1994). Competition for follicular niches excludes self-reactive B cells from the recirculating B cell repertoire. Nature *371*, 389–395.

Cyster, J. G., Healy, J. I., Kishihara, K., Mak, T. W., Thomas, M. L., and Goodnow, C. C. (1996). Regulation of B lymphocyte negative and positive selection by tyrosine phosphatase CD45. Nature *381*, 325–328.

Delibrias, C. C., Floettmann, J. E., Rowe, M., and Fearon, D. T. (1997). Downregulated expression of SHP-1 in Burkitt lymphomas and germinal center B lymphocytes. J. Exp. Med. *186*, 1575–1584.

DiSanto, J. P., Muller, W., Guy-Grand, D., Fischer, A., and Rajewsky, K. (1995). Lymphoid development in mice with a targeted deletion of the interleukin-2 receptor gamma chain. Proc. Natl. Acad. Sci. USA *92*, 377–381.

Doi, T. S., Takahashi, T., Taguchi, O., Takachika, A., and Obata, Y. (1997). NFκB RelA deficient lymphocytes: normal development of T cells and B cells, impaired production of IgA and IgG1 and reduced proliferative responses. J. Exp. Med. *185*, 953–961.

Doody, G. M., Dempsey, P. W., and Fearon, D. T. (1996). Activation of B lymphocytes: integrating signals from CD19, CD22 and FcγRIIb1. Curr. Opin. Immunol. *8*, 378–382.

Dougall, W. C., Glaccum, M., Charrier, K. et al. (1999). RANK is essential for osteoclast and lymph node development. Genes Dev. *13*, 2412–2424.

Durdik, J., Moore, M. W., and Selsing, E. (1984). Novel kappa light chain gene rearrangements in mouse lambda light chain producing B lymphocytes. Nature *307*, 409–452.

Ehlich, A., Schaal, S., Gu, H., Kitamra, D., Muller, W., and Rajewsky, K. (1993). Immunoglobulin heavy and light chain genes rearrange independently at early stages of B cell development. Cell *72*, 695–704.

Engel, P., Zhou, L.-J., Ord, D. C., Sato, S., Koller, B., and Tedder, B. F. (1995). Abnormal B lymphocyte development, activation, and differentiation in mice that lack or overexpress the CD19 signal transduction molecule. Immunity *3*, 39–50.

Fang, W., Mueller, D. L., Pennell, C. A., Rivard, J. J., Li, Y. S., Hardy, R. R., Schlissel, M. S., and Behrens, T. W. (1996). Frequent aberrant immunoglobulin gene rearrangements in pro-B cells revealed by a bcl-XL transgene. Immunity *4*, 291–299.

Ferradini, L., Gu, H., De Smet, A., Rajewsky, K., Reynaud, C.-A., and Weill, J.-C. (1996). Rearrangement enhancing element upstream of the mouse immunoglobulin kappa chain J cluster. Science *271*, 1416–1420.

Forster, R., Mattis, A. E., Kremmer, E., Wolf, E., Brem, G., and Lipp, M. (1996). A putative chemokine receptor, BLR1, directs B cell migration to defined lymphoid organs and specific anatomic compartments of the spleen. Cell *87*, 1037–1047.

Franzoso, G., Carlson, L., Poljak, L., Shores, E. W., Epstein, S., Leonardi, A., Grinberg, A., Tran, T., Scharton-Kersten, T., Anver, M., Love, P., Brown, K., and Siebenlist, U. (1998). Mice deficient in NFκB/p52 present with defects in humoral responses, germinal center reactions, and splenic architecture. J. Exp. Med. *187*, 147–159.

Franzoso, G., Carlson, L., Xing, L., Poljak, L., Shores, E. W., Brown, K. D., Leonardi, A., Tran, T., Boyce, B. F., and Siebenlist, U. (1997). Requirement for NFκB in osteoclast and B cell development. Genes Dev. *11*, 3482–3496.

Freitas, A. A., and Rocha, B. B. (1993). Lymphocyte lifespans: homeostasis, selection and competition. Immunol. Today *14*, 25–29.

Fulcher, D. A., and Basten, A. (1997). Influences on the lifespan of B cell populations defined by different phenotypes. Eur. J. Immunol. *27*, 1188–1199.

Gay, D., Saunders, T., Camper, S., and Weigert, M. (1993). Receptor editing: an approach by autoreactive B cells to escape tolerance. J. Exp. Med. *177*, 999–1008.

Gong, S., and Nussenzweig, M. (1996). Regulation of an early developmental checkpoint in the B cell pathway by Ig beta. Science *272*, 411–414.

Goodnow, C. C., Crosbie, J., Adelstein, S., Lavoie, T. B., Smith-Gill, S.-J., et al. (1988). Altered immunoglobulin expression and functional silencing of self-reactive B lymphocytes in transgenic mice. Nature *334*, 676–682.

Gorman, J. R., van der Stoep, N., Monroe, R., Cogne, M., Davidson, L., and Alt, F. W. (1996). The Igκ 3′ enhancer influences the ratio of Igκ versus Igλ lymphocytes. Immunity *5*, 241–252.

Grawunder, U., Leu, T. M., Schatz, D. G., Werner, A., Rolink, A., Melchers, F., and Winkler, T. H. (1995). Down-regulation of RAG1 and RAG2 gene expression in pre-B cells after functional immunoglobulin heavy chain rearrangement. Immunity *3*, 601–608.

Gu, H., Kitamura, D., and Rajewsky, K. (1991a). B cell development regulated by gene rearrangement: arrest of maturation by membrane bound Dμ protein and selection of DH reading frames. Cell *65*, 47–54.

Gu, H., Tarlinton, D., Muller, W., Rajewsky, K., and Forster, I. (1991b). Most peripheral B cells are ligand selected. J. Exp. Med. *173*, 1357–1371.

Gunn, M. D., Ngo, V. N., K.M., A., Ekland, E. H., Cyster, J. G., and Williams, L. T. (1998). A B-cell-homing chemokine made in lymphoid follicles activates Burkitt's lymphoma receptor-1. Nature *391*, 799–803.

Hardy, R. R., Carmack, C. E., Shinton, S. A., Kemp, J. D., and Hayakawa, K. (1991). Resolution and characteristization of pro-B abd pre-pro-B stages in normal mouse bone-marrow. J. Exp. Med. *172*, 1213–1225.

Hardy, R. R., Hayakawa, K., Haaijman, J., and Herzenberg, L. A. (1982). B-cell subpopulations identified by two colour fluorescence analysis. Nature *297*, 589–591.

Hardy, R. R., Hayakawa, K., Parks, D. R., and Herzenberg, L. A. (1983). Demonstration of B cell maturation in X-linked immunodeficient mice by simultaneous three colour immunofluorescence. Nature *306*, 270–272.

Hardy, R. R., Hayakawa, K., Parks, D. R., Herzenberg, L. A., and Herzenberg, L. A. (1984). Murine B cell differentiation lineages. J. Exp. Med. *159*, 1169–1188.

Hartley, S. B., Crosbie, J., Brink, R., Kantor, A. B., Basten, A., and Goodnow, C. C. (1991). Elimination from peripheral lymphoid tissues of self-reactive B lymphocytes recognizing membrane bound antigens. Nature *353*, 765–769.

Healy, J. I., Dometsch, R. E., Timmerman, L. A., Cyster, J. G., Thomas, M. L., Crabtree, G. R., Lewis, R. S., and Goodnow, C. C. (1997). Different nuclear signals are activated by the B cell receptor during positive versus negative signaling. Immunity *6*, 419–428.

Hendriks, R. W., deBruijn, M. F. T. R., Maas, A., Dingjan, G. M., Karis, A., and Grosveld, F. (1996). Inactivation of Btk by insertion of lacZ reveals defects in B cell development only past the pre-B stage. EMBO J. *15*, 4862–4872.

Hertz, M., and Nemazee, D. (1997). BCR ligation induces receptor editing in IgM^+IgD^- bone marrow B cells *in vitro*. Immunity *6*, 429–436.

Hibbs, M. L., Tarlinton, D. M., Armes, J., Graill, D., Hodgson, G., Maglitto, R., Stacker, S. A., and Dunn, A. R. (1995). Multiple defects in the immune system of Lyn deficient mice, culminating in autoimmune disease. Cell *83*, 301–311.

Huang, S. (1994). Blimp-1 is the murine homolog of the human transcriptional repressor PRD1-BF1. Cell *78*, 9.

Humbert, P. O., and Corcoran, L. M. (1997). Oct-2 gene disruption eliminates the peritoneal B-1 lymphocyte lineage and attenuates B-2 cell maturation and function. J. Immunol. *159*, 5273–5284.

Hunte, B. E., Hudak, S., Campbell, D., Xu, Y., and Rennick, D. (1996). flk2/flt3 ligand is a potent cofactor for the growth of primitive B cell progenitors. J. Immunol. *156*, 489–496.

Iritani, B. M., Forbush, K. A., Farrar, M. A., and Perlmutter, R. M. (1997). Control of B cell development by Ras-mediated activation of Raf. EMBO J. *16*, 7019–7031.

Jumaa, H., Wollscheid, B., Mitterer, M., Wienands, J., Reth, M., and Nielsen, P. J. (1999). Abnormal development and function of B lymphocytes in mice deficient for the signaling adaptor protein SLP-65. Immunity *11*, 547–554.

Karasuyama, H., Kudo, A., and Melchers, F. (1990). The proteins encoded by the VpreB and λ5 pre-B cell specific genes can associate with each other and with m heavy chain. J. Exp. Med. *172*, 969–972.

Karasuyama, H., Rolink, A., Shinkai, Y., Young, F., Alt, F. W., and Melchers, F. (1993). The expression of vPreB/λ5 surrogate light chain in early bone marrow precursor B cells of normal and B cell deficient mice. Cell *77*, 133–143.

Kee, B. L., and Paige, C. J. (1996). *In vitro* tracking of IL-7 responsiveness and gene expression during commitment of bipotent B cell/macrophage progenitors. Curr. Biol. *6*, 1159–1169.

Kerner, J. D., Appleby, M. W., Mohr, R. N., Chien, S. D. J. R., Maliszewski, C. R., Witte, O. N., and Perlmutter, R. M. (1995). Impaired expansion of mouse B cell progenitors lacking Btk. Immunity *3*, 301–312.

Keyna, U., Beck-Engeser, G. B., Jongstra, J., Applequist, S. E., and Jack, H.-H (1995). Surrogate light chain mediated selection of Ig heavy chain V regions. J. Immunol. *155*, 5536–5542.

Khan, W. N., Alt, F. W., Gerstein, R. M., Malynn, B. A., Larsson, I., Rathbun, G., Davidson, L., Muller, S., Kantor, A. B., Herzenberg, L. A., Rosen, F. S., and Sideras, P. (1995). Defective B cell development and function in Btk deficient mice. Immunity *3*, 283–299.

Kim, U., Qin, X. F., Gong, S., Stevens, S., Luo, Y., Nussenzweig, M., and Roeder, R. G. (1996). The B cell specific transcription coactivator OCA-B/OBF-1/Bob-1 is essential for normal production of immunoglobulin isotypes. Nature *383*, 542–547.

Kishihara, K., Penninger, J., Wallace. V. A., Kundig, T. M., Kawai, K., Wakeham, A., Timms, E., Pfeffer, K., Ohashi, P. S., and Thomas, M. L. (1993). Normal B lymphocyte development but impaired T cell maturation in CD45exon 6 protein tyrosine phosphatase deficient mice. Cell *74*, 143–156.

Kitamura, D., Kudo, A., Schaal, S., Muller, W., Melchers, F., and Rajewsky, K. (1992). A critical role of the λ5 protein in B cell development. Cell *69*, 823–831.

Kitamura, D., and Rajewsky, K. (1992). Targeted disruption of μ heavy chain membrane exon causes loss of heavy chain allelic exclusion. Nature *356*, 154–156.

Kitamura, D., Roes, J., Kuhn, R., and Rajewsky, K. (1991). A B cell deficient mouse by targeted disruption of the membrane exon of the immunoglobulin μ heavy chain. Nature *350*, 423–426.

Kong, Y.-Y., Yoshida, H., Sarosi, I., Tan, H.-L., Timms, E., Capparelli, C., Morony, S., Oliviera-dos-Santos, A. J., Van, G., Itie, A., Khoo, W., Wakeham, A., Dunstan, C. R., Lacey, D. L., Mak, T. W., Boyle, W. J., and Penninger, J. M. (1999). OPGL is a key regulator of osteoclastogenesis, lymphocyte development, and lymph-node organogenesis. Nature *397*, 315–323.

Kontgen, F., Grumont, R. J., Strasser, A., Metcalf, D., Li, R., Tarlinton, D., and Gerondakis, S. (1995). Mice lacking the c-Rel protooncogene exhibit defects in lymphocyte proliferation, humoral immunity and interleukin-2 expression. Genes Dev. *9*, 1965–1977.

Kopf, M., Brombacher, F., Hodgkin, P. D., Ramsay, A. J., Milbourne, E. A., Dai, W. J., Ovington, K. S., Behm, C. A., Kohler, G., Young, I. G., and Matthaei, K. I. (1996). IL-5 deficient mice have a developmental defect in $CD5^+$ B-1 cells and lack esoinophilia but have normal antibody and cytotoxic T cell responses. Immunity *4*, 15–24.

Kudo, A., and Melchers, F. (1987). A second gene, vPreB in the λ5 locus of the mouse, which appears to be selectively expressed in pre-B lymphocytes. EMBO J. *6*, 2267–2272.

Lam, K.-P., Kuhn, R., and Rajewsky, K. (1997). *In vivo* ablation of surface immunoglobulin on mature B cells by inducible gene targeting results in rapid death. Cell *90*, 1073–1083.

Lang, J., Arnold, B., Hammerling, G., Harris, A. W., Korsmeyer, S., Russell, D., Strasser, A., and Nemazee, D. (1997). Enforced Bcl-2 expression inhibits antigen-mediated clonal elimination of peripheral B cells in an antigen dose-dependent manner and promotes receptor editing in autoreactive, immature B cells. J. Exp. Med. *186*, 1513–1522.

Lassoued, K., Nunez, C. A., Billips, L., Kubagawa, H., Monteiro, R. C., LeBien, T. W., and Cooper, M. D. (1993). Expression of surrogate light chain receptors is restricted to a late stage in pre-B cell differentiation. Cell *73*, 73–86.

Legler, D. F., Loetscher, M., Roos, R. S., Clark-Lewis, I., Baggiolini, M., and Moser, B. (1998). B cell attracting chemokine 1, a human CXC chemokine expressed in lymphoid tissues, selectively attracts B lymphocytes via BLR1/CXCR5. J. Exp. Med. *187*, 655–660.

Lin, H., and Grosschedl, R. (1995). Failure of B cell differentiation in mice lacking the transcription factor EBF. Nature *376*, 263–267.

Loder, F., Mutschler, B., Ray, R. J., Paige, C. J., Sideras, P., Torres, R., Lamers, M. C., and Carsetti, R. (1999). B cell development in the spleen takes place in discrete steps and is determined by the quality of B cell receptor derived signals. J. Exp. Med. *190*, 75–89.

Loffert, D., Ehlich, A., Muller, W., and Rajewsky, K. (1996). Surrogate light chain expression is required to establish immunoglobulin heavy chain allelic exclusion during early B cell development. Immunity *4*, 133–144.

Lorenz, M., Jung, S., and Radbruch, A. (1995). Switch transcripts in immunoglobulin class switching. Science *267*, 1825–1828.

Ma, A., Pena, J. C., Chang, B., Margosian, E., Davidson, L., Alt, F. W., and Thompson, C. B. (1995). BclX regulates the survival of double positive thymocytes. Proc. Natl. Acad. Sci. USA *92*, 4763–4767.

Mackarehtschian, K., Hardin, J. D., Moore, K., Boast, S., Goff, S. P., and Lemischka, I. R. (1995). Targeted disruption of the flk2/flt3 gene leads to deficiencies in primitive hematopoietic progenitors. Immunity *3*, 147–161.

Medina, K. L., and Kincade, P. W. (1994). Pregnancy-related steroids are potential negative regulators of B lymphopoiesis. Proc. Natl. Acad. Sci. USA *91*, 5382–5386.

Melamed, D., Benschop, R. J., Cambier, J. C., and Nemazee, D. (1998). Developmental regulation of B lymphocyte immune tolerance compartmentalizes clonal selection from receptor selection. Cell *92*, 173–182.

Melamed, D., and Nemazee, D. (1997). Self-antigen does not accelerate immature B cell apoptosis but stimulates receptor editing as a consequence of developmental arrest. Proc. Natl. Acad. Sci. USA *94*, 9267–9272.

Minegishi, Y., Coustan-Smith, E., Wang, Y.-H., Cooper, M. D., Campana, D., and Conley, M. E. (1998). Mutations in the human λ5/14.1 gene result in B cell deficiency and agammaglobulinemia. J. Exp. Med. *187*, 71–77.

Minegishi, Y., Hendershot, L. M., and Conley, M. E. (1999a). Novel mechanisms control the folding and assembly of λ5/14.1 and vPreB to produce an intact surrogate light chain. Proc. Natl. Acad. Sci. USA *96*, 3041–3046.

Minegishi, Y., Rohrer, J., Coustan-Smith, E., Lederman, H. M., Pappu, R., Campana, D., Chan, A. C., and Conley, M. E. (1999b). An essential role for BLNK in human B cell development. Science *286*, 1954–1957.

Molina, H., Holers, V. M., Li, B., Fung, Y., Maraithasan, S., Goellner, J., Strauss-Schoenberger, J., Karr, R. W., and Chaplin, D. D. (1996). Markedly impaired immune response in mice deficient in complement receptors 1 and 2. Proc. Natl. Acad. Sci. USA *93*, 3357–3361.

Monroe, J. G. (1996). Tolerance sensitivity of immature stage B cells: can developmentally regulated B cell antigen receptor signal transduction play a role? J. Immunol. *156*, 2657–2660.

Mostoslavsky, R., Singh, N., Kirillov, A., Pelanda, R., Cedar, H., Chess, A., and Bergman, Y. (1998). Kappa chain monoallelic demethylation and the establishment of allelic exclusion. Genes Dev. *12*, 1801–1811.

Motoyama, N., Wang, F., Roth, K. A., Sawa, H., Nakayama, K., Negishi, I., Senju, S., Zhang, Q., Fujii, S., and Loh, D. Y. (1995). Massive cell death of immature hematopoietic cells and neurons in Bcl-x deficient mice. Science *267*, 1506–1510.

Nagasawa, T., Hirota, S., Tachibana, K., Takakura, N., Nishikawa, S., Kitamura, Y., Yoshida, N., Kikutani, H., and Kishimoto, T. (1996). Defects of B-cell lymphopoiesis and bone-marrow myelopoiesis in mice lacking the CXC chemokine PBSF-SDF-1. Nature *382*, 635–638.

Nagata, K., Nakamura, T., Kitamura, F., Kuramochi, S., Taki, S., Campbell, K. S., and Karasuyama, H. (1997). The Iga/Igb heterodimer on μ negative pro-B cells is competent for transducing signals to induce early B cell differentiation. Immunity *7*, 559–570.

Nakayama, K., Nakayama, K., Negishi, I., Kuida, K., Shinkai, Y, et al. (1993). Disappearance of the lymphoid system in Bcl-2 homozygous mutant chimeric mice. Science *261*, 1584–1588.

Nandurkar, H. H., Robb, L., Tarlinton, D., Barnett, L., Kontgen, F., and Begley, C. G. (1997). Adult mice with targeted mutation of the interleukin-11 receptor (IL11Rα) display normal hematopoiesis. Blood *90*, 2148–2159.

Nemazee, D., and Burki, K. (1989). Clonal deletion of of B lymphocytes in a transgenic mouse bearing anti MHC class I antibody genes. Nature *337*, 562–564.

Nielsen, P. J., Georgiev, O., Lorenz, B., and Schaffner, W. (1996). B lymphocytes are impaired in mice lacking the transcriptional co-activator Bob1/OCA-B/OBF-1. Eur. J. Immunol. *26*, 3214–3218.

Nishizumi, H., Taniuchi, I., Yamanashi, Y., Kitamura, D., Ilic, D., Mori, S., Watanabe, T., and Yamamoto, T. (1995). Impaired proliferation of peripheral B cells and indication of autoimmune disease in lyn-deficient mice. Immunity *3*, 549–560.

Norvell, A., and Monroe, J. G. (1996). Acquisition of surface IgD fails to protect from tolerance induction. J. Immunol. *156*, 1328–1332.

Nosaka, T., van Deursen, J. M. A., Tripp, R. A., Thierfelder, W. E., Wittuhn, B. A., McMickle, A. P., Doherty, P. C., Grosveld, G. C., and Ihle, J. N. (1995). Defective lymphoid development in mice lacking Jak3. Science *270*, 800–802.

Nunez, C., Nishimoto, N., Gartland, G. L., Billips, L. G., Burrows, P. D., Kubagawa, H., and Cooper, M. D. (1996). B cells are generated throughout life in humans. J. Immunol. *156*, 866–872.

Nussenzweig, M. C., Shaw, A. C., Sinn, E., Danner, D. B., Holmes, K. L., Morse, H. C. I., and Leder, P. (1988). Allelic exclusion in transgenic mice that express the membrane form of immunoglobulin μ. Science *236*, 816–819.

Nutt, S. L., Urbanek, P., Rolink, A., and Busslinger, M. (1997). Essential functions of pax5 (BSAP) in pro-B development: difference between fetal and adult B lymphopoiesis and reduced V-DJ recombination at the IgH locus. Genes Dev. *11*, 476–491.

Nutt, S. L., Heavey, B., Rolink, A. G., and Busslinger, M. (1999). Commitment to the B lymphocyte lineage depends on the transcription factor Pax5. Nature *401*, 556–562.

O'Keefe, T. L., Williams, G. T., Davies, S. L., and Neuberger, M. S. (1996). Hyperresponsive B cells in CD22 deficient mice. Science *274*, 798–801.

Oka, Y., Rolink, A. G., Andersson, J., Kamanaka, M., Uchida, J., Yasui, T., Kishimoto, T., Kikutani, H., and Melchers, F. (1996). Profound reduction of mature B cell numbers, reactivities and serum Ig levels in mice which simultaneously carry the Xid and CD40 genes. Int. Immunol. *8*, 1675–1685.

Oliver, A. M., Martin, F., Gartland, G. L., Carter, R. H., and Kearney, J. F. (1997). Marginal zone B cells exhibit unique activation, proliferative and immunoglobulin secretory responses. Eur. J. Immunol. *27*, 2366–2374.

Oritani, K., Kanakura, Y., Aoyama, K., Yokota, T., Copeland, N. G., Gilbert, D. J., Jenkins, N. A., Tomiyama, Y., Matsuzawa, Y., and Kincade, P. W. (1997). Matrix glycoprotein SC1/ECM2 augments B lymphopoiesis. Blood *90*, 3404–3413.

Osmond, D. G., and Nossal, G. J. V. (1974). Differentiation of lymphocytes in mouse bone marrow II. Kinetics of maturation and renewal of antiglobulin binding cells studies by double labeling. Cell. Immunol. *13*, 132–145.

Otipoby, K. L., Andersson, K. B., Draves, K. E., Klaus, S. J., Farr, A. G., Kerner, J. D., Perlmutter, R. M., Law, C. L., and Clark, E. A. (1996). CD22 regulates thymus independent responses and the lifespan of B cells. Nature *384*, 634–637.

Papavasiliou, F., Jankovic, M., and Nussenzweig, M. C. (1996). Surrogate or conventional light chains are required for membrane immunoglobulin μ to activate the precursor B cell transition. J. Exp. Med. *184*, 2025–2030.
Papavisiliou, F., Misulovin, Z., Suh, H., and Nussenzweig, M. C. (1995). The role of Igβ in precursor B cell transition and allelic exclusion. Science *268*, 409–411.
Pappu, R., Cheng, A. M., Li, B., Gong, Q., Chiu, C., Griffin, N., White, M., Sleckman, B. P., and Chan, A. C. (1999). Requirement for B cell linker protein (BLNK) in B cell development. Science *286*, 1949–1954.
Park, S. Y., Saijo, K., Takahashi, T., Osawa, M., Arase, H., Hirayama, N., Miyakie, K., Nakauchi, H., Shirasawa, T., and Saito, T. (1995). Developmental defects of lymphoid cells in JAK3 kinase deficient mice. Immunity *3*, 771–782.
Pelanda, R., Schaal, S., Torres, R. M., and Rajewsky, K. (1996). A premature expressed Igκ transgene, but not a VκJκ segment targeted into the Igκ locus can rescue B cell development in λ5 deficient mice. Immunity *5*, 229–239.
Pelanda, R., Schwers, S., Sonoda, E., Torres, R. M., and Rajewsky, K. (1997). Receptor editing in a transgenic mouse model: site, efficiency, and role in B cell tolerance and antibody diversification. Immunity *7*, 765–775.
Peschon, J. J., Morrissey, P. J., Grabstein, K. H., Ramsdell, F. J., Maraskovsky, E., Gliniak, B. C., Park, L. S., Ziegler, S. F., Williams, D. E., Ware, C. B., Meyer, J. D., and Davison, B. L. (1994). Early lymphocyte expansion is severely impaired in interleukin-7 deficient mice. J. Exp. Med. *180*, 1955–1960.
Pillai, S., and Baltimore, D. (1988). The omega and iota surrogate immunoglobulin light chains. Curr. Top. Microbiolo Immunol. *137*, 136–139.
Pillai, S., and Baltimore, D. (1987). The ω Ig light chain: an 18 kD protein that forms disulfide linked μ2ω2 tetramers in pre-B cells. Nature *329*, 172–174.
Radic, M. Z., Erikson, J., Litwin, S., and Weigert, M. (1993). B lymphocytes may escape tolerance by revising their antigen receptors. J. Exp. Med *177*, 1165–1173.
Rawlings, D. J., Saffran, D. C., Tsukada, S., Largaespada, D. A., Grimaldi, J. C., Cohen, L., Mohr, R. N., Bazan, J. F., Howard, M., Copeland, N. G., Jenkins, N. A., and Witte, O. N. (1993). Mutation of unique region of Bruton's tyrosine kinase in immunodeficient Xid mice. Science *261*, 358–361.
Reth, M. G., and Alt, F. W. (1984). Novel immunoglobulin heavy chains are produced from DJH gene segment rearrangements in lymphoid cells. Nature *312*, 418–423.
Reynaud, C. A., Garcia, C., Hein, W., and Weill, J.-C. (1995). Hypermutation generating the sheep immunoglobulin repertoire is an antigen-independent process. Cell *80*, 115–126.
Roark, J. H., Park, S. H., Jayawardena, J., Kavita, U., Shannon, M., and Bendelac, A. (1998). CD1.1 expression by mouse antigen presenting cells and marginal zone B cells. J. Immunol. *160*, 3121–3127.
Rolink, A., Karasuyama, H., Grawunder, U., Haasner, D., Kudo, A., and Melchers, F. (1993). B cell development in mice with a defective λ5 gene. Eur. J. Immunol. *23*, 1284–1288.
Rolink, A. G., Andersson, J., and Melechers, F. (1998). Characterization of immature B cells by a novel monoclonal antibody, by turnover and by mitogen reactivity. Eur. J. Immunol. *28*, 3738–3748.
Rolink, A. G., Brocker, T., Bluethmann, H., Kosco-Vilbois, M. H., Andersson, J., and Melchers, F. (1999). Mutations affecting either generation or survival of cells influence the pool size of mature B cells. Immunity *10*, 619–628.

Rolink, A. G., Winkler, T., Melchers, F., and Andersson, J. (2000). Precursor B-cell receptor-dependent B cell proliferation and differentiation does not require the bone marrow or fetal liver environment. J. Exp. Med. *191*, 23–32.

Sakaguchi, N., and Melchers, F. (1986). λ5, a new light chain related locus selectively expressed in pre-B lymphocytes. Nature *324*, 579–582.

Sandel, P. C., and Monroe, J. G. (1999). Negative selection of immature B cells by receptor editing or deletion is determined by site of antigen encounter. Immunity *10*, 289–299.

Sato, S., Miller, A. S., Inaoki, M., Bock, C. B., Jansen, P. J., Tang, M. L., and Tedder, T. F. (1996). CD22 is both a positive and negative regulator of B lymphocyte antigen receptor signal transduction: altered signaling in CD22 deficient mice. Immunity *5*, 551–562.

Scher, I., Ahmed, A., Strong, D. M., Steinberg, A. D., and Paul, W. E. (1975). X linked B lymphocyte defect in CBA/N mice. I. Studies of the function and composition of spleen cells. J. Exp. Med. *141*, 788–803.

Schilham, M. W., Oosterwegel, M. A. Moerer, P., Ya, J., deBoer, P. A., van de Wetering, M., Verbeek, S., Lamers, W. H., Kruisbeek, A. M., Cumano, A., and Clevers, H. (1996). Defect in cardiac outflow tract formation and pro-B lymphocyte expansion in mice lacking Sox-4. Nature *380*, 711–714.

Schittek, B., Rajewsky, K., and Forster, I. (1991). Dividing cells in bone marrow and spleen incorporate bromodeoxyuridine with high efficiency. Eur. J. Immunol. *21*, 235–238.

Schubart, D. B., Rolink, A., Kosco-Vilbois, M. H., Botteri, F., and Matthias, P. (1996). B cell specific coactivator OBF-1/OCA-B/Bob1 required for immune response and germinal centre formation. Nature *383*, 538–542.

Scott, E. W., Simon, M. C., Anastasi, J., and Singh, H. (1994). Requirement of transcription factor PU.1 in the development of multiple hematopoietic lineages. Science *265*, 1573–1577.

Sha, W. C., Hsiou-Chi, L., Toumanen, E. I., and Baltimore, D. (1995). Targeted disruption of the p50 subunit of of NF-κB leads to multifocal defects in immune responses. Cell *80*, 321–330.

Shachar, I., and Flavell, R. A. (1996). Requirement for invariant chain in B cell maturation and function. Science *274*, 106–108.

Shaffer, A. L., Peng, A., and Schlissel, M. S. (1997). *In vivo* occupancy of the κ light chain enhancers in primary pro- and pre-B cells: A model for κ locus activation. Immunity *6*, 131–143.

Shaffer, A. L., and Schlissel, M. S. (1997). A truncated heavy chain protein relieves the requirement for surrogate light chains in early B cell development. J. Immunol. *159*, 1265–1275.

Shaw, A. C., Swat, W., Ferrini, R., Davidson, L., and Alt, F. W. (1999). Activated ras signals developmental progression of RAG deficient pro-B lymphocytes. J. Exp. Med. *189*, 123–129.

Sigvardsson, M., O'Riordan, M., and Grosschedl, R. (1997). EBF and E47 collaborate to induce expression of the endogenous immunoglobulin surrogate light chain genes. Immunity *7*, 25–36.

Siminovitch, K. A., Bakhshi, A., Goldman, P., and Korsmeyer, S. J. (1985). A uniform deleting element mediates the loss of kappa genes in human B cells. Nature *316*, 260–262.

Snapper, C. M., Zelazowski, P., Rosas, F. R., Kehry, M. R., Tian, M., Baltimore, D., and Sha, W. C. (1996). B cells from p50/NFκB knockout mice have selective defects in proliferation, differentiation, germ-line CH transcription, and Ig class switching. J. Immunol. *156*, 183–191.

Solvason, N., Wu, W. W., Kabra, N., Lund-Johansen, F., Roncaraolo, M. G., Behrens, T. W., Grillot, D. A., Nunez, G., Lees, E., and Howard, M. (1998). Transgene expression of Bcl-XL permits anti immunoglobulin induced proliferation in Xid B cells. J. Exp. Med. *187*, 1081–1091.

Spanopoulou, E., Roman, C. A. J., Corcoran, L. M., Schlissel, M. S. Silver, D. P., Nemazee, D., Nussenzweig, M. C., Shinton, S. A., Hardy, R. R., and Baltimore, D. (1994). Functional immunoglobulin transgenes guide ordered B cell differentiation in Rag-1 deficient mice. Genes Dev. *8*, 1030–1042.

Sprent, J., and Bruce, J. (1984). Physiology of B cells in mice with X-linked immunodeficiency. III. Disappearance of xid B cells in double bone marrow chimeras. J. Exp. Med. *160*, 711–723.

Sprent, J., Bruce, J., Ron, Y., and Webb, S. R. (1985). Physiology of B cells in mice with X-linked immunodeficiency. I. Size, migratory properties, and turnover of the B cell pool. J. Immunol. *134*, 1442–1448.

Takagaki, Y., Seipelt, R. L., Peterson, M. L., and Manley, J. L. (1996). The polyadenylation factor CstF-64 regulates alternative processing of IgM heavy chain pre-mRNA during B cell differentiation. Cell *87*, 941–952.

Takeda, S., Shimizu, T., and Rodewald, H. (1997). Interactions between c-kit and stem cell factor are not required for B cell development. Blood *89*, 518–525.

Tarlinton, D., Strasser, A., McLean, M., and Basten, A. (1995). DH element reading frame selection is influenced by an Ig heavy chain transgene but not by bcl-2. J. Immunol. *154*, 3341–3350.

ten Boekel, E., Melchers, F., and Rolink, A. G. (1998). Precursor B cells showing H chain allelic inclusion display allelic exclusion at the level of pre-B receptor surface expression. Immunity *8*, 199–208.

Thomas, J. D., Sideras, P., Smith, C. I. E., Vorechovsky, I., Chapman, V., and Paul, W. E. (1993). Co-localization of X-linked agammaglobulinemia and X-linked immunodeficiency genes. Science *261*, 355–358.

Thomis, D. C., Gurniak, C. B., Tivol, E., Sharpe, A. H., and Berg, L. J. (1995). Defects in B lymphocyte maturation and T lymphocyte activation in micel lacking Jak3. Science *270*, 794–797.

Tiegs, S. L., Russell, D. M., and Nemazee, D. (1993). Receptor editing in self-reactive bone-marrow B cells. J. Exp. Med. *177*, 1009–1020.

Torres, R. M., Flaswinkel, H., Reth, M., and Rajewsky, K. (1996). Aberrant B cell development and immune response in mice with a compromised BCR complex. Science *272*, 1804–1808.

Tough, D. F., and Sprent, J. (1996). Measurement of T and B cell turnover with bromodeoxyuridine. In J. E. Coligan, A. M. Kruisbeek, D. H. Margulies, E. M. Shevach, and W. Strober, eds. (New York, NY.: John Wiley and Sons.).

Tsubata, T., and Reth, M. (1990). The products of the pre-B specific genes (λ5 and vPreB) and the immunoglobulin μ chain form a complex that is transported to the cell surface. J. Exp. Med. *172*, 973–976.

Tsukada, S., Saffran, D. C., Rawlings, D. J., Parolini, O., Allen, R. C., Klisak, I., Sparkes, R. S., Kubagawa, H., Mohandas, T., Quan, S., Belmont, J. W., Cooper, M. D., Conley, M. E., and Witte, O. N. (1993). Deficient expression of a B cell cytoplasmic tyrosine kinase in human X-linked agammaglobulinemia. Cell *72*, 279–290.

Turner, M., Gulbranson-Judge, A., Quinn, M. E. Walters, A. E., MacLennan, I. C. M., and Tybulewicz, V. L. J. (1997). Syk tyrosine kinase is required for the positive selection of immature B cells into the recirculating B cell pool. J. Exp. Med. *186*, 2013–2021.

Turner, M., Mee, P. J., Costello, P. S., Williams, O., Price, A. A., Duddy, L. P., Furlong, M. T., Geahlen, R. L., and Tybulewicz, V. (1995). Perinatal lethality and blocked B cell development in mice lacking the tyrosine kinaae Syk. Nature *378*, 298–303.

Urbanek, P., Wang, Z.-Q., Fetka, I., Wagner, E. F., Busslinger, M. (1994). Complete block of B cell differentiation and altered patterning of the posterior mid-brain in mire lacking Pax5/BSAP. Cell *79*, 901–912.

Veiby, O. P., Lyman, S. D., and Jacobsen, S. E. W. (1996). Combined signaling through interleukin-7 receptors and flt3 but not c-kit potently and selectively promotes B cell commitment and differentiation from uncommitted murine bone-marrow progenitor cells. Blood *88*, 1256–1265.

Veis, D. J., Sorenson, C. M., Shutter, J. R., and Korsmeyer, S. J. (1993). Bcl-2 deficient mice demonstrate fulminant lymphoid apoptosis, polycystic kidneys, and hypopigmented hair. Cell *75*, 229–240.

Vetrie, D., Vorechowsky, I., Sideras, P., Holland J., Davies, A., Flinter, F., Hammarstrom L., Kinnon, C., Levinsky, R., Bobrow, M., Smith, C. I. E., and Bentley, D. R. (1993). The gene involved in X-linked agammaglobulinemia is a member of the src-family of protein tyrosine kinases. Nature *361*, 226–233.

von Freeden-Jeffry, U., Vieira, P., Lucian, L. A., McNeil, T., Burdach, S. E., and Murray, R. (1995). Lymphopenia in I1–7 gene deleted mice identifies IL-7 as a non-redundant cytokine. J. Exp. Med. *181*, 1519–1526.

Waldschmidt, T. J., Kroese, F. G. M., Tygrett, L. T. Conrad, D. H. and Lynch, R. G. (1991). The expression of B cell surface receptors III. The murine low-affinity IgE receptor is not expressed on Ly-1 or Ly-1 like B cells. Int. Immunol. *3*, 305–314.

Wang, J., Koizumi, T., and Watanabe, T. (1996). Altered antigen receptor signaling and impaired Fas-mediated apoptosis of B cells in Lyn deficient mice. J. Exp. Med. *184*, 831–838.

Wang, J. H., Avitahl, N., Cariappa, A., Friedrich, C., Ikeda, T., Renold, A., Andrikopoulos, K., Liang, L., Pillai, S., Morgan, B. A., and Georgopoulos, K. (1998). Aiolos regulates B cell activation and maturation to effector state. Immunity *9*, 543–554.

Wasserman, R., Li, Y. S., Shinton, S. A., Carmack, C. E. Manser, T., Wiest, D. L., Hayakawa, K., and Hardy, R. R. (1998). A novel mechanism for B cell repertoire maturation based on response by B cell precursors to pre-B receptor assembly. J. Exp. Med. *187*, 259–264.

Weaver, D., Costantini, F., Imanishi-Kari, T., and Baltimore, D. (1985). A transgenic immunoglobulin μ gene prevents rearrangement of endogenous genes. Cell *42*, 117–127.

Weih, F., Carrasco, D., Durham, S. K., Barton, D. S., Rizzo, C. A., Ryseck, R.-P. Lira, S. A., and Bravo, R. (1995). Multiorgan inflammation and hematopoietic abnormalities in mice with a targeted disruption of RelB, a member of the NFκB/Rel family. Cell *80*, 331–340.

Yamagami, T., ten Boekel, E., Andersson, J., Rolink, A., and Melchers, F. (1999). Frequencies of multiple IgL chain gene rearrangements in single normal or κL chain deficient B lineage cells. Immunity *11*, 317–328.

Yamagami, T., ten Boekel, E., Schaniel, C., Andersson, J., Rolink, A., and Melchers, F. (1999). Four of Five RAG expressing JCκ$^{-/-}$ small pre-BII cells have no L chain gene rearangements. Detection by high efficiency single cell PCR. Immunity *11*, 309–316.

Yang, W., and Desiderio, S. (1997). BAP-135, a target for Bruton's tyrosine kinase in response to B cell receptor engagement. Proc. Natl. Acad. Sci. USA *94*, 604–609.

Ye, J., McCray, S. K., and Clarke, S. H. (1996). The transition of pre-BI to pre-BII cells is dependent on the V_H structure of the μ/surrogate L chain receptor. EMBO J. *15*, 1524–1533.

Yel, L., Minegishi, Y., Coustan-Smith, E., Buckley, R. H., Trubel, H., Pachman, L. M., Kitchingman, G. R., Campana, D., Rohrer, J., and Conley, M. E. (1996). Mutations in the μ heavy chain in patients with agammaglobulinemia. N. Engl. J. Med. *335*, 1486–1493.

Young, F., Ardman, B., Shinkai, Y., Lansford, R., Blackwell, T. K., Mendelsohn, M., Rolink, A., Melchers, F., and Alt, F. W. (1994). Influence of immunoglobulin heavy- and light chain expression on B cell differentiation. Genes. Dev. *8*, 1043–1057.

Zhuang, Y., Soriano, P., and Weintraub, H. (1994). The helix-loop-helix gene E2A is required for B cell formation. Cell *79*, 875–884.

8

Antigen-Induced B-Lymphocyte Differentiation

B cells can be triggered by two broad categories of antigen to differentiate along distinct pathways (Figure 8.1). Multivalent antigens, exemplified by bacterial polysaccharides, can directly trigger B cells with little T cell help. While this category of antigens can invoke an immune response and antibody secretion, cognate B cells are not generally triggered to go through the process of isotype switching, and affinity maturation of the immune response is not observed. As a result, the antibodies generated by T-independent antigens tend to be mainly of the immunoglobulin M (IgM) class and of relatively low affinity. Although some multivalent antigens can invoke the generation of memory, in general, memory B cells are generated in response to T-dependent antigens.

Protein antigens are internalized by B cells (by receptor-mediated endocytosis) cleaved into peptides, and specific peptide-MHC class II complexes are presented to cognate T cells. An activated T cell may trigger a B cell by a number of means. Critical receptors for this event on the B cell include CD40 and the IL-4 receptor, among others. T-dependent responses result in the induction of isotype switching and the formation of germinal centers in secondary lymphoid organs, and it is in these specialized expansions of B cells that the processes of somatic mutation and memory cell generation occur.

While we will discuss B cell activation in terms of this dichotomy between T-dependent and T-independent responses, keep in mind that these distinctions are probably much less sharp than they are often made out to be. Polysaccharide antigens or multivalent lipid antigens are generally not encountered in isolation but in the context of microbial infections, when they are seen in conjunction with protein antigens. T cells frequently modulate responses to polysaccharides. Polysaccharide antigens may often induce class switching and can also initiate germinal center formation and memory.

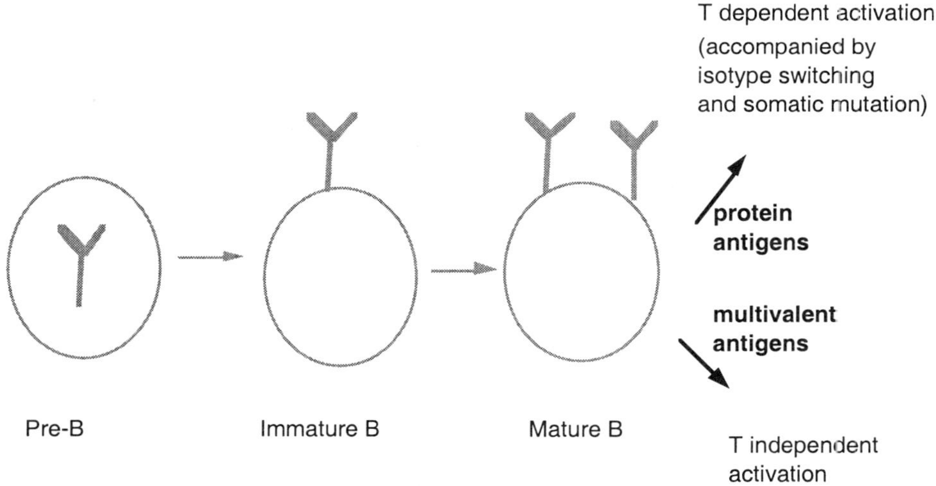

FIGURE 8.1. Two types of antigen can drive B cell differentiation.

T-Independent Immune Responses

Antibody responses that do not require T cell help are triggered primarily by polyvalent antigens known as type II T-independent antigens. These antigens tend to be macromolecules with multiple repeated determinants and are typically polysaccharides. They generally can fix complement and are poorly degraded. Type I T-independent antigens, on the other hand, are not really antigens in the classic mold but are essentially polyclonal mitogens that trigger innate immune responses. Molecules that trigger pattern recognition receptors on B cells include bacterial lipopolysaccharide (LPS), bacterial CpG DNA, and viral hemagglutinins. Vertebrate pattern recognition receptors on B cells homologous to the *Drosophila* Toll family include CD14, RP105, TLR2, and TLR4 (Toll-like receptors). CD14 is a glycosyl phosphatidylinosital GPI-anchored protein whose extracellular leucine-rich domain has homology to Toll and can bind LPS but cannot generate signals on its own. TLR2 and TLR4 may mediate responses to LPS and bacterial lipoprotein, leading to the induction of an IL-1R-like signaling pathway that contributes to NFκB activation. While many studies have been performed on signaling via RP105 on B cells, the ligand for this Toll family receptor remains unknown.

Much less is understood about T-independent immune responses than about responses to protein antigens. A mature follicular B cell or a marginal zone IgM^{hi} $IgD^{lo\ to-}$ B cell may encounter a cognate polyvalent antigen in a peripheral lymphoid organ and be triggered to secrete antibodies or occasionally to switch to the IgG3 (IgG2 in humans) or the IgA isotype. It is assumed that polysaccharide antigens, because of their polyvalent nature, can crosslink antigen receptors on B cells, thus triggering specific B cell clones. By virtue of their ability to fix complement by the alternative pathway, some polyvalent antigens stimulate B cells through the antigen receptor in conjunction with the CD21-CR2 co-receptor (discussed in Chapter 4), although this co-receptor is probably more critical in responses to protein antigens. It is assumed that free (as opposed to cell-bound) T-independent antigens can readily trigger the B cell receptor, although some evidence suggests that two categories of specialized macrophages in the spleen known as marginal zone macrophages and marginal zone (MZ) metallophilic macrophages may participate in the presentation of polysaccharide antigens. Indeed, it is well-recognized that the spleen might play a specialized role, both in mice and humans, in the induction of T-independent antibody responses. Genetic defects in molecules involved in B cell signaling (Tables 8.1 and 8.2) and splenectomy can independently contribute to an inability to respond to polysaccharide antigens. The spleen presumably plays an important role in sequestering and presenting these antigens to cognate B cells.

T-Independent "Memory" Responses and Isotype Switching

The vast majority of T-independent antigens do not induce the formation of germinal centers that are characteristically observed in the course of T-dependent responses. However, immunization with a native polysaccharide, α(1–6) dextran, has been reported to lead to the formation of germinal centers in mice. Most polysaccharide antigens do not elicit a memory response. Memory is generally induced in germinal centers in response to protein antigens. Memory responses to a small subset of polysaccharide antigens may be associated with the appearance of antigen-specific B cells in the MZ. Marginal zone B cells have been thought to represent a distinct lineage, and while this might prove to be the case, they might represent a population of B cells that have some memory characteristics and recognize and respond to polysaccharide antigens, possibly complement-coated polysaccharide antigens. MZ B cells may represent a subset of naive B cells which are triggered in a B cell receptor (BCR)-independent manner, perhaps by a macrophage-derived tumor neurosis factors (TNF) family ligand such as B cell-activating factor of the TNF family (BAFF), to assume a partially activated phenotype.

While we will discuss isotype switching in somewhat more detail in a later section of this chapter, it is of interest to note that this process can occur, albeit with a lower frequency, during responses to some T-independent antigens. These antigens may directly trigger B cells, T cells, natural killer (NK) cells, or macrophages to release cytokines by a poorly understood mecha-

TABLE 8.1. Mutations in selected molecules linked to B cell signaling

Target	Btk	Syk	Lyn	CD45	SHP-1	Vav	HS-1	PKCβ
ProB to PreB transition	Yes (mouse) No (human)	Blocked	Yes	Yes	Yes	Yes	Yes	Yes
Maintenance/ maturation	Defective	Defective	Defective	Defective	NE	Defective	NE	NE
Proliferation								
1. anti-IgM	Impaired	no B cells	Enh/Imp	Impaired	Enhanced	Impaired	Impaired	Impaired
2. CD40L	Normal (+/–)	No B cells	Impaired	Normal	Normal	Normal	Normal	Normal
3. LPS	Normal	No B cells	Impaired	Normal	Normal	Normal	Normal	Normal
T-independent responses	Impaired	No B cells	Impaired	Impaired	No defect	Impaired	Impaired	Impaired
B-1 cells	Absent	No B cells	Present	Absent	Increased	Absent	Do not apoptose	Absent
T-dependent responses	Yes	No B cells	Yes	Yes	Yes	Yes	Yes	Yes
Germinal centers	Yes	No B cells	Yes No PP	Yes	Yes	Yes	Yes	Yes
References	1–5	6, 7	8–10	11, 12	13, 14	15, 16	17	18

NE = not examined; PP = Peyer's patches.

References for Table 8.1: 1. Scher et al., 1975; 2. Khan et al., 1995; 3. Kerner et al., 1995; 4. Hendriks et al., 1996; 5. Cariappa et al., 1999; 6. Cheng et al., 1995; 7. Turner et al., 1997; 8. Hibbs et al., 1995; 9. Nishizumi et al., 1995; 10. Wang et al., 1996; 11. Kishihara et al., 1993; 12. Benatar et al., 1996; 13. Shultz and Green, 1976; 14. Cyster and Goodnow, 1995; 15. Tarakhovsky et al., 1995; 16. Zhang et al., 1995; 17. Taniuchi et al., 1995; 18. Leitges et al., 1996.

TABLE 8.2. Mutations in coreceptors and costimulators that influence B cell-receptor–driven signaling

Enockout	Phenotype	References
CD19$^{-/-}$	B-1a cells absent. Impaired T-dependent immune responses. Impaired germinal center (GC) formation. Little affinity maturation. No MZ B cells.	Rickert et al., 1995 Engel et al., 1995 Martin and Kearney, 2000
CD21/CD35$^{-/-}$	Impaired humoral responses to T-dependent antigens. Reduced size and numbers of GCs.	Ahearn et al., 1996 Croix et al., 1996 Molina et al., 1996
C3$^{-/-}$	Impaired humoral responses to T-dependent antigens. Reduced size and numbers of GCs.	Fischer et al., 1996
C4$^{-/-}$	Impaired humoral responses to T-dependent antigens. Reduced size and numbers of GCs.	Fischer et al., 1996
CD81$^{-/-}$	Reduced B1a cells. Normal T and B-2 cell development. Normal BCR signal transduction but decreased proliferation after BCR crosslinking.	Maecker and Levy, 1997 Miyazaki et al., 1997 Tsitsikov et al., 1997
CD22$^{-/-}$	Increased Fraction I cells ($IgD^{hi}IgM^{lo}$). Enhanced BCR signaling to T-independent antigens.	O'Keefe et al., 1996 Otipoby et al., 1996 Sato et al., 1996
FcγRIIb$^{-/-}$	Enhanced antibody and anaphylactic responses.	Takai et al., 1996
CD72$^{-/-}$	Increase in pre-B cells. Enhanced BCR signaling. Few bone marrow "recirculating" B cells	Pan et al., 1999

nism. Class switching generally requires activation of CD40 on B cells; "nonspecific" T–B interactions usually result in the Fas-mediated elimination of B cells, as will be discussed later in this chapter. It is unclear how CD40 is activated in selected T-independent responses, or if the requirement for the CD40L–CD40 interaction can be bypassed for T-independent antigens to induce class switching to IgG3 or IgA. Isotype switching induced by T-independent antigens remains very poorly understood.

Marginal Zone B Cells

Marginal zones mature long after the appearance of germinal centers in both mice and humans. In children, these anatomically defined entities are best seen in the spleen around 2 years of age, when the ability to respond to polysaccharides also matures. While the cells that are generally called MZ B cells are IgM^{hi} IgD^{-} $CD23^{-/lo}$ $CD21^{hi}$ and $CD1^{hi}$ (See Figures 7.15 and 7.16), some class switched "conventional" memory B cells of the IgG and IgA subclasses also colocalize in the marginal zone. A small proportion of the cells in this anatomically defined area are newly formed virgin B cells which are also $IgM^{hi}IgD^{lo/-}$ $CD23^{-}$, but which express low levels of CD21 and CD1. The major population of MZ B cells share some phenotypic features with B-1 cells (which are also IgM^{hi}, IgD^{-}, and $CD23^{-/lo}$). However, these two populations appear to be distinct. Mutations in Btk, for instance, lead to the disppearance of B-1 cells, but MZ B cell development is unaffected.

Marginal zones are found only in the spleen. However, in humans, MZ like B cells have been described in the gut. Many human "marginal zone" lymphomas are derived from MZ-like B cells of gut origin. Splenic MZs are perfused by blood sinusoids, which branch from marginal sinuses derived in turn from the central arteriole. The MZ is therefore the perfect location not just for B cells that recognize blood-borne T-independent antigens, but also for memory B cells that are specific for T-dependent or protein antigens. Memory B cells are believed to be located in "MZ-like" extrafollicular areas in lymph nodes, tonsils, and Peyer's patches—areas that encounter antigen soon after it enters a secondary lymphoid organ. When memory cells are stimulated by antigen, they migrate from marginal zones, or MZ-equivalent areas, to the T cell areas of these secondary lymphoid organs.

One of the few mouse mutants in which MZ B cell development is severely compromised is the $Aiolos^{-/-}$ mouse. Aiolos is a transcriptional regulator that is structurally related to Ikaros. In its absence, B cells are constitutively activated via the BCR and can be induced to proliferate with minimal crosslinking of the antigen receptor. In these mice, every naive B cell appears have undergone positive selection. Btk is constitutively activated in all B cells in the absence of Aiolos. Crossing in a Btk mutation completely abrogates the $Aiolos^{-/-}$ phenotype and restores the MZ B cell population. While strong Btk signals apparently impair MZ B cell development, some clones of MZ B cells fail to be selected in the absence of functional Btk. In the absence of CD19, MZ B cell development is completely abrogated. Low intensity BCR signals are probably required for MZ B cell development or survival while Btk may be required only for the positive selection of a small subset of MZ B cell clones.

MZ B cells appear to develop from a $CD21^{hi}$ precursor follicular B cell in Fraction II which may also be $CD1^{hi}$. Some of these cells apparently receive a separate signal which induces the development of MZ B cells. A survival or differentiation signal may be provided to MZ B cells or their precursors by a recently discovered TNF family ligand secreted by macrophages and dendritic cells. This protein called BAFF or BLys (B Lymphocyte stimulator) is cleaved off the macrophage cell surface by the furin protease. The generation of MZ B cells absolutely requires NFκBp50. It is likely that p50 containing NFκB complexes are activated by BAFF induced signal transduction (Figure 8.2).

Functionally MZ B cells can be very easily and rapidly triggered *in vitro* to proliferate and to express high levels of B7-1 and B7-2 or to differentiate into plasma cells. Stimuli that will trigger MZ B cells include LPS, CD40L and IL-4. MZ B cells may therefore represent a specialized B cell population with a distinct repertoire from follicular B cells which can be particularly easily triggered by T-dependent ligands and possibly by complement coated antigens (which could conceivably interact with CD21 on MZ B cells). Some activated MZ B cells may be particularly suited to present antigen, while others differentiate rapidly into antibody secreting plasma cells.

The B-1/CD5$^+$ Lymphocyte Lineage

B-1 cells were first decribed in 1983 and are most numerous in young animals. In the mouse, this subset of self-renewing B cells is found primarily in the peritoneal cavity; smaller numbers of B-1 lymphocytes populate the spleens and lymph nodes of adult mice. B-1 cells express the CD5/Ly-1 marker and may represent B lymphocytes that have been positively selected by self- or environmental antigens. This subset includes high frequencies of cells bearing receptors specific for multivalent self-antigens including phosphatidylcholine, Igs, DNA, membrane proteins on erythrocytes and thymocytes, as well as environmental antigens such as phosphorylcholine. B-1 cells represent lymphocytes that require BCR-derived signals for maintenance or for their genesis from mature B cell precursors, or for both these events. Although they might be described as members of a B cell subset that is more "intensely" positively selected than conventional naive B-2 cells, it is unclear whether B-1 lymphocytes should be considered to represent a "naive" B cell population.

B-1 cells secrete large quantities of IgM, IgG3, and IgA and are presumed to be responsible for "natural" immunity. They are also involved, both in mice and humans, in the generation of some disease-associated autoantibodies. The level of natural antibodies may be regulated by controlling the numbers of these intrinsically activated, self-regenerating B-1 cells. These cells are categorized as either B-1a cells, which express CD5, or B-1b cells, which are very similar to B-1a cells but do not express CD5. Both B-1a and B-1b cells are phenotypically IgM^{hi} IgD$^-$ CD23$^-$ and Mac-1$^+$. B-1a cells develop primarily from fetal liver-derived hematopoietic stem cells (HSCs), while B-1b cells develop from HSCs from both the fetal liver and the bone marrow.

A widely held view, sometimes referred to as the lineage hypothesis, is that B-1 cells represent a separate lineage, derived from committed pre-

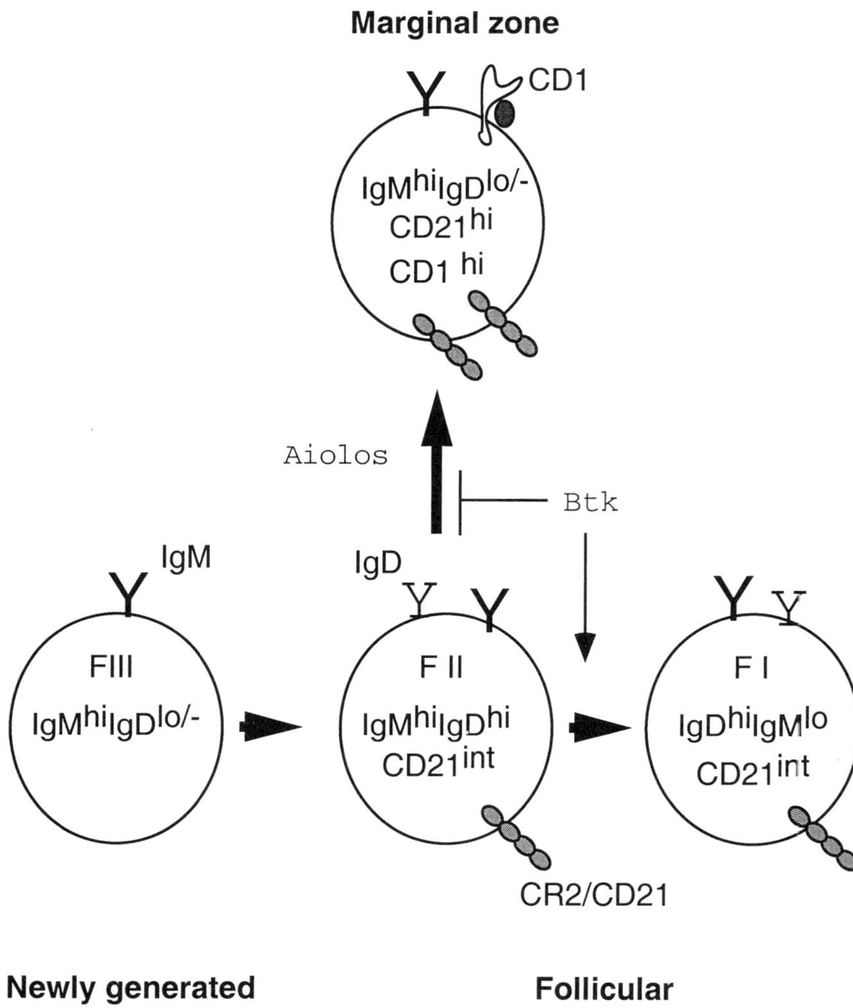

FIGURE 8.2. Marginal zone B cells are derived from follicular B cells which may require low level BCR signaling during development or for survival. Aiolos may be required for MZ B cell generation by its ability to attenuate Btk derived signals. $IgM^{hi}IgD^{hi}$ CD21 ^{hi} Fraction II follicular B cells are presumed precursors of MZ B cells. These cells may receive signals delivered by BAFF which may be expressed on MZ macrophages. An "intermediate" follicular stage in which some follicular cells express CD1 has also been described. The CD1 molecule is an MHC class I like protein, which is discussed in Chapter 9 in connection with NT (natural killer-like T) cells.

cursors largely of fetal liver origin that cannot give rise to conventional or B-2 lymphocytes. An alternative model, sometimes called the induced differentiation hypothesis, suggests that B-1 cells may be derived from B-2 precursor cells that have been triggered by polyvalent T-independent-type

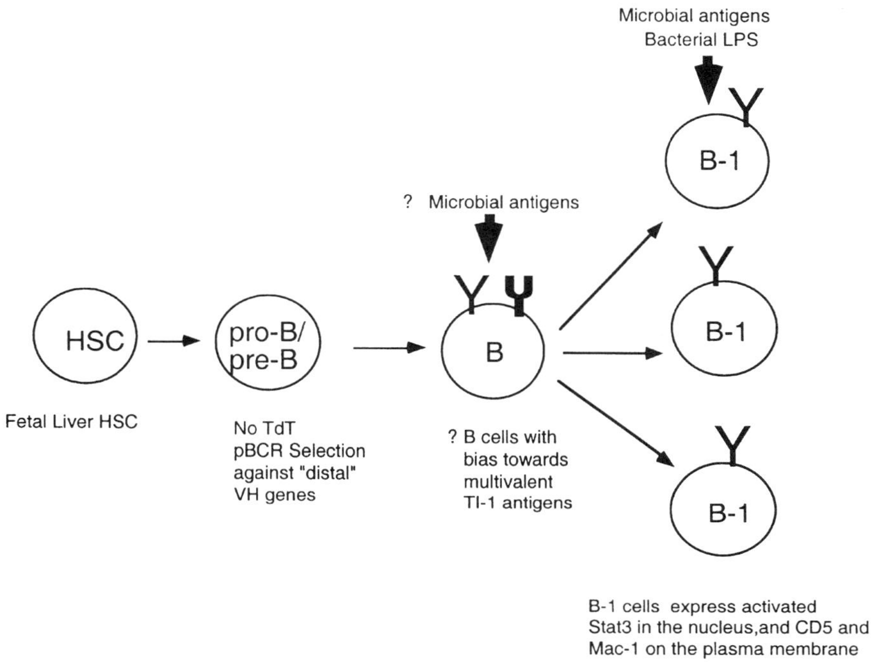

FIGURE 8.3. One model for the origin and maintenance of B-1 cells. B-1 cells develop primarily from fetal liver-derived HSCs, and their antigen receptors may be biased toward multivalent, macromolecular antigens. Maintenance of B-1 cells may depend on BCR-derived signals. The self-renewing capacity may be an inherent property of this lineage but may well be induced by exposures to environmental mitogens such as LPS and might be linked to their constitutive phosphorylation of nuclear STAT3.

2 antigens. Although the field seems divided between the proponents of the lineage hypothesis and those who insist that B-1 cells are "made" and not "born," the distance between these views may be smaller than imagined (Figure 8.3). Fetal liver HSCs presumably give rise to precursors with a receptor bias toward multivalent self- or environmentally derived antigens. These cells are apparently neither clonally deleted nor altered by receptor editing during ontogeny. This may be because they are not susceptible to these processes because of a hardwiring difference at the chromatin level or because the inciting antigens are not available at the site of early development, which is most often the fetal liver. These B-1 precursors may then be positively selected by multivalent self- or environmental antigens and induced to express CD5.

The CD5 protein is a member of the scavenger receptor family of cell surface proteins that is expressed on T cells and is known to negatively regulate T cell-receptor (TCR) mediated signaling. Interestingly, CD5 is upregulated during the TCR-dependent "induction" step of positive selection in $CD4^+/CD8^+$ double-positive thymocytes. CD5 is also induced when naive B-2 cells are triggered via their antigen receptors. CD5 induction may therefore be a marker of antigen-receptor–mediated positive selection in both B and T cells. Clearly naive peripheral B-2 cells do not express high levels of CD5 and, unlike B-1 lymphocytes, show little biochemical evidence of having gone through a selection event. This might reflect the fact that B-1 cell genesis or maintenance involves a much more intense triggering event by a multivalent TI-2 antigen. If self-antigen–mediated positive selection of immature B-2 cells actually does occur, the inciting antigen may be of low affinity or avidity—perhaps even protein antigens might play a role in such a "tickling" process. In the case of B-1a cells, it appears that the presence of the CD5 protein plays a critical role in downregulating BCR signals, possibly because it can recruit the SH2 containing PTP-1 (SHP-1) tyrosine phosphatase to the vicinity of the crosslinked receptor (CD5 has been shown to be capable of complexing with the BCR by a poorly understood mechanism). In B-1b cells, which lack CD5, negative regulation of antigen-receptor signaling might be mediated by other proteins that can recruite SHP-1, such as CD22, or other B cell surface proteins whose cytosolic tails contain ITIM motifs. Examples of such proteins include the paired Ig-like receptors (PIRs), the leukocyte Ig receptor (LIR), and the leukocyte-associated Ig-like receptor (LAIR-1). However, CD5 itself lacks immunoreceptor tyrosine inhibitory motifs (ITIMs), and no evidence indicates a direct interaction between CD5 and SHP-1.

From a developmental standpoint, it is of interest that most B-1 cells develop from fetal liver progenitors and therefore have a restricted repertoire, biased toward the recognition of multivalent, naturally occurring, macromolecules. As discussed in Chapter 7, fetal liver stem cells do not express terminal deoxynucleotidyl transferase (TdT), and as a result, rearranged heavy chain genes in B cells derived from these stem cells lack N regions and are of restricted diversity. Pre-B receptor signaling in fetal liver-derived HSCs may also help bias the repertoire toward incorporating selected V_H genes, mainly from proximally located V_H families, as described in Chapter 7.

Proteins Involved in BCR Signaling Are Critical for B-1 Cell Development or Maintenance

A number of gene-targeting experiments (most of which are summarized in Tables 8.1 and 8.2) have revealed that certain defects in B cell signaling impair T-independent and T-dependent responses as well as the generation of B-1 cells. In general, the failure to generate a strong crosslinking signal via the BCR and the CD21 co-receptor complex results in the absence

TABLE 8.3. Factors required for B-1 cell generation

Knockout	B-1 cells	References
Xid and $Btk^{-/-}$	Absent	Herzenberg et al., 1986; Khan et al., 1995; Kerner et al., 1995; Hendriks et al., 1996
$PKC\beta^{-/-}$	Absent	Leitges et al., 1996
$p85\alpha^{-/-}$	Absent	Suzuki et al., 1999; Fruman et al., 1999
$CD45^{-/-}$	Absent	Byth et al., 1996
$Ig\alpha^{-/-}$	Absent	Torres et al., 1996
$CD19^{-/-}$	Absent	Rickert et al., 1995; Engel et al., 1995
$CD21/CD35^{-/-}$	Reduced	Ahearn et al., 1996; Molina et al., 1996
$CD81^{-/-}$	Reduced	Maecker and Levy, 1997; Miyazaki et al., 1997; Tsitsikov et al., 1997
$Vav^{-/-}$	Absent	Tarakhovsky et al., 1995; Zhang et al., 1995
$Oct\text{-}2^{-/-}$	Absent	Corcoran and Karvelas, 1994; Humbert and Corcoran, 1997
$Aiolos^{-/-}$	Decreased	Wang et al., 1998
$IL\text{-}5^{-/-}$	Transient decrease	Kopf et al., 1996
$IL\text{-}5R\alpha^{-/-}$	Transient decrease	Yoshida et al., 1996

of B-1 cells. The exact role of the BCR in B-1 cells, however, is unclear. Whether the BCR is required only for the activation of precursors by multivalent antigens or whether the antigen receptor is primarily needed for maintenance is not known. BCR-ligated peritoneal B-1 cells fail to proliferate and undergo apoptosis. A list of the genes known to be important for B-1 cell generation is provided in Table 8.3.

If antigen-receptor signaling leads to the apoptosis of B-1 cells, can this be reconciled with the possibility that cognate antigens (and therefore the BCR) most likely contribute to the maintenance and proliferation of these self-renewing cells, as well as the induction of antibody production? It is possible that self/environmental antigens do not induce apoptosis of B-1 cells because of "costimulation" from gut microflora-derived LPS or other signals. B-1 cells, but not unstimulated conventional B cells, express a constitutively phosphorylated form of the STAT3 transcriptional regulator. The phosphorylation of STAT3 in B-1 cells may or may not be linked to BCR activation. The pattern of STAT3 phosphorylation induced by antigen-receptor ligation *in vitro* represents a subset of the modifications noted constitutively in B-1 cells. STAT3 activation may depend on some unknown intrinsic mechanism or may require a combination of signals that may include inputs from the antigen receptor. A knockout of STAT3 resulted in an embryonal lethal phenotype, and the importance of STAT3 for B-1 cell proliferation or survival remains to be fully evaluated. There is a transient deficit of B-1 cells in mice in which IL-5 or the IL-5R have been knocked out, suggesting that IL-5 may be a preferred but nonessential cytokine for B-1 cell maintenance (see Table 8.3). Antibodies to IL-10 can block the development or maintenance of B-1 cells, but the IL-10 and IL-10R knockout mice do not have major deficits in B-1 cell numbers.

In the *Xid* and *Btk* null mice, as well as in mice that lack CD45, PKCβ, CD19, and Vav, B-1 cells are not seen. CD45 is a tyrosine phosphatase required for Src-family kinase activation, and Src-family kinases are critical for the activation of Btk. In contrast, in *motheaten viable* mice that lack functional SHP-1, BCR signaling is enhanced and B-1 cells accumulate.

B-1 cells are not seen in the absence of Oct-2, a transcription factor that has been implicated as a downstream effector of BCR-mediated signaling (and is discussed later in this chapter and briefly in Chapter 2). There is a significant decrease in B-1 cells seen in the Aiolos knockout mouse, and this might reflect the inability of these mice to efficiently respond to T-independent antigens, as discussed above. B-1 cells are diminished in mice lacking CD81 (which complexes with CD21 and CD19 as a co-receptor for BCR-mediated signaling).

Btk is critical at least at three distinct stages in B cell development. In humans, but not in mice, Btk is required for B cell generation, presumably because of its requirement in preBCR signaling. In the mouse, Btk provides a maintenance signal for peripheral $IgD^{hi}IgM^{lo}$ Fraction I B cells, as discussed in Chapter 7. In addition, Btk is required for the activation of peripheral B cells by multivalent antigens, an event that is probably critical for T-independent immune responses and B-1 cell generation. This latter function of Btk may also be dependent on the activation of PKCβ. Mice in which the PKCβ gene has been knocked out, are, like *Xid* and *Btk* null mice, unable to proliferate in response to T-independent antigens and lack B-1 cells. Although physical association between PKCβ and Btk has been noted, a more likely functional link between these two enzymes may depend on the fact that Btk may be required for the activation of PLC-γ following BCR crosslinking. PLC-γ in turn, by generating diacyl glycerol (DAG) and inositol 1,4,5 trisphosphate (IP3), is most likely required for the activation of protein kinase C β (PKCβ) following BCR ligation. While it has been suggested that PKCβ might be relevant upstream of Btk, it is quite likely that Btk activation actually leads to the activation of PKCβ. Some critical substrate or substrates of PKCβ and/or of Btk is/are presumably required for B cell proliferation and B-1 cell generation. This Btk/PKCβ pathway may be required for the induction in activated B cells of survival factors such as c-Rel (an NFκB family member) and Bcl-X_L. No hard evidence at present links the putative Btk/PKCβ pathway to c-Rel induction, but a connection between Btk signaling and Bcl-X_L has been suggested.

CD21 or CR2 is a receptor for the C3d fragment of the C3 complement component. CD21 associates with CD19 and CD81 (TAPA-1) to form a signaling complex. Antigens coated with C3d can activate BCR signaling far more efficiently than uncoated antigens. As discussed in Chapter 4, the BCR can be activated by crosslinking as well as by the simultaneous ligation of the CD21 co-receptor; proximity presumably allows Src-family kinases associated with the cytoplasmic portions of the receptor and co-receptor to phosphorylate each other and initiate signal transduction. Vav, as also discussed in Chapter 4, is a guanine nucleotide exchange factor for Rac that can contribute to the activation of the JNK/SAPK pathway. Vav is tyrosine phosphorylated after BCR activation and can also be recruited to CD19. Mice lacking CD21, CD19, or Vav all lack B-1 cells. In the absence of CD81/TAPA-1, there is a reduction in the abundance of B-1 cells.

In summary, the absence of CD45, Btk, PKCβ, CD19, or Oct-2 can individually lead to a lack of B-1 cells. CD21 and CD19 contribute as part of the same co-receptor complex to reducing the threshold for BCR signaling. Vav is an exchange factor for Rac that is downstream of BCR and CD19 signaling.

Modulators of BCR Responses to T-Dependent and T-Independent Antigens

A large proportion of proteins encountered by the immune system may exist as oligomers, which might have the ability to crosslink B cell receptors. Although soluble proteins that are truly monomeric might not be expected to be able to crosslink the B cell receptor, it is apparent that once such proteins are covalently linked to C3d, they may activate BCR signaling by bringing the CR2/CD21 coreceptor and the antigen receptor in close proximity. The CD21/CD19 coreceptor complex is required for optimal responses to T-dependent antigens.

As discussed in Chapter 4, a few receptors qualify as modulators of BCR signaling. The phenotypes of mice that lack CD21, C3, C4 (which is required for cleavage of C3), CD19, CD81, CD22, and FcγRIIb are summarized in Table 8.2.

T-Dependent Antigens Induce Class Switching, Memory Generation, and Somatic Mutation

T-dependent signals contribute to B cell clonal expansion, the initiation of isotype switching, the formation of germinal centers, somatic mutation, and the generation of memory. Signals delivered via CD40 are required for all of these phenomena. In patients with the X-linked hyper IgM syndrome, T–B collaboration is abrogated because of an inherited absence or defect in the CD40 ligand (Figure 8.4), which is normally induced in activated helper T cells. Early in an immune response, the interaction of an antigen-triggered B cell with a helper T cell in the outer T cell zone in the spleen may initiate the formation of a focal B cell aggregate at the boundary between the T cell zone and the red pulp. Similar focal aggregates may form

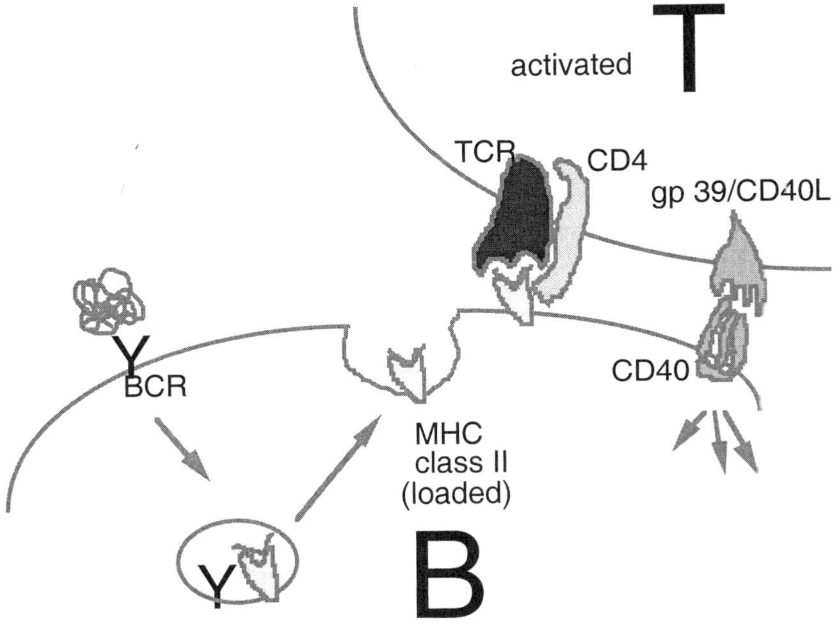

FIGURE 8.4. The B cell receptor internalizes cognate protein antigens for antigen presentation on MHC class II molecules to specific T cells. The CD40L on activated T cells provides growth and differentiation signals to B cells. Signals may also be provided by the B cell receptor, the CD21 coreceptor complex, cytokine receptors on the B cell, and other costimulatory receptors as well.

in lymph nodes. Later in an immune response, T–B collaboration may lead to the initiation of a germinal center reaction. Isotype switching may occur in these focal aggregates before germinal center formation but can also occur after somatic mutation has been completed within the germinal center, as will be discussed below.

T-Dependent Protein Antigens Distract B Cells in T Cell Zones and Facilitate T–B Collaboration in the Outer PALS

The T cell zone in the spleen is called the periaarteriolar lymphoid sheath (PALS), and the place where specific B and T cells get to know each other is called the outer PALS. A naive recirculating antigen-specific B cell enters secondary lymphoid organs and traverses the outer T cell zone on its way to a primary lymphoid follicle (which lacks germinal centers) or possibly to the follicular mantle zone of a secondary follicle. If this B cell encounters a cognate antigen, it receives antigen-receptor–derived signals, which basically distract the cell and prevent it from homing to the follicle. Antigen-receptor ligation instructs B cells to remain in the outer PALS close to the interface between B and T cell areas. If a B cell is activated elsewhere, either in the red pulp, the marginal zone, or in the follicle itself, it receives signals to migrate to the interface between T and B cell areas. This B cell will present peptides from the internalized and processed protein antigen in an MHC-class II context on the cell surface (see Figure 8.4). This cell looks around for help, which it might receive from a cognate helper T cell when that latter cell makes its rounds through this particular T cell zone (and is also directed to move to the outer PALS as a consequence of TCR stimulation and costimulation). During the course of an infection, the chances that such an encounter will happen are probably quite high, since the protein antigen in question will be presented by professional antigen-presenting cells (APCs), which activate and expand specific helper T cell clones. The CD40 ligand (also known as CD40L or gp39) on activated T cells delivers a growth signal to B cells via the CD40 molecule on the latter. Additional signals are provided by T cell-derived cytokines and possibly by other less well-studied costimulatory receptors.

Early in an immune response, an activated B cell may form an expanded collection of clonal B cells in the outer T cell zone close to the T cell zone–red pulp interface (Figure 8.5) in the spleen or close to the medullary area in lymph nodes. Cells in these focal extrafollicular foci or aggregates may undergo isotype switching and differentiate into short-lived plasma cells, which, unlike germinal center-derived "long-lived" plasma cells, probably do not receive signals to leave secondary lymphoid organs and seed the bone marrow. Once activated B cells in these extrafollicular aggregates and/or plasma cells secrete antibodies, local immune complex formation with circulating antigen is facilitated. These immune complexes can then be presented by follicular dendritic cells in B cell follicles, and about 48 hours after the first appearance of extrafollicular foci can trigger the formation of follicular germinal centers. It is in these germinal centers that somatic mutation occurs, following which, the B cells with the highest affinity for antigen are selected, a process known as affinity maturation. The further development of germinal centers and the differentiation of B cells at these sites will be discussed a little later in this chapter.

Isotype Switching

The process of isotype switching serves to functionally diversify the immune response. While signals delivered by CD40 are essential for this process, specific signals from distinct cytokine receptors may influence switching from IgM to selected Ig classes. The generation of IL-4 by helper T cells,

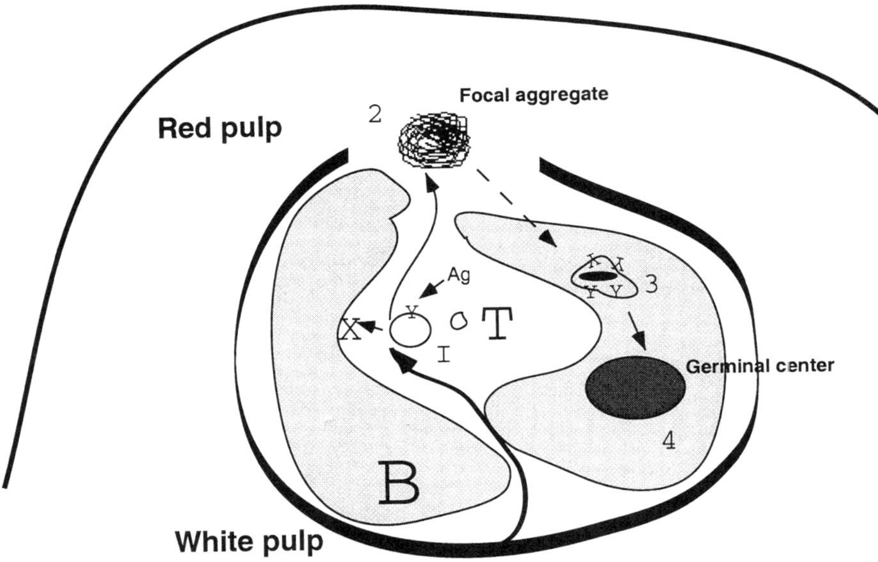

FIGURE 8.5. Some of the steps involved in T-dependent focal aggregate and germinal center formation by antigen-specific B cells. (1) When an antigen-specific B cell is triggered by a cognate T-dependent antigen in the outer T cell zone, it is prevented from entering the B cell follicular area, and cognate T–B collaboration is facilitated. Alternatively, stimulated follicular B cells are induced to move to the boundary between B and T cell areas (2) T–B collaboration results in the formation of focal aggregates. (3) A little later, immune complexes form on follicular dendritic cells (FDCs) in the follicle. (4) FDCs presenting immune complexes to antigen and T cell-activated B cells facilitate the formation of germinal centers.

for instance, may help drive class switching to IgE, while a high local concentration of transforming growth factor β (TGF-β) may drive switching to IgA. Signal transduction leads to the transcriptional induction of DNA-binding proteins. These proteins identify switch sites upstream of heavy chain constant regions and help mediate a deletional recombination event, which brings the rearranged V(D)J unit from its location adjacent to the μ constant regions to a site next to the heavy chain isotype targeted for switching (Figure 8.6). Based on the nature and location of the antigenic provocation, cytokines may be generated, which help trigger accessibility in B cells of recombinationally relevant regions upstream of a desired Ig isotype. Cytokines may also help drive the switching process by means other than the enhancement of accessibility, as will be discussed below. These signals permit the switch recombinase to facilitate the deletion and joining process. Based on the rearrangement of artificially generated substrates introduced into a range of cell lines, it appears that switch recombinase activity is B cell specific, being found only in pre-B and B cell lines. Very

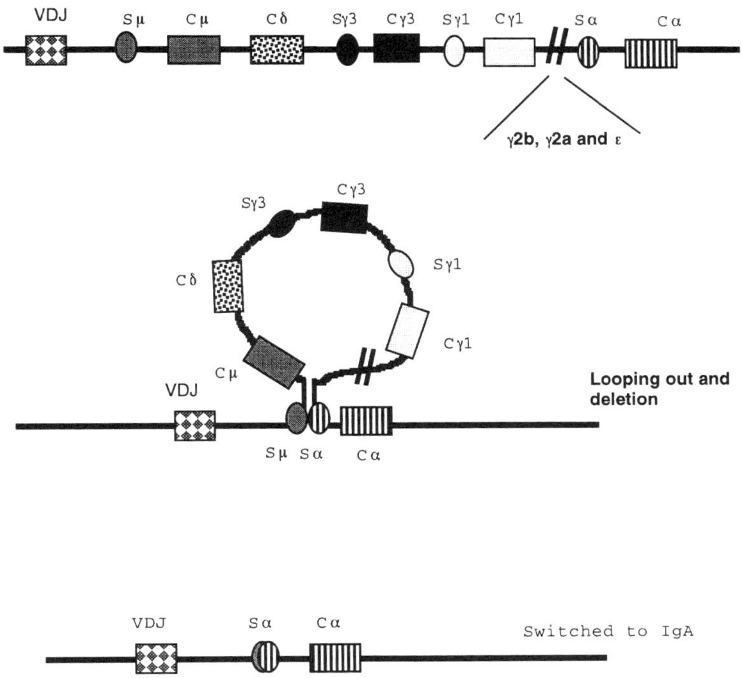

FIGURE 8.6. A schematic view of the process of isotype switching. Cell surface signals initiate the process, as will be considered in Figure 8.7. The process involves looping out and deletion of DNA between switch regions, followed by double-strand break repair which requires DNA-PKcs, Ku 70, and Ku80.

little is known at present about this recombinase. It has been shown that progenitor B cells from *Rag-2* null mice have the ability to mediate class switching, but progenitor B cells from *Scid* mice, which carry a mutant DNA-PK gene, are unable to switch. The Ku70/Ku80/DNA-PK complex participates in V(D)J recombination (see Chapter 3) as part of the machinery involved in double-strand break repair. It is likely that this complex serves a similar function during the process of switch recombination. Switch recombination, like V(D)J recombination, involves the generation of double-stranded breaks at distant sites, deletion of an intervening region, and ligation of the broken ends. Rearranged "knocked-in" Ig loci have been engineered into mice that lack Ku70 and Ku80, and an impairment of class switch recombination has been noted in these mice. B cells in these mice can respond to cytokines and induce switch transcripts but are unable to complete the recombination process. A very interesting connection has been made between CD40 and Ku. It has been shown that a substantial portion of Ku is non-nuclear and that Ku proteins associate with the membrane proximal region of the cytoplasmic tail of CD40. Ligation of the

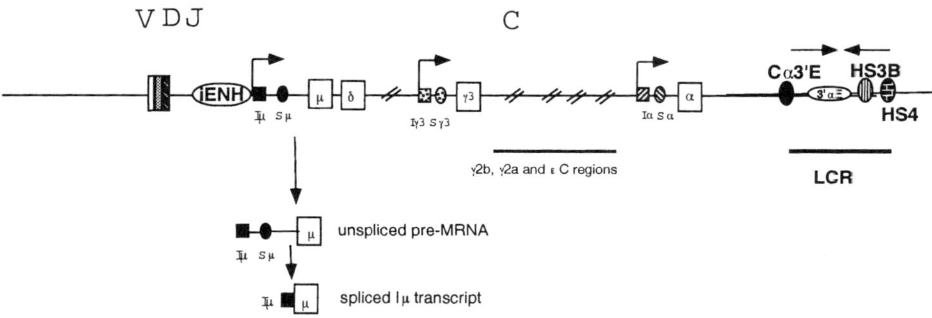

FIGURE 8.7. A schematic view of the Ig heavy chain locus emphasizing the structural features of switch transcripts. It has been speculated that the spliced switch transcript may have a structural role in recombination. Alternatively, the spliced out intron might form an RNA-DNA triple helix in the switch region. Spliceosome components might be involved in class switching. The 3′LCR, discussed in Chapter 2, has been implicated in the regulation of switch accessibility as discussed in the text.

receptor leads to the translocation of Ku into the nucleus by an unknown mechanism. Ku proteins are tyrosine phosphorylated, but it is unclear if this modification is linked to the nuclear translocation of Ku during the T-dependent activation of B cells.

Upstream of each constant region is a region of DNA known as a switch region. Switch regions contain multiple repeats of GAGCT and TGGGG and may range from about 1 to 10 kb in length. Just upstream of each switch region is an I region promoter and an I exon (Figure 8.7). When a heavy chain isotype is targeted for class switching, a transcript is initiated from the I region 5′ to the relevant constant region. The transcript reads through the I exon, the switch region, and the constant region exons and is terminated and polyadenylated at the 3′ end of the constant regions. This transcript is spliced (removing the switch region and other introns). In general, switch transcripts contain multiple-stop codons and cannot be translated into protein. There is one report of a specific switch transcript that was capable of being translated, but this appears to represent a rare exception. Replacement of an I region promoter with a constitutively active promoter can make class switching cytokine independent.

Are switch transcripts merely markers of accessibility, or do they have a specific structural or functional role? The structure of the spliced switch RNA requires to be conserved in order for isotype switching to proceed optimally. There may be a need both for some conservation of the structure of the I exon as well as for the I region transcript to be properly spliced for it to facilitate switching. This has led to the suggestion that the spliced switch transcript might itself be a structural component of the switch recombinase.

It has also been suggested that the intronic "spliced-out" portions of switch transcripts (which include the switch regions) might generate RNA-DNA triplex structures that could serve as intermediates prior to class switching. The switch sequence specific RNA-DNA hybrids might facilitate the recombination process by facilitating the formation of nuclease sensitive short stem-loop structures involving tandem repeat sequences in the switch regions. An alternative explanation for the need to splice switch region transcripts is that the splicing machinery may be essential for class switching. It is possible that components of the spliceosome are co-opted into the recombinational process during the splicing of switch region transcripts.

Some progress has been made in attempting to identify components of the switch recombination machinery by biochemical purification of factors that may participate in the recombination of switch substrates. A number of RNA and DNA binding proteins have been identified. Two of these proteins, nucleophosmin and nucleolin, are involved in the processing of ribosomal RNA in the nucleus. This supports the growing view that components important for splicing might participate in switch recombination.

Regulation of Isotype Switching

Isotype switching is regulated by signals initiated at the B cell surface from receptors such as CD40 or from cytokine receptors. The process is regulated at the level of accessibility and also at other levels, perhaps by regulating the levels of proteins that are required for the cutting or joining processes. In a general sense, there is a requirement for the intronic H-chain enhancer to induce accessibiity upstream of the constant regions of μ (Figure 8.7; see also Chapter 2). Interference with the function of the *3′H chain LCR* region (by insertion of a neomycin resistance cassette into the *3′αE* region) repressed most germline heavy chain transcripts. This regulatory region probably cooperates with specific I region promoters to regulate accessibility at most H-chain loci. A large number of studies have revealed the importance of specific cytokines in inducing switching to specific isotypes. The influence of these cytokines on switching has generally been correlated with increased switch region transcripts from the isotype of interest. Exposing B cells to LPS and IL-4 results in increased *Iγ1* and *Iε* transcription and switching to *γ1* and ε. Exposure of these cells to IFN-γ results in a decrease in *Iγ1* and *Iε* transcripts, but an increase in *Iγ2a* transcripts and class switching is directed toward IgG2a. TGFβ can enhance the ability of cells to switch to *Cα*, and isotype switching can also be influenced by IL-5 and IL-10. Many of the influences of cytokines on class switching, particularly those listed above, can be explained in terms of the accessibility model, since they result in the activation or induction of DNA-binding proteins that bind to specific I region promoters. A number of the effects of cytokines, however, are quite complex and may not involve an increase in accessibility alone.

Germline transcripts can be induced in resting B cells by exposure to the appropriate cytokines, but these cells will not switch unless they have also been induced to enter the cell cycle. A direct role for DNA synthesis in the switching process has been suggested but remains speculative. Stimulated B cells apparently need to go through at least three cell cycles before switched isotypes appear on their surfaces. Why exactly B cells need to enter cycle to class switch remains to be determined. Another interesting correlate of locus accessibility both for V(D)J recombination as well as switch recombination is demethylation. It has been suggested that cis-acting factors may regulate methylation of the locus and that undermethylation results in increased accessibility for switching.

Following activation of B cells via the B cell receptor, CD40, or cytokines such as IL-4, a number of transcription factors are either induced or activated. Some of these factors play important roles in maintenance, proliferation, germinal center formation, isotype switching, and the process of differentiation into plasma cells. We will discuss some of these transcription factors as a group later in this chapter. In the context of class switching, it is worth mentioning the important roles ascribed to factors such as NFκB1, and Pax-5/BSAP.

B cells from the *NFκB1*$^{-/-}$ (p50 null) mouse exhibit specific but limited defects in isotype switching (apart from also having defects in cell maintenance, as will be discussed later in this chapter). There are a number of sites for p50 in the heavy chain locus, in the *Iγ3*, *Iγ1*, and *Iε* regions as well as in the 3′ locus control region (*3′LCR*). p50 Binding sites have also been found in *Sγ3*, *Sγ2b*, and *Sγ1* switch regions close to sites where recombination occurs. Defects in switching to IgG3 and IgE correlate with a decrease in germline transcripts generated by the Iγ3 and Iε promoters. However, there is also a defect in switching to IgA, although transcripts from the *Iα* promoter are not decreased.

Pax-5/BSAP may also be involved in isotype switching. Since targeted disruption of this gene results in a block in early B cell development, it has been difficult to assess its role later in development. Antisense studies on cultured cells have suggested a role for Pax-5 in isotype switching. It is unclear whether this role in switching is mediated by Pax-5 transcriptionally activating I region transcripts or because Pax-5 is required for B cell proliferation.

Another transcriptional regulator that is involved in early B cell development but which is also induced after BCR ligation is E2A. The use of a dominant negative form of E2A in mature B cells has revealed that E2A is not required for BCR induced proliferation, but is absolutely required for isotype switching.

Transcriptional Regulators in Activated B Cells

As discussed in Chapter 4, many of the signals delivered by the B cell receptor, CD40, bacterial LPS, and cytokine receptors such as the IL-4 receptor culminate in the induction or activation of transcriptional regulators. The Ras/MAPK (mitogen-activated protein kinase) pathway, for instance, eventually leads to the phosphorylation by the ERK (extracellular signal-regulated kinase) MAPK of transcription factors such as Elk-1. Both ERK and JNK (Jun N-terminal kinase) MAPKs contribute to the activation of components of the AP-1 transcription factor (made up of Jun and Fos proteins, as discussed in Chapter 4). The increase of intracellular calcium that follows antigen-receptor signaling activates calcineurin, which dephosphorylates members of the NFAT family of transcription factors and thus contributes to the activation of target genes. Signal transduction by both the B cell receptor and by CD40 can activate members of the NFκB family. Numerous other transcriptional regulators are no doubt critical during B cell proliferation and differentiation.

Transcription factors contribute to B cell maintenance, maturation, proliferation, isotype switching, germinal center formation, plasma cell generation, and undoubtedly to memory cell generation as well. Some transcription factors thought to be involved in developmental events such as tissue-specific Ig gene expression have been discovered to be critical components of B cell signaling after they have been knocked out and may not have yet been connected to any specific signaling pathway. A number of transcription factors are believed to cooperate during later events in B cell differentiation, particularly in activated B cells undergoing class switch recombination and gearing up to become plasma cells.

Table 8.4 lists some of the functions of transcription factors in the later stages of B cell development.

Members of the NFκB Family

Members of the NFκB family such as p50/NFκB1, c-Rel, and RelA, which are expressed at high levels in hematopoietic cells, as well as proteins such as Rel B, NFκB2/p52, and Bcl-3, which are expressed at high levels primarily in antigen-presenting cells, may play a critical role in B cell development and activation. These proteins have been discussed in Chapter 4, and Table 4.4 lists the phenotypes of mice in which genes of this family have been disrupted. Some of this information is repeated for the sake of convenience in Table 8.4.

NFκB1/p50 is required for the maintenance of immature and mature naive B cells. The life span of B cells is considerably reduced in the absence of p50. This reduction of life span affects both immature and mature B cells. B cell proliferation in response to LPS is defective, although proliferation

TABLE 8.4. Transcription factors and later events in B cell development (incorporates some material from Table 4.4)

		References
NFκB1 (p50)$^{-/-}$	Defective B cell proliferation in response to LPS; normal response to CD40L and anti-IgM; defects in B cell maturation; defects in isotype switching (IgG3 and IgE reduced); defective T cell proliferation.	Sha et al., 1995 Snapper et al., 1996a
NFκB2 (p52)$^{-/-}$	Reduction in B cell numbers; defects in humoral immune responses, germinal center formation, and splenic microarchitecture; lack FDC networks and MZ metallophilic macrophages.	Franzoso et al., 1998 Caamano et al., 1998
NFκB1$^{-/-}$/ *NFκB2*$^{-/-}$	Defective B cell and osteoclast development.	Franzoso et al., 1997
RelA (p65)$^{-/-}$	Death at day 16 of embryogenesis due to massive liver degeneration; rescued B and T cells have defects in proliferation; defects in isotype switching to IgG1 and IgA.	Beg et al., 1995 Doi et al., 1997
c-Rel$^{-/-}$	Defective B cell proliferation in response to LPS, CD40, and anti-IgM; defects in isotype switching (IgG1 and IgE reduced); defective T cell proliferation and cytokine generation.	Kontgen et al., 1995
RelB$^{-/-}$	Reduced thymic epithelial and dendritic cells; myeloid hyperplasia; inflammatory infiltrate in many organs; defective B cell proliferation.	Weih et al., 1995 Burkly et al., 1995 Snapper et al., 1996b
NFκB1$^{-/-}$/ *RelA*$^{-/-}$	Mice die at day 12 of embryogenesis due to massive liver necrosis; stem cell rescue studies reveal a failure of lymphopoiesis due to defect in stromal cells.	Horwitz et al., 1997
RelB$^{-/-}$/*NFκB1*$^{-/-}$	Defects in B cell proliferation, thymic medullary development, and dendritic cells, as in individual KOs; inflammation more severe.	Weih et al., 1997
Bcl-3$^{-/-}$	Defects in germinal center formation, splenic and lymph node microarchitecture; humoral immune responses defective; phenotype resembles NKκB2 KO.	Franzoso et al., 1997 Schwarz et al., 1997
Oct-2$^{-/-}$	Neonatal lethality; peripheral B cell maintenance defect; proliferation in response to anti-IgM and LPS impaired. No B1 cells. T-dependent and T-independent immune responses defective.	Corcoran et al., 1993 Corcoran and Karvelas, 1994 Humbert and Corcoran, 1997
OBF-1/OCA-B/ Bob-1$^{-/-}$	Defect in B cell maintenance and anti-IgM–induced proliferation. No germinal centers. Production of switched isotypes impaired at transcriptional level.	Kim et al., 1996 Schubart et al., 1996 Nielsen et al., 1996

TABLE 8.4. (*Cont.*)

		References
Pip/IRF4/ LSIRF$^{-/-}$	Reduced Fraction I follicular B cells. Impaired proliferation in response to anti-IgM and LPS. No germinal center B cells.	Mittrucker et al., 1997
Bcl-6$^{-/-}$	Dysregulated cytokine gene expression. Inflammation. No germinal centers.	Dent et al., 1997 Ye et al., 1997
Spi-B$^{-/-}$	Defective anti-IgM–induced proliferation smaller germinal centers due to maintenance defect.	Su et al., 1997
Aiolos$^{-/-}$	Defective B cell maturation, reduced B-1 cells, and absence of marginal zone B cells. Enhanced proliferation of antigen-receptor–ligated B cells. Spontaneous germinal center formation and enhanced T-dependent immune responses.	Wang et al., 1998

in response to antigen-receptor crosslinking and CD40L exposure is unaffected. This transcription factor, as mentioned above, also participates in the process of isotype switching, and levels of IgG3 and IgE are reduced in p50-deficient mice. In the absence of p50, T-dependent immune responses are impaired, but there is no defect in germinal center formation.

The proliferation of B cells in response to anti-IgM crosslinking, CD40 ligation, and LPS is impaired in mice that lack c-Rel, another member of the NFκB family. These mice have defects in isotype switching to IgG1 and IgE and also have defects in T cell proliferation and cytokine induction.

In the absence of c-Rel or of the p50/NFκB1 protein, B cells undergo apoptosis, and progression through the G1 phase of the cell cycle does not occur in response to antigen-receptor ligation or mitogens. However, the failure of cell cycle progression is not merely due to increased apoptosis. When apoptosis is inhibited by forced expression of Bcl-2, these B cells nevertheless fail to respond to proliferative signals, suggesting that other critical genes downstream of NFκB must be activated. Although studies in cell lines had suggested that *c-myc* may be a target of NFκB, in *c-Rel* and *p50* null mice, *c-myc* is normally induced in G1. Overall, it appears that yet to be identified targets of p50/NFκB1 are essential for survival of G0 cells, and that other genes downstream of p50 and c-Rel may need to be induced to ensure G1 progression.

Although the RelA/p65-deficient mouse dies at day 16 due to massive hepatic necrosis, reconstitution of the lymphoid compartment reveals defects in B and T lymphocyte proliferation as well as defects in isotype switching to IgG1 and IgA. In a *p50*$^{-/-}$/*p65*$^{-/-}$ double-deficient mouse, death is observed at embryonic day 12 due to massive liver necrosis. Reconstitution of the hematopoietic sytem reveals a defect in lymphopoiesis due to a defect in a nonlymphoid "supporting" cell of bone marrow origin.

The phenotypes of mice lacking NFκB2/p52 and Bcl-3 are very similar. As mentioned in Chapter 4, Bcl-3 is a member of the IκB family that actually serves as an activator of transcription in collaboration with p50 or p52 (see Chapter 4). In the absence of p52 or Bcl-3, FDC networks are not seen, MZ metallophilic macrophages are absent, and germinal center formation is compromised. The defects in splenic and lymph node microarchitecture that are seen in the absence of these transcriptional regulators mimic the defects seen in the absence of some members of the TNF family, as discussed in Chapter 6.

RelB is expressed only in lymphoid tissue and is primarily expressed in interdigitating dendritic cells. The primary defect in the absence of this transcription factor is in the generation of CD8α^- myeloid-type dendritic cells (as opposed to CD8α^+ lymphoid dendritic cells, which can develop normally) in the thymus and peripheral lymphoid organs. The absence of these cells compromises thymic medullary development and results in defective antigen presentation, compromised cell-mediated immunity, splenomegaly, and multiorgan inflammation. B cell proliferation in response to antigen-receptor ligation, LPS, or triggering of CD40 is defective, but there is no defect in the capacity of these B cells to undergo isotype switching or Ig secretion.

Oct-2 and OBF-1/OCA-B/Bob1

Oct-1 and Oct-2 are transcription factors that bind to the octamer motif in Ig promoters and enhancers, as discussed in Chapter 2. While Oct-2 is expressed in a B cell-specific manner, Oct-1 is also found in B cells, and the B cell-specific transcriptional activation of target genes has not been easy to explain, since Oct-1 is ubiquitous. Tissue-specific expression of octamer motif-containing genes is achieved by the association of the co-activator OBF-1/OCA-B/BOB-1, as discussed in Chapter 2.

In the absence of Oct-2, mice die in the neonatal stage. The production of B lineage cells in the bone marrow occurs normally, but mature B cells are reduced in the periphery. Reconstitution experiments in *Rag-1*$^{-/-}$ mice has revealed that Oct-2 is required not only for the maintenance of peripheral B cells but also for the proliferation of B cells in response to antigen-receptor crosslinking or to LPS. Proliferation of B cells in response to CD40 is unaffected. However, an impairment of both T-independent and T-dependent immune responses was noted in Oct-2-deficient reconstituted mice. The absence of B-1 cells in *Oct-2*$^{-/-}$ mice has been discussed earlier.

Targeted disruption of *Obf-1* has revealed that it is required for B cell maintenance and for B cell proliferation following antigen-receptor crosslinking. Although *Oct-2*$^{-/-}$ mice have no B-1 cells, there is no defect in B-1 cell generation in the *Obf-1*$^{-/-}$ mouse. OBF-1 is induced in primary B cells by signals from the antigen receptor, CD40, and cytokine receptors. OBF-1 is highly expressed in germinal centers and is absolutely required for germinal center formation. In the absence of this transcription factor, primary B cell follicles, MZs, and plasma cells are seen, but there is a very specific defect in development of germinal center B cells. There is a reduction in the number of recirculating mature B cells presumably because of a compromised maintenance function. In *Obf-1*-deficient mice, the production of IgG1, IgG2a, IgG2b, IgG3, IgA, and IgE are reduced. This reduction is not the result of a defect in the isotype switching process but because transcription from the switched loci is reduced in the absence of OBF-1.

Bcl-6

Another transcription factor that is required for germinal center formation is Bcl-6. This transcriptional repressor is expressed at high levels in germinal center B cells and is required for germinal center development. Bcl-6 has an N-terminal transcriptional repressor domain, which includes a protein–protein interaction motif known as a POZ domain. The C-terminal half of this protein contains six Kruppel-type Zinc finger motifs, which are involved in DNA binding. Bcl-6 was initially discovered as a gene that is frequently mutated in an aggressive type of human non-Hodgkin's B cell lymphoma known as diffuse large cell lymphoma. Bcl-6 null mice have dysregulated cytokine gene expression, inflammation, and the absence of germinal centers. The molecular targets that apparently must be repressed by Bcl-6 in germinal center B cell precursors in order for these structures to develop remain to be identified. Intriguingly, B cell receptor-mediated ERK MAP kinase activation leads to the phosphorylation of Bcl-6 and its subsequent degradation by a ubiquitin-proteasomal pathway. Whether this degradation of Bcl-6 is important during memory cell generation is unclear.

Pip/IRF4/LSIRF

Pip is a transcription factor of the interferon response factor (IRF) family and is also known as IRF4 or lymphocyte-specific IRF (LSIRF). Pip binds to sites adjacent to PU-1 binding and may cooperate with PU.1 in regulating light chain gene recombinational accessibility and rearrangement during early B cell development. A targeted mutation of Pip, however, has revealed that this gene is dispensable for early B cell development or the expression of κ light chain genes but is required for B cell maturation, for germinal center formation, and for B cell proliferation in response to anti-IgM stimulation and in response to bacterial LPS. IgD^{hi} IgM^{lo} B cells (Fraction I) are diminished, and $CD23^{hi}$ IgM^{hi} cells fail to develop, suggesting a major defect in follicular B cell maturation. This maturational defect may account for the defect in B cell activation and the absence of germinal center B cells.

Spi-B

Spi-B is a lymphoid-specific Ets family member that is very closely related to PU.1. Spi-B binds to most putative PU.1 target genes but unlike PU.1 is not expressed in myeloid cells. Spi-B is expressed in T cells as well as in B cells, and its expression increases with B cell maturation. In the absence of Spi-B, B cells proliferate poorly in response to anti-IgM stimulation and die (although they proliferate readily in response to LPS). T-dependent secondary immune responses are defective *in vivo*, and germinal centers form but are not maintained.

Aiolos

Aiolos is a transcriptional regulator of the Ikaros family that dampens B cell activation and proliferation in response to T-dependent antigens. It also participates in B cell maturation, the generation of B-1 cells, and in the generation of MZ B cells (as discussed earlier). The major phenotypic changes in the Aiolos mouse results from a dysregulation of Btk signaling. Btk is constitutively activated in every B cell in the absence of Aiolos. While the absence of MZ B cells in the absence of Aiolos may be linked to the constitutive activation of Btk, the absence of B-1a cells in *Aiolos*$^{-/-}$ mice is paradoxical and has not yet been explained.

Pax-5

Pax-5, as discussed in Chapter 2 and 7, is critical early in development, but it is also important during the final stages of B cell activation. Pax-5 may also participate in B cell proliferation and isotype switching, as discussed above. Pax-5 is expressed in the pro-B, pre-B, and mature B stages but most Pax-5 isoforms, except for one isoform called Pax-5b, are not expressed in plasma cells. The inactivation of Pax-5 expression may be critical for the differentiation of antigen-activated B cells into plasma cells.

E2A

The importance of E2A during early B cell development has been stressed in Chapters 2 and 7. As mentioned above, E2A is induced after BCR ligation and is absolutely required for isotype switching.

Blimp-1

Blimp-1 (B lymphocyte-induced maturation protein-1) is a transcription factor that might be involved in the differentiation of antigen-activated B cells into plasma cells. Blimp-1 contains four Kruppel-like Zinc fingers and is induced in mature B cells by IL-2 and IL-5. Transfection of Blimp-1 into an appropriate B cell line (BCL-1) results in the transcriptional induction of J chain synthesis, the secretion of IgM, and the expression of Syndecan-1.

When B cells mature into plasma cells, Blimp-1 levels rise and c-Myc levels fall. Blimp-1 can transcriptionally repress c-Myc expression and may induce the differentiation of B cells into plasma cells. Since c-Myc may control both proliferation and apoptosis, it is interesting that in a B cell line (WEHI 231) that resembles immature B cells, Blimp-1 overexpression results in apoptosis.

Affinity Maturation Depends on Somatic Mutation Followed by Selection in Germinal Centers

In the germinal center, the process of B cell ontogeny is virtually re-created, and activated lymphocytes go through successive events of somatic mutation, the selection of cells with the highest affinity, a second opportunity for isotype switching, and eventually differentiation into memory B cells or plasma cells. It has been estimated that any given germinal center might represent the clonal outgrowth of from one to three antigen-activated B cells. Somatic mutation, as we will discuss in detail below, is coupled to cell selection processes. Clones with the highest affinity are selected and expand (Figures 8.8 and 8.9). The selection of mutated cells based on affinity is referred to as *affinity maturation*.

Factors Involved in Germinal Center Formation

Germinal centers are remarkably organized structures that are induced during an immune response and in which there exists an exquisite symmetry between structure and function. The formation of these structures depends on a number of sequential events.

1. A lymphoid follicle must first be generated during normal development prior to antigen exposure, and this must contain a network of follicular dendritic cells (FDCs). B cells must migrate into follicles for the FDC network to be created. This process has been described in Chapter 6.

2. A B cell that is activated by T-dependent antigens must migrate to the T cell area and be induced by cognate T cell help to form a focal extrafollicular aggregate, as described in Figure 8.5.

3. Once antibodies are generated and immune complexes are formed, these complexes may be presented by FDCs within follicles, and activated B cells are in some unexplained manner now targeted to follicles.

4. Activated B cells that have been brought into the follicles expand clonally. It is generally assumed that they divide rapidly as centroblasts in

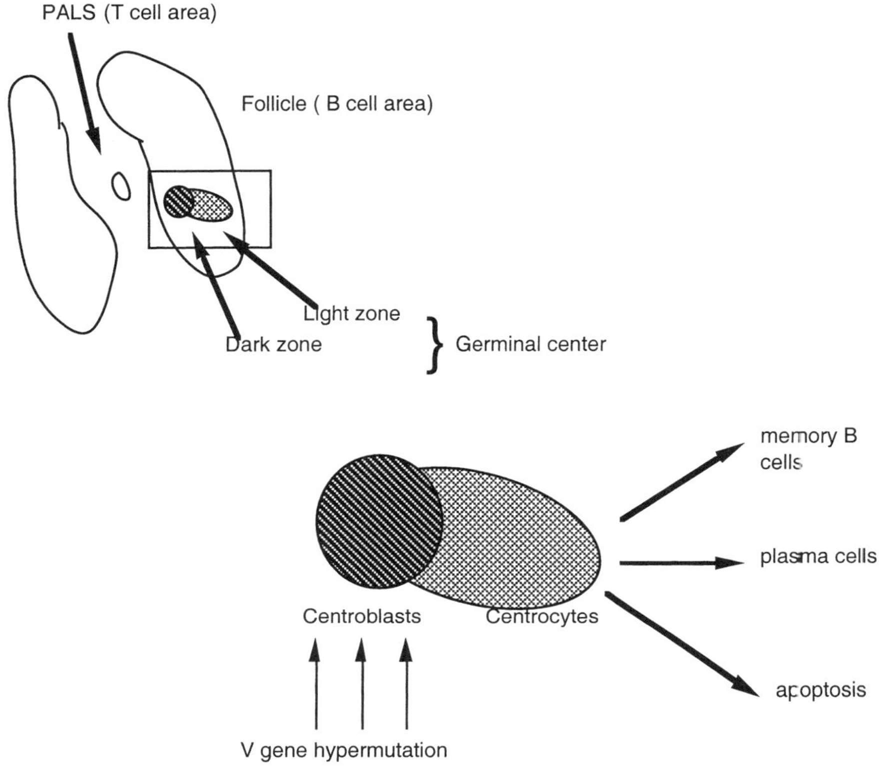

FIGURE 8.8. Somatic mutation is linked to the selection and maturation of activated B cells.

the dark zone, where they undergo somatic mutation. They then migrate into the basal light zone as centrocytes and may be selected by antigen in immune complexes on FDCs. Selected centrocytes may receive further T cell help, undergo isotype switching, and differentiate into memory cells or plasma cells in the apical light zone (Figure 8.9). Germinal centers, however, are dynamic entities, and the formation of dark and light zones represents a transient phase in their genesis and subsequent decline.

The generation of lymphoid follicles requires two types of signals depending on the type of secondary lymphoid tissue being considered (see Chapter 6 for a more extensive discussion). For the generation of primary follicles in lymph nodes, a signal is delivered via LTαβ2 heterotrimers and the LTβ receptor, which is found on accessory cells of non-bone marrow origin in these organs. Lymphotoxin α (LTα), a member of the TNF family of cytokines, and/or TNF trigger TNFRI tumor neurosis factor receptor type I on accessory cells in the spleen and in Peyer's patches (both these

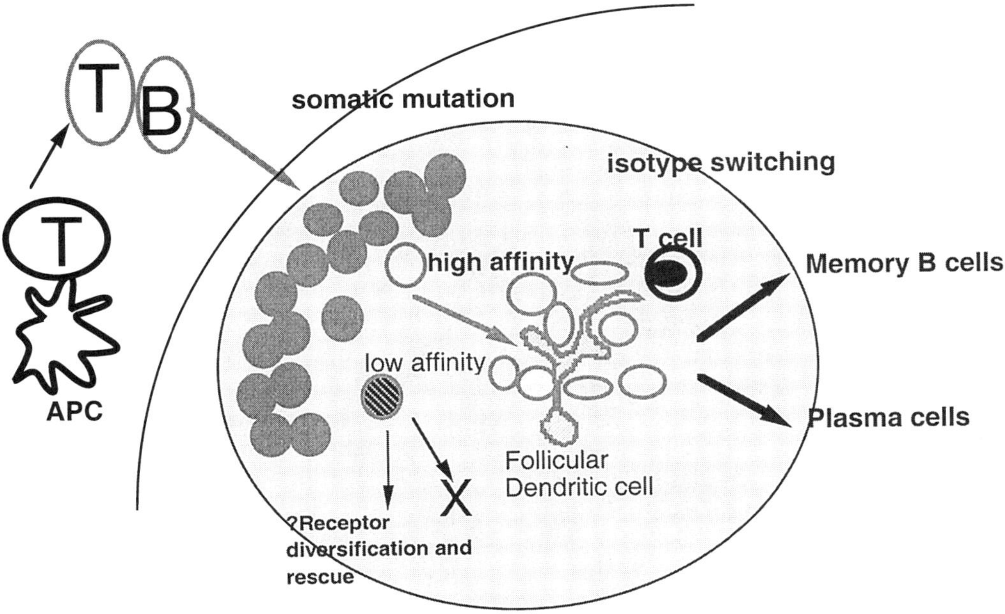

FIGURE 8.9. B cells responding to T-dependent antigens undergo expansion, isotype switching, and somatic mutation. Following the mutational process, cells with the highest affinity for antigen are selected and expand, a process known as affinity maturation. Although most cells with lower affinity for antigen are probably lost by apoptosis, some may undergo further rearrangement and receptor diversification—a process that may involve re-expression of *Rag* genes. Somatic mutation and the selection processes that result in affinity maturation occur only in germinal centers. The process of isotype switching does not require germinal centers, is initiated at the boundary of the T and B cell zones, but may also occur in germinal centers following the process of somatic mutation.

cytokines bind to the same receptor the p55 TNFRI; p70TNFRII is not required for follicle generation). Essential signals are also delivered by the osteoprotegerin ligand (OPGL) to its receptor, RANK. While TNF and LTα are secreted proteins, LTβ subunits (which form the LTαβ2 ligand) are transmembrane polypeptides. The nonhematopoietic cells being triggered are probably FDCs themselves or poorly characterized FDC precursors. Formation of the FDC network requires these above cytokine signals and also the presence of B cells themselves, which provide trophic factors that may include TNF and LT. During fetal life, a specialized $CD4^{lo}$ $CD3^{-}$ cell described in Chapter 6 is the source of LTβ, and this cell triggers nonlymphoid cells in the lymph node anlage. Ligation of TNFR family receptors on FDCs may result in the activation of specific NFκB family members such as p52/NFκB2. Entry of B cells into follicles at different sites requires either the activation by the BLC chemokine of CXCR5 on B cells or the activation of CXCR4 by SDF-1. CXCR5 knockout mice lack inguinal lymph nodes and Peyer's patches and have defective splenic primary follicles. SDF-1 is probably required for migration of B cells into other sites, but this has not yet been examined in a knockout situation because a "conventional" SDF-1 knockout leads to compromised pro-B cell development (see Chapter 7). Genes involved in germinal center formation are listed in Table 8.5.

Formation of germinal centers requires that B cells be triggered by T-dependent antigens and by cognate helper T cells. Mice in which CD19 has been targeted lack germinal centers. Mice deficient in CD21, CD81, the C3 component of complement or the C4 component all exhibit a decrease in the number and size of germinal centers. This speaks to the role played by complement and the CR2 coreceptor in the activation of B cells by T-dependent antigens. IL-6 is believed to participate in germinal center formation as well, perhaps because of its role in increasing extrahepatic "local" C3 production by macrophages in secondary lymphoid organs.

MHC class II deficient mice (which lack T helper cells) and mice lacking CD40 or the CD40L are defective in germinal center formation. Mice that lack CD28 on T cells or B7-1 and B7-2 on B cells and APCs also are unable to form germinal centers. This is probably because CD28-dependent costimulation is required for the induction of CD40L on activated T cells.

Apart from NFκB2 and Bcl-3, which may be required in accessory cells for the formation of FDC networks, a number of transcription factors expressed in B cells, described a little earlier in this chapter, participate in germinal center formation. These include OCA-B/OBF-1/Bob-1, Bcl-6, and Pip/IRF4/LSIRF. In addition, Spi-B may contribute to the maintenance of germinal centers.

Somatic Mutation

Somatic hypermutation, also known as somatic mutation, involves the specific generation of mutations at an extraordinarily high rate in rearranged *IgV* genes in B cells responding to protein antigens. The process is site-specific, lineage-specific, and differentiation stage-specific; that is, it occurs only in the *IgV* genes and their flanking regions, in antigen-activated B cells. Somatic mutations occur in activated B cells primarily in a compartment of the germinal center made up of extremely rapidly dividing cells known as the dark zone. Centroblasts in this zone are estimated to have a doubling time of about 6 to 8 hours. As a result of this process, a number of B cell clones with altered affinity for the original immunogen are generated. Cells that emerge from the dark zone into the basal light zone are known as centrocytes. The antigen that initiated germinal center formation is presented as a part of immune complexes by FDCs. These immune complexes usually

TABLE 8.5. Mutant mice that fail to develop germinal centers (See Tables 4.4 and 6.3 for additional details)

Defects	References
Mice with defects in B cell activation and T–B collaboration	
Defects in B cell surface molecules	
CD19$^{-/-}$	(Rickert et al., 1995; Engel et al., 1995)
CD40$^{-/-}$	(Kawabe et al., 1994)
B7-1$^{-/-}$, *B7-2*$^{-/-}$	(Borriello et al., 1997)
B7-2$^{-/-}$	(Borriello et al., 1997)
Defects in T cell surface molecules	
CD28$^{-/-}$	(Shahinian et al., 1993; Ferguson et al., 1996)
CD40L$^{-/-}$	(Renshaw et al., 1994; Xu et al., 1994)
Defects in intracellular regulators in B cells	
Bcl-6$^{-/-}$	(Dent et al., 1997; Ye et al., 1997)
OBF-1/OCA-1/Bob-1$^{-/-}$	(Schubart et al., 1996; Kim et al., 1996)
Pip/IRF4/LSIRF$^{-/-}$	(Mittrucker et al., 1997)
Spi-B$^{-/-}$	(Su et al., 1997)
Defects in T helper cell development	
MHC class II$^{-/-}$	(Grusby et al., 1991)
Defects in lymphoid architecture	
Defects in chemokine receptors	
BLR-1$^{-/-}$	(Forster et al., 1996)
Defects in signals to accessory cells	
LT$\alpha^{-/-}$	(DeTogni et al., 1994; Banks et al., 1995; Matsumoto et al., 1996)
LT$\beta^{-/-}$	(Alimzhanov et al., 1997; Koni et al., 1997)
LTβ*R*$^{-/-}$	(Futterer et al., 1998)
OPGL$^{-/-}$	(Kong et al., 1999)
TNF$^{-/-}$	(Pasparakis et al., 1996)
TNFRI$^{-/-}$ *(p55*$^{-/-}$*)*	(Pfeffer et al., 1993; Rothe et al., 1993; Le Hir et al., 1996; Neumann et al., 1996; Tkachuk et al., 1998)
NFκ*B2*$^{-/-}$ *(p52*$^{-/-}$*)*	(Franzoso et al., 1998; Caamano et al., 1998)
Bcl-3$^{-/-}$	(Franzoso et al., 1997; Schwarz et al., 1997)
Aly/Aly$^{-/-}$ *(NIK mutant)*	(Shinkura et al., 1996, 1999)
Mutations in complement components and CR2 lead to impaired germinal center formation	
CD21/CD35$^{-/-}$	(Ahearn et al., 1996)
C3$^{-/-}$ *and C4*$^{-/-}$	(Fischer et al., 1996)
IL-6$^{-/-}$	(Kopf et al., 1994)

contain complement components and are presented by CD21 on FDCs to centrocytes that have emerged from the dark zone in many mutant forms. Cells that have high-affinity receptors are selected and expanded.

Somatic mutation largely involves point substitutions, although occasionally deletions and insertions may also be seen. Although the mechanisms involved in this fascinating process remain poorly understood, the following features of the process have led to some interesting models:

1. The rate of mutation of *IgV* genes in germinal centers is estimated to be around 10^{-4} to 10^{-3}/base pair/generation, which is about four to six orders of magnitude higher than the estimated rate of mutation in other vertebrate genes.
2. Single-nucleotide substitutions are added in a stepwise manner.
3. The mutational process involves rearranged (and therefore expressed) *V* genes.
4. Mutations start from the 5′ end of the rearranged *V* gene (downstream of the promoter). However, the 3′ border of the segment that is mutated is variable, and mutation can spread into the adjacent intron.
5. There is a strand bias in the process favoring the transcribed strand; transitions are seen more often than transversions.
6. The rate of transcription of transgenic substrates directly influences the frequency of somatic mutation.
7. Studies using transgenes have indicated that there are hot spots for mutation.

Two broad categories of models have been proposed for somatic hypermutation. One category of models suggests the specific and localized repression of postreplicative repair or of DNA polymerase proofreading activity or of both following semiconservative DNA replication as a means of achieving somatic mutation. Other models invoke the need for a specific mutational polymerase. These latter models that do not invoke the repression of postreplicative repair in general and involve a short patch of nonreplicative DNA synthesis. This process may involve the generation of single-strand or double-strand breaks in the V-region followed by gap formation and error prone-repair or, as has been suggested recently, may involve a transcription coupled repair process.

Transcription has been linked mechanistically to somatic mutation in a number of studies. Intronic enhancers and matrix-attachment regions (MARs) are required for this process to occur. In the case of κ transgenes, the 3′ enhancer is not required for somatic mutation to occur, and other promoters can replace the *V* gene promoter, though perhaps with varying efficiency. However, the presence of the intronic κ enhancer is essential for mutation to be seen. Engineering a promoter just upstream of the constant region of a transgene led to the generation of mutations in the constant region. The need for transcription led to the formulation of a transcription coupled repair model. In such a model, a yet to be identified mutation-activating gene or a mutation factor drives the process.

In one version of such a model, the unidentified "mutation factor" loads onto a transcription complex. A specific polymerase moves along the transcription bubble, and the activity of this polymerase is dependent on the mutation factor. DNA replication then fixes the mutations, and the muta-

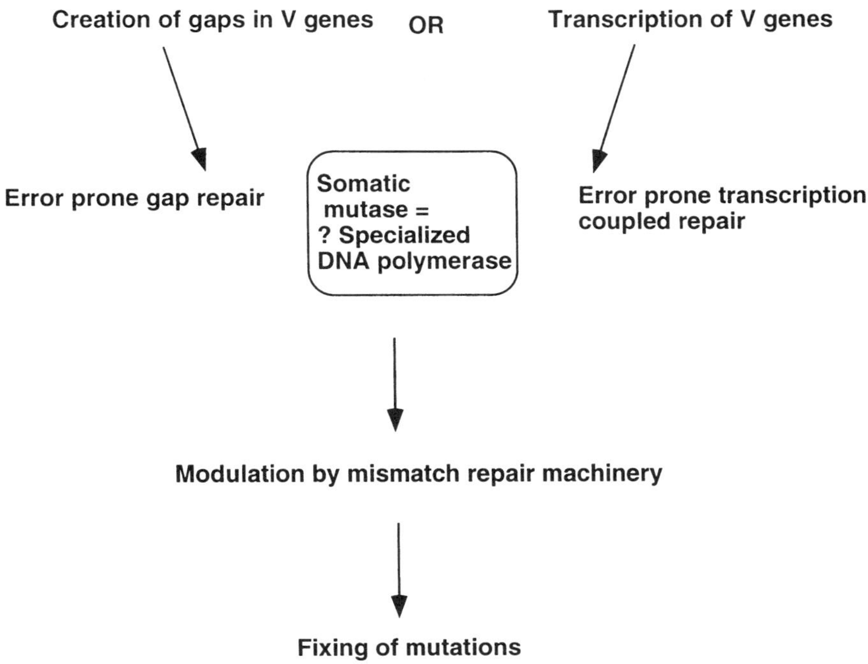

FIGURE 8.10. Some current models for somatic hypermutation. An error-prone gap repair or transcription-coupled repair process may be involved. This process probably utilizes a highly specialized DNA polymerase expressed in germinal center B cells.

tion factor falls off or is removed. In such a model, the strand bias for mutation favoring the transcribed strand may be easily explained, and hotspots may be assumed to reflect sites at which the unidentified polymerase has a propensity to stall. Clearly, many other equally speculative models may be considered to explain the existing state of knowledge in the field (Figure 8.10).

It has long been speculated that DNA repair mechansims may have something to do with somatic mutation. Either specific repair mechanisms might be shut off to permit mutation to occur, or specific repair machinery may be modified and used to generate mutations. Repair mechanisms could also modulate the mutational process. A mouse mismatch repair gene *Pms2* has recently been invoked as being involved in somatic mutation, but discrepant results from different laboratories using similar mice still need to be reconciled. *Pms2* is a homologue of the *E. coli mutL* gene, which is involved in the *mutS/mutL* mismatch repair pathway. Human homologues of *mutS* and *mutL* are involved in the pathogenesis of hereditary nonpolyposis colorectal cancer. In *Pms2* null mice, the frequency of somatic mutation was found to be dramatically reduced, suggesting that mismatch repair may be required for somatic mutation rather than being shut off to permit such a process to proceed. *Pms2* represents a gene that may be involved in somatic mutation, but given that some laboratories note an increase or no change in the mutation frequency, this view remains to be established. It is possible that the repair process corrects some mutations rather than induces them, and it may well be that DNA repair serves a modulating or "editing" function in this process. In mice lacking the *Msh2* gene (a *mutS* homologue), the frequency of somatic mutation is unchanged, but there is a marked preference for mutations derived from G and C residues.

It has been postulated that some unknown mutation activating gene may be involved in generating point substitutions in rearranged V genes. Mismatch repair genes may then be co-opted to "correct" the sequence of the normal strand using the mutated strand as a template, thus fixing the mutations. Somatic mutation remains one of the most intriguing and poorly understood issues in lymphocyte development.

A fairly intriguing link may be made between the machinery involved in the proliferation of B cells, somatic mutation, and the genesis of lymphomas. The vast majority of human lymphomas have their genesis in germinal center B cells. Somatic mutation may generate single-stranded and double-stranded breaks in Ig genes, which may predispose them to undergo translocations. The mutational process itself may target a translocated gene such as *c-Myc* or *Bcl-6* and functionally alter the translocated protein. Indeed, although somatic mutation is generally specific for the V regions of Ig genes, for some reason, nontranslocated *Bcl-6* is also targeted by the mutational process, and this might contribute on occasion to the process of lymphomagenesis.

Receptor Diversification and Cell Selection in Germinal Centers

Activated germinal center B cells known as centroblasts undergo somatic mutation, and a proportion of these cells also die by apoptosis. However, the highest rates of apoptosis in the germinal center are observed in centrocytes of the basal light zone close to the dark zone. It is assumed that cells that have mutated their V genes, and are either no longer able to bind the original antigen or do so with low affinity, may die by apoptosis. The propensity of germinal center B cells to undergo apoptosis may be explained by the fact that these cells express low levels of Bcl-2, and high levels of Bax, c-Myc, and p53. While germinal center cells may intrinsically be prone to die and are "waiting to be saved" by high-affinity antigen if they make the "right" kind of receptors, it appears that there might be a rescue pathway that is designed to give cells that have failed to make high-affinity receptors another chance. In a manner that is reminiscent of the receptor-editing phenomenon seen in immature B cells, it has been suggested that germinal center B cells also re-express *Rag-1* and *Rag-2*. A similar phenomenon has been described to occur in B-1 cells. This issue is extremely controversial; some studies suggest that all Rag-expressing cells in the spleen might represent immature cells that have recently emigrated from the bone marrow and have yet to shut off *Rag* gene expression. It has been suggested, and this view is now a matter of debate, that cells that cannot be positively selected by antigen in germinal centers may be induced to rearrange their light chain loci to generate new B cells with diversified receptors. Unlike the situation in the bone marrow, where antigen-receptor ligation of immature B cells induces receptor editing, it has been suggested that germinal center B cells that cannot be triggered by antigen are induced to re-express *Rag-1* and *Rag-2*. It has been suggested that Rag-1 and Rag-2 might even have a role to play in the process of somatic mutation, perhaps by their ability to make nicks in DNA. If *Rag* gene reexpression really does occur at this stage, a teleological reason suggested to explain this phenomenon is that centrocytes that are likely to apoptose are prevented from doing so and are provided with the opportunity to contribute to the immune repertoire again in a useful manner. There are conflicting views at present as to whether this *Rag* gene reexpression can or cannot be induced.

Cells that express high-affinity receptors are positively selected in the light zone by antigens contained in immune complexes and presented by FDCs to centrocytes. FDCs express an alternatively spliced form of CD21 that gives rise to a longer protein than that found on B cells. It is unclear whether the slightly longer extracellular domain of this protein allows it to function in a different way on FDCs. CD21 on FDCs may not have a signaling function but may need to be able to hold onto immune complexes on the cell surface for extended periods of time. This positive selection process leads to the expansion of high-affinity B cells and provides them with the opportunity to class switch with T cell help in the apical light zone. These cells may receive further CD40L signals and opt to become memory B cells. Alternatively, they may receive a distinct set of signals that permit them to become plasma cells.

Fas-Mediated Elimination of Bystander B Cells That Are Promiscuously Activated

In secondary lymphoid organs, activated helper T cells express high levels of CD40L. Since most follicular B cells express CD40, this obviously represents a situation where the potential for the promiscuous activation of bystander B cells exists. How is specificity maintained? Activated T cells also express high levels of FasL, which contributes to the suicidal process of activation-induced cell death (AICD) that eliminates most of these T cells. Specific and nonspecific B cells are respectively "intentionally" or "promiscuously" triggered by CD40, and this also induces their levels of Fas expression. As a result, these B cells not only receive signals via CD40 but they also receive death signals via Fas. "Promiscusously" activated B cells are induced to undergo apoptosis downstream of Fas signaling. The "intentionally" activated or antigen-specific B cells are "saved" and do not die. Death is held at bay by signals generated by the antigen receptor, which is triggered by specific antigen. These signals probably activate ERK, which in turn induces the transcriptional activation of Flice Inhibitory Protein (FLIP). As explained in Chapter 5, FLIP structurally resembles caspase-8 or Fas-linked interleukin-1 cleaving enzyme (FLICE) but lacks caspase activity. It thus competes with caspase-8 for recruitment by Fas-associated death domain (FADD), inhibits Fas signaling, and abrogates the induction of apoptotic death (Figure 8.11). Bcl-X induction by anti-IgM signaling has been postulated to be involved in the rescue process.

An intriguing role has been suggested for the SLAM/CDw150 (signaling lymphocyte activation molecule) protein in enhancing the activation of the Fas pathway in nonspecifically activated B cells. SLAM is a member of the CD2 family and is a costimulatory molecule that functions in both B and T cells. The cytoplasmic tail of SLAM associates with the product of the X-linked lymphoproliferation gene or SLAM-associated protein (SAP). SAP consists of an SH2 domain with a minuscule tail. It is expressed in T cells but not in B cells. While the possibility that SLAM enhances the apoptosis of nonspecifically stimulated B cells is intriguing, it appears that the absence of SAP compromises T cell-mediated antiviral immune responses and that the increase in B cell proliferation in patients with X-linked lymphoproliferation (XLP) reflects the failure to clear Ebstein-Barr virus (EBV).

The Generation of Memory B Cells and Plasma Cells

How does a B cell that has undergone affinity maturation choose to become a memory B cell or a plasma cell? Very little is known as to how these differentiation events are orchestrated. In the absence of CD40 or CD40L, IgM plasma cell formation is not impaired. From *in vitro* studies, it appears that stimulation with CD40L favors memory cell generation and actually serves to inhibit the process of differentiation into plasma cells. CD40 signaling inhibits J chain synthesis and IgM secretion and prevents the induction of Blimp-1. Plasma cells do not appear in germinal centers, and it is probably important that centrocytes that are being selected on the basis of their affinity for antigen on FDCs should not secrete antibody. Even a small amount of secreted antibody could potentially compromise the ability of antigen to select the appropriate high-affinity B cells.

Triggering of cells with cytokines such as IL-6, IL-5, and IL-2 in the absence of CD40L stimulation may favor plasma cell generation

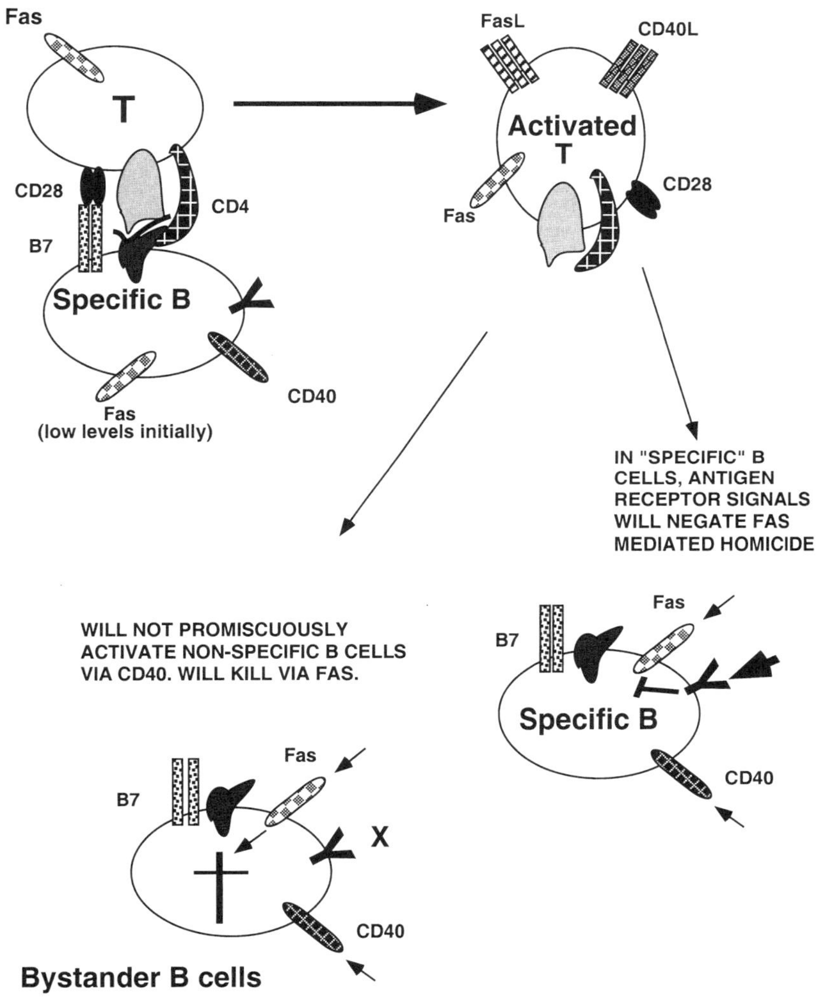

FIGURE 8.11. Fas signaling eliminates promiscuously activated bystander B cells. Specific B cells that are triggered via their antigen receptors are protected. Antigen-receptor signaling presumably activates ERK signaling and the induction of FLIP, which inhibits caspase-8 recruitment to FADD.

(Figure 8.12). One mechanism may involve the induction of *Blimp-1*, discussed earlier in this chapter. Blimp-1 is capable of inducing J chain synthesis, syndecan expression, and Ig secretion. It can also transcriptionally repress *c-Myc* and may thereby contribute to the inhibition of proliferation and the concomitant induction of terminal differentiation. It has been suggested on the basis of studies in cell line models for plasma cell

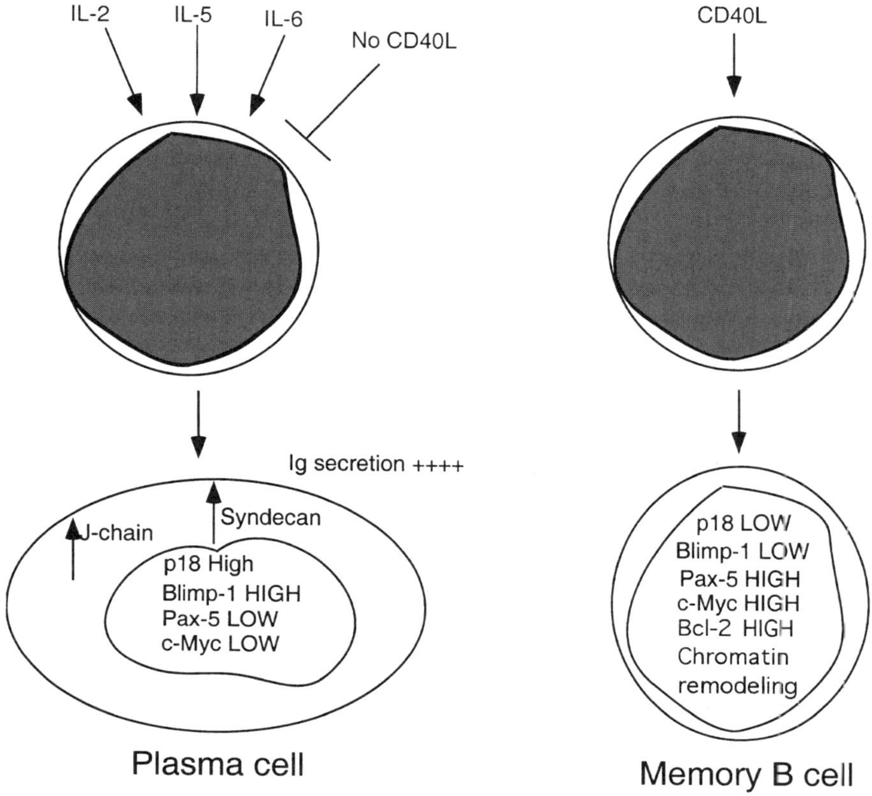

FIGURE 8.12. Models for plasma cell and memory B cell generation. Terminal differentiation is inhibited by CD40 signaling. Cytokines induce Blimp-1, which can repress c-Myc and induce J-chain synthesis, IgM secretion, and syndecan expression. A drop in Pax-5 levels and the induction of the CDK inhibitor, p18, may contribute to plasma cell differentiation.

differentiation, that IL-6 might transcriptionally activate the p18 CDK inhibitor and cause activated B cells to cease proliferating and to secrete Igs. This cessation of proliferation may also be permissive for the induction of the plasma cell differentiation pathway. The enhanced synthesis of secretory Igs depends not merely on high-level transcription but on regulation at the level of RNA processing. Selective polyadenylation and splicing (described in Chapter 7) are involved in the switch from the generation of largely membrane Ig heavy chains to high levels of the secretory form. Plasma cells may downregulate the expression of genes encoding B cell receptor-associated proteins such as Igα. Surface expression of the BCR is markedly diminished.

Does plasma cell generation occur only in secondary lymphoid tissues? What are the circumstances under which an activated B cell receives cytokine-mediated signals but is not activated by CD40L? Does plasma cell generation require that no CD40 signaling occur, or do other receptors counter the effects of CD40-mediated signals? Clear answers are not available to these questions. Although plasma cell generation probably occurs primarily in lymphoid tissues, it is now becoming clear that at least two different categories of plasma cells might be generated during an immune response. Short-lived plasma cells that have a life span of a few days or weeks may arise from focal aggregates and probably remain in the red pulp of the spleen or in the medullary chords of lymph nodes. Plasma cells that are generated from memory cells that move away from the germinal centers, perhaps away from T cells that can continue to trigger them via a CD40L–CD40 interaction, eventually migrate to the bone marrow as long-lived plasma cells whose life span may be as long as that of the animal. These long-lived plasma cells presumably contribute to the extended secretion of specific antibody long after antigen has been cleared. CD30 is a TNF family receptor expressed on an outer rim of germinal center B cells. CD30L is expressed on activated T cells. CD30 signaling can antagonize CD40 signaling and may thus contribute to the shutting off of isotype switching in promiscuously activated B cells; it might contribute to the terminal differentiation process that generates plasma cells. Plasma cells are often found in sites of chronic inflammation. It is not known whether these represent short-lived or long-lived plasma cells or whether they are generated from memory cells that differentiate in tissue sites.

Membrane Ig heavy chains other than those of IgM and IgD all have relatively long cytoplasmic tails. It appears that the tail of IgG1 and IgE at least, and probably of all downstream isotypes, are required for the expansion and propagation of memory cells. The activation process involving the antigen receptor and CD40 signaling presumably primes memory B cells so that a number of their genes are induced to be in an open chromatin cofiguration, ensuring that they can respond rapidly and with high amplitude when they are restimulated by high-affinity antigens. These altered B cells must be kept alive so that they are around when the host is once again exposed to the activating foreign antigen. It is possible that memory B cells are constantly "tickled" by low-affinity self-antigens (which might have positively selected these B cells when they first emerged from the bone marrow) and that this tickling process is responsible for the induction or modification of survival factors. Long-lived cells in general express high levels of Bcl-2.

Summary

B cells may be activated by T-independent multivalent antigens or by T-dependent protein antigens. T-independent antigens may trigger fetal liver-derived B cells and thus give rise to the B-1 lineage. Most B-1 cells express CD5 (when they are known as B-1a cells) and seed the peritoneum. These cells are self-renewing and synthesize most natural antibodies. The generation of B-1 cells requires proteins involved in antigen-receptor signaling as well as nuclear factors that are probably downstream of these signals. $IgM^{hi}IgD^{-}CD23^{-}CD1^{hi}$ marginal zone cells in the spleen may represent a population of partially activated naive B cells that develop from follicular B cells.

T-dependent antigens are internalized by B cells, processed, and presented in an MHC class II context to specific T cells at the interface of T and B cell areas in secondary lymphoid organs. Activated T cells express CD40L, and the triggering of B cells via CD40 is critical for the induction of isotype switching, proliferation, germinal center formation, and the induction of somatic hypermutation. Some of the activated B cells in focal aggregates may be induced to undergo isotype switching and to differentiate into plasma cells. Other activated B cells migrate into follicles and initiate the formation of germinal centers. The generation of germinal center B cells requires antigen-receptor signaling, CD40 activation, the establishment of the appropriate lymphoid architecture, the proper formation of FDC networks, and the activation of critical nuclear factors both in accessory cells as well as in B cells. Somatic hypermutation alters the V regions of Igs in rapidly dividing germinal center B cells known as centroblasts, which express very little surface Ig and transiently form the dark zone of the germinal center. After the mutational process, high-affinity B cells are selected by antigen presented on FDCs in the light zone of the germinal center. Nonselected B cells may die or make new receptors. Memory B cells emerge from germinal centers following CD40 triggering. In the absence of CD40 triggering, some antigen-specific cells are induced by cytokines to express high levels of Blimp-1 and to differentiate into plasma cells.

Selected Reviews

Camacho, S. A., Kosco-Vilbois, M. H., and Berek, C. (1998). The dynamic structure of the germinal center. Immunol. Today *19*, 511–514.

Carroll, M. C. (1998). The role of complement and complement receptors in induction and regulation of immunity. Annu. Rev. Immunol. *16*, 545–568.

Doody, G. M., Dempsey, P. W., and Fearon, D. T. (1996). Activation of B lymphocytes: integrating signals from CD19, CD22 and FcγRIIb1. Curr. Opin. Immunol. *8*, 378–382.

Fearon, D. T., and Carter, R. H. (1995). The CD19/CR2/TAPA-1 complex of B lymphocytes: linking natural to acquired immunity. Annu. Rev. Immunol. *13*, 127–150.

Finkelman, F. D., Holmes, J., Katona, I. M., Urban, J. F. J., Beckmann, M. P., Park, L. S., Schooley, K. A., Coffman, R. L., Mosmann, T. R., and Paul, W. E. (1990). Lymphokine control of *in vivo* immunoglobulin isotype selection. Annu. Rev. Immunol. *8*, 303–333.

Foy, T. M., Aruffo, A., Bajorath, J., Buhlmann, J. E., and Noelle, J. (1996). Immune regulation by CD40 and its ligand gp39. Annu. Rev. Immunol. *14*, 591–617.

Haughton, G., Arnold, L. W., Whitmore, A. C., and Clarke, S. H. (1993). B-1 cells are made, not born. Immunol. Today *14*, 84–87.

Herzenberg, L. A., Stall, A. M., Lalor, P. A., Sidman, C., Moore, W. A., Parks, D., and Herzenberg, L. A. (1986). The Ly-1 B cell lineage. Immunol. Rev. *93*, 81–102.

Kantor, A. B., and Herzenberg, L. A. (1993). Origin of murine B cell lineages. Annu. Rev. Immunol. *11*, 501–538.

Kelsoe, G. (1996). Life and death in germinal centers (redux). Immunity *4*, 107–111.

Kim, N., and Storb, U. (1998). The role of DNA repair in somatic hypermutation of immunoglobulin genes. J. Exp. Med. *187*, 1729–1733.

Kraal, G. (1992). Cells in the marginal zone of the spleen. Int. Rev. Cytol. *132*, 31–74.

Lorenz, A., and Radbruch, A. (1996). Developmental and molecular regulation of immunoglobulin class switch recombination. Curr. Top. Microbiol. Immunol. *217*, 151–169.

MacLennan, I. C. M., and Gray, D. (1986). Antigen driven selection of virgin and memory B cells. Immunol. Rev. *91*, 61–85.

McHeyzer Williams, M. G., and Ahmed, R. (1999). B cell memory and the long-lived plasma cell. Curr. Opin. Immunol. *11*, 172–179.

Michaelson, J. M., Singh, M., and Birshtein, B. K. (1996). B cell lineage-specific activator protein (BSAP). A player at multiple stages of B cell development. J. Immunol. *156*, 2349–2351.

Mowat, A. M., and Viney, J. L. (1997). The anatomical basis of intestinal immunity. Immunol. Rev. *156*, 145–166.

Neuberger, M. S., and Milstein, C. (1995). Somatic hypermutation. Curr. Opin. Immunol. *7*, 248–254.

Nussenzweig, M. C. (1998). Immune receptor editing: revise and select. Cell *95*, 875–878.

Sha, W. C. (1998). Regulation of immune responses by NFκB/Rel transcription factors. J. Exp. Med. *187*, 143–146.

Shannon, M., and Weigert, M. (1998). Fixing mismatches. Science *279*, 1159–1160.

Smith, C. A., Farrah, T., and Goodwin, R. G. (1994). The TNF receptor superfamily of cellular and viral proteins: activation, co-stimulation, and death. Cell *76*, 959–962.

Snapper, C. M., Marcu, K. B., and Zelazowski, P. (1997). The immunoglobulin clas switch: beyond accessibility. Immunity *6*, 217–223.

Stavnezer, J. (1996). Antibody class switching. Adv. Immunol. *61*, 79–146.

Storb, U. (1996). The molecular basis of somatic hypermutation of immunoglobulin genes. Curr. Opin. Immunol. *8*, 206–214.

Tarakhovsky, A. (1997). Bar-Mitzvah for B-1 cells: how will they grow up? J. Exp. Med. *185*, 981–984.

van Kooten, C., and Banchereau, J. (1997). Functions of CD40 on B cells, dendritic cells and other cells. Curr. Opin. Immunol. *9*, 330–335.

Wabl, M., Cascalho, M., and Steinberg, C. (1999). Hypermutation in antibody affinity maturation. Curr. Opin. Immunol. *11*, 186–190.

Some Selected References

Achatz, G., Nitschke, L., and Lamers, M. C. (1997). Effect of transmembrane and cytoplasmic domain of IgE on the IgE response. Science *276*, 409–411.

Ahearn, J. M., Fischer, M. B., Croix, D., Goerg, S., Ma, M., Xia, J., Zhou, X., Howard, R. G., Rothstein, T. L., and Carroll, M. C. (1996). Disruption of the Cr2 locus results in a reduction in B-1a cells and in an impaired B cell response to T-dependent antigen. Immunity *4*, 251–262.

Alimzhanov, M. B., Kuprash, D. V., Kosco-Vilbois, M. H., Luz, A., Turetskaya, R. L., Tarakhovsky, A., Rajewsky, K., Nedospasov, S. A., and Pfeffer, K. (1997). Abnormal development of secondary lymphoid organs in lymphotoxin beta deficient mice. Proc. Natl. Acad. Sci. USA *94*, 9302–9307.

Banks, T. A., Rouse, B. T., Kerley, M. K., Blair, P. J., Godfrey, V. L., Kuklin, N. A., Bouley, D. M., Thomas, J., Kanangat, S., and Mucenski, M. L. (1995). Lymphotoxin alpha deficient mice. Effects on secondary lymphoid organ formation and humoral immune responsiveness. J. Immunol. *155*, 1685–1693.

Beg, A. A., Sha, W. C., Bronson, R. T., Ghosh, S., and Baltimore, D. (1995). Embryonic lethality and liver degeneration in mice lacking the RelA component of NF κB. Nature *376*, 167–170.

Benatar, T., Carsetti, R., Furlonger, C., Kamalia, N., Mak, T., and Paige, C. J. (1996). Immunoglobulin mediated signal transduction in B cells from CD45 deficient mice. J. Exp. Med. *183*, 329–334.

Bertocci, B., Quint, L., Delbos, F., Garcia, C., Reynaud, C.-A., and Weill, J.-C. (1998). Probing immunoglobulin gene hypermutation with microsatellites suggests a non-replicative short patch DNA synthesis. Immunity *9*, 257–265.

Betz, A., Milstein, C., Gonzalez-Fernandez, A., Pannell, R., Larson, T., and Neuberger, M. S. (1994). Elements regulating somatic hypermutation of an immunoglobulin κ gene: critical role for the intron enhancer/matrix attachment region. Cell *77*, 239–248.

Bikah, G., Carey, J., Ciallella, J. R., Tarakhovsky, A., and Bondada, S. (1996). CD5-mediated negative regulation of antigen receptor growth signals in B-1 β cells. Science *274*, 1906–1909.

Borggrefe, T., Wabl, M., Akhmedov, A. T., and Jessberger, R. (1998). A B cell specific DNA recombination complex. J. Biol. Chem. *273*, 17025–17035.

Borriello, F., Sethna, M. P., Boyd, S. D., Schweitzer, A. N., Tivol, E. A., Jacoby, D., Strom, T. B., Simpson, E. M., Freeman, G. J., and Sharpe, A. H. (1997). B7-1 and B7-2 have overlapping, critical roles in immunoglobulin class switching and germinal center formation. Immunity *6*, 303–314.

Bottaro, A., Lansford, R., Xu, L., Zhang, J., Rothman, P., and Alt, F. W. (1994). S region transcription (per se) promotes basal IgE class switch recombination but additional factors regulate the efficiency of the process. EMBO J. *13*, 665–674.

Bottaro, A., Young, F., Chen, J., Serwe, M., Sablitzky, F., and Alt, F. W. (1998). Deletion of the IgH intronic enhancer and associated matrix-attachment regions decreases but does not abolish class switching at the mu locus. Int. Immunol. *10*, 799–806.

Brenner, S., and Milstein, C. (1966). Origin of antibody variation. Nature *211*, 242–243.

Burkly, L., Hession. C., Ogata, L., Reilly, C., Marconi, L. A., Olson, D., Tizard, R., Cate, R., and Lo, D. (1995). Expression of RelB is required for the development of thymic medulla and dendritic cells. Nature *373*, 531–536.

Byth, K. F., Conroy, L. A., Howlett, S., Smith, A. J. H., May, J., Alexander, D. R., and Holmes, N. (1996). CD45-null transgenic mice reveal a positive regulatory role for CD45 in early thymocyte development, in the selection of $CD4^+$ $CD8^+$ thymocytes, and in B cell maturation. J. Exp. Med. *183*, 1707–1718.

Caamano, J. H., Rizzo, C. A., Durham, S. K., Barton, D. S., Raventos-Suarez, C., Snapper, C. M., and Bravo, R. (1998). NFκB2 (p100/p52) is required for normal splenic microarchitecture and B cell mediated immune responses. J. Exp. Med. *187*, 185–196.

Cariappa, A., Kim T. J., and Pillai, S. (1999). Accelerated emigration of B lymphocytes in the Xid mouse. J. Immunol. *162*, 4417–4423.

Cascalho, M., Wong, J., Steinberg, C., and Wabl, M. (1998). Mismatch repair co-opted by hypermutation. Science *279*, 1207–1210.

Casellas, R., Nussenzweig, A., Wuerffel, R., Pelanda, R., Reichlin, A., Suh, H., Qin, X., Besmer, E., Kenter, A., Rajewsky, K., and Nussenzweig, M. C. (1998). Ku 80 is required for immunoglobulin isotype switching. EMBO J. *17*, 2404–2411.

Cerutti, A., Schaffer, A., Shah, S., Zan, H., Liou, H. C., Goodwin, R. G., and Casali, P. (1998). CD30 is a CD40-inducible molecule that negatively regulates CD40 mediated immunoglobulin class switching in non-antigen-selected human B cells. Immunity *9*, 247–256.

Cheng, A. M., Rowley, B., Pao, W., Hayday, A., Bolen, J. B., and Pawson, T. (1995). Syk tyrosine kinase required for mouse viability and B cell development. Nature *378*, 303–307.

Clarke, S. H., and Arnold, L. W. (1998). B-1 cell development: evidence for an uncommitted Immunoglobulin (Ig)M^+ B cell precursor in B cell differentiation. J. Exp. Med. *187*, 1325–1334.

Cogne, M., Lansford, R., Bottarp, A., Zhang, J., Gorman, J., Young, F., Cheng, H. L., and Alt, F. W. (1994). A class switch control region at the 3′ end of the immunoglobulin heavy chain locus. Cell *77*, 737–747.

Corcoran, L. M., and Karvelas, M. (1994). Oct-2 is required early in T-independent B cell activation for G1 progression and for proliferation. Immunity *1*, 635–645.

Corcoran, L. M., Karvelas, M., Nossal, G. J. V., Ye, Z., Jacks, T., and Baltimore, D. (1993). Oct-2, although not required for early B cell development, is critical for later B cell maturation. Genes Dev. *7*, 570–581.

Croix, D. A., Ahearn, J. M., Rosengard, A., Han, S.-H., Ma, M., and Carroll, M. C. (1996). Antibody response to a T-dependent antigen requires B cell expression of complement receptors. J. Exp. Med. *183*, 1857–1864.

Cyster, J. G., and Goodnow, C. C. (1995). Protein tyrosine phosphatase 1C negatively regulates antigen receptor signaling in B lymphocytes and determines thresholds for negative selection. Immunity *2*, 13–24.

De Togni, P., et al. (1994). Abnormal development of peripheral lymphoid organs in mice deficient in lymphotoxin. Science *264*, 703–707.

Dent, A. L., Hu-Li, J., W. E. P., and Staudt, L. M. (1998). T helper type 2 inflammatory disease in the absence of interleukin 4 and transcription factor STAT6. Proc. Natl. Acad. Sci. USA *95*, 13823–13828.

Dent, A. L., Shaffer, A. L., Yu, X., Allman, D., and Staudt, L. M. (1997). Control of inflammation, cytokine expression and germinal center formation by Bcl-6. Science *276*, 589–592.

Doi, T. S., Takahashi, T., Taguchi, O., Takachika, A., and Obata, Y. (1997). NF-κB RelA deficient lymphocytes: normal development of T cells and B cells, impaired production of IgA and IgG1 and reduced proliferative responses. J. Exp. Med. *185*, 953–961.

Doody, G. M., Justemet, L. B., Delibrias, C. C., Matthews, R. J., Lin, J., Thomas, M. L., and Fearon, D. T. (1995). A role in B cell activation for CD22 and the protein tyrosine phosphatase SHP. Science *269*, 242–244.

Engel, P., Zhou, L.-J., Ord, D. C., Sato, S., Koller, B., and Tedder, B. F. (1995). Abnormal B lymphocyte development, activation, and differentiation in mice that lack or overexpress the CD19 signal transduction molecule. Immunity *3*, 39–50.

Erikson, J., Radic, M. Z., Camper, S. A., Hardy, R. R., Carmack, C., and Weigert, M. (1991). Expression of anti-DNA immunoglobulin transgenes in non-autoimmune mice. Nature *349*, 331–334.

Ferguson, S. E., Han, S. H., Kelsoe, G., and Thompson, C. B. (1996). CD28 is required for germinal center formation. J. Immunol. *156*, 4576–4581.

Fischer, M. B., Goerg, S., Shen, L. M., Prodeus, A. P., Goodnow, C. C., Kelsoe, G., and Carroll, M. C. (1998). Survival of germinal center B cells is dependent on expression of CD21/CD35. Science *280*, 582–585.

Fischer, M. B., Ma, M., Goerg, S., Zhou, X., Xia, J., Finco, O., Han, S., Kelsoe, K., Howard, R. G., Rothstein, T. L., Kremmer, E., Rosen, F. S., and Carroll, M. C. (1996). Regulation of the B cell response to T-dependent antigens by classical pathway complement. J. Immunol. *157*, 549–556.

Foote, L. C., Marshak-Rothstein, A., and Rothstein, T. L. (1998). Tolerant B lymphocytes acquire resistance to fas mediated apoptosis after treatment with IL-4 but not after treatment with specific antigen unless a surface immunoglobulin threshold is reached. J. Exp. Med. *187*, 147–153.

Forster, R., Mattis, A. E., Kremmer, E., Wolf, E., Brem, G., and Lipp, M. (1996). A putative chemokine receptor, BLR1, directs B cell migration to defined lymphoid organs and specific anatomic compartments of the spleen. Cell *87*, 1037–1047.

Franzoso, G., Carlson, L., Poljak, L., Shores, E. W., Epstein, S., Leonardi, A., Grinberg, A., Tran, T., Scharton-Kersten, T., Anver, M., Love, P., Brown, K., and Siebenlist, U. (1998). Mice deficient in NFκB/p52 present with defects in humoral responses, germinal center reactions, and splenic architecture. J. Exp. Med. *187*, 147–159.

Franzoso, G., Carlson, L., Scharton-Kersten, T., Shores, E. W., Epstein, S., Grinberg, A., Tran, T., Shacter, E., Leonardi, A., Anver, M., Love, P., Sher, A., and Siebenlist, U. (1997). Critical roles for the Bcl-3 oncoprotein in T cell mediated immunity, splenic microarchitecture and germinal center reactions. Immunity *6*, 479–490.

Frey, S., Bertocci, B., Delbos, F., Quint, L., Weill, J. C., and Reynaud, C. A. (1998). Mismatch repair deficiency interferes with the accumulation of mutations in chronically stimulated B cells and not with the hypermutation process. Immunity *9*, 127–134.

Fruman, D. A., Snapper, S. B., Yballe, C. M., Davidson, L., Yu, J. Y., Alt, F. W., and Cantley, L. C. (1999). Impaired B cell development and proliferation in absence of phosphoinositide 3 kinase p85α. Science *283*, 393–397.

Futterer, A., Mink, K., Luz, A., Kosco-Vilbois, M. H., and Pfeffer, K. (1998). The lymphotoxin β receptor controls organogenesis and affinity maturation in peripheral lymphoid tissues. Immunity *9*, 59–70.

Garside, P., Ingulli, E., Merica, R. R., Johnson, J. J., Noelle, R. J., and Jenkins, M. K. (1998). Visualization of specific B and T lymphocyte interactions in the lymph node. Science *281*, 96–99.

Goossens, T., Klein, U., and Kuppers, R. (1998). Frequent occurrence of deletions and duplications during somatic hypermutation: implication for oncogene translocations and heavy chain disease. Proc. Natl. Acad. Sci. USA *95*, 2463–2468.

Goyenechea, B., Klix, N., Yelamos, J., Williams, G. T., Riddell, A., Neuberger, M. S., and Milstein, C. (1997). Cell strongly expressing Igκ transgenes show clonal recruitment of hypermutation: a role for both MAR and the enhancers. EMBO J. *16*, 3987–3994.

Grusby, M. J., Johnson, R. S., Papaioannou, V. E., and Glimcher, L. H. (1991). Depletion of CD4⁺ T cells in major histocompatibility complex class II-deficient mice. Science *253*, 1417–1420.

Han, S., Dillon, S. R., Zheng, B., Shimoda, M., Schlissel, M. S., and Kelsoe, G. (1997). V(D)J recombinase activity in a subset of germinal center B lymphocytes. Science *278*, 298–301.

Han, S., Zheng, B., Schatz, D. B., Spanopoulou, E., and Kelsoe, G. (1996). Neoteny in lymphocytes: rag 1 and rag 2 expression in germinal center B cells. Science *274*, 2094–2097.

Hardy, R. R., and Hayakawa, K. (1991). A developmental switch in B lymphopoiesis. Proc. Natl. Acad. Sci. USA *88*, 11550–11554.

Harriman, G. R., Bradley, A., Das, S., Rogers-Fani, P., and Davis, A. C. (1996). IgA class switch in Iα exon deficient mice. J. Clin. Invest. *97*, 477–485.

Hayakawa, K., Asano, M., Shinton, S. A., Gui, M., Allman, D., Stewart, C. L., Silver, J., and Hardy, R. R. (1999). Positive selection of natural autoreactive B cells. Science *285*, 113–116.

Hayakawa, K., Carmack, C. E., Hyman, R., and Hardy, R. R. (1990). Natural autoantibodies to thymocytes: origin, VH genes, fine specificities, and the role of Thy-1 glycoprotein. J. Exp. Med. *172*, 869–878.

Hayakawa, K., Hardy, R. R., Honda, M., Herzenberg, L. A., Steinberg, A. D., and Herzenberg, L. A. (1984). Ly-1 B cells: functionally distinct lymphocytes that secrete IgM autoantibodies. Proc. Natl. Acad. Sci. USA *81*, 2494–2498.

Hayakawa, K., Hardy, R. R., Parks, D. R., and Herzenberg, L. A. (1983). The "Ly-1 B" cell subpopulation in normal, immunodefective and autoimmune mice. J. Exp. Med. *157*, 202–218.

Hein, K., Lorenz, M. G. O., Siebenkotten, G., Petry, K., Rainer, C., and Radbruch, A. (1998). Processing of switch transcripts is required for targeting of antibody class switch recombination. J. Exp. Med. *188*, 2369–2374.

Hendriks, R. W., deBruijn, M. F. T. R., Maas, A., Dingjan, G. M., Karis, A., and Grosveld, F. (1996). Inactivation of Btk by insertion of lacZ reveals defects in B cell development only past the pre-B stage. EMBO J. *15*, 4862–4872.

Hertz, M., Kouskoff, V., Nakamura, T., and Nemazee, D. (1998). V(D)J recombinase induction in splenic B lymphocytes is inhibited by antigen receptor signaling. Nature *394*, 292–295.

Hibbs, M. L., Tarlinton, D. M., Armes, J., Graill, D., Hodgson, G., Maglitto, R., Stacker, S. A., and Dunn, A. R. (1995). Multiple defects in the immune system of Lyn deficient mice, culminating in autoimmune disease. Cell *83*, 301–311.

Hikida, M., Mori, M., Takai, T., Tomochika, K., Hamatani, K., and Ohmori, H. (1996). Reexpression of Rag-1 and Rag-2 genes in activated mouse mature B cells. Science *274*, 2092–2094.

Hodgkin, P. D., Lee, J. H., and Lyons, A. B. (1996). B cell differentiation and isotype switching is related to cell cycle number. J. Exp. Med. *184*, 277–281.

Horwitz, B. H., Scott, M. L., Cherry, S. R., Bronson, R. T., and Baltimore, D. (1997). Failure of lymphopoiesis after adoptive transfer of NFκB deficient fetal liver cells. Immunity *6*, 765–772.

Huang, S. (1994). Blimp-1 is the murine homolog of the human transcriptional repressor PRD1-BF1. Cell *78*, 9.

Humbert, P. O., and Corcoran, L. M. (1997). Oct-2 gene disruption eliminates the peritoneal B-1 lymphocyte lineage and attenuates B-2 cell maturation and function. J. Immunol. *159*, 5273–5284.

Iwasato, T., Shimizu, A., Honjo, T., and Yamagishi, H. (1990). Circular DNA is excised by immunoglobulin class switch recombination. Cell *62*, 143–149.

Jacobs, H., Fukita, Y., van der Horst, G. T. L., de Boer, J., Weeda, G., Esser, J., deWind, N., Engelward, B. P., Samson, L., and Verbeek, S., et al. (1998). Hypermutation of immunoglobulin genes in memory B cells of DNA repair deficient mice. J. Exp. Med. *187*, 1735–1743.

Jung, S., Rajewsky, K., and Radbruch, A. (1993). Shutdown of class switch recombination by deletion of a switch region control element. Science *259*, 984–987.

Kaisho, T., Schwenk, F., and Rajewsky, K. (1997). The roles of γ1 heavy chain membrane expression and cytoplasmic tail in IgG1 responses. Science *276*, 412–415.

Karras, J. G., Wang, Z., Huo, L., Howard, R. G., Frank, D. A., and Rothstein, T. L. (1997). STAT3 is constitutively activated in normal, self-renewing B-1 cells, but only inducibly expressed in conventional B-1 lymphocytes. J. Exp. Med. *185*, 1035–1042.

Kawabe, T., Naka, T., Yoshida, K., Tanaka, T., Fujiwara, H., Suematsu, S., Yoshida, N., Kishimoto, T., and Kikutani, H. (1994). The immune repsonses in CD40 deficient mice: impaired immunoglobulin class switching and germinal center formation. Immunity *1*, 167–178.

Kerner, J. D., Appleby, M. W., Mohr, R. N., Chien, S. D. J. R., Maliszewski, C. R., Witte, O. N., and Perlmutter, R. M. (1995). Impaired expansion of mouse B cell progenitors lacking Btk. Immunity *3*, 301–312.

Khan, W. N., Alt, F. W., Gerstein, R. M., Malynn, B. A., Larsson, I., Rathbun, G., Davidson, L., Muller, S., Kantor, A. B., Herzenberg, L. A., Rosen, F. S., and Sideras, P. (1995). Defective B cell development and function in Btk deficient mice. Immunity *3*, 283–299.

Kim, U., Qin, X. F., Gong, S., Stevens, S., Luo, Y., Nussenzweig, M., and Roeder, R. G. (1996). The B cell specific transcription coactivator OCA-B/OBF-1/Bob-1 is essential for normal production of immunoglobulin isotypes. Nature *383*, 542–547.

Kishihara, K., Penninger, J., Wallace, V. A., Kundig, T. M., Kawai, K., Wakeham, A., Timms, E., Pfeffer, K., Ohashi, P. S., and Thomas, M. L. (1993). Normal B lymphocyte development but impaired T cell maturation in CD45exon 6 protein tyrosine phosphatase deficient mice. Cell *74*, 143–156.

Kong, Y.-Y., Yoshida, H., Sarosi, I., Tan, H.-L., Timms, E., Capparelli, C., Morony, S., Oliviera-dos-Santos, A. J., Van, G., Itie, A., Khoo, W., Wakeham, A., Dunstan, C. R., Lacey, D. L., Mak, T. W., Boyle, W. J., and Penninger, J. M. (1999). OPGL is a key regulator of osteoclastogenesis, lymphocyte development, and lymph-node organogenesis. Nature *397*, 315–323.

Koni, P. A., Sacca, R., Lawton, P., Browning, J. L., Ruddle, N. H., and Flavell, R. A. (1997). Distinct roles in lymphoid organogenesis for lymphotoxins α and β revealed in lymphotoxin β deficient mice. Immunity *6*, 491–500.

Kontgen, F., Grumont, R. J., Strasser, A., Metcalf, D., Li, R., Tarlinton, D., and Gerondakis, S. (1995). Mice lacking the c-Rel protooncogene exhibit defects in lymphocyte proliferation, humoral immunity and interleukin-2 expression. Genes Dev. *9*, 1965–1977.

Kopf, M., Baumann, H., Freer, G., Freudenberg, M., Lamers, M., Kishimoto, T., Zinkernagel, R., Buethmann, H., and Kohler, G. (1994). Impaired immune and acute-phase responses in interleukin-6 deficient mice. Nature *368*, 39–342.

Kopf, M., Brombacher, F., Hodgkin, P. D., Ramsay, A. J., Milbourne, E. A., Dai, W. J., Ovington, K. S., Behm, C. A., Kohler, G., Young, I. G., and Matthaei, K. I. (1996). IL-5 deficient mice have a developmental defect in $CD5^+$ B-1 cells and lack eosinophilia but have normal antibody and cytotoxic T cell responses. Immunity *4*, 15–24.

Lansford, R., Manis, J., Sonoda, E., Rajewsky, K., and Alt, F. (1998). Ig heavy chain class switching in rag-deficient mice. Int. Immunol. *10*, 325–332.

Le Hir, M., Bluethmann, H., Kosco-Vilbois, M. H., Muller, M., di Pardova, F., Moore, M., Ryffel, B., and Eugster, H. P. (1996). Differentiation of follicular dendritic cells and full antibody responses require tumor necrosis factor receptor-1 signaling. J. Exp. Med. *183*, 2367–2372.

Leitges, M., Schmedt, C., Guinamard, R., Davoust, J., Schaal, S., Stabel, S., and Tarakhovsky, A. (1996). Immunodeficiency in protein kinase C beta deficient mice. Science *273*, 988–991.

Lin, Y., Wong, K., and Calame, K. (1997). Repression of c-myc transcription by Blimp-1, an inducer of terminal B cell differentiation. Science *276*, 596–599.

Lorenz, M., Jung, S., and Radbruch, A. (1995). Switch transcripts in immunoglobulin class switching. Science *267*, 1825–1828.

Maeeker, H. T., and Levy, S. (1997). Normal lymphocyte development but delayed humoral immune response in CD81-null mice. J. Exp. Med. *185*, 1505–1510.

Manis, J. P., Gu, Y., Lansford, R., Sonoda, E., Ferrini, R., Davidson, L., Rajewsky, K., and Alt, F. W. (1998). Ku70 is required for late B cell development and immunoglobulin heavy chain class switching. J. Exp. Med. *197*, 2081–2089.

Manis, J. P., van der Stoep, N., Tian, M., Ferrini, R., Davidson, L., Bottaro, A., and Alt, F. W. (1998). Class switching in B cells lacking 3′ immunoglobulin heavy chain enhancers. J. Exp. Med. *188*, 1421–1431.

Martin, F., and Kearney, J. F. (2000). Positive selection from newly formed to marginal zone B cells depends on the rate of clonal production, CD19, and Btk. Immunity *12*, 39–49.

Matsumoto, M., Maraiathasan, S., Nahm, M. H., Baranyay, F., Peschon, J. J., and Chaplin, D. D. (1996). Role of lymphotoxin and the type I TNF receptor in the formation of germinal centers. Science *271*, 1289–1291.

Matsuoka, M., Yoshida, K., Maeda, T., Usuda, S., and Sakano, H. (1990). Switch circular DNA formed in cytokine treated mouse splenocytes: evidence for intramolecular DNA deletion in immunoglobulin class switching. Cell *62*, 135–142.

Mercolino, T. J., Arnold, L. W., Hawkins, L. A., and Haughton, G. (1998). Normal mouse peritoneum contains a large number of $Ly\text{-}1^+$ (CD5) B cells that recognize phosphatidyl choline. Relationship to cells that secrete hemolytic antibody specific for autologous erythrocytes. J. Exp. Med. *168*, 687–698.

Messika, E. J., Lu, P. S., Sung, Y.-J., Yao, T., Chi, J.-T., Chien, Y.-h., and Davis, M. M. (1998). Differential effect of B-lymphocyte-induced maturation protein (Blimp-1) expression on cell fate during B lymphocyte development. J. Exp. Med. *188*, 515–525.

Mittrucker, H. W., Matsuyama, T., Grossman, A., Kundig, T. M., Potter, J., Shahinian, A., Wakeham, A., Patterson, B., Ohashi, P. S., and Mak, T. W. (1997). Requirement for the transcription factor LSIRF/IRF4 for mature B and T lymphocyte function. Science *275*, 540–543.

Miyazaki, T., Muller, U., and Campbell, K. S. (1997). Normal development but differentially altered proliferative responses of lymphocytes in mice lacking CD81. EMBO J. *16*, 4217–4225.

Molina, H., Holers, V. M., Li, B., Fung, Y., Maraithasan, S., Goellner, J., Strauss-Schoenberger, J., Karr, R. W., and Chaplin, D. D. (1996). Markedly impaired immune response in mice deficient in complement receptors 1 and 2. Proc. Natl. Acad. Sci. USA *93*, 3357–3361.

Monroe, R. J., Seidl, K. J., Gaertner, F., Han, S., Chen, F., Sekiguchi, J., Wang, J., Ferrini, R., Davidson, L., Kelsoe, G., and Alt, F. W. (1999). RAG2:GFP knockin mice reveal novel aspects of RAG2 expression in primary and peripheral lymphoid tissues. Immunity *11*, 201–212.

Morio, T., Hanissian, S. H., Bacharier, L. B., Teraoka, H., Nonoyama, S., Seki, M., Kondo, J., Nakano, H., Lee, S.-K, Geha, R. S., and Yata, J.-I. (1999). Ku in the cytoplasm associates with CD40 in human B cells and translocates into the nucleus following incubation with IL-4 and anti-CD40 mAb. Immunity *11*, 339–348.

Morse, L., Chen, D., Franklin, D., Xiong, Y., and Chen-Kiang, S. (1997). Induction of cell cycle arrest and B cell terminal differentiation by CDK inhibitor p18 (INK4c) and IL-6. Immunity *6*, 47–56.

Murakami, M., Tsubata, T., Okamoto, M., Shimizu, A., Kumagai, S., Imura, H., and Honjo, T. (1992). Antigen induced apoptotic death of Ly-1 b cells responsible for autoimmune disease in transgenic mice. Nature *357*, 77–80.

Neumann, B., Luz, A., Pfeffer, K., and Holzmann, B. (1996). Defective Peyer's patch organogenesis in mice lacking the 55 kD receptor for tumor necrosis factor. J. Exp. Med. *184*, 259–264.

Neurath, M. F., Max, E. E., and Strober, W. (1995). Pax-5 (BSAP) regulates the murine immunoglobulin 3′ enhancer by suppressing binding of NF-aP a protein that controls heavy chain transcription. Proc. Natl. Acad. Sci. USA *92*, 5336–5340.

Nielsen, P. J., Georgiev, O., Lorenz, B., and Schaffner, W. (1996). B lymphocytes are impaired in mice lacking the transcriptional co-activator Bob1/OCA-B/OBF-1. Eur. J. Immunol. *26*, 3214–3218.

Nishizumi, H., Taniuchi, I., Yamanashi, Y., Kitamura, D., Ilic, D., Mori, S., Watanabe, T., and Yamamoto, T. (1995). Impaired proliferation of peripheral B cells and indication of autoimmune disease in lyn-deficient mice. Immunity *3*, 549–560.

Niu, H., Ye, B. H., and Della-Favera, R. (1998). Antigen receptor signaling induces MAP kinase-mediated phosphorylation and degradation of the BCL6 transcription factor. Genes Dev *12*, 1953–1961.

O'Keefe, T. L., Williams, G. T., Davies, S. L., and Neuberger, M. S. (1996). Hyperresponsive B cells in CD22 deficient mice. Science *274*, 798–801.

Oliver, A. M., Martin, F., Gartland, G. L., Carter, R. H., and Kearney, J. F. (1997). Marginal zone B cells exhibit unique activation, proliferative and immunoglobulin secretory responses. Eur. J. Immunol. *27*, 2366–2374.

Oliver, A. M., Martin, F., and Kearney, J. F. (1999). IgM high CD 21 high lymphocytes enriched in the splenic marginal zone generate effector cells more rapidly than the bulk of follicular B cells. J. Immunol. *162*, 7198–7207.

Otipoby, K. L., Andersson, K. B., Draves, K. E., Klaus, S. J., Farr, A. G., Kerner, J. D., Perlmutter, R. M., Law, C. L., and Clark, E. A. (1996). CD22 regulates thymus independent responses and the lifespan of B cells. Nature *384*, 634–637.

Pan, C., Baumgarth, N., and Parnes, J. R. (1999). CD72-deficient mice reveal nonredundant roles of CD72 in B cell development and activation. Immunity *11*, 495–506.

Papavisiliou, F., Casellas, R., Suh, H., Qin, X. F., Besmer, E., Pelanda, R., Nemazee, D., Rajewsky, K., and Nussenzweig, M. C. (1997). V(D)J recombination in mature B cells: a mechanism for altering antibody responses. Science *278*, 298–301.

Pasparakis, M., Alexopoulou, L., Episkopou, V., and Kollias, G. (1996). Immune and inflammatory responses in TNF α deficient mice: a critical requirement for TNF α in the formation of primary B cell follicles, follicular dendritic cell networks and germinal centers, and in the maturation of the humoral immune response. J. Exp. Med. *184*, 1397–1411.

Peters, A., and Storb, U. (1996). Somatic hypermutation of immunoglobulin genes is linked to transcription initiation. Immunity *4*, 57–65.

Pfeffer, K., Matsuyama, T., Kundig, T. M., Wakeham, A., Kishihara, K., Shahinian, A., Wiegmann, K., Ohashi, P. S., Kronke, K., and Mak, T. W. (1993). Mice deficient for th 55 kd tumor necrosis factor receptor are resistant to endotoxic shock, yet succumb to *L. monocytogenes* infection. Cell *73*, 457–467.

Phung, Q. H., Winter, D. B., Cranston, A., Tarone, R. E., Bohr, V. A., Fishel, R., and Gearhart, P. J. (1998). Increased hypermutation at G and C nucleotides in immunoglobulin variable genes from mice deficient in the MSH2 mismatch repair protein. J. Exp. Med. *187*, 1–7.

Qin, X.-F., Reichlin, A., Luo, Y., Roeder, R. G., and Nussenzweig, M. C. (1998). OCA-B integrates B cell antigen receptor-, CD40L-, and IL-4 mediated signals for the germinal center pathway of B cell development. EMBO J. *17*, 5066–5075.

Qin, X.-F., Schwers, S., Yu, W., Papavasiliou, F., Suh, H., Nussenzweig, A., Rajewsky, K., and Nussenzweig, M. C. (1999). Secondary V(D)J recombination in B-1 cells. Nature *397*, 355–359.

Quong, M. W., Harris, D. P., Swain, S. L., and Murre, C. (1999). E2A activity is induced during B cell activation to promote immunoglobulin class switch recombination. EMBO J. *18*, 6307–6318.

Rada, C., Ehrenstein, M. R., Neuberger, M. S., and Milstein, C. (1998). Somatic hypermutation in MSH2-deficient mice is more focused on intrinsic hot spots suggesting hypermutation targeting to be a two-staged process. Immunity *9*, 135–141.

Randall, T. D., Heath, A. W., Santos-Argumedo, L., Howard, M. C., Weissman, I. L., and Lund, F. E. (1998). Arrest of B lymphocyte terminal differentiation by CD40 signaling: mechanism for lack of antibody secreting cells in germinal centers. Immunity *8*, 733–742.

Rathmell, J. C., Cooke, M. P., Ho, W. Y., Grein, J., Townsend, S. E., Davis, M. M., and Goodnow, C. C. (1995). CD95(Fas) dependent elimination of self-reactive B cells upon interaction with $CD4^+$ T cells. Nature *376*, 181–183.

Renshaw, B. R., Fanslow, W. C., Armitage, R. J., Campbell, K. A., Liggitt, D., Wright, B., Davison, B. L., and Maliszewski, C. R. (1994). Humoral immune responses in CD40 ligand deficient mice. J. Exp. Med. *180*, 1889–1900.

Reynaud, C. A., Garcia, C., Hein, W., and Weill, J.-C. (1995). Hypermutation generating the sheep immunoglobulin repertoire is an antigen-independent process. Cell *80*, 115–126.

Rickert, R. C., Rajewsky, K., and Roes, J. (1995). Impairment of T cell dependent B cell responses and B-1 cell development in CD19 deficient mice. Nature *376*, 352–355.

Roark, J. H., Park, S. H., Jayawardena, J., Kavita, U., Shannon, M., and Bendelac, A. (1998). CD1.1 expression by mouse antigen presenting cells and marginal zone B cells. J. Immunol. *160*, 3121–3127.

Rolink, A., Melchers, F., and Andersson, J. (1996). The SCID but not the Rag-2 gene product is required for Sμ-Sε heavy chain class switching. Immunity *5*, 319–330.
Rothe, J., Lesslauer, W., Lotscher, H., Lang, Y., Koebel, P., Kontgen, F., Althage, A., Zinkernagel, R., Steinmetz, M., and Bluethmann, H. (1993). Mice lacking the tumor necrosis factor receptor 1 are resistant to TNF mediated toxicity but highly susceptible to infection by *Listeria monocytogenes*. Nature *364*, 798–802.
Rothstein, T. L., Wang, J. K., Panka, D. J., Foote, L. C., Wang, Z., Stanger, B., Cui, H., Ju, S. T., and Marshak-Rothstein, A. (1995). Protection against Fas-dependent Th1-mediated apoptosis by antigen receptor engagement in B cells. Nature *374*, 163–165.
Sato, S., Miller, A. S., Inaoki, M., Bock, C. B., Jansen, P. J., Tang, M. L., and Tedder, T. F. (1996). CD22 is both a positive and negative regulator of B lymphocyte antigen receptor signal transduction: altered signaling in CD22 deficient mice. Immunity *5*, 551–562.
Scher, I., Ahmed, A., Strong, D. M., Steinberg, A. D., and Paul, W. E. (1975). X linked B lymphocyte defect in CBA/N mice. I. Studies of the function and composition of spleen cells. J. Exp. Med. *141*, 788–803.
Schubart, D. B., Rolink, A., Kosco-Vilbois, M. H., Botteri, F., and Matthias, P. (1996). B cell specific coactivator OBF-1/OCA-B/Bob1 required for immune response and germional centre formation. Nature *383*, 538–542.
Schwarz, E. M., Krimpenfort, P., Berns, A., and Verma, I. M. (1997). Immunological defects in mice with a targeted disruption in Bcl-3. Genes Dev. *11*, 187–197.
Seipelt, R. L., Spear, B. T., Snow, E. C., and Peterson, M. L. (1998). A nonimmunoglobulin transgene and the endogenous immunoglobulin μ gene are coordinately regulated by alternative RNA processing during B cell maturation. Mol. Cell. Biol. *18*, 1042–1048.
Sha, W. C., Hsiou-Chi, L., Toumanen, E. I., and Baltimore, D. (1995). Targeted disruption of the p50 subunit of of NFκB leads to multifocal defects in immune responses. Cell *80*, 321–330.
Shahinian, A., Pfeffer, K., Lee, K. P., Kundig, T. M., Kishihara, K., Wakeham, A., Kawai, K., Ohashi, P. S., Thompson, C. B., and Mak, T. W. (1993). Differential T cell costimulatory requirements in CD28 deficient mice. Science *261*, 609–612.
Shen, H. M., Peters, A., Baron, B., Zhu, X., and Storb, U. (1998). Mutation of Bcl-6 gene in normal B cells by the process of somatic hyperemutation of Ig genes. Science *280*, 1750–1752.
Shinkura, R., Kitada, K., Matsuda, F., Tashiro, K., Ikuta, K., Suzuki, M., Kogishi, K., Serikawa, T., and Honjo, T. (1999). Alymphoplasia is caused by a point mutation in the mouse gene encoding NFκB inducing kinase. Nature Genet. *22*, 74–77.
Shinkura, R., Matsuda, F., Sakiyama, T., Tsubata, T., Hiai, H., Paumen, M., Miyawaki, S., and Honjo, T. (1996). Defects in somatic mutation and class switching in alymphoplasia (aly) mutant mice. Int. Immunol. *8*, 1067–1075.
Shultz, L. D., and Green, M. C. (1976). Motheaten, an immunodeficient mutant of the mouse. II. Depressed immune competence and elevated serum immunoglobulins. J. Immunol. *116*, 936–943.
Singh, M., and Birshtein, B. K. (1996). Concerted repression of an immunoglobulin heavy chain enhancer 3′αE (hs1,2). Proc. Natl. Acad. Sci. USA *93*, 4392–4397.
Snapper, C. M., Rosas, F. R., Zelazowski, P., Moorman, M. A., Kehry, M. R., Bravo, R., and Weih, F. (1996a). B cells lacking RelB are defective in proliferative responses, but undergo normal B cell maturation to Ig secretion and Ig class switching. J. Exp. Med. *184*, 1537–1541.

Snapper, C. M., Zelazowski, P., Rosas, F. R., Kehry, M. R., Tian, M., Baltimore, D., and Sha, W. C. (1996b). B cells from p50/NFκB knockout mice have selective defects in proliferation, differentiation, germ-line CH transcription, and Ig class switching. J. Immunol. *156*, 183–191.

Solvason, N., Wu, W. W., Kabra, N., Wu, X., Lees, E., and Howard, M. C. (1996). Induction of cell cycle regulatory proteins in anti-immunoglobulin stimulated mature B lymphocytes. J. Exp. Med. *184*, 407–417.

Stavnezer-Nordgren, J., and Sirlin, S. (1986). Specificity of immunoglobulin heavy chain class switch correlates with activity of germline heavy chain genes prior to switching. EMBO J. *5*, 95–102.

Strubin, M., Newell, J. W., and Matthias, P. (1995). OBF-1, a novel B cell specific coactivator that stimulates immunoglobulin promoter activity through association with octamer-binding proteins. Cell *80*, 497–506.

Su, G. H., Chen, H. M., Muthusamy, N., Garrett-Sinha, L. A., Baunoch, D., Tenen, D. G., and Simon, M. C. (1997). Defective B cell receptor-mediated responses in mice lacking the Ets protein, SpiB. EMBO J. *16*, 7118–7129.

Suzuki, H., Terauchi, Y., Fujiwara, M., Aizawa, S., Yazaki, Y., Kadowaki, T., and Koyasu, S. (1999). Xid like immunodeficiency in mice with disruption of the p85α subunit of phosphoinositide 3-kinase. Science *283*, 390–392.

Takagaki, Y., Seipelt, R. L., Peterson, M. L., and Manley, J. L. (1996). The polyadenylation factor CstF-64 regulates alternative processing of IgM heavy chain pre-mRNA during B cell differentiation. Cell *87*, 941–952.

Takai, T., Ono, M., Hikida, M., Ohmori, H., and Ravetch, J. V. (1996). Augmented humoral and anaphylactic responses in FcγRII deficient mice. Nature *379*, 346–349.

Taniuchi, I., Kitamura, D., Maekawa, Y., Fukuda, T., Kishi, H., and Watanabe, T. (1995). Antigen receptor induced clonal expansion and deletion of lymphocytes are impaired in mice lacking HS1 protein, a substrate of the antigen-receptor coupled tyrosine kinases. EMBO J. *14*, 3664–3678.

Tarakhovsky, A., Turner, M., Schaal, S., Mee, P. J., Duddy, L. P., Rajewsky, K., and Tybulewicz, V. L. J. (1995). Defective antigen receptor mediated proliferation of B and T cells in the absence of of Vav. Nature *374*, 467–470.

Tew, J. G., Wu, J., Qin, D., Helm, S., Burton, G. F., and Szakal, A. K. (1997). Follicular dendritic cells and presentation of antigen and costimulatory signals to B cells. Immunol. Rev. *156*, 39–52.

Tkachuk, M., Bolliger, S., Ryffel, B., Pluschke, G., Banks, T. A., Herren, S., Gisler, R. H., and Kosco-Vilbois, M. H. (1998). Crucial role of tumor necrosis factor receptor 1 expression on nonhematopoietic cells for B cell localization within the splenic white pulp. J. Exp. Med. *187*, 469–477.

Torres, R. M., Flaswinkel, H., Reth, M., and Rajewsky, K. (1996). Aberrant B cell development and immune response in mice with a compromised BCR complex. Science *272*, 1804–1808.

Tsitsikov, E. N., Gutierrez-Ramos, J. C., and Geha, R. S. (1997). Impaired CD19 expression and signaling, enhanced antibody repsonses to type II T independent antigen and reduction of B-1 cells in CD81-deficient mice. Proc. Natl. Acad. Sci. USA *94*, 10844–10849.

Tumas-Brundage, K., and Manser, T. (1997). The transcriptional promoter regulates hypermutation of the antibody heavy chain locus. J. Exp. Med. *185*, 239–250.

Turner, M., Gulbranson-Judge, A., Quinn, M. E., Walters, A. E., MacLennan, I. C. M., and Tybulewicz, V. L. J. (1997). Syk tyrosine kinase is required for the positive selection of immature B cells into the recirculating B cell pool. J. Exp. Med. *186*, 2013–2021.

von Schwedler, U., Jack, H. M., and Wabl, M. (1990). Circular DNA is a product of the immunoglobulin class switch rearrangement. Nature *345*, 452–456.

Wang, J., Koizumi, T., and Watanabe, T. (1996). Altered antigen receptor signaling and impaired Fas-mediated apoptosis of B cells in Lyn deficient mice. J. Exp. Med. *184*, 831–838.

Wang, J. H., Avitahl, N., Cariappa, A., Friedrich, C., Ikeda, T., Renold, A., Andrikopoulos, K., Liang, L., Pillai, S., Morgan, B. A., and Georgopoulos, K. (1998). Aiolos regulates B cell activation and maturation to effector state. Immunity *9*, 543–554.

Weigert, M. G., Cesari, I. M., Yonkovich, S. J., and Cohn, M. (1970). Variability in the lambda light chain sequences of mouse antibody. Nature *228*, 1045–1047.

Weih, F., Carrasco, D., Durham, S. K., Barton, D. S., Rizzo, C. A., Ryseck, R.-P., Lira, S. A., and Bravo, R. (1995). Multiorgan inflammation and hematopoietic abnormalities in mice with a targeted disruption of Rel B, a member of the NFκB/Rel family. Cell *80*, 331–340.

Weih, F., Warr, G., Yang, H., and Bravo, R. (1997). Multifocal defects in immune responses in relB deficient mice. J. Immunol. *158*, 5211–5218.

Wilson, P. C., de Boutellier, O., Liu, Y.-J., Potter, K., Banchereau, J., Capra, J. D., and Pascual, V. (1998). Somatic mutation introduces insertions and deletions into immunoglobulin V genes. J. Exp. Med. *187*, 59–70.

Wu, H. Y., and Kaartinen, M. (1995). The somatic mutation activity of a follicular lymphoma links to large insertions and deletions of immunoglobulin genes. Scand. J. Immunol. *42*, 52–59.

Xu, J. C., Foy, T. M., Laman, J. D., Elliott, E. A., Dunn, J. J., Waldschmidt, T. J., Elsemore, J., Noelle, R. J., and Flavell, R. A. (1994). Mice deficient for CD40 ligand. Immunity *1*, 423–431.

Ye, B. H., Cattoretti, G., Shen, Q., Zhang, J., Hawe, N., de Waard, R., Leung, C., Nouri-Shirazi, M., Orazi, A., Chaganti, R. S., Rothman, P., Stall, A. M., Pandolfi, P. P., and Della-Favera, R. (1997). The Bcl-6 proto-oncogene controls germinal center formation and Th2 type inflammation. Nat. Genet. *16*, 161–170.

Yelamos, J., Klix, B., Goyenechea, B., Lozano, F., Chui, Y. L., Gonzalez-Fernandez, A., Pannell, R., Neuberger, M. S., and Milstein, C. (1995). Targeting of non-Ig sequences in place of the V segment by somatic hypermutation. Nature *376*, 225–229.

Yoshida, T., Ikuta, K., Sugaya, H., Maki, K., Takagi, M., Kanazawa, H., Sunaga, S., Kinashi, T., Yoshimura, K., Miyazaki, J.-i., Takaki, S., and Takatsu, K. (1996). Defective B-1 cell development and impaired immunity against *Angiostrongylus cantonensis* in IL-5Rα deficient mice. Immunity *4*, 483–494.

Yu, W., Nagaoka, H., Jankovic, M., Misulovin, Z., Suh, H., Rolink, A., Melchers, F., Meffre, E., and Nussenzweig, M. C. (1999). Continued RAG expression in late stages of B cell development and no apparent re-induction after immunization. Nature *400*, 682–686.

Zhang, R., Alt, F. W., Davidson, L., Orkin, S. H., and Swat, W. (1995). Defective signaling through the T and B cell antigen receptors in lymphoid cells lacking the vav proto-oncogene. Nature *374*, 470–473.

9

T Lymphocyte Development

In this chapter we will discuss the development of naive T lymphocytes. The activation and antigen-driven differentiation of T cells will be considered in Chapter 10. T cell progenitors develop from fetal liver-derived stem cells *in utero* and from bone marrow progenitors in adult life. These precursors migrate to the thymus, where most T cells are generated. Once progenitors seed the thymus via the blood stream, they begin to develop in this central organ going through a number of cellular stages that are similar, at least in broad outline, to the developmental stages considered during B cell development. The thymus provides stromal signals, both cell-bound and soluble, that drive T cell development, and this organ is uniquely required for a "compartmentalization" function that is critical for the development of most T cells, particularly cells of the αβ lineage. T cell development can also occur at extrathymic sites, and this aspect of T cell generation will also be considered in this chapter.

There are two major T lymphoid lineages, the αβ and the γδ lineages. About 90% of all T cells are αβ T cells. Most αβ T cells mature into either CD4- or CD8-expressing (single-positive, SP) T cells. Cells of a "third" T lymphocyte lineage express αβ T cell receptors, may either express CD4, or may lack both CD4 and CD8, and generally express the (NK1.1) marker (characteristically found on NK natural killer cells). These cells are referred to as natural T (NT) cells (they are also called NK T cells or NK 1.1^{+}T cells). A developing T cell commits to mature either as a γδ or an αβ cell. Exactly how the decision is made for a progenitor cell to commit to one or the other lineage remains to be established. It is unclear whether the process is stochastic or instructive, and current models to explain this event will be discussed in this chapter. In general, as we shall see, instructive, but "leaky" commitment processes best explain how decisions are made during T cell development. Early T cells of the αβ lineage are positively selected if they productively rearrange the T cell-receptor (TCR) β chain, a process known as positive selection I, or β selection. This event is analogous to positive selection I in the B lineage, and signals generated during β selection may contribute to a number of important processes, one of which is the generation of accessibility for rearrangement at the *TCRα* locus. Pre-TCR signaling may in a sense represent the instructive part of commitment to the αβ lineage in the course of a gradual but "leaky" instructive process. αβ T cells that have completed the program of receptor gene rearrangement subsequently undergo a second positive selection event (positive selection II, commonly referred to as "positive selection" by most people in the field) and may also be negatively selected at the $CD4^+CD8^+$ double-positive (DP) stage. In the T lineage, negative selection leads to the deletion of self-reactive cells. Although growing evidence suggests that cells are given "another chance" if they encounter self-antigen during B cell development, at present no evidence suggests that receptor editing occurs early in T cell development (although a similar process might occur in the periphery). Positive selection II in T cells is also closely linked to a lineage commitment decision—to make SP $CD4^+$ or $CD8^+$ cells. A simplified overview of T cell development (primarily of αβ T cells) is provided in Figure 9.1, and a slightly more detailed picture is provided in Figure 9.2.

Homing of Progenitors to the Thymus

During fetal life, T cell precursors leave the fetal liver in waves and colonize the rudimentary thymus at precise stages of development starting from day 11 of gestation (dE11). The first wave of precursors that seed the thymus before dE13 give rise to all the T cells that are produced by the thymus during gestation and the first week after birth. A second wave of progenitors from the fetal liver seeds the thymus after dE13, and this produces T cells that begin to dominate the repertoire from about day 7 after birth. After birth, however, a steady exodus of progenitors leaves the bone marrow and homes to the thymus. Exactly how waves of precursors are generated during fetal life is unclear. Homing of progenitors might require specific exit mechanisms from the bone marrow as well as specific entry mechanisms that assist in binding to thymic endothelium and entry into the thymus. While α6 integrins and the hyaluronan receptor, CD44, may be required for thymic entry, the specificity of thymic progenitors for the thymus is poorly explained. A molecule on perivascular thymic stromal cells called vanin-1 has been demonstrated to be important for thymic entry. Vanin-1 is attached to the plasma membrane of thymic stromal cells by a glycosylphosphatidyl inositol anchor, and the extracellular portion of Vanin-1 is homologous to biotinidases. Vanin-1 is probably responsible for a postendothelial migration step during thymic homing. Exactly how it works and what it recognizes on the surface of progenitor T cells are not known.

Undoubtedly, specific chemokines will be shown to be involved in the homing of progenitors to the thymus. At the present time, the best, but

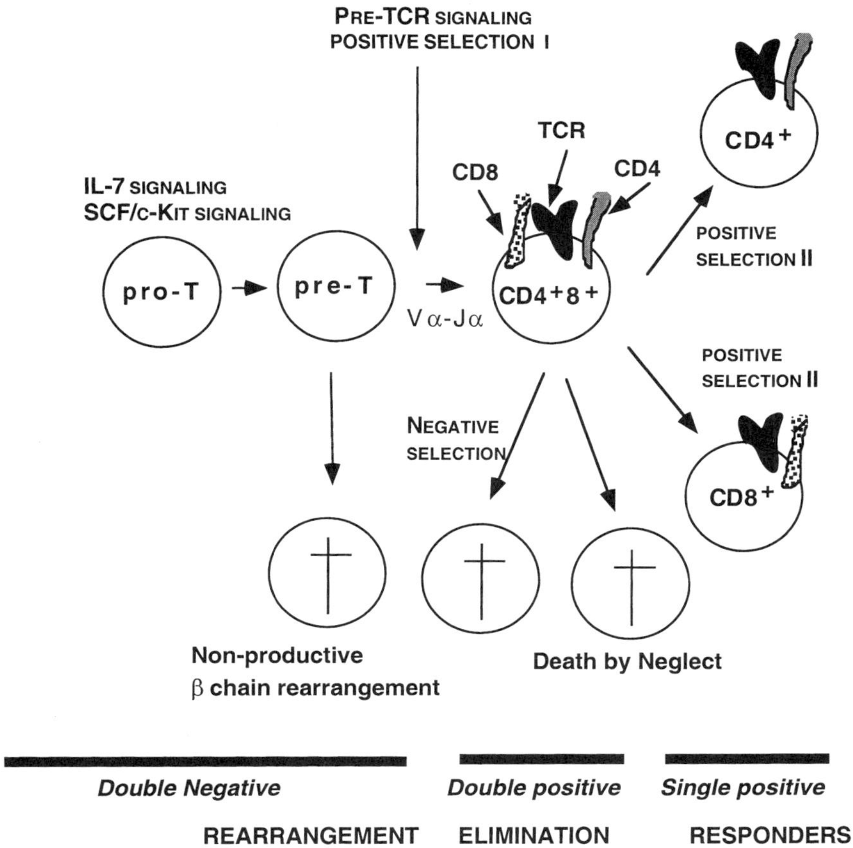

FIGURE 9.1. A simplified overview of T cell development. T cell progenitors lacking CD4 and CD8 (double-negative cells, DN) rearrange their TCR β, γ, and δ genes at the pre-T stage. These early pro-T and pre-T cells express very low levels of CD3 and are frequently referred to as triple-negative (TN) cells as well. Cells committed to the αβ lineage that productively rearrange their β chain genes receive signals via the pre-T receptor (positive selection I/β selection). The pre-T receptor provides survival and progression signals, contributes to allelic exclusion of the β chain of the TCR, and is required for the generation of accessibility at the TCRα locus. Pre-T receptor signaling is required for the DN to DP transition. DP T cells may "die by neglect" if their TCRs do not recognize self-MHC, may be eliminated if their TCRs have a high affinity for self-MHC–peptide complexes, and may be positively selected if their TCRs have a low-to-moderate affinity for self-MHC–peptide complexes. Positive selection of DP T cells (positive selection II) is accompanied by a lineage commitment event whereby SP cells restricted to self-MHC class II express the CD4 coreceptor, and cells bearing T cell receptors restricted to self-MHC class I express the CD8 coreceptor.

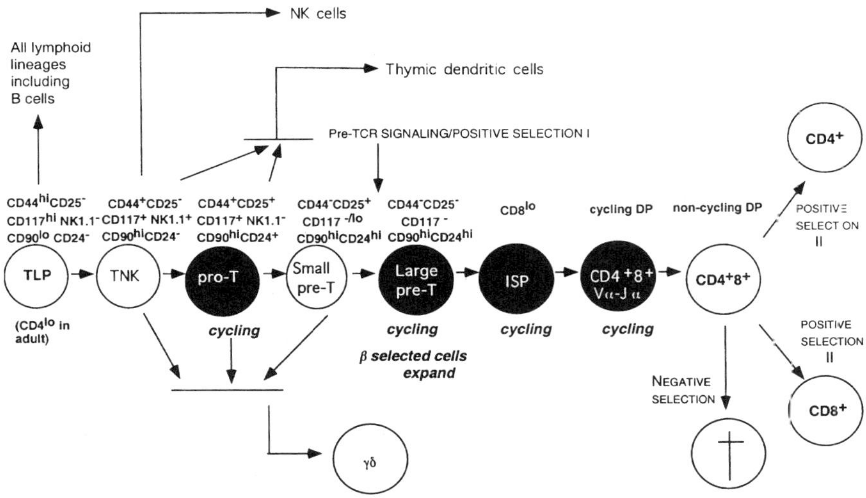

FIGURE 9.2. A slightly more detailed overview of T cell development. Pro- and pre-T cells are generally referred to as either triple negative (TN, $CD3^-$, $CD4^-$, $CD8^-$) or DN T cells. Multipotent thymic lymphoid progenitors (TLPs), which are $CD117^+CD44^+CD25^-CD90^+$ (CD117 is c-Kit, CD90 is Thy-1, CD24 is HSA) still have the potential to differentiate into B cells, NK cells, and lymphoid dendritic cells. They are functionally similar to bone marrow common lymphoid progenitors (CLPs). TLPs differentiate into TNK cells (bipotential T/NK cells; FTNK in fetal thymus), which have similar surface characteristics to TLPs (but also express NK 1.1, and do not express CD4). These latter cells have the potential to develop into αβ or γδ T cells, NK cells, thymic dendritic cells, or NT cells (a separate lineage of αβ T cells, not shown in this figure), but not into B cells. TNK cells may be induced by thymic signals to mature into pro-T cells ($CD44^+CD25^+$), which expand and differentiate into small pre-T cells. Small pre-T cells ($CD44^-CD25^+$) may initiate and complete rearrangement of *TCR β, γ,* or *δ* genes. αβ lineage T cells that make productive rearrangements at the β chain locus may be selected via pre-T receptor signaling and later undergo positive and negative selection at the $CD4^+CD8^+$ DP stage. Cells that make productive rearrangements at the γ and δ loci may commit to the γδ lineage (the commitment process is discussed later). ISP refers to immature single-positive cells that express low levels of CD8 (in most mouse trains; in humans these cells may express low levels of CD4). This classification is based in part on Godfrey et al., 1993; Shortman and Wu, 1996; and Carlyle and Zuniga-Pflucker, 1998.

poorly studied, candidate is the C-C chemokine TECK, which is made by thymic dendritic cells and may facilitate the entry of progenitors into the thymus—although this issue has not yet been systematically addressed.

Why Do T Cells Develop Primarily in the Thymus and Not Elsewhere?

If any organ has been deified by immunologists it is the thymus. It has been described as the "maestro" or the "master orchestrator" of the immune system. Some of the implied "intelligence" attributed to selection processes in the thymus, particularly the process of major histocompatibility complex (MHC) restriction, may need to be thought through from a strictly developmental, but perhaps teleologically less exciting, viewpoint (see discussion below). Why exactly do we need the thymus? Are there any unique trophic factors in this organ that do not exist at other anatomical sites? Is there any reason why stromal cells in the bone marrow or other cells elsewhere cannot induce positive and negative selection? One potentially critical inductive influence could be the presence of ligands on thymic epithelial cells for Notch1. It is clear that Notch1 signaling permits the development of T cells in the thymus at the expense of B cells. In mice in which Notch1 deletion is induced, T cell development is blocked, and B cell development in the thymus is enhanced. However, while Notch 1 is critical for T cell development, transmembrane ligands for Notch such as Jagged 1, Jagged 2, and Delta have not been shown to be thymus specific, although more definitive information on ligands and their expression patterns is awaited. Moreover, while overexpression of Notch1 results in proliferation of T cell precursors in the bone marrow, they are unable to mature into single-positive T cells, suggesting some other more specialized and specific role for the thymus.

The unique architecture of the thymus appears to be essential during lymphocyte development. While the thymus does provide the signals for commitment and differentiation as described below, recent studies suggest that T cell development cannot proceed efficiently in extrathymic sites, in part because of a feedback inhibition effect. It has been demonstrated that recently generated mature T cells inhibit the further development of precursors. While bone marrow precursors can give rise to mature SP $\alpha\beta$ T cells in the bone marrow itself, mature cells provide feedback signals that prevent the efficient maturation of precursors within this compartment. In *in vitro* studies, when DN T cell precursors in the bone marrow are separated from mature cells (largely of recirculating origin), these precursors can quite efficiently generate SP $CD4^+$ and $CD8^+$ cells. In the presence of mature T cells, *in vitro* T cell maturation is relatively inefficient. It has been suggested that perhaps because T cells and their cytokine products can potently influence hematopoiesis, it is advantageous for T cell production not to occur in the immediate vicinity of blood cell generation. Accordingly, a negative feedback loop may have evolved in vertebrates by which mature T cells inhibit the generation of T cells from precursors. Although such a scheme of things is totally speculative, these observations lend support to the view that the underlying design of the thymus is to provide an environment that overcomes the process of feedback inhibition. One major feature of "thymus design" is the compartmentalization of precursors in the cortex and of more mature cells in the medulla. It is likely that in such a setup, more mature T cells are physically segregated from immature T cells and are thus unable to negatively influence T cell development.

Thymocyte Precursors Can Develop in the Absence of the Thymus: Lineage Commitment Is Generally Completed in the Thymus

Apart from the "compartmentalization" function, the thymus appears to be required to efficiently complete the process of T lineage commitment. Common lymphoid progenitors (CLPs) in the fetal liver and bone marrow can give rise to all lymphoid lineages including T, B, NK, and lymphoid dendritic cells. Lymphoid dendritic cells may be distinguished from dendritic cells of myeloid origin by the fact that they express CD8α. Precursors of these CLPs, and CLPs themselves, may home to the thymus, and cells have been identified in the fetal thymus that can give rise to all the lymphoid lineages. These CLP-like cells in the thymus are called thymic lymphoid progenitors (TLPs). They express c-Kit/CD117 (the receptor tyrosine kinase that responds to stem cell factor, SCF) and CD44 but do not express NK1.1 (CD161) or CD25 (see Figure 9.2). They also express low levels of CD4 in the adult mouse. CD44 is the hyaluronan receptor and is discussed in Chapter 6. NK1.1 is expressed on NK cells and a subset of T cells known as NT cells and will be discussed later in this chapter and in Chapter 11. CD25 is the α chain of the IL-2 receptor (see Chapter 4) and is a useful marker for T cell commitment.

Both extrathymic CLPs as well as TLPs can differentiate into a more committed progenitor called a TNK cell. This TNK cell (so called because it is a precursor for both T and NK cells) can no longer give rise to B cells but has the potential to develop into a T cell, an NK cell, or a lymphoid dendritic cell. Preferential development of this TNK cell presumably requires Notch1 signals and occurs at the expense of commitment to the B lineage. This progenitor still does not express CD25 but expresses NK1.1. In adult mice, CD4 expression is lost at this stage. TNK cells are not seen in the fetal liver but may be observed in fetal spleen and fetal blood in addition to the fetal thymus. They are observed in the fetal circulation even in the absence of the thymus (in nude mice). Committed $CD25^+$ pro-T cells are not observed in the circulation in normal or athymic mice but are efficiently generated in the thymus in normal mice. Differentiation up to the TNK stage therefore does not require the thymus. However, apart from its compartmentalization function, the thymus does contribute in a major way to providing signals for completing the T lineage commitment process and for further differentiation.

Cortical and medullary thymic epithelial cells might have specific and distinct functions during T cell development, as discussed below.

The Development of the Thymus

The thymus is made up of a highly organized network of epithelial cells, mesenchymal cells, and T lineage cells and their progenitors. The thymic epithelium, fibroblasts, dendritic cells, and macrophages together make up the thymic stroma. The latter two cell types are of bone marrow origin. Dendritic cells in the thymus, however, may be of myeloid or lymphoid origin, as discussed above, and thymic macrophages may also be fairly diverse. The early, relatively undeveloped, thymic primordium is formed by the fusion of the endoderm from the third pharyngeal pouch with the ectodem of the third branchial cleft. The development of this thymic anlage depends on the inductive influence of the surrounding neural crest-derived mesenchyme. In the thick, it has been suggested that pharyngeal pouch endodermal cells alone can give rise to thymic epithelial cells. In the mouse, both the ectoderm and the endoderm are believed to contribute to the development of these epithelial cells. In the *nude* mouse, discussed below, ectodermally derived epithelial cells fail to develop. The importance of the neural crest-derived mesenchyme in inducing the development of the thymic primordium has been supported by experimental ablation studies. In humans, the DiGeorge syndrome (discussed in Chapter 12) might result from a defect in neural crest-derived mesenchyme. Mice in which the *Hox-a3/Hox 1.5* gene has been knocked out have athymic embryos along with other craniofacial abnormalities. Whether the underlying defect is in cells of ectodermal, endodermal, or mesenchymal origin remains to be established. However, the similarity of the thymic defect to that noted in the DiGeorge syndrome suggests a defect at the level of the mesenchymal induction of thymus development. A list of some of the known factors in thymus development is provided in Table 9.1, later in this chapter.

The thymic primordium is invaded by blood vessels and associated mesenchyme of mesodermal origin. Once the blood supply to this organ has been established, bone marrow-derived macrophages, thymic dendritic cells, and thymic progenitors can enter. Further development of the thymus into distinguishable cortical and medullary regions requires the development of early T cells, as discussed below. Immunohistochemical markers in a fully developed thymus permit the delineation of a number of distinct types of thymic epithelial cells. These include cortical epithelial cells, medullary epithelial cells, capsular cells that line the mesenchymal capsule of the thymus, and distinguishable epithelial cells at the corticomedullary junction.

A number of genes have been identified that specifically affect the development of thymic epithelial cells. Genes that are critical for epithelial cell development include the *whn* gene and *pax-1*. E-cadherin also probably contributes to proper thymic architecture and function. Normal hair growth and thymic development are disrupted in *nude* mice and rats. These rodents carry a mutation in the *winged helix* or *whn* gene. The Whn protein is a member of the winged-helix or forkhead group of transcription factors. These proteins contain an N-terminal presumptive DNA-binding winged-helix domain and a negatively charged, proline-rich C-terminal domain that might participate in transcriptional activation. Mutations in the winged-helix domain in members of this family induce homeotic transformations in *Drosophila*. A member of this family of proteins is critical for vulval development in *Caenoshabditis elegans*. In the *nude* mouse, mutations are found in the winged-helix (DNA binding) domain of the *whn* gene (this gene is also defective in the *nude* rat). Mice in which *whn* was knocked out by homologous recombination have a phenotype identical to the *nude* mouse. The *whn* gene is normally expressed by thymic epithelial cells starting from embryonic day 10, about 2 days before T cell precursors begin to enter the thymus. The primitive thymic anlage is formed in the absence of *whn*, but the thymic epithelium does not develop the ability to support the entry or survival of T cell precursors. The Whn protein probably plays a role in the differentiation or maturation of thymic epithelial cells and is believed to be critical for the further differentiation of epithelial cells of ectodermal origin.

The *pax-1* gene is a murine homolog of the *Drosophila paired* and *gooseberry* genes and encodes a transcriptional regulator. This gene contributes to the development of cortical epithelial cells that are derived from the endoderm of the third pharyngeal pouch. Mice with *pax-1* mutations (*undulated* mice) have defects in vertebral column development and T cell generation, the latter resulting from abnormalities in the development of the thymic cortex. E-cadherin in thymic epithelial cells may be required for homotypic interactions and the generation of proper epithelial architecture. In experiments involving the use of antibodies in a fetal thymic organ culture system, blocking E-cadherin–dependent homotypic interactions led to a marked impairment in T cell generation.

Early T cell commitment and development occur in the thymic cortex. Interestingly, cortical epithelial cells in the primitive thymic stroma are essential for T cell development, and in turn, the development of the cortex and medulla are both influenced by developing T cells. Cortical epithelial cells promote thymocyte survival and the initiation of V(D)J rearrangement of TCR genes. These cells provide stem cell factor (SCF; the c-Kit ligand) and IL-7, which apparently influence similar molecular events in developing T cells, although SCF initiates signals via the c-Kit receptor tyrosine kinase and IL-7 activates Jak-STAT signaling (as discussed in Chapter 4). These cells also express ligands for Notch1 and may secrete specific Wnt proteins such as Wnt 1 and Wnt 4. Although T cell development can proceed in the absence of either γc (which is essential for IL-7 signaling) or c-Kit, T cell development is abrogated in mice that lack both c-Kit and γc. In mice expressing a CD3ε transgene, there is a severe defect in the development of the thymic cortex, and T cell development is blocked at the TLP stage. This underscores the important role of the thymic cortex in T cell commitment and the initiation of TCR gene rearrangement. It should be emphasized that CD3 components including CD3ε have no known function prior to the formation of the pre-T receptor, and the overexpressed CD3ε presumably interferes with signaling via pathways that are relevant earlier, such as those downstream of c-Kit or the IL-7 receptor. This engineered block in thymus development also suggests that early thymocytes may contribute to the development of thymic epithelial cells. This particular defect in cortical development can be corrected by providing HSCs early in life but not later, suggesting that developing thymocytes provide the appropriate signals for cortical development (Figure 9.3).

Development of the thymic medulla is regulated both by genes expressed in medullary epithelial cells as well as by trophic influences provided by more mature thymocytes. In the absence of the *relB* gene (a member of the NFκB family), thymic medullary epithelial cells that are UEA-1$^+$ (UEA-1 is a lectin that identifies this subset of epithelial cells) are not seen, and dendritic cell differentiation is defective. CD8α$^-$ myeloid dendritic cells are not produced, because of a cell intrinsic requirement for RelB. CD8α$^+$ lymphoid dendritic cells are not observed in the thymus, primarily because of the defect in development of thymic medullary epithelium. These lymphoid dendritic cells do not intrinsically require RelB; if *relB*$^{-/-}$ progenitors are transferred to a normal thymic environment, myeloid dendritic cells do not develop, but the development of lymphoid dendritic cells is unimpaired.

Dendritic cells and thymic medullary epithelial cells participate in antigen presentation and are particularly important in the presentation of peptides involved in negative selection. In these *relB*$^{-/-}$ mice, commitment to the T lineage and the development of thymocytes is generally unaffected, although the mice are likely to have deficits in negative selection (an issue that requires further analysis). In *rag-1*$^{-/-}$ or *rag-2*$^{-/-}$ mice as well as in SCID mice, T cell development is blocked at the CD44$^-$CD25$^+$ stage because of the absence of pre-T receptor–derived signals (discussed below). In these mice, cortical development is intact but the medullary compartment fails to develop. Medullary areas are also missing in the *TCRα*$^{-/-}$ mouse in which development is blocked at the double-positive CD4$^+$CD8$^+$ stage. This medullary defect can be corrected by the introduction of normal mature T cells.

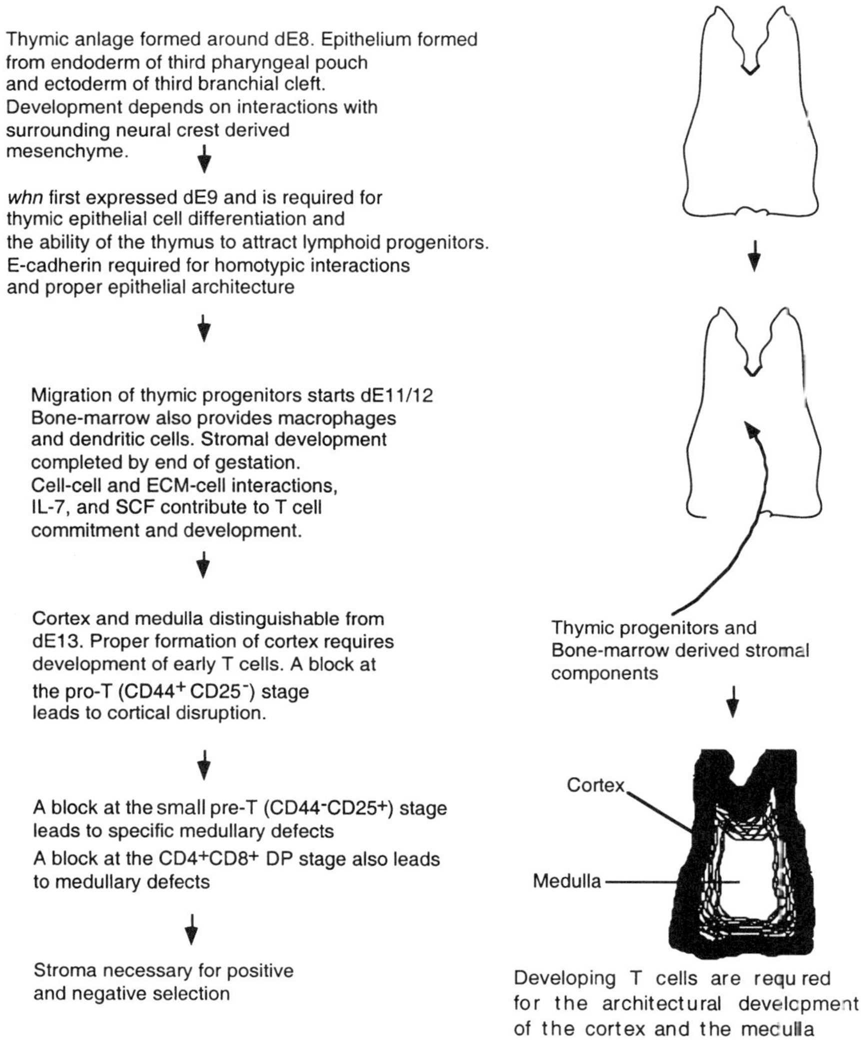

FIGURE 9.3. Thymus development is a two-way street. The thymic cortex provides signals for thymocyte development. Trophic influences from developing thymocytes are required for the generation of the thymic cortex as well as the medulla.

Thymic Function and T Cell Production Persists Late in Adult Life

The thymus is known to involute with age. It is often assumed that most T cells are generated in young animals and that new T cell generation does not occur late in adult life. However, it should be kept in mind that some thymic tissue has been noted in humans autopsied at a fairly advanced age (as old as 107 years). Studies looking for deleted circles during TCRα gene rearrangement have revealed that new T cells are produced late in life. During deletional V(D)J rearangement, the deleted DNA is joined at the signal ends of recombination signal sequences (RSSs) (see Chapter 3) and is maintained in cells as an episome, though it is not duplicated during mitosis. The presence of these circles in naive T cells can be used as a measure of recent thymic emigrants. There is a decline in recent thymic emigrants with age and a drastic reduction after thymectomy. Using this approach, it has been demonstrated that patients with human immunodeficiency virus (HIV) infection can also generate new peripheral T cells after effective antiviral therapy.

The issue of new T cell generation with age has been conveniently examined in the chick because of the discovery of a cell-surface marker that is expressed only on recent thymic emigrants. These studies have revealed that while fewer new T cells are generated with age, T cell generation continues at some level throughout adult life.

Some T cell function persists after thymectomy, and it is possible that a small number of new T cells are generated after thymectomy. It is possible that new T cells could be generated in extrathymic sites (discussed below) such as the bone marrow and intestinal cryptopatches. It is equally likely that no thymectomy is ever perfect; perhaps some residual thymic tissue persists after the most meticulous extirpation.

Differences Between the Development of T Cells From Fetal Liver and Bone Marrow-Derived Progenitors

As noted in Chapter 2, fetal liver-derived hemopoietic stem cells (HSCs) are functionally distinct from bone marrow-derived HSCs. The thymus is colonized by waves of progenitors in embryonic life. Progenitors that seed the thymus between dE10 and dE13 generate all the T cells produced by the thymus during fetal development and for the first week after birth. Precursors that seed the thymus after dE13 provide a second wave of T cells that begin to emerge after the first week of life.

All Vγ3 γδ T cells that are found in the skin of the adult mouse (Thy-1^+ dendritic epidermal γδ cells, DECs) are derived from fetal liver-derived HSCs responsible for the the first wave of T cells during embryonic life. Adult bone marrow-derived HSCs are incapable of giving rise to Vγ3 rearrangements, and all DECs are therefore of fetal origin. Why fetal HSCs and a fetal thymic environment are both necessary for the generation of Vγ3^+ DECs and why adult stem cells cannot generate these rearrangements remain unanswered questions.

As discussed in Chapter 2, fetal liver-derived stem cells do not give rise to terminal deoxynucleotidyl transferase (TdT)-containing T and B cell progenitors. Fetal HSC-derived T cells therefore lack N regions. Adult HSCs contribute to more $CD4^+$ than $CD8^+$ cells compared to fetal HSCs, presumably because the presence of N regions biases positive selection driven by MHC class II molecules.

Early T Cell Development

Apart from CLPs, which are seen in the fetal liver and bone marrow, two distinct-type progenitors derived from HSCs may also exist. A progenitor has been described that can give rise to macrophages, T cells, and B cells. This progenitor may be identical to a B/macrophage progenitor described many years ago, whose T cell potential had not been previously explored. This T/B/macrophage progenitor may well represent a stage of development that just precedes the development of the CLP, although it might also represent a distinct type of precursor that does not go through the CLP stage. Similarly, the existence of a T/macrophage precursor has also been suggested, and it remains unclear if this represents a distinct stage or a less fully characterized precursor of a CLP.

The earliest thymic T cell progenitor is also known as a thymic lymphoid progenitor (TLP) because it can give rise to all lymphoid lineages. It is functionally similar to bone marrow CLPs, but may have distinct surface characteristics (TLPs may express higher levels of CD43 and CD44 than CLPs, for instance). TLPs in adult mice are distinguished from other $CD44^+CD25^-$ cells by their $CD3^-CD4^{lo}CD8^-Thy\text{-}1^{lo}$ phenotype (see Figures 9.2 and 9.4). Why low levels of CD4 are expressed by these progenitors is unclear. It is unlikely that the CD4 molecule plays a functional role at this stage, and indeed, T cell development through these stages is not affected in any way in mice lacking CD4. In the embryonic thymus, the earliest identifiable progenitor cell does not express CD4. These $CD4^{lo}$ bone marrow-derived progenitors seen in the adult thymus are capable of differentiating into other lymphoid lineages, but they become progressively "more" committed to the T lineage as they mature in the thymus (Figure 9.4). Committed T cell progenitors either may develop outside the thymus and migrate to this organ, or they may be generated in the thymus. Most T cells are probably derived from precursors that commit to the T lineage in the thymus itself.

In Chapter 2, we examined the role of transcription factors such as SCL/Tal-1, AML1, c-Myb, GATA-2, PU.1, and Ikaros in the generation of progenitors of B and T lymphocytes. HES1 is a bHLH protein that is activated downstream of Notch signaling (see Chapter 4). In mice that lack HES1, development of the thymus is compromised. The defect in these mice lies at the level of the generation of the earliest T cell precursors, presumably because inductive signals from Notch are not acted upon. Another transcription factor that plays a critical role in the development of T cells,

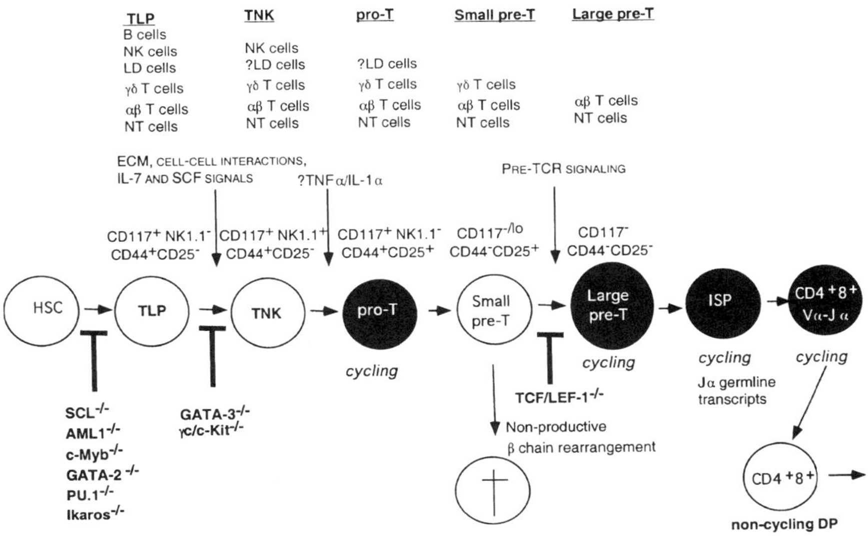

FIGURE 9.4. An overview of some of the early events in T cell development. Filled circles represent cycling cells. TLPs ($CD44^+CD25^-$) are committed to all lymphoid lineages, but TNK cells (also $CD44^+CD25^-$; they also express NK 1.1) cannot give rise to B cells. Pro-T cells ($CD44^+CD25^+$) are unable to generate B or NK cells, and early pre-T ($CD44^-CD25^+$) cells are no longer able to give rise to thymic dendritic/lymphoid dendritic (LD) cells. After pre-TCR signaling, developing T cells will infrequently give rise to γδ T cells but cannot do so after TCRα gene rearrangement. Signals generated via the common γ chain and c-Kit are essential for thymocyte development. The γc is a component of the receptors for many cytokines, including IL-7. Cortical epithelial cells trigger T cell progenitors by triggering Notch1, secreting IL-7 and SCF, and by less well understood cell–cell interactions. IL-7 provides survival signals and probably indirectly initiates V(D)J rearrangement at the β, δ, and γ loci in early pre-T cells ($CD44^-CD25^+$). In cells that commit to the αβ lineage, pre-TCR signaling leads to the successive generation of cycling large pre-T ($CD44^-CD25^-$), immature single positive (ISP), and $CD4^+$ $CD8^+$ double positive (DP) cells. Eventually, noncycling DP cells differentiate further into $CD4^+$ and $CD8^+$ single positive T cells, as decribed elsewhere in this chapter. Exactly when the capacity to generate LD cells is lost is unclear. Based in part on Godfrey et al., 1993, Shortman and Wu, 1996, and Carlyle and Zuniga-Pflucker, 1998.

but not of other lineages is GATA-3. Development of the earliest committed thymic T lineage progenitors is abrogated in the absence of GATA-3. This zinc-finger transcription factor is expressed in the developing kidney and nervous system as well as in thymocytes and T cells. *Gata-3* null mice die at day 12 of embryogenesis. However, using chimeric mice in which the hematopoietic stem cells were *Gata-3* null, it has been demonstrated that while development of all other hematopoietic lineages is normal, T cell development is abrogated and CD3$^-$CD4$^-$CD8$^-$ T cell progenitors in the thymus are not seen. As we shall see in Chapter 10, GATA-3 is also required later in T cell development. Table 9.1 summarizes critical signals for T cell development prior to the pre-TCR checkpoint.

TLPs can be distinguished from most other TN cells by their c-Kit$^+$CD44$^+$CD25$^-$ phenotype. c-Kit$^+$CD44$^+$CD25$^-$ cells are heterogeneous, and TLPs (which in adult mice express low levels of CD4) develop into TNK cells, which do not express CD4. Both the common lymphoid progenitor and TNK cells express Sca-2 and low levels of Thy-1. Sca-2 is not found on bone marrow HSCs and is a glycosylphosphatidylinositol-anchored protein of the Ly-6 family. It is important to define TLPs and TNK progenitors as cells that express c-Kit in addition to being CD44$^+$CD25$^-$. TN thymocytes have been described that are CD44$^+$CD25$^-$ but are c-Kit$^-$. These cells are not progenitors but include more mature αβ T cells that lack CD4 and CD8. TNK cells are fairly plastic and have the ability to differentiate into αβ or γδ T cells, cells, NK cells, NT cells and probably thymic dendritic (lymphoid dendritic, LD) cells as well. TNK cells expresses *rag-1* and *rag-2*, but their β, γ, and δ TCR genes remain in germline configuration. Although in Figures 9.2 and 9.4, it is assumed that pro-T cells may retain the capacity to generate LD cells, it is unclear exactly when during development the ability to generate these cells is lost.

In studies employing cytokines and blocking antibodies in a fetal thymic organ culture system, TNF and IL-1α have been shown to be able to significantly influence the conversion of TNK cells (CD44$^+$CD25$^-$) into CD44$^+$CD25$^+$ pro-T cells. Although no inhibition of differentiation has been observed in either TNF or IL-1α null mice, these mice will need to be bred together to determine whether there is any impairment of early T cell development in the combined absence of these cytokines. The CD44$^+$CD25$^-$ pro-T cell represents a proliferative stage of thymocyte development. These cells may receive signals from IL-7 and SCF and probably from other poorly defined thymic stromal factors as well. IL-7 provides a Bcl-2-dependent survival signal and may directly or indirectly influence accessibility for V(D)J rearrangement, while SCF probably provides proliferative signals. A transcriptional regulator that might be important downstream of IL-7R or c-Kit signaling is the δEF1 protein. This transcriptional regulator contains two Kruppel-type zinc fingers and a homeodomain. It is the murine homolog of the human ZEB protein (briefly discussed in Chapter 2) and was originally identified as a factor that binds the enhancer of the chicken δ1 crystallin gene. The δEF1 protein binds to CACCT motifs and may generally function as a competitive transcriptional repressor of bHLH proteins. Although the actual targets of δEF1 are not known, mice in which the δEF1 gene has been targeted have major defects in thymocyte development with a decrease in cells noted from the TNK stage on. Total thymic cellularity is reduced to about 1% of normal. This gene may affect development at a numbers of stages during T cell ontogeny.

TABLE 9.1. Signals that drive T cell development prior to the pre-TCR checkpoint

Factor	Comments	References
Factors involved in thymic epithelial development		
Winged helix/nude	Thymic epithelial defect in *nude* or *whn*$^{-/-}$ mice. No development of T cell precursors.	Nehls et al., 1994, 1996
Hox-a3/Hox 1.5$^{-/-}$	*hox-*$\alpha 3^{-/-}$ mice have athymic embryos and craniofacial abnormalities. Hox-a3, Hox-b3 and Hox-d3 contribute to the migration of the thymic anlage.	Manley and Capecchi, 1995, 1998
Pax-1	*pax-1* mutant (*undulated*) mice have defects in thymic cortical epithelial and T cell generation as well as vertebral column defects.	Wallin et al., 1996
RelB	In *relB* null mice, UEA-1^{+} medullary epithelial cells and thymic myeloid-related CD8α^{-} DC fail to mature. Lack of CD8α^{+} lymphoid DC is secondary to thymic epithelial defect. No defect in commitment to T lineage. Impaired negative selection later in thymic development.	Burkly et al., 1995 Wu et al., 1998
E-cadherin	The use of antibodies in fetal thymic organ culture suggests that E-cadherin is required for homotypic interactions between epithelial cells and thus indirectly for thymocyte development.	Muller et al., 1997
Other factors involved in early T cell development		
Notch 1	Drives T cell commitment at expense of B cells. Induced deletion leads to absence of T cells.	Radtke et al., 1999 Pui et al., 1999
HES1	Absence of HES1 leads to abrogation of development of earliest T cell precursors.	Tomita et al., 1999
IL-7 and SCF	In combined γc/c-Kit null mice, pro-T cell expansion is markedly impaired.	Rodewald et al., 1997
$\alpha 4\beta 1$ integrin	In the absence of $\alpha 4$ integrins, fetal T cell development is unaffected, but adult T cell development does not occur.	Arroyo et al., 1996
GATA-3	In the absence of GATA-3, T cell progenitors are unable to commit to the T lineage.	Ting et al., 1996

Rearrangement of *TCR β*, *δ*, and *γ* genes is initiated and completed at the CD44⁻CD25⁺ small pre-T stage. Induction of *TCR β* gene rearrangement requires that the β chain locus be made accessible for recombination. It is known from germline deletion studies that rearrangement at this locus requires the TCRβ enhancer (Eβ). It was generally assumed that Eβ is required for the generation of accessibility in the vicinity of the *TCR β* gene. More recent studies examining accessibility in the presence and absence of this enhancer suggest that, while Eβ is required for rearrangement of the β locus, it may not be a critical contributor to accessibility. It has been suggested that Eβ might be essential for the recruitment of a general DNA repair factor. It is not clear which specific motifs in Eβ (Figure 9.5) are required for the facilitation of recombination. When an αβ T cell completes the program of β chain gene rearrangement, it no longer expresses c-Kit (pre-T cells are CD117⁻). At this stage, cells become dependent on a signal from the pre-T receptor for survival and further differentiation.

Induction of *TCR δ* gene rearrangement requires the TCR δ enhancer (Eδ). As seen in Figure 9.6, the *TCR α* and *δ* genes share a common locus. The reason they share a common locus may have to do with the fixing of

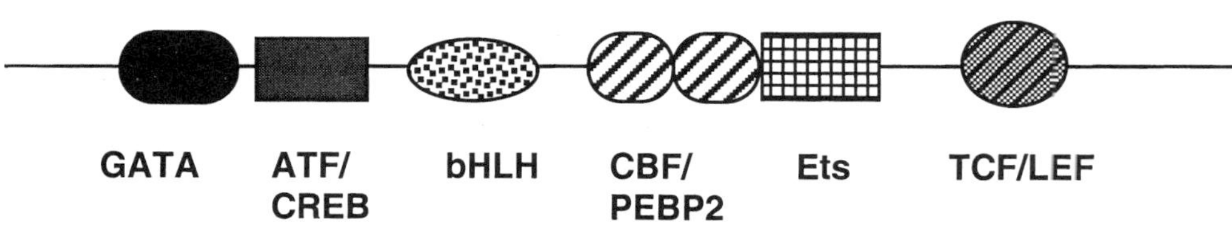

FIGURE 9.5. Some of the presumed regulatory motifs in the TCRβ core enhancer region. While this enhancer is required for the rearrangement process, it has been suggested that it is not required for accessibility but might contribute to the recruitment of some essential DNA repair factor.

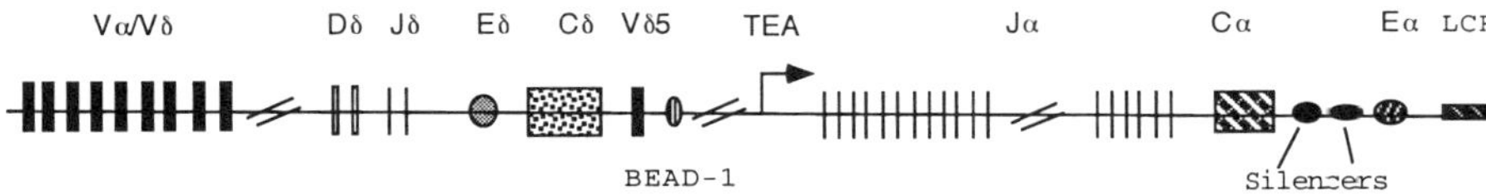

FIGURE 9.6. The *Tcr α/δ* locus. Vα/Vδ segments are located at the 5′ end of this locus. The δ locus is rearranged early in T cell development, and the process is nonordered (recombination may start as a V to Dδ event or a Dδ to Jδ event). The BEAD-1 element is an insulator, which presumably prevents Eδ (the enhancer that regulates accessibility and rearangement at the δ locus) from opening up the Jα region early in T cell development. After pre-TCR signaling, in an αβ committed T cell, rearrangement at the α locus deletes any rearranged TCR δ unit, thus ensuring that commitment is irreversible. Regulatory regions in the α locus include an enhancer (Eα), two silencer elements, and an LCR.

the commitment process in αβ T cells (by the deletion of rearranged δ genes), as discussed in the following section. *TCR δ* gene rearrangement occurs early in development, and *α* gene rearrangement is initiated later, following pre-TCR signaling. *Vα/Vδ* gene segments are located at the 5′ end of the locus, and early in development, Eδ is activated, presumably as a result of signals received in part via the IL-7R and c-Kit. This enhancer is required for the generation of local accessibility prior to *TCR δ* gene rearrangement, which (unlike *TCR β* or *IgH* gene rearrangement) is a nonordered process and may in a given cell first involve a V to Dδ recombination event or a Dδ to Jδ rearrangement. As evidenced from the Eδ knockout mouse, Eδ is absolutely required for Dδ to Jδ joining, but some other unidentified regulatory element may be necessary for V to Dδ recombination. This enhancer contains a c-Myb site adjacent to a CBF2 site and also contains GATA sites. The c-Myb and CBF2 sites have been shown to be important for the induction of *TCR δ* gene recombination in studies performed with mutant transgenes. Sites in Eδ that regulate recombinational accessibility at this locus remain to be identified. Mice lacking c-Myb and CBF2 have significant hematopoietic defects (as discussed in Chapter 2), and conditional stage-specific deletion of these genes will have to be undertaken to elucidate their role in T lymphopoiesis. In addition, "knock-in" experiments to mutate endogenous c-Myb and CBF2 sites in Eδ will help confirm the importance of these transcriptional regulators in inducing V(D)J recombination at the *TCR δ* locus.

A very interesting facet of gene regulation has been revealed from an analysis of *TCR δ* gene rearrangement. While Eδ contributes to accessibility at the δ locus, it is critical that this enhancer not "open-up" the adjacent Jα segments at the time that TCR δ rearrangement occurs. As seen in Figure 9.6, there is a regulatory element known as BEAD-1 (for boundary element Alpha Delta-1) that lies between the δ locus and downstream Jα segments. This boundary element functions as an insulator, preventing Eδ from opening up the Jα region even as it makes the upstream δ region accessible for recombination at the pro-T stage. The function of BEAD-1 was originally examined in stable transfectants in cell lines. However BEAD-1 is not required for the regulation of murine *TCR α* and *δ* gene expression. In the cell-line context, this insulator element can be functionally replaced by *Drososphila* insulator elements. It is assumed that insulators bind specific proteins and impede the spread of chromatin accessibility in a poorly understood fashion.

While nomenclatures vary, it would be appropriate to define cells of the αβ lineage that have productively rearranged their *TCR β* genes, but have yet to rearrange their *TCR α* genes, as pre-T cells. These include T cells that are $CD44^{-}CD25^{+}$ and that have just rearranged their *TCR β* genes, proliferating cells that are $CD44^{-}CD25^{-}$, and also proliferating cells at the immature single positive (ISP) stage, which generally express low levels of CD8 (or CD4 in humans, and in some strains of mice). The pre-T cell receptor (pre-TCR) is generally described as driving cells through the double-negative to double-positive transition. While this is correct in a broad sense, in actual fact, in the total absence of pre-T receptor signaling, cells remain frozen at the DN $CD44^{-}CD25^{+}$ stage and do not increase in size and begin to proliferate or make it to the DN $CD44^{-}CD25^{-}$ stage. Depending on the "severity" of the block, mutations in different genes can block pre-TCR–dependent development at different steps prior to the DP stage. In wild-type mice, the pre-TCR is expressed at low levels on cycling DP T cells that are actively rearranging their *TCR α* loci. Pre-TCR signaling will be discussed in a later section.

Commitment to the αβ or the γδ Lineage

A developing double-negative T cell may rearrange its *TCR β*, *γ*, and *δ* genes around the same time. What drives a cell to develop as an αβ T cell or a γδ T cell? How is the process regulated? γδ T cells may develop in the thymus or extrathymically, as discussed below. They are not restricted by classical MHC class I or class II molecules and do not develop via an intermediate DP stage. γδ Cells either express no CD4 or CD8, or they may sometimes express CD8. However, unlike αβ CD8 T cells that express CD8αβ heterodimers, γδ T cells express CD8αα homodimers. While we appreciate that αβ and γδ T cells develop quite differently, and have some understanding of how commitment to the αβ lineage may be sustained (once it has occurred), fundamental questions about the commitment process remain unanswered. As has been stressed elsewhere in this book, it is most useful to think of most commitment events as being gradual and stepwise, with true irreversibility being attained late in the day. Both the αβ versus γδ commitment event, as well as the CD4 versus the CD8 lineage commitment decision, which we will discuss later in this chapter, do not occur in a sudden and precipitate manner. Simplified models, while of heuristic value, do not necessarily fit these processes in an entirely convincing manner.

Instructive models suggest that the productive rearrangement of the appropriate TCR genes generates signals that drive a developing T cell to commit to the lineage represented by the rearranged genes. *Stochastic models* suggest that progenitor T cells randomly commit to becoming either γδ or αβ cells before *TCR* gene rearrangement is initiated. If a cell commits to the γδ lineage, for instance, stochastic models suggest that such a committed cell will be incapable of developing into an αβ T cell, even if one of its *TCR β* loci is productively rearranged (Figure 9.7).

In instructive models, productive rearrangement of both the δ and γ chain genes in a given cell might commit it to the γδ lineage. Commitment to the αβ lineage presumably occurs if an in-frame rearrangement at the *TCR β* locus precedes the generation of in-frame rearrangements at the γ and δ loci. Some support for this model has been obtained from studies that suggest that very few productive Vγ to Jγ rearrangements are seen in αβ T

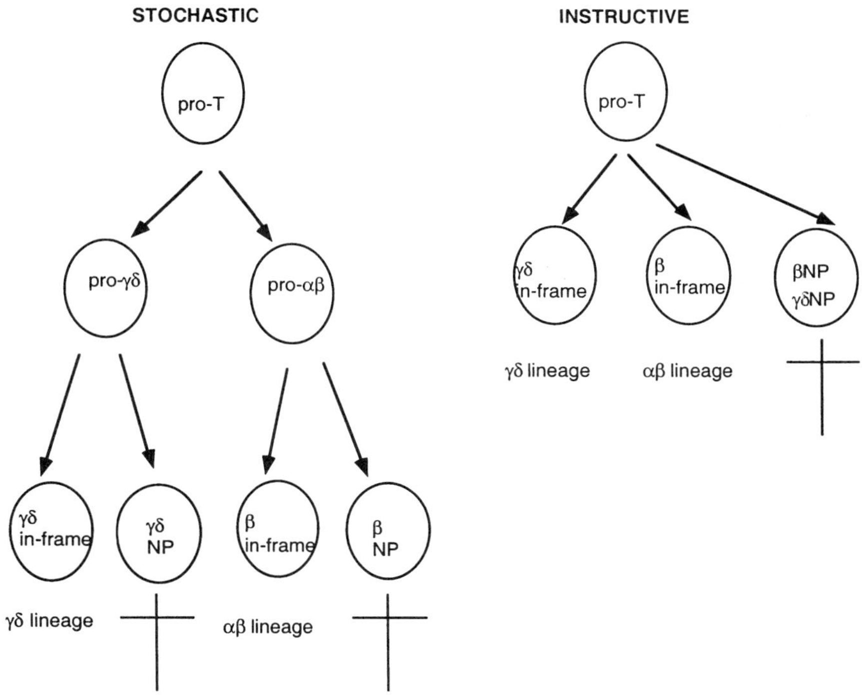

FIGURE 9.7. Instructive and stochastic models for the αβ versus γδ lineage choice. In instructive models, based on which of their TCR genes are productively rearranged, developing T cells receive signals to commit to the appropriate T lineage. In stochastic models, cells commit to a lineage prior to the rearrangement process. Committed cells that make the "right" rearrangements then receive survival signals. The αβ versus γδ decision might occur stochastically, although lineage choice is best explained by a "leaky instructive" process.

cells. In-frame V-DJβ rearrangements, however, are sometimes observed in γδ T cells. It appears that, even if a pro-T cell rearranges its *TCR β* chain productively, it may sometimes commit to the γδ lineage; however, single-cell polymerase chain reaction (PCR) analyses suggest that most often instructive signals are delivered following productive TCRβ chain rearrangement, and these result in commitment to the αβ lineage. Similarly, productive rearrangement of the *TCR δ* chain does not necessarily commit a cell to the γδ lineage. However, some more recent analyses suggest that expression of the TCR γ and δ genes in developing T cells does impede, at least in a statistical sense, these cells from developing into αβ T cells. This provides further support, though unfortunately not in a black and white manner, for instructional models.

αβ T cell development involves the generation of DP T cells, whereas during γδ T cell development, cells retain the DN phenotype. Occasionally, productive γ and δ rearrangements might occur in the absence of a productive β rearrangement. In such a cell, the $\gamma\delta$ TCR may provide signals for movement to the DP stage (which is characteristic of $\alpha\beta$ T cells but not $\gamma\delta$ T cells), as witnessed in transgenic experiments. Productively rearranged TCR γ chains may also associate with the preTα chain, albeit fairly inefficiently, and form an atypical pre-TCR. However, expression of the TCRγ chain tends to be downregulated when cells differentiate along the $\alpha\beta$ developmental pathway, and these "surrogate" pre-TCRs, which contain the TCR γ chain, are not very effective in driving development along the $\alpha\beta$ pathway. While some signals may well be generated in a stochastic way, many molecular events probably need to be completed to irreversibly commit a cell to the $\alpha\beta$ or $\gamma\delta$ lineage. The available data are consistent with a "leaky" instructive model, which takes into account that commitment is a gradual process. The single event that appears to be critical for irreversible commitment to the $\alpha\beta$ lineage may be the assembly of the pre-T receptor.

Details of the molecular changes that are necessary for a cell to commit to the $\alpha\beta$ lineage remain unknown. It has been suggested that the random induction of expression of the pTα component of the pre-T receptor may commit a developing T cell to the $\alpha\beta$ lineage, though no experimental data support such a model. However, when a cell productively rearranges its TCR β chain, signaling via the pre-TCR may fix the commitment of such a cell to the $\alpha\beta$ lineage. Pre-TCR signaling probably induces accessibility at the *TCR α* locus, as discussed below. Since the *TCR α* and *δ* loci are interwoven (see Figure 9.6), a V to Jα rearrangement will delete any rearranged V-DJδ unit in the same locus, thus ensuring irreversible commitment to the $\alpha\beta$ lineage.

While at present no evidence suggests that pTα expression is selectively induced in some early thymocytes and not in others, it seems very likely that the productive rearrangement of both the TCR γ and δ genes in early T lineage cells may drive development along the $\gamma\delta$ lineage. The data, however, are equivocal and at best consistent, as suggested above, with a leaky instructive model. A major stumbling block to the acceptance of an instructive model, at least for $\gamma\delta$ T cell development, is the fact that $\gamma\delta$ transgenic mice do make $\alpha\beta$ T cells. It is likely therefore that signals from the pre-TCR are "dominant" and may play a major instructive role in a T cell's decision to commit.

The nature of the signal generated by $\gamma\delta$ receptors during T cell development is poorly understood. It is clear, however, that such a signal might be capable of influencing commitment even as late as the small pre-T (noncycling $CD44^-CD25^+$) stage. This signal, unlike that generated by the pre-TCR, does not induce selected cells to proliferate in a significant way but may influence commitment and further differentiation along the $\gamma\delta$ pathway. We will return to the issue of $\gamma\delta$ T cell development later in this chapter.

It has been suggested that the Notch pathway may be involved in driving the αβ versus γδ decision. Studies examining T cell development in mice expressing a constitutively active Notch transgene suggest that Notch signaling might favor αβ T cell development. The ability of an activated Notch allele to influence αβ T cell development does not necessarily suggest that Notch is involved in commitment to the αβ lineage but could be consistent with a role for Notch signaling downstream of or in parallel with the pre-T receptor.

Positive Selection I and Allelic Exclusion Are Mediated by the pre-T Receptor

The pre-TCR is made up of the β chain of the T cell receptor, a surrogate α chain known as pre-Tα/gp33 (pTα), and the associated CD3 complex (Figure 9.8). pTα is a 33-kd type I membrane glycoprotein which is expressed in pro-T and pre-T stages prior to *TCR α* chain gene rearrangement. This protein has a fairly long cytoplasmic tail that could potentially participate in signaling events. In the absence of pTα, positive selection I in the T lineage is abrogated, T cells are frozen at the DN stage, and allelic exclusion fails to occur at the *TCR β* locus. This defect can be rescued by the transgenic introduction of a truncated pTα chain, which lacks most of the cytoplasmic tail. This suggests that the cytoplasmic tail of pTα is not essential for pre-TCR signaling. However, keep in mind that the rescue experiment involved a very high degree of overexpression of the tail-deleted *pTα* transgene. It is possible that a signaling role for the cytoplasmic tail of pTα may be revealed by appropriately engineering the endogenous gene by a "knock-in" approach, but this remains to be demonstrated. However, since the cytoplasmic tails of murine and human pTα are not conserved, it is likely that this portion of the protein is not critical during T cell development.

Signal transduction by the pre-TCR not only mediates a survival function and therefore promotes the DN to DP T cell transition but also contributes to allelic exclusion at the β chain locus of the T cell receptor. All the issues that we have considered in connection with the pre-B receptor and how it functions (in Chapter 7) may be equally relevant when discussing the pre-TCR. This receptor is expressed at low levels on the surface of DN T cells. No ligand has been found for the pre-TCR, and it is very likely that preantigen receptors signal constitutively and do not require a ligand in the classical sense. Deletion of the extracellular portions of the TCR β chain and of pTα does not compromise the ability of the pre-TCR to drive differentiation or to mediate allelic exclusion. As discussed in Chapter 7, preantigen receptors, unlike most other cell surface receptors, are designed to monitor whether the right kind or rearrangement has been

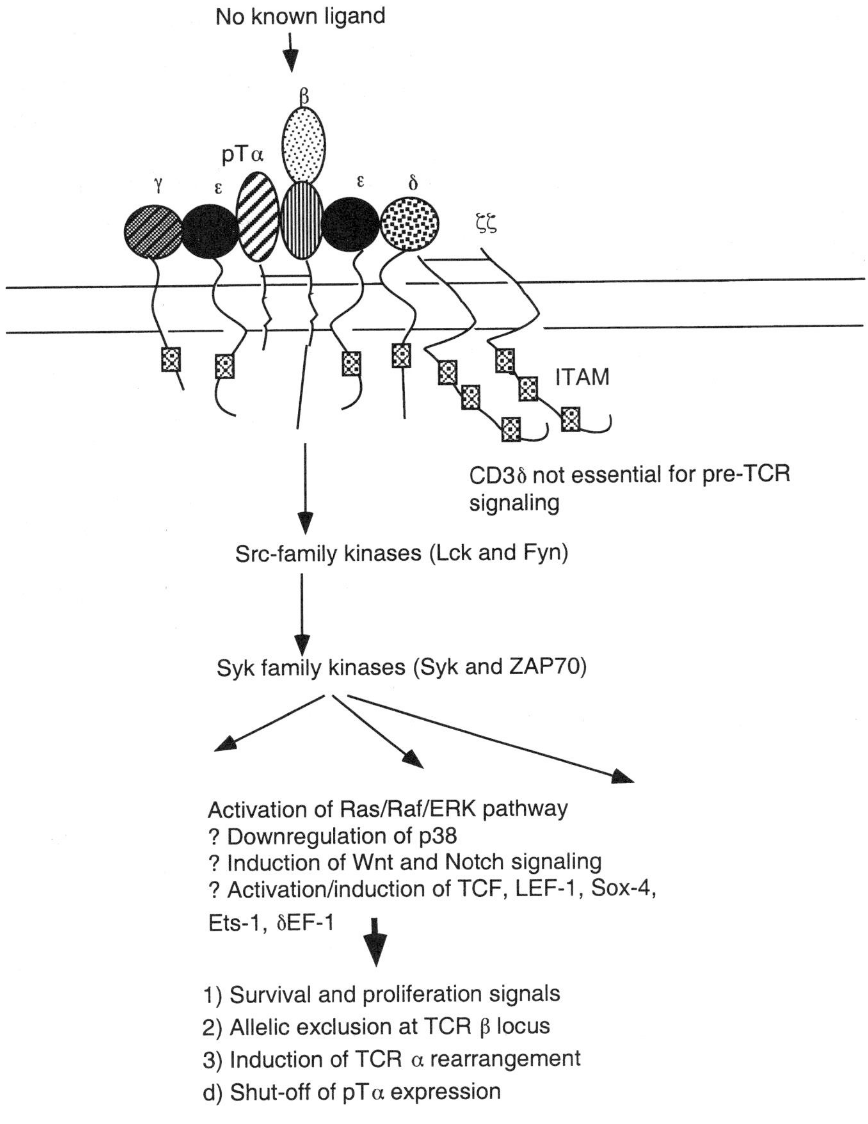

FIGURE 9.8. The pre-T receptor induces multiple signaling programs during the double negative to double positive thymocyte transition.

made and are not really designed to respond to external cues. Nevertheless, pre-T receptors that are engineered to be intracellularly retained (and therefore might associate more readily with chaperones than with signaling proteins) fail to provide differentiation signals. While these results appear to suggest a need for the pre-TCR to be expressed on the cell

TABLE 9.2. Mutations that block development at the pre-TCR checkpoint

1. Mutations that abrogate TCR β rearrangement	
a. *Rag-1*$^{-/-}$	Mombaerts et al., 1992a
b. *Rag-2*$^{-/-}$	Shinkai et al., 1992
c. *Ku80*$^{-/-}$	Nussenzweig et al., 1996; Zhu et al., 1996
d. *DNA-PK*$^{-/-}$	Bosma and Carroll, 1991; Danska et al., 1996
2. Mutations in pre-T receptor components or in their expression	
a. $TCR\beta^{-/-}$	Mombaerts et al., 1992b
b. $E\beta^{-/-}$	Bories et al., 1996; Bouvier et al., 1996
c. $pT\alpha^{-/-}$	Fehling et al., 1995
d. $CD3\varepsilon^{-/-}$	Malissen et al., 1995
e. $CD3\gamma^{-/-}$	Haks et al., 1998
f. $CD3\zeta/h^{-/-}$ (partial block)	Love et al., 1993; Malissen et al., 1993; Ohno et al., 1993; Ardouin et al., 1998; Shores et al., 1998
g. $CD3\delta^{-/-}$ (no block)	Berger et al., 1997
3. Signaling molecules	
a. *Lck*$^{-/-}$ (partial block)	Molina et al., 1992
b. *Fyn*$^{-/-}$ (no block)	Appleby et al., 1992
c. *Lck*$^{-/-}$, *Fyn*$^{-/-}$	Groves et al., 1996; van Oers et al., 1996
d. *Syk*$^{-/-}$ (no block)	Turner et al., 1995; Cheng et al., 1995
e. *Zap-70*$^{-/-}$ (no block)	Negishi et al., 1995; Wiest et al., 1997
f. *Syk*$^{-/-}$, *Zap-70*$^{-/-}$	Cheng et al., 1997
g. *Vav*$^{-/-}$ (Block is minimal)	Turner et al., 1997
h. *SLP-76*$^{-/-}$	Clements et al., 1998; Pivniouk et al., 1998
i. *LAT*$^{-/-}$	W. Zhang et al., 1999
4. Transcriptional regulators	
a. *Tcf*$^{-/-}$ (block in adult)	Verbeek et al., 1995
b. *Tcf*$^{-/-}$, *LEF*$^{-/-}$	Okamura et al., 1998
c. *Sox-4*$^{-/-}$	Schilham et al., 1997
d. ? *Ets-1*$^{-/-}$	Bories et al., 1995; Muthusamy et al., 1995
e. ? $\delta EF\text{-}1^{-/-}$	Higashi et al., 1997
f. ? *Egr-1*	Miyazaki, 1997
5. Others	
a. *OPGL*$^{-/-}$	Kong et al., 1999
b. ? *Notch*—(Activated Notch in (T cells)	Robey et al., 1996; Washburn et al., 1997

surface, they do not prove that this is the case. The level of expression of the pre-T receptor on the cell surface is miserable! If signaling occurs constitutively, without the need for an external ligand, it is more than likely that signals may actually be transduced from intracellular sites soon after assembly of the receptor.

The pre-T receptor drives the DN to DP transition. A summary of mutations that block T cell development at the pre-TCR checkpoint is provided in Table 9.2. In the absence of a rearranged *TCR* β chain gene or of $pT\alpha$, of most members of the CD3 complex, or of critical signaling components downstream of the pre-T receptor, cells remain frozen at the $CD44^{-}CD25^{+}$

stage. In mice lacking Rag-1, Rag-2, or Ku80, development does not proceed beyond the CD44$^-$CD25$^+$ stage, presumably because these mice cannot generate a rearranged TCR β chain.

While the CD3 γ and ε chains are required at the pre-TCR checkpoint, the CD3 δ chain is dispensable. Absence of the CD3 associated ζ/η chain leads to a partial abrogation of the DN to DP transition. Mice lacking these immunoreceptor tyrosine-based activation motif (ITAM)-containing molecules do reach the DP stage but unlike wild-type DP cells these cells continue to express CD25. They do initiate *TCRα* gene rearrangement but fail to mediate allelic exclusion at the TCRβ locus. These results suggest that signaling via the CD3ζ/η proteins serves to strengthen the signals generated via the pre-TCR and that these more potent signals may be absolutely required to initiate allelic exclusion. In mice that lack Csk, T cells complete the DN to DP transition even in the absence of the pre-TCR, indicating that Src-family kinases are of critical importance at this checkpoint. The absence of Lck in knockout mice also contributes to a partial abrogation of the DN to DP transition, while the absence of Fyn alone has no discernible phenotype. However, in the absence of both Lck and Fyn, pre-T receptor signaling is not initiated and T cells remain frozen at the DN CD44$^-$CD25$^+$ stage. In the absence of either Syk or Zap-70, the DN to DP transition is not compromised, but when both Syk and Zap-70 are deficient, pre-T receptor signaling fails to occur and development freezes at the DN stage. All this evidence supports the view that pre-T receptor signaling proceeds, as expected, via the activation of Src and Syk family kinases and that CD3 γ, ε, ζ, and η chains participate in the recruitment and activation of these tyrosine kinases. The requirement for CD45 at the pre-T stage is not absolute suggesting that the magnitude of Src and Syk family kinase activation required at this transition is relatively limited. The absence of Vav also does not markedly compromise pre-T receptor mediated positive selection. The Ras/Raf/ERK pathway is also required for pre-TCR signaling and for driving DN T cells into the DP stage (see Figure 9.8). Expression of an activated form of the p38 MAP kinase can inhibit the DN to DP transition, and it has been therefore assumed that p38 negatively regulates this transition. There is, however, no reason to believe that the pre-TCR actually produces signals that downregulate p38 *in vivo*.

It is likely that pre-TCR signaling contributes to the induction or activation of a subset of transcription factors that drive the cell through the DN to DP transition. In the absence of lymphoid enhancer factor 1 (LEF-1) and TCF (T cell factor 1), both members of the HMG group of transcription factors, T cell development is blocked at the ISP stage in fetal as well as adult T cells. TCF and LEF-1 are both expressed in a number of tissues during embryogenesis; in the adult, TCF is expressed only in T cells while LEF-1 is expressed both in T cells and immature B cells. In the absence of LEF-1 alone, both fetal and adult T cell development is unaffected. In the absence of TCF alone, fetal thymic development is relatively normal, but adult thymopoiesis is blocked at the ISP stage. These transcription factors function downstream of the Wnt/Wingless signaling pathway (discussed in Chapter 4), and it is possible, therefore, that pre-T receptor and Wnt signaling pathways intersect. Wnt signaling involves the interaction of Wnt-1 with its receptor, Frizzled, which is a member of the serpentine receptor family, as discussed in Chapter 4. Activation of this receptor leads to the activation in turn of a cytosolic protein known as Disheveled. Disheveled in turn inhibits the activity of a cytosolic kinase, GSK3β. In the absence of Wnt signaling, GSK3β constitutively phosphorylates and destabilizes β-catenin. When GSK3β is inactivated, β-catenin forms a complex with transcription factors such as TCF and LEF-1, enters the nucleus, and transcriptionally activates (or represses) specific target genes. One potential target is the TCR α locus, as discussed below. However β-catenin is not required for TCF activity in the content of the *TCR-α* enhance. TCF/LEF-1 sites are found in both the TCRβ core enhancer as well as the TCRα core enhancer, but these sites are not likely to be critical for the DN to DP transition (since expression of TCRα is not required for this transition to occur). TCF-1 and LEF-1 presumably act as activators or repressors of critical genes that may be regulated downstream of the pre-TCR.

Another member of the HMG family, Sox-4, is also required, for the DN to DP transition and may, like TCF-1 and LEF-1 be relevant downstream of the pre-TCR. *Sox-4*$^{-/}$ mice have been discussed in Chapter 7 in the context of B cell development.

Interestingly, Notch signaling might also influence pre-T receptor-driven events. An activated Notch transgene can drive the DN to DP transition in TCR β-deficient mice, suggesting that Notch signaling may be downstream of pre-T receptor signaling. While Notch signaling (see Chapter 4) is known to influence Wnt-derived signals during invertebrate development, it is conceivable that these pathways intersect in pre-T cells as well. Although it is conceivable that Wnt signaling may be involved in T cell development, evidence for the involvement in development of both the Wnt and Notch pathways at the pre-T stage remains quite indirect. Wnt isoforms are expressed in the thymus and may well be involved in T cell development. The nature of vertebrate Notch ligands and the identity of the cells on which they are expressed also remain issues of conjecture at this time. In *Drosophila*, the Notch ligand, Delta—for long thought to exist exclusively as a membrane anchored protein—is cleaved from the cell surface and is released as a soluble molecule. Notch ligands in vertebrates may also exist as soluble proteins. The putative role of Notch in positive selection and lineage commitment will be discussed a little later in this chapter.

Apart from TCF-1, LEF-1, and Sox-4, it is likely that many other nuclear factors are targets of pre-TCR signaling. A potential target might be the Ets-1 transcription factor. Sites for Ets family protein binding exist in both the TCR β enhancer and the TCR α enhancer. From gene-targeting experiments, it is apparent that Ets-1 might play an important role in T cell survival. It is likely that Ets-1 is important not only downstream of the pre-TCR but possibly also at later stages of development downstream of the TCR itself. T cell survival at multiple checkpoints might be compromised in the absence of other transcriptional regulators as well. Other regulators that fit into this somewhat nebulous category include the bHLH E12 and E47 protein products of the *E2A* gene (discussed in Chapter 7; these proteins are far more critical during B cell development), and also the δEF1 transcriptional regulator, which was considered earlier in this chapter.

The osteoprotegerin ligand (OPGL) has been discussed in Chapters 6 and 7. *OPGL*$^{-/-}$ mice have no peripheral lymph nodes and have a major block in pre-B cell development. While splenic B cell numbers are markedly reduced in the absence of OPGL, splenic T cell numbers are not significantly decreased. However, thymic cellularity is considerably reduced, primarily because of a block in development from the CD44$^-$CD25$^-$ to the CD44$^-$CD25$^-$ stage of development. This developmental block is intrinsic to bone marrow-derived cells—the introduction of normal bone marrow into *OPGL*$^{-/-}$ mice restores normal thymic development. It is presumed that OPGL signaling is in some unexplained way required for pre-TCR signaling.

As has been briefly considered in Chapter 3, signals from the pre-TCR are also responsible for ensuring that only one *TCRβ* allele is productively rearranged during ontogeny. Signals from this receptor drive proliferation of CD25$^-$ pre-T cells and thus contribute to cyclin-dependent kinase (CDK)-induced phosphorylation, ubiquitination, and degradation of Rag-2. This post-translational regulation of Rag-2 may contribute to allelic exclusion at the *TCR β* locus; it probably plays an important role in transiently shutting off rearrangement (some experiments suggest that Rag-2 degradation may not be relevant for allelic exclusion, as discussed in Chapter 3). Pre-TCR signaling must also contribute to chromatin remodeling at the *TCRβ* locus to render this gene inaccessible to Rag-mediated recombination, especially while the cell is rearranging its *TCRα* genes. Although the TCRβ enhancer is required for the induction of *TCRβ* rearrangement, the cis-acting sequences in this locus that are responsible for the generation of inaccessibility (to ensure allelic exclusion) have yet to be identified.

Pre-TCR signaling is required for the initiation of rearrangement at the *TCRα* locus. It is likely that nuclear events downstream of the pre-TCR are responsible for the opening up of the *TCRα* locus. Accessibility to the α locus may depend on a number of elements including an LCR at the 3′ end of the locus, an enhancer in the α locus (Eα), and the T early α promoter (TEA). Eα plays a critical role in ensuring accessibility for *TCRα* rearrangement. This enhancer contains a number of regulatory sites, as depicted in Figure 9.9. Interestingly, one of the sites is a TCF/LEF-1 site. Deletion of this regulatory region results in the loss of germline transcription and rearrangement of the *TCRα* gene. The TCR α enhancer

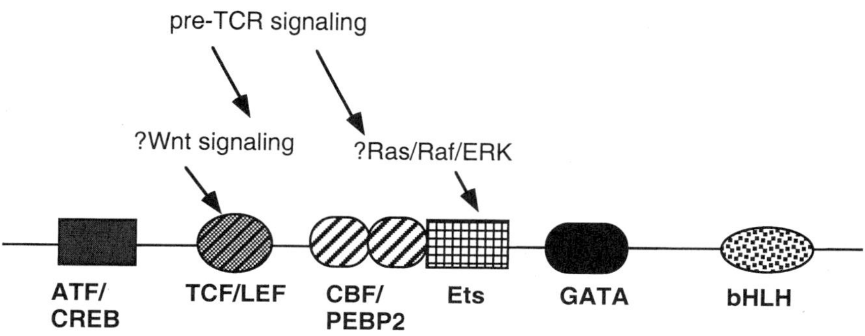

FIGURE 9.9. The core α enhancer (Eα) may be a target of pre-TCR signaling. Eα is required for the initiation of germline transcription at the α locus and for TCR α rearrangement and expression. While Eα is not required for rearrangement at the δ locus, it does contribute to the optimal transcriptional expression of the *TCR δ* gene in mature γδ T cells.

also contributes to *TCRδ* gene expression in mature T cells of the γδ lineage.

The TEA promoter is located just 5′ to the Jα cluster. Targeted deletion of the TEA compromises the rearrangement of the most *5′TCR Jα* gene segments. The developmental roles played by the locus control region (LCR) and of the two silencer elements in the *TCR α* locus remain unclear at this time, though these elements probably contribute to the regulation of accessibility and expression of *TCRα*.

Positive and Negative Selection at the Double-Positive T Cell Stage

A number of models have been proposed to explain how a developing T cell makes the choice between positive and negative selection at the DP stage. Some information is now available about cytosolic effectors that mediate positive versus negative selection downstream of the TCR. For instance, distinct members of the mitogen activated protein (MAP) kinase family are required for positive and negative selection. Nevertheless, it is unclear how the same cell may engage different cytosolic effectors when the TCR is triggered. Qualitatively or quantitatively distinct signal outputs must be generated by the same receptor during these two selection processes. Some of the categories of models proposed to explain this phenomenon are listed below:

1. Affinity models suggest that T cells bearing receptors with no affinity for self-MHC complexed with peptide "die by neglect," cells bearing TCRs with

low affinity for self-ligands are positively selected, and cells with high-affinity TCRs are eliminated.

2. Avidity models suggest that low-avidity interactions lead to positive selection, and high-avidity interactions lead to tolerance. High avidity may be generated by higher concentrations of ligand or by cooperative interactions that depend on coreceptors.

3. Peptide-specificity models suggest that certain types of peptides that behave as agonists mediate negative selection if they are of self-origin. Variants of these peptides that do not trigger the T cell receptor as efficiently are described as "antagonists." Antagonist peptides of self-origin have been postulated to be responsible for positive selection.

4. Cell-specificity models have suggested that different cells mediate positive and negative selection.

a. In one version of this model, cells responsible for positive selection express distinct subsets of self-proteins and therefore present different peptides when compared with cells responsible for negative selection.
b. In another version of this type of model, the antigen-processing machinery is assumed to be different in cells responsible for positive and negative selection, and therefore, distinct subsets of self-peptides might particpate in the two selection processes.
c. Another version of the cell specificity model (which is consistent with one type of avidity model) suggests that the cells that mediate positive and negative selection express different coreceptors, thus ensuring that distinct signals are generated during positive and negative selection.
d. It has also been postulated that in the absence of glucocorticoids, TCR signaling initiated even by moderate- to low-affinity peptide–MHC ligands may provoke apoptosis, while collaboration with glucocorticoids might prevent death. Glucocorticoid receptor activation may occur only in cells mediating positive selection because these steroids may be synthesized only in the thymic cortex.

Versions of the *affinity model* are now widely accepted as being best able to explain most of the available data on positive and negative selection. If a DP T cell expresses an antigen receptor that can be triggered by a high-affinity MHC–peptide combination in the thymus (and which is therefore "strongly" self-reactive), it receives signals that lead to its elimination (negative selection). If a DP T cell expresses no receptor whatsoever, or if it synthesizes an antigen receptor that does not recognize self-MHC (see discussion below, as to whether such cells exist), it is allowed to "die by neglect." If a DP T cell recognizes self-MHC–peptide complexes with moderate or low affinity, it receives trophic signals that allow it to survive. It is this category of positively selected T cells (undergoing positive selection II) that contributes to the naive T cell repertoire in the periphery.

Although the affinity model is well supported by a large body of evidence, an alternative model that has received some support is the *avidity* model. Avidity of a TCR–MHC interaction may be influenced by ligand density. The avidity of the interaction between the TCR on a T cell and an MHC molecule on an antigen-presenting cell may also be influenced by adhesion molecules that hold these disparate cells together, thus increasing the likelihood that appropriate intermolecular collisions will occur. The peptides involved in positive selection are fairly promiscuous. This is perhaps not too surprising given that small peptides are likely to assume many different conformations even when restrained by one or two invariant anchor residues within the MHC groove. When considering MHC–peptide complexes that interact with TCRs with relatively low affinity, promiscuity may be all the more likely. A number of ingenious approaches to examine positive selection in the context of a single peptide or a small number of peptides have all confirmed that a few peptides may be involved in the positive selection of a relatively large number of distinct T cells. It has been argued, based on positive selection studies performed in fetal thymic organ culture models, that a low concentration of a high-affinity peptide–MHC ligand (a high-affinity peptide–MHC complex is sometimes considered an "agonist" ligand) may provide the same kind of positive selection signal through the TCR as is provided by a high concentration of a low affinity MHC–peptide complex. The latter may be a so-called "antagonist" ligand if it can compete with an agonist ligand for the MHC groove; a low-affinity MHC–peptide complex that associates with the same TCR, however, is far more likely to involve a peptide that has no discernible structural similarity to the "agonist" peptide. All the evidence to date on the functional behavior of T cells selected by low concentrations of agonists (antigenic peptides!) indicates that such T cells are incapable of mounting functional responses. High-affinity self at low concentrations has not been shown to positively select functional T cells, and there is therefore at present limited support for avidity models. However, it is likely that the avidity of the interaction between the TCR and MHC–peptide ligands is influenced by a number of means, particularly by adhesion molecules or other costimulatory interactions.

Positively and negatively selecting peptide ligands were once postulated to be selectively synthesized in different cell types. This is not likely to be the case for the vast majority of peptides, since for the most part the same peptides appear to be presented by MHC molecules expressed in thymic cortical epithelial cells and in bone marrow-derived (largely medullary) cells. Some differences may exist in the antigen-processing machinery found in cells expressed in the thymic cortex as opposed to the medulla; distinct cathepsins (lysosomal/late endosomal proteases responsible for most of the cleavage that occurs during antigen processing in the MHC class II pathway) have been shown to be differentially expressed in the thymic cortex and the medulla. As discussed below, positive and negative selection appear to be mediated in a sequential fashion by cells in the thymic cortex and medulla, respectively, and the reason why different types of cell may play different roles during thymic selection may relate to their differential ability to contribute to the avidity of TCR–MHC interactions or to induct costimulatory responses.

Acceptance of the basic premise of the affinity model suggests that a DP T cell can tell when it has received a "strong" signal that leads to elimination and also recognize a moderate/weak signal that has a trophic function and allows for positive selection. Some of the pathways that must be invoked to ensure positive selection and signaling pathways that mediate negative selection will be considered below. DP T cells are hard-wired differently from mature SP T cells, and as a result, these former cells are eliminated by the kind of high-affinity MHC ligand that would drive an immune response in the periphery. The nature of this hard-wiring difference is now being explored. DP T cells refuse to enter the cell cycle when stimulated by the same MHC–ligand complexes that will trigger the proliferation of SP T cells in the periphery.

Does Positive Selection Cull the T Cell Repertoire?

It was once generally believed that a large number of newly generated αβ T cells express receptors that do not recognize MHC-like shapes and that positive selection at the DP stage was basically designed to cull out those T cells that recognized MHC–self-peptide complexes with low-to-moderate affinity. The "need" for positive selection was to ensure MHC restriction by getting rid of the T cells that recognize "garbage" or non-MHC shapes. It has now become clear that all TCRs are intrinsically biased toward the recognition of MHC-like shapes. There are probably no T cells generated whose antigen receptors recognize "garbage," and there is probably no real reason to "cull" the repertoire toward MHC molecules in general, since all αβ T cells are already MHC directed. However, it clearly helps to cull the repertoire toward *self*-MHC, as discussed below.

It has been shown that if TCRs are tested against a wide range of MHCs, virtually all newly formed T cells will prove to be alloreactive. In studies using TCR transgenics, preferential positive selection by self-MHC molecules, compared with a small subset of nonself-MHC proteins, has been noted. Nevertheless, it is quite likely that some other nonself-MHC–peptide combinations that were not tested might potentially trigger the test TCR more strongly than self-MHC. If all newly generated T cells are potentially alloreactive against some nonself-MHC molecule, why is positive selection necessary? It could be argued that an alloreactive T cell that cannot see self-MHC + self-peptide should nonetheless have the potential to recognize self MHC + some foreign peptide pretty well and should therefore be worth keeping around. It is probably critical for T cells to be *self*-MHC restricted because they cannot survive unless they interact with MHC-like shapes, and the only ones that they are going to see are self-MHC shapes. Indeed, as we shall see, peripheral maintenance of T cells requires that they be constantly "tickled" by self-MHC–peptide complexes. The "goal" of positive selection may primarily be to provide trophic signals to T cells that express complete antigen receptors and to ensure that only T cells that can survive in response to these trophic signals be invited to stick around.

Sequential Positive and Negative Selection Events in the Thymic Cortex and Medulla: The Role of Costimulators in Thymic Education

Although both positive and negative selection might potentially be mediated by a number of cell types and tissues, it appears that positive selection primarily occurs in the thymic cortex and is mediated by thymic cortical epithelial cells. Negative selection probably occurs both in the cortex and in the medulla. In the latter site, it might be mediated by bone marrow-derived thymic dendritic cells and macrophages and also by medullary epithelial cells. When MHC class II molecules were targeted specifically to the thymic cortical epithelium in MHC class II-deficient mice (by using a portion of a keratinocyte specific promoter to drive a transgene), positive selection was unaffected but negative selection was either incomplete or was not actually initiated. In these mice, MHC class II molecules were not expressed either by bone marrow-derived medullary cells or by medullary epithelial cells, suggesting that these latter cell types might be critical for the negative selection process. Both these latter cell types are specifically lost in RelB knockout mice, suggesting that these otherwise distinct cells may be functionally related. It is likely that thymic cortical epithelium supports early T cell maturation and positive selection but that it is relatively inefficient in initiating negative selection.

Is there anything special about thymic cortical epithelial cells? Apart from making goodies such as IL-7 and SCF to drive the process of T cell differentiation, these MHC class II-bearing cells may functionally differ from medullary cells in a couple of important ways. It is known that thymic cortical epithelial cells and medullary cells express distinct cathepsins. It is possible that cortical cells may process proteins to yield distinct peptides (which primarily mediate positive selection) from those that are generated in the medulla and are involved in negative selection. While the difference in the pattern of cathepsin expression has been demonstrated, no direct evidence suggests that cortical and medullary cells present distinct subsets of MHC-bound peptides. It has been suggested that while both cortical and medullary cells express the intercellular adhesion molecule-1 (ICAM-1) costimulatory/adhesion molecule, B7-1 is not expressed in cortical cells but is expressed in the medulla. Signaling in the cortex may therefore be qualitatively distinct or relatively incomplete when compared with TCR/costimulatory signaling in the medulla. It is therefore likely that T cell progenitors that arrive from the bone marrow first migrate to the cortex, where they receive signals for commitment, pass through the pre-TCR checkpoint, and then are positively selected. While negative selection may be initiated in the cortex, the bulk of the deletional process may occur after positive selection is complete, and this may occur in the medulla.

It has been shown that signaling through the glucocorticoid receptor (GR), a transcription factor of the nuclear hormone receptor family, can synergize with TCR signaling to prevent death. MHC–peptide complexes with low-to-moderate affinity for a specific TCR might actually provoke death in the absence of GR signals but mediate positive selection in the presence of glucocorticoids. Glucocorticoids are synthesized by thymic cortical cells and shutting off endogenous glucocorticoid synthesis by a transgenic antisense approach has been shown to make thymocytes more prone to elimination following TCR ligation. In addition, glucocorticoids themselves, in the absence of any TCR signals, may provoke apoptosis of thymocytes. Glucocorticoids in the cortex may therefore be responsible for the "background level" of "death by neglect." A combination of GR and TCR signaling in the cortex may be responsible for positive selection. TCR signaling unopposed by GR in the medulla might mediate negative selection (Figure 9.10).

Positive Selection and the CD4 Versus CD8 Lineage Commitment Decision

As we discussed in the previous sections, DP T cells expressing T cell receptors that recognize MHC–peptide complexes with low or moderate affinity are positively selected and help generate the T cell repertoire. Positive selection identifies cells that have made in-frame rearrangements at both the *TCR* α and β loci and also identifies cells with receptors that are MHC restricted. In addition, this selection event commits cells to the CD4 or CD8 lineage.

For a DP αβ T cell to be positively selected, signaling through the TCR is essential. Just as a T cell must decide to become an αβ or a γδ cell, αβ cells themselves must make a decision to commit either to the CD4 or the CD8 lineage (Figures 9.11 and 9.12). Since TCRs on CD4 cells recognize MHC class II–peptide complexes, and TCRs on CD8 cells recognize MHC class I–peptide complexes, instructive models were proposed to explain this developmental switch (Figure 9.11). These models postulated that an MHC class II–peptide complex might engage CD4 on a DP T cell and thus induce the shutoff of CD8 expression in a DP T cell to give rise to a CD4 T cell. Similarly, MHC class I–peptide complexes might instruct a DP T cell to shut

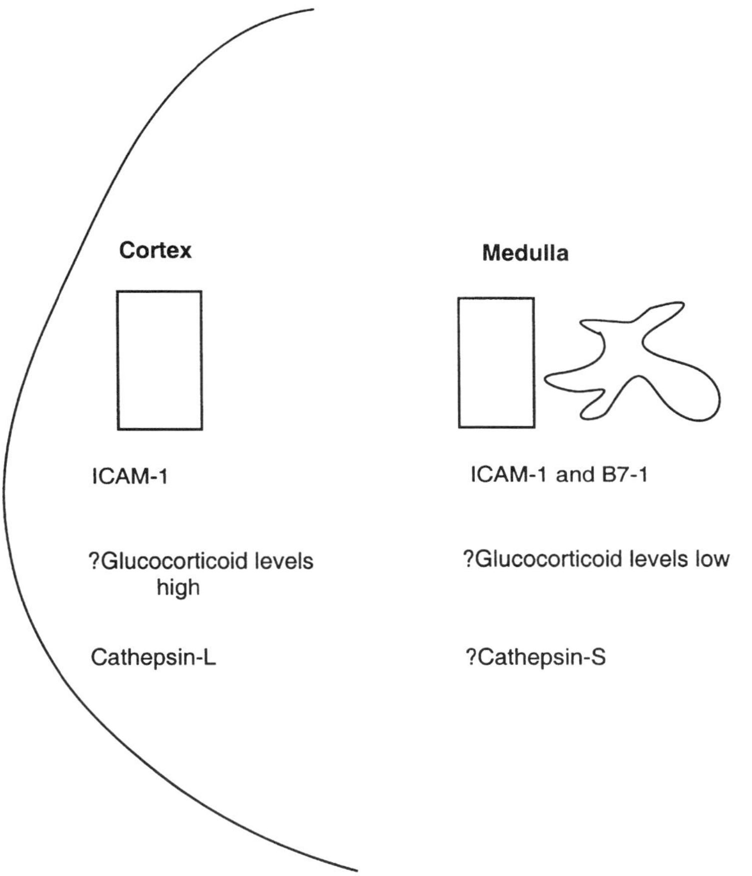

FIGURE 9.10. Positive and negative selection occur sequentially in the cortex and medulla. Cortical cells mediate positive selection, while negative selection may be incomplete in the absence of medullary cells. Cortical cells might present distinct peptides from medullary cells or may activate different signals because they lack B7.1. Glucocorticoids might be available in the cortex but not in the medulla.

off CD4 synthesis and thus allow a DP T cell to mature into a CD8 T cell. These models assume that a DP T cell will "know" that its CD4 coreceptor (or its CD8 coreceptor, as the case may be) has been engaged and as a result will shut off expression of its other type of coreceptor.

Alternative stochastic models (see Figure 9.12) suggest that a given DP T cell may either switch off CD4 or CD8 prior to commitment to a SP lineage. If a stochastically committed CD4 T cell has a receptor that recognizes self-MHC class II–self-peptide complexes with low affinity, it may be positively selected. If such a cell expresses a TCR that can recognize self-MHC class I–self-peptide complexes, it will not receive a co-receptor signal

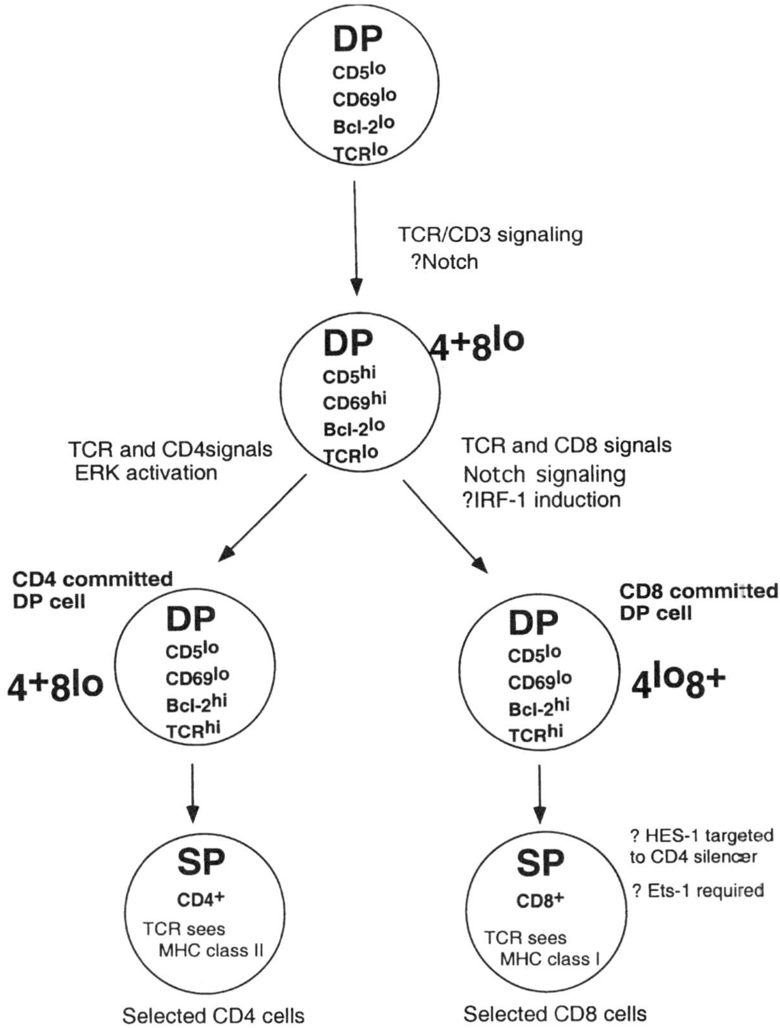

FIGURE 9.11. An instructive model for CD4/CD8 lineage commitment. DP cells that initially express low levels of the TCR and Bcl-2 are first induced to express activation markers such as CD5 and CD69. They commit during the DP stage to the CD4 or CD8 lineages at a point when TCR and Bcl-2 levels are elevated. TCR signaling may induce extracellular signal-regulated kinase (ERK) and provide instructions for CD4 commitment. Notch signaling might contribute to positive selection and also to the switch to the CD8 lineage. Interferon-regulated factor 1 (IRF-1) may also contribute to the $CD8^+$ lineage commitment decision. Notch signaling may induce hairy enhancer of split-1 (HES-1), which may be targeted to the CD4 silencer in CD8 cells.

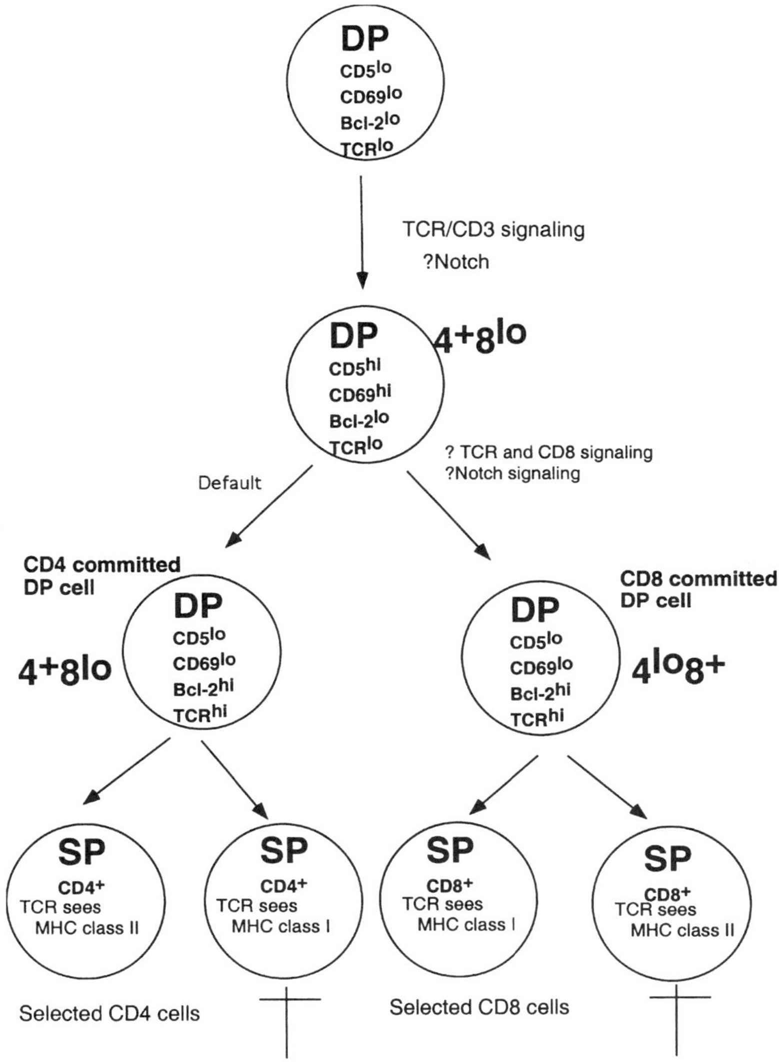

FIGURE 9.12. A stochastic model for CD4/CD8 lineage commitment. DP cells that initially express low levels of the TCR and Bcl-2 are first induced to express activation markers such as CD5 and CD69. They commit during the DP stage to the CD4 or CD8 lineages at a point when TCR and Bcl-2 levels are elevated. CD4 commitment might represent a default process. Notch signaling might contribute to positive selection and also to the switch to the CD8 lineage. In the "asymmetric commitment" model, $CD4^+$ cell differentiation is considered to be stochastic, and signals may be involved in $CD8^+$ cell generation. If the process of lineage commitment is actually stochastic, cells will still need to be selected on the basis of coreceptors expressed (or on the basis of the strength of TCR signaling) by MHC class I or MHC class II ligands.

(since it expresses only CD4) and will be allowed to "die by neglect." Purely stochastic models for commitment incorporate the need for the selection of committed cells that express the "right" type of coreceptor by the matched MHC class of ligand.

To help confuse the picture, it is now apparent, primarily from knockouts of the CD4 gene and of the CD8α gene, that commitment to the CD4 and CD8 lineages does not, in an absolute sense, require the generation of signals via the respective coreceptors. Instructive signals that are quantitatively distinct and have the potential to influence lineage choice may be provided by the TCR in the absence of coreceptors. In most T cells (in normal mice), coreceptors undoubtedly participate in the selection process that determines whether a TCR recognizes MHC class I or MHC class II complexes.

What is the best evidence to support an instructional model? One study using a dominant gain of function mutation of the *erk2* gene has suggested that activation of extracellular signal-regulated kinase (ERK) in the T lineage favors CD4 T cell generation. Since TCR-signaling activates ERK, this suggests that T cells are instructed to become CD4 cells. A similar result has not been observed in somewhat different approaches to examining the Ras/Raf/ERK pathway. In the absence of Csk, $CD8^+$ cells are not generated, suggesting that enhanced signaling via Src-family kinases such as Lyn and Fyn may drive cells down the $CD4^+$ route. Commitment to the $CD8^+$ lineage may require specific signals that may be randomly delivered by Notch (discussed below) and/or may depend on quantitatively or qualitatively distinct instructive TCR signals in cells triggered by MHC class I ligands. The interferon regulated transcription factor (IRF-1) might also be required in a cell-intrinsic manner for $CD8^+$ cell differentiation. The signals involved in positive selection will be discussed in more depth below.

It has been suggested that commitment to the $CD4^+$ lineage may represent a default process and might occur in the absence of MHC molecules, while $CD8^+$ development may be dependent on signals. This partly instructive, partly "spontaneous" process has been referred to as the "asymmetric commitment" model. None of the existing data are sufficiently convincing to unequivocally state that commitment it purely stochastic or instructionally driven.

In summary, TCR signaling by low- to moderate-affinity MHC–peptide ligands drives positive selection. DP cells that initially express low levels of the TCR and of Bcl-2 are first induced to express high levels of CD5 and CD69 (these should be considered as markers of T cell activation). Commitment to the CD4 lineage may not require additional signals and may represent the default pathway (the asymmetric commitment model), or it might be instructional and depend on Src-family kinase and ERK activation. Differentiation to the CD8 lineage may depend on CD8 signals or on the triggering of the Notch pathway. During the process of lineage commitment, DP T cells need to selectively shut off either CD4 expression or CD8 expression.

Signals Involved in Lineage Commitment: ERK and Notch May Influence the Process

We will discuss the signals involved in positive and negative selection together in the next section. An important aspect of positive selection is lineage commitment, and the signals that may be involved in this process are considered here. It has been suggested that a "default pathway" induced by the triggering of the TCR in conjunction with coreceptors in some "generic" way may drive cells that receive trophic signals to differentiate into $CD4^+$ cells (as opposed to cells that receive no signal whatsoever and "die by neglect"). The evidence that ERK may be required for commitment to the CD4 lineage, however, suggests that CD4 differentiation is "instructional," and this may not really represent a "default" pathway.

Cells that receive a qualitatively distinct TCR signal, perhaps in conjunction with a Notch-derived signal, may commit to the CD8 lineage. The presumed role of Notch signaling for CD8 commitment is based in part on experiments involving the forced expression of an activated form of Notch in the T lineage. More direct evidence will undoubtedly be made available in the coming years to resolve whether or not Notch signaling is involved in lineage commitment, and if so, which Notch ligands participate in the process. Notch is also believed to cooperate with the pre-T receptor in positive selection I, as we have discussed earlier. In a separate study, Notch has been implicated in providing a survival signal during positive selection II, as opposed to playing a role in lineage commitment. It has been argued that activated Notch transgenes in some studies might have been inadvertently expressed at higher levels in CD8 cells. It remains to be resolved whether Notch actually participates in lineage commitment, though most of the available evidence is quite compelling. As discussed in Chapter 4, the HES1 protein (hairy/enhancer of split) may be downsteam of Notch. Activated Notch in T cells induces *Hes1* transcription, and HES1 has been shown to be capable of functionally interacting with the CD4 silencer (see below).

IRF-1 binds to the promoters of a number of interferon-regulated genes and positively influences their transcription. These genes include H-$2K^b$, the transporter associated with antigen processing, TAP-1, and the proteasomal subunit, LMP-2. In the absence of IRF-1, MHC class I expression on a number of cell types is impaired and IL-15 secretion by stromal cells is compromised. IRF-1–deficient mice have defects in the development of a number of lymphoid lineages including NK cells, NT cells, and intraepithelial $\gamma\delta$ T cells. The reduced numbers of $CD8^+$ cells observed in the thymus of *IRF-1*$^{-/-}$ mice could potentially be accounted for by the requirement of this gene in non-T cells to induce either class I expression or IL-15 secretion. However, in studies involving *in vivo* bone marrow chimeras and *in vitro* reaggregation cultures, it has been demonstrated that the critical defect in $CD8^+$ T cell development in these mice reflects the absence of IRF-1, not in the thymic environment, but in developing T cells themselves. IRF-1 levels are relatively high in mature T cells but are dramatically induced in DP T cells that are triggered through their TCRs. The induction of IRF-1 appears to be important during both positive and negative selection and is apparently required for lineage commitment of $CD8^+$ cells.

Even if DP T cells commit to the CD4 or CD8 lineage stochastically, they need to be informed whether their TCRs are directed toward MHC class II or MHC class I complexes and will therefore be selected appropriately by an instructive process. Cells that express CD4 and whose receptors recognize MHC class II–peptide complexes must be distinguished from cells that express CD4 but whose receptors recognize MHC class I–peptide complexes. While many suggestions have been made as to how signaling via CD4-linked TCRs may be distinguished from signals transduced by CD8-linked TCRs, no convincing explanations have been provided regarding how these different signals are generated or sensed. It has been suggested that CD4 may associate more tightly with Lck than CD8 and that, as a result, the signals produced via TCR–CD4 complexes might be quantitatively distinct from the signals generated by TCR–CD8 signaling. There is limited support for this explanation because mutant CD4 molecules that cannot bind Lck seem quite capable of mediating positive selection.

A critical component of all models for lineage commitment is the need to shut off *CD8* expression in a $CD4^+$ cell and *CD4* expression in a $CD8^+$ cell. Clearly, a complete understanding of the transcriptional regulation of both the *CD4* and *CD8* genes will help fill in many of the blanks in the molecular understanding of the lineage commitment process. At present, the transcriptional regulation of the *CD4* gene during the lineage commitment process is better understood than that of *CD8*. A T lineage-specific enhancer is located about 13 kb upstream of the *CD4* gene, and this enhancer directs *CD4* expression in both $CD4^+$ and $CD8^+$ cells. Specific expression of this gene in SP $CD4^+$ cells depends on the activity of silencer elements located in the first intron of the CD4 gene. Somehow, during lineage commitment, this silencer is turned "on" only in $CD8^+$ cells. Two factors that bind to this silencer have been identified. One is a novel homeodomain protein called SAF, and the other is HES-1. Mutating the binding sites for SAF and HES-1 abrogate silencing.

Regulation of *CD8* gene expression is far more complex. The *CD8α* and *CD8β* genes are separated by about 36 kb on mouse chromosome 6, and these two genes are regulated both coordinately and independently in different cell types and during the process of T cell differentiation. Most $CD8^+$ T cells of the αβ lineage express heterodimners of CD8α and CD8β. Intestinal intraepithelial lymphocytes (IELs) that develop extrathymically (discussed later in this chapter), some NK cells, and dendritic cells of lymphoid origin express the CD8αα homodimer. At least four distinct enhancers have been identified in the murine CD8 locus, and one of these appears to be specifically required for the expression of CD8αα homodimers in IELs. How exactly the expression of CD8α as well as of CD8β is shut off during lineage commitment is still being worked out.

Signals Involved in Positive and Negative Selection

Most DP T cells are postmitotic cells. TCR signals cannot induce these cells to proliferate. However, negative selection ligands can induce the induction of apoptotic processes in these cells. Positive selection of these cells presumably involves the generation of survival signals and the specific targeting by signal transduction of regulatory motifs in the CD4 locus and the CD8 locus to ensure the selected transcriptional downregulation of either CD4 expression (in cells that commit to the CD8 lineage) or of CD8 expression (in CD4-committed T cells).

Both positive and negative selection involve activation of the TCR; in the absence of TCR/CD3 signaling, one would expect both these processes to be compromised. However, since the TCR distinguishes between trophic self-MHC ligands and death-inducing self-MHC ligands, it would be expected that the TCR induces very different cellular signaling processes during positive and negative selection. TCR triggering leads to the activation of ERK, JNK, and p38 MAP kinases. The ERK MAP kinase is required for the DN to DP transition (positive selection I) as mentioned earlier and also participates in positive selection II and lineage commitment. An activated *erk2* transgene has been shown to contribute to the enhanced generation of CD4 T cells. The ERK MAP kinase, however, is not required for negative selection. The use of a transgenically expressed dominant negative (kinase dead) JNK1 protein has suggested that the JNK pathway is involved in negative selection. Activated forms of MKK1 (which activates ERK), and MKK6 (which activates p38), have also been introduced into thymocytes in fetal thymic organ culture. Pharmacological inhibitors of specific MAP kinases have also been used in these culture systems. These studies have revealed that while ERK is required for positive selection, p38, like JNK, is required for negative selection.

Positive selection is abrogated in the absence of MHC class I or class II ligands. Mice that lack β2-microglobulin, that carry mutations that compromise MHC class I heavy chain expression, that have engineered defects in the TAP1 or TAP2 peptide transporters and in proteasomal subunits such as LMP2, all present with defects in the positive selection of $CD8^+$ T cells. Similarly, mice that lack MHC class II proteins, that do not express the MHC class II associated invariant chain, that have defects in transcriptional regulators of MHC class II gene expression, or mice that have been engineered to lack one of the chains of the H-2M heterodimer, all exhibit defects in the positive selection of CD4 T cells. H-2M is the murine equivalent of the human MHC class II-like human leukocyte antigen-DM (HLA-DM) heterodimer. These proteins facilitate the exchange process that permits exogenously derived processed peptides to displace the invariant chain-derived CLIP peptide (class II associated invariant chain derived peptide) in a late endosomal compartment. Transcriptional regulators that are responsible for inherited defects in MHC class II expression in humans are discussed in Chapter 12. Mutations that involve components of the TCR/CD3 complex can also prevent positive selection of DP T cells if they involve proteins that did not contribute to pre-TCR signaling or only partially did so. These include the knockout of TCRα and of Eα.

TCR signaling at the double positive stage may induce Id3, an inhibitor of E2A proteins such as E47 and E12 (discussed in Chapter 2). Although in B cells E47 homodimers bind DNA, in thymocytes E47 forms a heterodimer with another helix-loop-helix protein called HEB (Hela E box-binding protein). Inhibition of E47 expression and activity promotes positive selection in the T lineage. In $Id3^{-/-}$ mice positive selection of $CD4^{+}$ T cells is very severely compromised, but the positive selection of $CD8^{+}$ T cells is on marginally perturbed.

Apart from the requirement for p38 and JNK, negative selection probably involves the induction of the orphan steroid receptor gene *Nur 77* (discussed in Chapter 5). Exactly why *Nur 77* is required for negative selection is unclear. While there is a vast body of information supporting a role for Fas-FasL signaling in activation-induced cell death (AICD) in the periphery, it has been suggested that Fas is also relevant for negative selection in the thymus, particularly in the context of self-antigens that are expresssed at relatively high levels. Fas is induced in developing thymocytes, and Fas signaling may well be critical in overriding positive selection. Fas-mediated apoptosis via caspase-8 dependent pathways may lead to the caspase-mediated degradation of antiapoptotic proteins such as Bcl-2, thus dominantly suppressing positive selection signals that induce these factors. A member of the TNF receptor family, CD30, has also been implicated in negative selection. Negative selection is impaired in $CD30^{-/-}$ mice. It is possible that CD30 is transiently induced on DP T cells that avidly recognize self-MHC. How exactly CD30 facilitates negative selection is unclear. E2F-1 is a proapoptotic transcription factor that is released from the Rb pocket following CDK-mediated phosphorylation in the G1 phase of the cell cycle. This factor might be involved in negative selection as has been suggested by *in vitro* studies on thymocytes obtained from an E2F-1 knockout mouse. The relevance of IRF-1 for both positive and negative selection in the $CD8^{+}$ lineage was discussed in the previous section.

Mutations that block T cell development at the TCR-driven positive selection checkpoint are listed in Table 9.3.

TABLE 9.3. Signals involved in positive selection at the DP T cell stage

Mutations involving TCR/CD3 signaling	
TCRα$^{-/-}$	Mombaerts et al., 1992b, Philpott et al., 1992
Eα$^{-/-}$	Sleckman et al., 1997
CD3δ$^{-/-}$	Dave et al., 1997
CD3ζ/h$^{-/-}$	Love et al., 1993; Malissen et al., 1993; Ohno et al., 1993; Ardouin et al., 1998; Shores et al., 1998
Lck$^{-/-}$	Molina et al., 1992
Vav$^{-/-}$	Turner et al., 1997
CD45$^{-/-}$	Byth et al., 1996
Zap-70$^{-/-}$	Negishi et al., 1995; Wiest et al., 1997
IRF-1$^{-/-}$	Matsuyama et al., 1993; Reis et al., 1994; Penninger et al., 1997)
Id3$^{-/-}$	Rivera et al., 2000
Mutations involving the MHC class I pathway	
β2-microglobulin$^{-/-}$	Zijlstra et al., 1990; Koller et al., 1990
MHC class I H-chain	Tanchot et al., 1997
TAP transporters	Van Kaer et al., 1992; De la Salle et al., 1994
LMP2$^{-/-}$	Van Kaer et al., 1994
IRF-1$^{-/-}$	Matsuyama et al., 1993; Reis et al., 1994; Penninger et al., 1997
Mutations involving MHC class II pathway	
MHC class II$^{-/-}$	Cosgrove et al., 1991; Grusby et al., 1991
MHC class II transcriptional regulators	Mach et al., 1996
H2-M$^{-/-}$	Fung-leung et al., 1996; Martin et al., 1996; Miyazaki et al., 1996
Invariant chain$^{-/-}$	Viville et al., 1993; Bikoff et al., 1993

A Brief Comment on Allelic Exclusion at the TCR α Locus

While allelic exclusion at the TCR β locus is well documented and clearly occurs at the level of TCR β gene rearrangement, most T cell rearrange both TCR α alleles. In studies on mouse T cells that carry two in-frame Vα-Jα rearrangements and express two mRNAs, more often than not, only one TCR α chain has been shown to be expressed on the cell surface. In many, but not all, T cells, allelic exclusion probably does occur at the TCR α locus, but the mechanism appears to involve a post-translational event. This is in contrast to the B lineage where allelic exclusion at the Igκ locus is regulated at the level of gene expression.

In some studies, the expression of two different α chains on the surface of human T cells has been reported. It may be reasoned that allelic exclusion is not really required at the *TCRα* locus. It could be argued that if there were in fact two expressed α chains in a single T cell, even if one of these chains formed part of a receptor that could be positively selected by self-

MHC, it would be extremely unlikely that the other αβ combination on the same cell would also be self-MHC restricted. By this reasoning, allelic exclusion at the β locus alone would suffice to ensure clonal specificity in a functional sense. Notwithstanding these arguments, it does appear that post-translational mechanisms involving competition between expressed TCRα chains ensure the cell surface expression of only one α chain on the surface of individual murine T cells.

It has been reported that during mouse T cell development, allelic exclusion is not observed in immature thymocytes (which may express two different α chains on the cell surface) but that this process occurs at the time of thymocyte TCR upregulation during positive selection (see Figure 9.11, for instance). In immature thymocytes, the TCRα chain is rapidly degraded. If it complexes with the β chain, it is protected from degradation. These immature cells accordingly have a large excess of TCRβ chains, and under these circumstances, competition between two α chains for association with an excess of β chain is likely to be minimal. TCR^{lo} $CD69^{lo}$ T cells transcriptionally upregulate the expression of their rearranged α chain genes during positive selection and mature into TCR^{hi} $CD69^{hi}$ cells. The increase in α chain protein expression at this transition ensures that TCRβ chains are not in large excess, and therefore competition for β chain binding between the α chains ensues. As a result, the α chain with a higher affinity for the specific β chain wins out, and the other α chain is rapidly degraded and functionally excluded.

Maturation and Export of Naive Peripheral T Cells

Although it has been estimated that approximately 95% of developing T cells die in the thymus and that only a small proportion of newly generated T cells is exported to the periphery, estimates of cell death are probably fairly inaccurate and reveal little information about when during development death actually occurs. What proportion of dying thymocytes represent cells that fail to go through β selection, or cells that "die by neglect" at the DP stage, or cells that are actually negatively selected at the DP stage is not known. It has been estimated that only about 1×10^6 newly formed αβ T cells are actually released from the thymus each day in young mice. In the case of γδ T cells, it is estimated that about 2×10^4 newly formed T cells are exported from the thymus each day. A significant proportion of the daily γδ T cell output is probably derived, as we will discuss below, from extrathymic sources such as intestinal cryptopatches.

Once SP cells enter the medulla, they mature into peripheral SP T cells and are exported into the periphery. As SP T cells mature, they express lower levels of HSA/CD24 and express higher levels of L-selectin, Qa-2, and high molecular restricted (R) isoforms of the CD45 tyrosine phosphatase (CD45 RA, B, and C). The higher levels of L-selectin probably contribute to the ability of newly formed T cells to home to lymph nodes. There appears to be a highly regulated, but poorly understood, export process required for peripheral T cell homeostasis. This ensures that peripheral cell numbers are maintained in part by the process of regulated export of newly formed SP T cells from the medulla to the periphery.

Maintenance of Naive Peripheral T Cells

Apart from positive selection in the thymic cortex, the maintenance and survival of mature peripheral T cells depend upon frequent, if not continuous, stimulation of the TCR on "resting" peripheral T lymphocytes by MHC molecules. The maintenance of resting peripheral CD8 T cells specific for the H-Y male antigen was shown to require the presence of the appropriate self-MHC class I molecule (but not the specific male antigen). Activation of these cells requires both the MHC class I molecule and the H-Y antigen. (In contrast, in similar experiments, memory T cells could be maintained even by a nonspecific MHC class I molecule, and while the activation of memory T cells required specific MHC class I molecules, it did not require the specific antigen. Issues related to memory T cells will be considered in the next chapter).

Resting peripheral T cells therefore require "tickling" by specific MHC molecules in the periphery for them to survive. The nonmitotic signals delivered via the TCR probably induce survival pathways. These survival pathways may, as in B cells, require Tec kinases. As summarized in Table 4.2 in Chapter 4, engineered mutations in two Tec kinases, Itk and Rlk, can lead to a defect in peripheral T cell maintenance and consequently also of T cell proliferation in response to TCR ligation. The $Itk^{-/-}/Rlk^{-/-}$ mouse has a T cell phenotype that resembles the B cell phenotype seen in *Xid or* $Btk^{-/-}$ mice.

It has been suggested that transcriptional regulation might contribute to the maintenance of the "resting" state. LKLF (lung Kruppel-like factor) is a transcription factor that is induced during the maturation of quiescent SP T cells and that its expression is extinguished when these cells are activated. In LKLF null mice, SP peripheral T cells apoptose at a high rate and appear to take on an activated phenotype and express FasL before they die. Another transcription factor that is required for the maintenance of peripheral T cells is Ets-1.

It is possible that low-level triggering of the TCR in the periphery contributes to the induction of Ets-1 and possibly LKLF that may be required for the survival of SP peripheral T cells. The maintenance of the quiescent state may require quantitatively or qualitatively distinguishable TCR signals upstream of LKLF induction. LKLF might prevent resting T cells from being activated by nonantigenic trophic factors that are undoubtedly freely available in the circulation.

Thymic Versus Extrathymic Development and the Generation of γδ T Cells

In an earlier section, we have discussed how the thymus may be required for a compartmentalization function that facilitates the generation of T cells by inhibiting a negative feedback loop. Most conventional αβ T cells develop in the thymus, probably because SP αβ T cells that are generated at any other extrathymic location significantly inhibit T cell maturation. T cells can develop at extrathymic sites, albeit less efficiently than in the thymus, and these sites include the bone marrow, the gut, and the liver. While all T lineages can develop in the thymus, a large number of γδ T cell actually develop in other sites, particularly in the intestine. Natural T (NT)/NK1.1 T cells develop mainly in the thymus. Unconventional and rare DN αβ T cells that do not resemble NT cells may develop in the bone marrow itself.

γδ T cells constitute a small proportion of the T lineage. These cells have a less restricted repertoire than αβ T cells but serve an important function in host defense, particularly at mucosal sites. A large number of γδ T cells are located in the gut epithelium itself as intra-epithelial T cells and may be held there by interactions between αE integrins and E-cadherin. IELs may be either αβ or γδ cells and often express the CD8 co-receptor. Many γδ T cells express no coreceptors, and some express CD8αα homodimers. We have already discussed commitment to the γδ lineage. In the absence of γc, γδ T cell development is completely blocked in adult mice and is very severely compromised in the fetus as well. IL-7/γc signaling is probably required for the survival of early precursors prior to TCR γ gene rearrangement, although it has also been argued that this cytokine might be required for the induction of the rearrangement process itself. Given that the γc knockout compromises γδ development more severely than αβ or NT development, it is possible that γc is required in a particularly critical fashion by precursors that have committed to the γδ lineage. Further development of these cells does not depend on the generation of DP intermediates. Signaling from γδ TCRs probably contributes to the development of cells of this lineage and may represent a form of positive selection. Issues of positive and negative selection for this lineage will be considered below. A large proportion of γδ T cells may develop extrathymically, and most intestinal IELs appear to develop in the gut. The IL-15 R provides either developmental or survival signals to γδ IELs. In the absence of IL-15Rα or of IRF-1 (which is required for the transcription of IL-15 by stromal cells), IEL development is markedly compromised.

T cell precursors that mature into both αβ and γδ T cells appear to accumulate in the gut in regions known as *intestinal cryptopatches*. Bone marrow precursors migrate to the gut and Lin^- cells expressing c-Kit, the IL-7R, and Thy-1 accumulate under the epithelial cells in the crypt regions of intestinal villi. Each of these patches of developing T lymphocytes consists of about a thousand cells. These precursors can give rise to both αβ and γδ T cells when transferred into immunodeficient recipients. These gut-derived precursors give rise to αβ T cells in mesenteric lymph nodes as well as to αβ and γδ-expressing IELs.

How exactly self–nonself-discrimination is achieved during the development of γδ cells is unclear. These cells have very restricted repertoires. It is possible that they do not need to be negatively selected, since they are less likely to be triggered by self-antigens under normal circumstances. In humans, two main populations of γδ T cells are present. Vγ2/Vδ2 T cells are largely circulating cells, and they recognize soluble bacterial antigens. Vδ1 γδ T cells make up about 70% to 90% of γδ IELs, and they recognize the MHC class I-like MICA (MHC class I-related chain A) and MICB (MHC class I-related chain B) molecules, particularly when their expression is induced by epithelial cells that are "under stress" (usually because they are infected). These MHC class I-like proteins lack β2-microglobulin and are not known to be loaded with peptides by the MHC class I pathway. Activation of these γδ IELs may contribute to the elimination of infected epithelial cells. It is possible that a weak interaction with MICA or MICB in noninfected or nonstressed epithelial cells may provide trophic/positive selection signals to cognate γδ T cells. Some aspects of γδ T cell development are summarized in Figure 9.13.

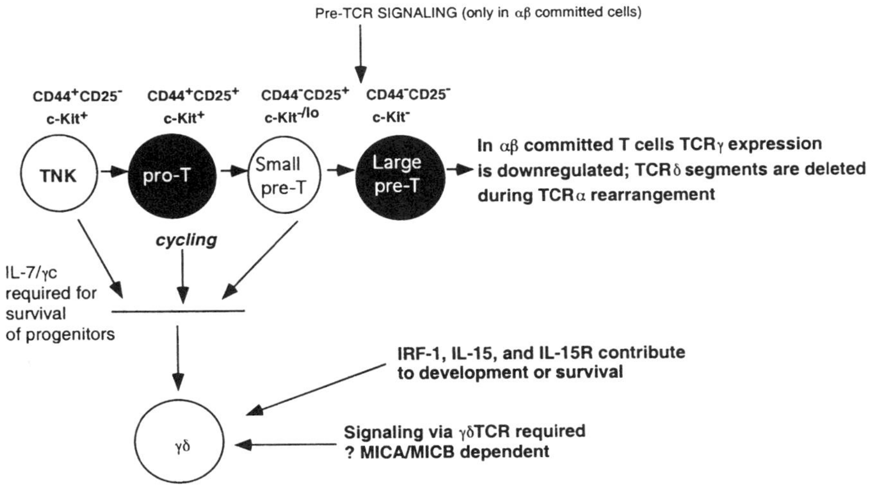

FIGURE 9.13. The development of γδ T cells. The development of this lineage requires IL-7/γc signaling. Trophic signals are also provided by the γδ T cell receptor. MHC class I-like proteins such as MICA and MICB might be required to trigger the TCR on these cells.

Some γδ T cells, particularly those that home to the epidermal or vaginal epithelium, have invariant γδ T cell receptors. These T cells could conceivably have been selected and expanded by antigen. Alternatively, they could have been generated by the regulation of V(D)J rearrangement at the level of accessibility in fetal precursors of these γδ T cells. The use of a rearrangement competent transgene with a frameshift mutation in the V region resulted in canonical Vγ3-Jγ1 junctions in epidermal γδ cells indicating that selection is not required to generate these T cells. By some poorly understood means, specific Vγ and Jγ genes appear to be targeted for rearrangement during fetal development. The absence of TdT in fetal liver stem cells is permissive for the formation of short homology repeats during the rearrangement process (as discussed in Chapter 3). However, the absolute requirement for a fetal thymic environment, as discussed earlier in this chapter, for the development of invariant γδ T cells suggests that some external ligand found only in the fetal thymus generates the specfic accessibility of target gene segments during γδ T cell development.

Development of Cells of the NT Lineage

As mentioned, NT cells are also known as NK1.1 T cells. These DN or CD4-expressing T cells secrete high levels of IL-4 and IFN-γ, express an invariant TCR α chain, Vα14-Jα281, preferentially utilize TCR Vβ8, Vβ7, and Vβ2 chains, express high levels of CD44 and Ly6C, and in some mouse strains express a protein known as NK1.1, found generally on NK cells, but widely used to help identify the NT population. NT cells also express intermediate levels of the αβ TCR (as compared with high levels on conventional αβ T cells) and like NK and γδ T cells, express high levels of the IL-2Rβ chain (conventional αβ T cells express low levels of this chain shared by the IL-2 and IL-15 receptors). The development or survival of NT cells is dependent in part on signals via the IL-15R. NT cell generation is impaired in mice that lack IL-15Rα, as well as in mice deficient in IRF-1, which is required for the synthesis of IL-15. The requirement for IL-15 signaling during NT cell generation is not as critical as it is in NK cells (see Chapter 11).

NT cells express both Ly49 and NKR-P1 family proteins. The NK 1.1 marker is a protein of the NKR-P1 family (which, like the Ly49 family, includes activating and inhibitory NK cell receptors). These receptors (see Chapter 11) are all type II membrane proteins that contain C-type lectin domains and form disulfide linked dimers. With the existing serological reagents, NK1.1 has been conclusively demonstrated to be present only on the surface of NT and NK cells from C57/B6, New Zealand Black (NZB), and SJL mice. While NT cells probably serve as a first line of defense against infection and possibly modulate T cell responses to exogenous antigens by virtue of their ability to secrete cytokines, the actual function of these cells remains to be determined.

NT cells recognize and are activated by CD1 molecules. These MHC class I like molecules contain a β2-microglobulin moiety but present antigens—including both peptide and lipid antigens, depending on the type of CD1 molecule—in a TAP-independent fashion. In humans, there are five different CD1 isotypes: CD1a, b, c, d, and e. Human CD1b molecules present lipid antigens that complex with the CD1 molecule in a late endosome-like compartment. Mice inherit two CD1 genes, which encode the CD1.1 and CD 1.2 proteins that are very similar in structure. Mouse CD1 molecules are structurally similar to human CD1d proteins. NT cell development has mainly been studied in the B6 mouse that expresses only the CD1.1 protein. CD 1.1 is expressed on the surface of cortical thymocytes (but not of epithelial cells) and is also expressed in the liver and to a lesser extent in the spleen and lung.

NT cells are extremely infrequent at birth but gradually accumulate, with levels peaking at about 6 to 8 weeks after birth. Their major site of synthesis is the thymus, and most available evidence suggests that they are generated by a process of selection rather than by the directed rearrangement of a precommitted precursor. NT cells can be generated from both fetal and adult-type HSCs. A T cell that has committed to the αβ lineage and that has randomly made the appropriate Vα14-Jα281 rearrangement is presumably triggered by CD1 expressed on the surface of other cortical thymocytes. These NT cells are probably positively selected at the DP stage, and cells that express CD8 might be negatively selected. Exactly how DN NT cells are generated is unclear. Although no direct evidence indicates that these cells pass through the DP stage, this is considered likely, since this is the stage at which *TCRα* rearrangement occurs. Moreover, evidence from transgenic mice suggests that CD8-expressing NT cells are negatively selected. Following maturation, these cells receive signals for further expansion and constitute multiple oligoclonal pools sharing the same TCRα chain and a limited variety of β chains. The development of NT cells is severely compromised in mice in which the CD1 gene has been knocked out. A somewhat incomplete model for NT cell generation is provided in Figure 9.14.

The NK 1.1 marker is first expressed on oligopotential TNK precursors ($CD44^+CD25^-$). It is not expressed, however, during the subsequent stages of αβ T cell development. Exactly when the NK 1.1 molecule is induced again in NT cells is unclear, although it has been speculated that this molecule may be turned on in response to TCR signaling and that it serves as an attenuator or modulator of signal transduction based on the presence of immunoreceptor tyrosine inhibitory motif (ITIM) in its cytoplasmic tail. NK cell markers, however, are not expressed on NT cells that develop in γc-deficient mice. Interestingly, mice that lack γc absolutely fail to generate NK cells, γδ T cells, and intraepithelial T cells, but conventional αβ T cell or NT cell development is not significantly compromised. The development of NT cells in the thymus does not require the expression of NK cell markers.

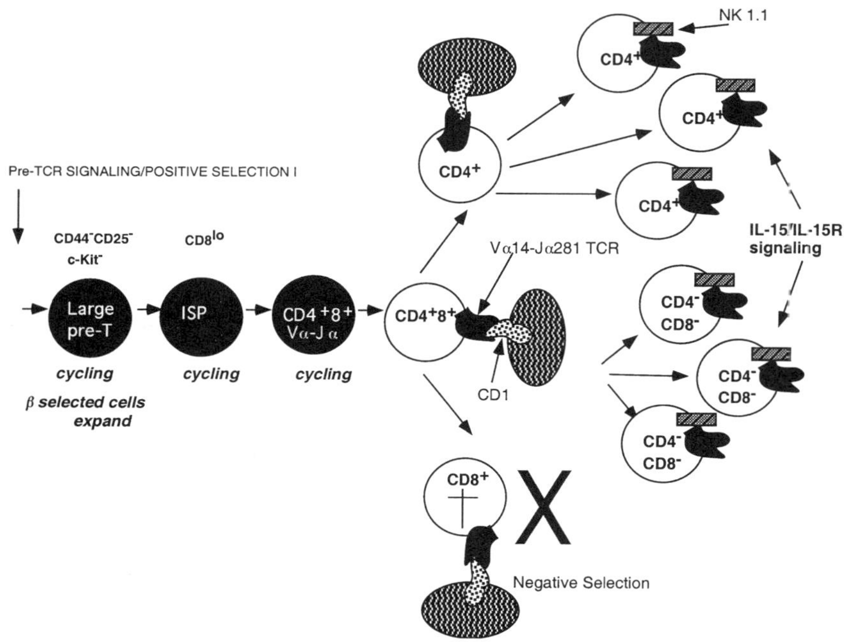

FIGURE 9.14. The development of NT cells in the thymus from αβ precursors expressing invariant TCRs. CD1.1 is expressed on cortical thymocytes and influences the selection and development of NT cells. It is presumed that DP T cells expressing Vα14-Jα281 TCRs are triggered by CD1 and may receive signals to induce the expression of NK complex genes (Ly49 and NKPR1). Exactly how DN NT cells are generated is not known. $CD8^+$ cells are negatively selected. The development or survival of NT cells is compromised in mice that lack IRF-1 and the IL-15Rα chain. Exactly when IL-15 signaling influences NT cell development is unclear.

Some cells that have all the characteristics of NT cells but that do not express NK 1.1 have been described (even in mice strains that do express NK 1.1 on most NT and NK cells). These cells probably develop in a similar fashion to other NT cells.

Some NT cells express CD3 ζ/ FcεRIγ heterodimers. However, NT cell generation is not compromised in mice lacking FcεRIγ. Like mainstream T cells, development of NT cells does not proceed in the absence of CD3ζ. Interestingly the development and expansion of NT cells apparently does not require the Ras/Raf/ERK pathway that is needed during the positive selection of conventional αβ T cells. This indirectly supports the view that NT cells are selected by CD1 at the DP stage of αβ T cell development.

A separate population of αβ lineage DN cells has been described in the bone marrow. These cells do not express the canonical Vα14-Jα 281 rerrangement. They have been considered to represent a "natural suppressor" population that can inhibit graft-versus-host disease. Similar DN αβ T cell populations have been described in normal lymph nodes and liver and have also been described to accumulate in *lpr*$^{-/-}$ mice and in human patients with the autoimmune lymphoproliferative syndrome (discussed in Chapter 12). In the absence of Fas-mediated AICD, activated SP T cells that should normally be eliminated may downregulate coreceptor expression, possibly by recapitulating events that occur during the lineage commitment decision. The development and functions of all these minor T cell populations, however, remains poorly understood.

Summary and Perspective

HSCs generate partially committed progenitors that migrate to the thymus and progressively become more committed to the T lineage. While intermediate stages of commitment may be defined by surface markers, little is understood of the actual commitment process itself. GATA-3 represents a transcription factor required for entry into the T lineage. Most T cells belong to the αβ lineage and develop in the thymus. Commitment to the αβ or γδ lineages may involve a "leaky" instructive mechanism, but it remains possible that stochastic events drive this developmental decision. A critical function of the thymus is to compartmentalize and separate precursor cells from more mature cells that can inhibit the maturation process. Precursors develop in the thymic cortex. Cortical epithelial cells provide poorly understood cell surface and extracellular matrix ligands for receptors on T cell progenitors. They also secrete IL-7 and SCF, which are required for the survival and proliferation of T lineage precursors and either directly or indirectly influence the commitment process. TCRβ gene rearrangement is induced in cortical thymocytes, and while Eβ (the enhancer in the TCR β locus) is required, the actual cell surface signals that trigger accessibility factors, and the specific cis-acting sequences that bind to these unknown accessibility factors, remain to be identified. The coming years will probably see the identification of the extracellular ligands and receptors that might drive commitment, and gene-targeting experiments will undoubtedly be conducted to establish which specific cis-acting motifs in the TCRβ locus are required for the initiation of V(D)J rearrangement at this locus. TCRβ chains contribute to the formation of the pre-TCR, which probably signals constitutively in a ligand-independent fashion to select cells that have made in-frame rearrangements at the TCRβ locus. The pre-TCR is required for survival and subsequent proliferation of selected pre-T cells and for the initiation of allelic exclusion, probably both at the level of the transient induction of Rag-2 degradation and the subsequent loss of recombinational accessibility. Pre-TCR signals also probably contribute to the induction of TCR α accessibility in an Eα-dependent manner, although the specific cis-acting motifs and accessibility factors remain to be identified.

Following β selection, DP T cells may be successively positively and negatively selected. Positive selection depends on the induction of TCR signaling by low- to moderate-affinity MHC–peptide ligands. High-affinity MHC–peptide ligands induce deletion of T cells, and T cells that do not recognize MHC–like shapes probably "die by neglect." Positive selection is initiated by MHC–peptide ligands on cortical epithelial cells, and negative selection is completed in the medulla by MHC–peptide complexes that bind with high affinity to cognate TCRs. Differences in costimulators, processing enzymes, and the levels of glucocorticoids in the cortex and medulla could conceivably contribute to the temporal and physical separation of positive and negative selection in the cortex and medulla, respectively (this separation is not absolute). Positive selection is accompanied by a lineage commitment decision whereby DP cells shut off expression either of CD4 or CD8. Lineage commitment may involve instructive processes or may be achieved by an asymmetric commitment mechanism that is stochastic in part. Even if the lineage commitment process is completely stochastic (and this is unlikely), selection processes must subsequently dictate the survival of cells with TCRs being matched, based on the class of MHC recognized, with either the CD4 or CD8 coreceptors.

A significant proportion of γδ (and some αβ) T cells may develop extrathymically; an important site for IEL generation is intestinal cryptopatches. Invariant γδ T cells that develop in a fetal thymic environment from fetal HSCs are not selected by antigen but are actually programmed to initiate very specific patterns of TCR γ and δ gene rearrangement. NT cells express canonical Vα14-Jα281 rearrangements but appear to emerge primarily in the thymus by the selection of specific αβ T cells driven by CD1 ligands expressed on cortical thymocytes.

Selected Reviews

Anderson, G., Moore, N. C., Owen, J. J. T., and Jenkinson, E. J. (1996). Cellular interactions in thymocyte development. Annu. Rev. Immunol. *14*, 73–99.

Bosma, M. J., and Carroll, A. M. (1991). The SCID mouse mutant: definition, characterization, and potential uses. Annu. Rev. Immunol. *9*, 323–350.

Cresswell, P. (1998). Proteases, processing, and thymic selection. Science *280*, 394–395.

Fehling, H. J., and von Boehmer, H. (1997). Early αβ T cell development in the thymus of normal and genetically altered mice. Curr. Opin. Immunol. *9*, 263–275.

Fink, P. J., and Bevan, M. J. (1995). Positive selection of thymocytes. Adv. Immunol. *59*, 99–133.

Fischer, A., and Malissen, B. (1998). Natural and Engineered disorders of lymphocyte development. Science *280*, 237–243.

Freitas, A. A., and Rocha, B. (1999). Peripheral T cell survival. Curr. Opin. Immunol. *11*, 152–156.

Glimcher, L. H., and Singh, H. (1999). Transcription factors in lymphocyte development—T and B cell get together. Cell *96*, 13–23.

Kisielow, P., and von Boehmer, H. (1995). Development and selection of T cells: facts and puzzles. Adv. Immunol. *58*, 87–209.

Krangel, M. S., Hernandez-Munain, C., Lauzurica, P., McMurry, M., Roberts, J. L., and Zhong, X. P. (1998). Developmental regulation of V(D)J recombination at the TCR α/δ locus. Immunol. Rev. *165*, 131–147.

Kuo, C. T., and Leiden, J. M. (1999). Transcriptional regulation of T lymphocyte development and function. Annu. Rev. Immunol. *17*, 149–187.

Mach, B., Steimle, V., Martinez-Soria, E., and Reith, W. (1995). Regulation of MHC class II genes: Lessons from a disease. Annu. Rev. Immunol. *14*, 301–331.

Marrack, P., and Kappler, J. (1997). Positive selection of thymocytes bearing αβ T cell receptors. Curr. Opin. Immunol. *9*, 250–255.

Miller, J. F. A. P. (1994). The thymus: maestro of the immune system. BioEssays *16*, 509–512.

Poussier, P., and Julius, M. (1994). Thymic independent T cell development and selection in the intestinal epithelium. Annu. Rev. Immunol. *12*, 521–553.

Shortman, K., and Wu, L. (1996). Early T lymphocyte progenitors. Annu. Rev. Immunol. *14*, 29–47.

Robey, E., and Fowlkes, B. J. (1998). The αβ versus γδ T cell lineage choice. Curr. Opin. Immunol. *10*, 181–187.

Rodewald, H. R. (1995). Pathways from hematopoietic stem cells to thymocytes. Curr. Opin. Immunol. *7*, 176–187.

von Boehmer, H. (1996). CD4/CD8 lineage commitment: back to instruction? J. Exp. Med. *183*, 713–715.

von Boehmer, H., Aifantis, I., Feinberg, J., Lechner, O., Saint-Ruf, C., Walter, U., Buer, J., and Azogui, O. (1999). Pleiotropic changes controlled by the pre-T receptor. Curr. Opin. Immunol. *11*, 135–142.

Zuniga-Pflucker, J. C., and Lenardo, M. J. (1996). Regulation of thymocyte development from immature progenitors. Curr. Opin. Immunol. *8*, 215–224.

Some Relevant References

Akashi, K., Kondo, M., von Freeden-Jeffry, U., Murray, R., and Weissman, I. L. (1997). Bcl-2 rescues T lymphopoiesis in interleukin-7 receptor deficient mice. Cell *89*, 1033–1041.

Alberola-Ila, J., Forbush, K. A., Seger, R., Krebs, E. G., and Perlmutter, R. M. (1995). Selective requirement for MAP kinase activation in thymocyte differentiation. Nature *373*, 620–623.

Alberola-Ila, J., Hogquist, K. A., Swan, K. A., Bevan, M. J., and Perlmutter, R. M. (1996). Positive and negative selection invoke distinct signaling pathways. J. Exp. Med. *184*, 9–18.

Amakawa, R., Hakem, A., Kundig, T. M., Matsuyama, T., Simard, J. J. et al. (1996). Impaired negative selection of T cells in Hodgkin's disease antigen CD30-deficient mice. Cell *84*, 551–562.

Anderson, G., Anderson, K. L., Conroy, L. A., Hallam, T. J., Moore, N. C., Owen, J. J. T., and Jenkinson, E. J. (1995). Intracellular signaling events during positive and negative selection of $CD4^+$ $CD8^+$ thymocytes *in vitro*. J. Immunol. *154*, 3636–3643.

Appleby, M. W., Gross, J. A., Cooke, M. P., Levin, S. D., Qian, X., and Perlmutter, R. M. (1992). Defective T cell receptor signaling in mice lacking the thymic isoform of p59fyn. Cell *750*, 751–763.
Ardavin, C., Wu, L., Li, C. L., and Shortman, K. (1993). Thymic dendritic cells and T cells develop simultaneously within the thymus from a common precursor population. Nature *362*, 761–763.
Ardouin, L., Ismaili, J., Malissen, B., and Malissen, M. (1998). The CD3-γδε and CD 3 ζ/η modules are each essential for allelic exclusion at the T cell receptor β locus but are both dispensable for the initiative of V to (D)J recombination at the T cell receptor β, γ and δ loci. J. Exp. Med. *187*, 105–116.
Arroyo, A. G., Yang, J. T., Rayburn, H., and Hynes, R. O. (1996). Differential requirements for α4 integrins during fetal and adult hematopoiesis. Cell *85*, 997–1008.
Asarnow, D. M., Cado, D., and Raulet, D. H. (1993). Selection is not required to produce invariant T cell receptor γ-gene junctional sequences. Nature *362*, 158–160.
Ashton-Rickardt, P. G., Bandeira, A., Delaney, J. R., Van Kaer, L., Pircher, H. P., Zinkernagel, R. M., and Tonegawa, S. (1994). Evidence for a differential avidity model of T cell selection in the thymus. Cell *76*, 651–663.
Bain, G., Quong, M. W., Soloff, R. S., Hedrick, S. M., and Murre, C. (1999). Thymocyte maturation is regulated by the activity of the helix-loop-helix protein, E47. J. Exp. Med. *190*, 1605–1616.
Baker, J. E., Cado, D., and Raulet, D. H. (1998). Developmentally programmed rearrangement of T cell receptor Vγ genes is controlled by sequences immediately upstream of the Vγ genes. Immunity *9*, 159–168.
Barton, G. M., and Rudensky, A. Y. (1999). Requirement for diverse low-abundance peptides in positive selection of T cells. Sciences *283*, 67–70.
Berger, M. A., Dave, V., Rhodes, M. R., Bosma, G. C., Bosma, M. J., Kappes, D. J., and Wiest, D. L. (1997). Subunit composition of pre-T cell receptor complexes expressed by primary thymocytes: CD3 δ is physiscally associiated but not functionally required. J. Exp. Med. *186*, 1461–1467.
Bhandoola, A., Cibotti, R., Punt, J. A., Granger, L., Adams, A. J., Sharrow, S. O., and Singer, A. (1999). Positive selection as a developmental progression initiated by αβ TCR signals that fix TCR specificity prior to lineage commitment. Immunity *10*, 301–311.
Bikoff, E. K., Huang, L. Y., Episkopu, V., van Meerwijk, J., Germain, R. N., and Robertson, E. J. (1993). Defective major histocompatibility class II assembly, transport, peptide acquisition, and $CD4^+$ T cell selection in mice lacking invariant chain expression. J. Exp. Med. *177*, 1699–1712.
Bories, J. C., Demengeot, J., Davidson, L., and Alt, F. W. (1996). Gene-targeted deletion and replacement mutations of the T cell receptor β-chain enhancer: the role of enhancer elements in controlling V(D)J recombination accessibility. Proc. Natl. Acad. Sci. USA *93*, 7871–7876.
Bories, J. C., Willerford, D. M., Grevin, D., Davidson, L., Camus, A., Martin, P., Stehelin, D., and Alt, F. W. (1995). Increased T cell apoptosis and terminal B cell differentiation induced by inactivation of the Ets-1 proto-oncogene. Nature *377*, 635–638.
Bouvier, G., Watrin, F., Nespetti, M., Verthuy, C., Naquet, P., and Ferrier, P. (1996). Deletion of the mouse T-cell receptor β gene enhancer blocks αβ T cell development. Proc. Natl. Acad. Sci. USA *93*, 7877–7881.

Burkly, L., Hession, C., Ogata, L., Reilly, C., Marconi, L. A., Olson, D., Tizard, R., Cate, R., and Lo, D. (1995). Expression of RelB is required for the development of thymic medulla and dendritic cells. Nature *373*, 531–536.

Byth, K. F., Conroy, L. A., Howlett, S., Smith, A. J., May, J. Alexander, D. R., and Holmes, N. (1996). CD45 null transgenic mice reveal a positive regulatory role for CD45 in early thymocyte development, in the selection of $CD4^{+}CD8^{+}$ thymocytes and B cell maturation. J. Exp. Med. *183*, 1707–1718.

Cao, X., Shores, E. W., Hu-Li, J., Anver, M. R., Kelsall, B. L., Russel, S. M., Drago, J., Noguchi, M., Grinberg, A., and Bloom, E. T., et al. (1995). Defective lymphoid development in mice lacking expression of the common cytokine receptor γ chain. Immunity *2*, 223–238.

Carlyle, J. R., and Zuniga-Pflucker, J. C. (1998). Requirement for the thymus in $\alpha\beta$ T lymphocyte development. Immunity *9*, 187–198.

Castro, J. E., Listman, J. A., Jacobson, B. A., Wang, Y., Lopez, P. A., Ju, S., Finn, P. W., and Perkins, D. L. (1996). Fas modulation of apoptosis during negative selection of thymocytes. Immunity *5*, 617–627.

Chen, Y. H., Chiu, N. M., Mandal, M., Wang, N., and Wang, C. R. (1997). Impaired $NK1^{+}$ T cell development and early IL-4 production in CD1 deficient mice. Immunity *6*, 459–467.

Cheng, A. M., Negishi, I., Anderson, S. J., Chan, A. C., Bolen, J., Loh, D. Y., and Pawson, T. (1997). The Syk and ZAP-70 SH2 containing tyrosine kinases are implicated in pre-T cell receptor signaling. Proc. Natl. Acad. Sci. USA *94*, 9797–9801.

Cheng, A. M., Rowley, B., Pao, W., Hayday, A., Bolen, J. B., and Pawson, T. (1995). Syk tyrosine kinase required for mouse viability and B cell development. Nature *378*, 303–307.

Cibotti, R., Punt, J. A., Dash, K. S., Sharrow, S. O., and Singer, A. (1997). Surface molecules that drive T cell development *in vitro* in the absence of thymic epithelium and in the absence of lineage-specific signals. Immunity *6*, 245–264.

Clements, J. L., Yang, B., Ross-Barta, S. E., Eliason, S. L., Hrstka, R. F., Williamson, R. A., and Koretzky, G. A. (1998). Requirement for the leukocyte specific adapter protein SLP-76 for normal T cell development. Science *281*, 416–419.

Cosgrove, D., Gray, D., Dierich, A., Kaufman, J., Lemeur, M., Benoist, C., and Mathis, D. (1991). Mice lacking MHC class II molecules. Cell *66*, 1051–1066.

Crompton, T., K. C., G., and Owen, M. J. (1996). The MAP kinase pathway controls differentiation from double negative to double positive thymocytes. Cell *86*, 243–251.

Danska, J. S., Holland, D. P., Mariathasan, S., Williams, K. M., and Guidos, C. (1996). Biochemical and genetic defects in the DNA protein kinase in murine scid lymphocytes. Mol. Cell. Biol. *16*, 5507–5517.

Dave, V. P., Cao, Z., Browne, C., Alarcon, B., Fernandez-Miguel, G., Lafaille, J., de la Hera, A., Tonegawa, S., and Kappes, D. J. (1997). CD3 δ deficiency arrests development of the $\alpha\beta$ but not the $\gamma\delta$ lineage. EMBO J. *16*, 1360–1370.

de la Salle, H., Hanau, D., Fricker, D., Urlacher, A., Kelly, A., Salamero, J., Powis, S. H., and Donato, L., et al. (1994). Homozygous human TAP peptide transporter mutation in HLA class I deficiency. Science *265*, 237–241.

Deftos, M. L., He, Y. W., Ojala, E. W., and Bevan, M. J. (1998). Correlating notch signaling with thymocyte maturation. Immunity *9*, 777–786.

Donda, A., Schulz, M., Buerki, K., DeLibero, G., and Uematsu, Y. (1996). Identification and characterization of a human CD4 silencer. Eur. J. Immunol. *26*, 493–500.

Douek, D. C., McFarland, R. D., Keiser, P. H., Gage, E. A., Massey, J. M., Haynes, B. F., Polis, M. A., Haase, A. T., Feinberg, M. B., Sullivan, J. L., Jamieson, B. D., Zack, J. A., Picker, L. I., and Koup, R. A. (1998). Changes in thymic function with age and during the treatment of HIV infection. Nature *396*, 690–695.

Dudley, E. C., Petri, H. T., Shah, L. M., Owen, M. J., and Hayday, A. C. (1994). T cell receptor β chain gene rearrangement and selection during thymocyte development in adult mice. Immunity *1*, 83–93.

Elliott, E. A., Drake, J. R., Amigorena, S., Elsemore, J., Webster, P., Mellman, I., and Flavell, R. A. (1994). The invariant chain is required for intracellular transport and function of major histocompatibility complex class II molecules. J. Exp. Med. *179*, 681–694.

Ellmeier, W., Sunshine, M. J., Losos, K., and Littman, D. R. (1998). Multiple developmental stage-specific enhancers regulate CD8 expression in developing thymocytes and in thymus-independent T cells. Immunity *9*, 485–496.

Ernst, B., Lee, D.-S., Chang, J. M., Sprent, J., and Surh, C. D. (1999). The peptide ligands mediating positive selection in the thymus control T cell survival and homeostatic proliferation in the periphery. Immunity *11*, 173–182.

Fehling, H. J., Krotkova, A., Saint-Ruf, C., and von Boehmer, H. (1995). Crucial role of the pre-T receptor a gene in development of αβ but not γδ T cells. Nature *375*, 795–798.

Fischer, K.-D., Zmuidzinas, A., Gardner, S., Barbacid, M., Bernstein, A., and Guidos, C. (1995). Defective T cell receptor signalling and positive selection of Vav deficient $CD4^+CD8^+$ thymocytes. Nature *374*, 474–478.

Fung-Leung, W. P., Surh, C. D., Liljedahl, M., Pang, J., Leturcq, D., Peterson, P. A., Webb, S. R., and Karlson, L. (1996). Antigen presentation and T cell development in H2-M deficient mice. Science *271*, 1278–1281.

Garcia-Ojeda, M. E., Dejbakhsh-Jones, S., Weissman, I. L., and Strober, S. (1998). An alternate pathway for T cell development supported by the bone marrow microenvironment: recapitulation of thymic maturation. J. Exp. Med. *187*, 1813–1823.

Georgopoulos, K., Bigby, M., Wang, J. H., Molnar, A., Wu, P., Winandy, S., and Sharpe, A. (1994). Early arrest in lymphocyte differentiation in Ikaros mutant mice. Cell *78*, 143–156.

Godfrey, D. I., Kennedy, J., Suda, T., and Zlotnik, A. (1993). A developmental pathway involving four phenotypically and functionally distinct subsets of CD3-CD4-CD8-triple negative adult mouse thymocytes defined by CD44 and CD25 expression. J. Immunol. *155*, 4244–4252.

Goldrath, A. W., and Bevan, M. J. (1999). Low affinity ligands for the TCR drive proliferation of mature $CD8^+$ cells in lymphopenic hosts. Immunity *11*, 183–190.

Groettrup, M., Ungeweiss, K., Azogui, O., Palacios, R., Owen, M. J., Hayday, A. C., and von Boehmer, H. (1993). A novel disulfide linked heterodimer on pre-T cells consists of the T cell receptor β chain and a 33 kd glycoprotein. Cell *75*, 283–294.

Groh, V., Steinle, A., Bauer, S., and Spies, T. (1998). Recognition of stress-induced MHC molecules by intestinal epithelial γδ T cells. Science *279*, 1737–1740.

Groves, T., Smiley, P., Cooke, M. P., Forbush, K., Perlmutter, R. M., and Guidos, C. J. (1996). Fyn can partially substitute for Lck in T lymphocyte development. Immunity *5*, 417–428.

Grusby, M. J., Johnson, R. S., Papaioannou, V. E., and Glimcher, L. H. (1991). Depletion of $CD4^+$ T cells in major histocompatibility complex class II-deficient mice. Science *253*, 1417–1420.

Haks, M. C., Krimpenfort, P., Borst, J., and Kruisbeek, A. M. (1998). The CD3 γ chain is essential for the development of both the TCR αβ and TCR γδ lineages. EMBO J. *17*, 1871–1882.

Higashi, Y., Moribe, H., Takagi, T., Sekido, R., Kawakami, K., Kikutani, H., and Kondoh, H. (1997). Impairment of T cell development in dEF1 mutant mice. J. Exp. Med. *185*, 1467–1480.

Hoffman, E. S., Passoni, L., Crompton, T., Leu, T. M. J., Schatz, D. G., Koff, A., Owen, M. J., and Hayday, A. C. (1996). Productive T cell receptor β-chain gene rearrangement: coincident regulation of cell cycle and clonality during development *in vivo*. Genes Dev. *10*, 948–962.

Hostert, A., Garefalaki, A., Mavria, G., Tolaini, M., Roderick, K., Norton, T., Mee, P. J., Tybulewicz, V. L. J., Coles, M., and Kioussis, D. (1998). Hierarchical interactions of control elements determine CD8α gene expression in subsets of thymocytes and peripheral T cells. Immunity *9*, 497–508.

Jamieson, B. D., Douek, D. C., Killian, S., Hultin, L. E., Scripture-Adams, D. D., Giorgi, J. V., Marelli, D., Koup, R. A., and Zack, J. A. (1999). Generation of functional thymocytes in the human adult. Immunity *10*, 569–576.

Kang, J., Coles, M., Cado, D., and Raulet, D. H. (1998). The developmental fate of T cells is critically influenced by TCR γδ expression. Immunity *8*, 427–438.

Kawai, K., and Ohashi, P. S. (1995). Immunological function of a defined T cell population tolerized to low affinity self antigens. Nature *374*, 68–69.

Kawamoto, H., Ohmura, K., and Katsura, Y. (1997). Direct evidence for the commitment of hematopoietic stem cells to T, B, and myeloid lineages in murine fetal liver. Int. Immunol. *9*, 1011–1019.

Kim, H. K., and Siu, G. (1998). The notch pathway intermediate HES-1 silences CD4 gene expression. Mol. Cell. Biol. *18*, 7166–7175.

Kishimoto, H., Cai, Z., Brunmark, M., Jackson, M., Peterson, P. A., and Sprent, J. (1996). Differing roles for B7 and ICAM 1 in negative selection of thymocytes. J. Exp. Med. *184*, 531–537.

Kishimoto, H., Surh, C. D., and Sprent, J. (1998). A role for Fas in negative selection of thymocytes *in vivo*. J. Exp. Med. *187*, 1427–1438.

Koller, B. H., Marrack, P., Kappler, J. W., and Smithies, O. (1990). Normal development of mice deficient in β2 microglobulin, MHC class I proteins, and $CD8^+$ T cells. Science *248*, 1227–1230.

Kong, Y.-Y., Yoshida, H., Sarosi, I., Tan, H.-L., Timms, E., Capparelli, C., Morony, S., Oliviera-dos-Santos, A. J., Van, G., Itie, A., Khoo, W., Wakeham, A., Dunstan, C. R., Lacey, D. L., Mak, T. W., Boyle, W. J., and Penninger, J. M. (1999). OPGL is a key regulator of osteoclastogenesis, lymphocyte development, and lymph-node organogenesis. Nature *397*, 315–323.

Krotkova, A., von Boehmer, H., and Fehling, H. J. (1997). Allelic exclusion in preTα deficient mice: no evidence for cell surface expression of two T cell receptor β chains but less efficient inhibition of endogenous Vβ to DJβ rearrangements in the presence of a functional TCR β transgene. J. Exp. Med. *186*, 767–775.

Kuo, C. T., Veselits, M. L., and Leiden, J. M. (1997). LKLF: a transcriptional regulator of single-positive T cell quiescence and survival. Science *277*, 1986–1990.

Lacaud, G., Carlsson, L., and Keller, G. (1998). Identification of a fetal hematopoietic precursor with B cell, T cell, and macrophage potential. Immunity *9*, 827–838.

Lalli, E., Sassone-Corsi, P., and Ceredig, R. (1996). Block of T lymphocyte differentiation by activation of the cAMP-dependent signal transduction pathway. EMBO J. *15*, 528–537.

Lantz, O., Sharara, L. I., Tilloy, F., Andersson, A., and DiSanto, J. P. (1997). Lineage relationships and differentiation of Natural Killer (NK) T cells: intrathymic selection and interleukin-4 production in the absence of NKR-P1 and Ly49 molecules. J. Exp. Med. *185*, 1395–1401.

Levelt, C. N., Mombaerts, P., Wang, B., Kohler, H., Tonegawa, S., Eichmann, K., and Terhorst, C. (1995). Regulation of thymocyte development through CD3: functional dissociation between p56lck and CD3ζ in early thymic selection. Immunity *3*, 215–222.

Love, P. E., Shores, E. W., Johnson, M. D., Tremblay, M. L., Lee, E. J., Grinberg, A., Huang, S. P., Singer, A., and Westphal, H. (1993). T cell development in mice that lack the ζ chain of the T cell antigen receptor complex. Science *261*, 918–921.

Malissen, M., Gillet, A., Ardouin, L., Bouvier, G., Trucy, J., Ferrier, P., Vivier, E., and Malissen, B. (1995). Altered T cell development in mice with a targeted mutation of the CD3 ε gene. EMBO J. *14*, 4641–4653.

Malissen, M., Gillet, A., and Rocha, B., et al. (1993). T cell development in mice lacking the CD3 ζ/ε gene. EMBO J. *12*, 4347–4355.

Malissen, M., Pereira, P., Gerber, D. J., Malissen, B., and DiSanto, J. P. (1997). The common cytokine receptor γ chain controls survival of γδ T cells. J. Exp. Med. *186*, 1277–1285.

Mallick, C. A., Dudley, E. C., Viney, J. L., Owen, M. J., and Hayday, A. C. (1993). Rearrangement and diversity of T cell receptor β chain genes in thymocytes: a critical role for the β chain in development. Cell *75*, 513–519.

Manley, N. R., and Capecchi, M. R. (1998). Hox group 3 paralogs regulate the development and migration of the thymus, thyroid and parathyroid glands. Dev. Biol. *195*, 1–15.

Manley, N. R., and Capecchi, M. R. (1995). The role of Hox-a3 in mouse thymus and thyroid development. Development *121*, 1989–2003.

Maraskovsky, E., O'Reilly, L. A., Teepe, M., Corcoran, L. M., Peschon, J. J., and Strasser, A. (1997). Bcl-2 can reduce T lymphocyte development in interleukin-7 receptor deficient mice but not in mutant rag-1–/– mice. Cell *89*, 1011–1019.

Martin, W. D., Hicks, G. G., Mendiratta, S. K., Leva, H. I., Ruley, H. E., and Van Kaer, L. (1996). H2-M mutant mice are defective in the peptide loading of class II molecules, antigen presentation, and T cell repertoire selection. Cell *84*, 543–550.

Matsuyama, T., Kimura, T., and Kitagawa, M., et al. (1993). Targeted disruption of IRF-1 and IRF-2 results in abnormal type I IFN gene induction and aberrant lymphocyte development. Cell *75*, 83–97.

Matechak, E. O., Killeen, N., Hedrick, S. M., and Fowlkes, B. J. (1996). MHC class II-specific T cells can develop in the CD8 lineage when CD4 is absent. Immunity, *4*, 337–347.

Mendiratta, S., Martin, W. D., Hong, S., Boesteanu, A., Joyce, S., and Van Kaer, L. (1997). Cd1d1 mutant mice are deficient in natural T cells that promptly produce Il-4. Immunity *6*, 469–477.

Miyazaki, T. (1997). Two distinct steps during thymocyte maturation from CD4$^-$CD8$^-$ to CD4$^+$CD8$^+$ distinguished in the early growth response (Egr)-1 transgenic mice with a recombinase activating gene deficient background. J. Exp. Med. *186*, 877–885.

Miyazaki, T., Wolf, P., Tourne, S., Waltzinger, C., Dierich, A., Barois, N., Ploegh, H., Benoist, C., and Mathis, D. (1996). Mice lacking H2-M complexes, enigmatic elements of the MHC class II peptide-loading pathway. Cell *84*, 531–541.

Molina, T. J., Kishihara, K., Siderovski, D. P., et al. (1992). Profound block in thymocyte development in mice lacking p56lck. Nature *357*, 161–164.

Mombaerts, P., Clarke, A. R., Rudnicki, M. A., Iacomini, J., Itohara, S., Lafaille, J. J., Ichikawa, Y., Jaenisch, R., Hooper, M. L., and Tonegawa, S. (1992a). Mutations in T cell receptor genes α and β block thymocyte development at different stages. Nature *360*, 225–231.

Mombaerts, P., Iacomini, J., Johnson, R. S., Herrup, K., Tonegawa, S., and Papaioannou, V. E. (1992b). Rag-1 deficient mice have no mature T and B lymphocytes. Cell *68*, 869–877.

Monroe, R. J., Sleckman, B. P., Monroe, B. C., Khor, B., Claypool, S., Ferrini, R., Davidson, L., and Alt, F. W. (1999). Developmental regulation of TCRδ locus accessibility and expression by the TCRδ enhancer. Immunity *10*, 503–514.

Muller, K. M., Luedecker, C. J., Udey, M. C., and Farr, A. G. (1997). Involvement of E-cadherin in thymus organaogenesis and thymocyte maturation. Immunity *6*, 257–264.

Muthusamy, N., Barton, K., and Leiden, J. M. (1995). Defective activation and survival of T cells lacking the Ets-1 transcription factor. Nature *377*, 639–642.

Nakagawa, T., Roth, W., Wong, P., Nelson, A., Farr, A., Villadabgos, J. A., Ploegh, H., Peters, C., and Rudensky, A. Y. (1998). Cathepsin L; critical role in Ii degradation and CD4 T cell selection in the thymus. Science *280*, 450–453.

Negishi, I., Motoyama, N., Nakayama, K., Nakayama, K., Senju, S., Hatakeyama, S., Zhang, Q., Chan, A. C., and Loh, D. Y. (1995). Essential role for ZAP-70 in both positive and negative selection of thymocytes. Nature *376*, 435–438.

Nehls, M., Kyewski, B., Messerle, M., Waldschutz, R., Schuddekopf, K., Smith, A. J. H., and Boehm, T. (1996). Two genetically separable steps in the differentiation of thymic epithelium. Science *272*, 886–889.

Nehls, M., Pfeifer, D., Schorpp, M., Hedrich, H., and Boehm, T. (1994). New member of the winged-helix protein family in mouse and rat nude mutations. Nature *372*, 103–107.

Nussenzweig, A., Chen, C., daCosta Soares, V., Sanchez, M., Sokol, K., Nussenzweig, M. C., and Li, G. C. (1996). Requirement for Ku80 in growth and immunoglobulin V(D)J recombination. Nature *382*, 551–555.

Ohno, H., Aoe, T., Taki, S., Kitamura, F., Ishida, Y., Rajewsky, K., and Saito, T. (1993). Developmental and functional impariment of T cells in mice lacking CD3 ζ chains. EMBO J. *12*, 4357–4366.

Ohteki, T., Yoshida, H., Matsuyama, T., Duncan, G. S., Mak, T. W., and Ohashi, P. S. (1998). The transcription factor interferon regulatory factor-1 (IRF-1) is important during the maturation of natural killer 1.1$^+$ T cell receptor αβ$^+$ (NK1$^+$T) cells, natural killer cells and intestinal intraepithelial cells. J. Exp. Med. *187*, 967–972.

Okamura, R. W., Sigvardsson, M., Galceran, J., Verbeek, S., Clevers, H., and Grosschedl, R. (1998). Redundant regulation of T cell differentiation and TCRα gene expression by the transcription factors IEF-1 and TCF-1. Immunity *8*, 11–20.

Oosterwegel, M. A., Haks, M. C., Jeffry, U., Murray, R., and Kruisbeek, A. M. (1997). Induction of TCR gene rearrangements in uncommitted stem cells by a subset of IL-7 producing MHC class-II expressing thymic stromal cells. Immunity *6*, 351–360.

Penninger, J. M., Sirard, C., Mittrucker, H.-W., Chidgey, A., Kozieradzki, I., Nghiem, M., Boyd, R., Taniguchi, T., Matsuyama, T., and Mak, T. W. (1997). The interferon regulatory transcription factor IRF-1 controls positive and negative selection of $CD8^+$ thymocytes. Immunity *7*, 243–254.

Peschon, J. J., Morrissey, P. J., Grabstein, K. H., Ramsdell, F. J., Maraskovsky, E., Gliniak, B. C., Park, L. S., Ziegler, S. F., Williams, D. E., and Ware, C. B., et al. (1994). Early lymphocyte expansion is severely impaired in interleukin-7 deficient mice. J. Exp. Med. *180*, 1955–1960.

Petrie, H. T., Livak, F., Burtrum, D., and Mazel, S. (1995). T cell receptor gene recombination patterns and mechanisms: cell death, rescue, and T cell production. J. Exp. Med. *182*, 121–127.

Philpott, K. L., Viney, J. L., Kay, G., Rstan, S., Gardner, E. M., Chae, S., Hayday, A. C., and Owen, M. J. (1992). Lymphoid development in mice congenitally lacking T cell receptor αβ expressing cells. Science *256*, 1448–1452.

Pivniouk, V., Tsitsikov, E., Swinton, P., Rathbun, G., Alt, F. W., and Geha, R. S. (1998). Impaired viability and profound block in thymocyte development in mice lacking the adaptor protein SLP-76. Cell *94*, 229–238.

Pui, J. C., Allman, D., Xu, L., DeRocco, S., Karnell, F. G., Bakkour, S., Lee, J. Y., Kadesch, T., Hardy, R. R., Aster, J. C., and Pear, W. S. (1999). Notch 1 expression in early lymphopoiesis influences B versus T lineage determination. Immunity *11*, 299–308.

Radtke, F., Wilson, A., Stark, G., Bauer, M., van Meerwijk, J., MacDonald, H. R., and Aguet, M. (1999). Deficient T cell fate specification in mice with an induced inactivation of Notch 1. Immunity *10*, 547–558.

Reis, L. F., Ruffner, H., Stark, G., Aguet, M., and Weismann, C. (1993). Mice devoid of interferon regulatory factor-1 (IRF-1) show normal expression of type I interferon genes. EMBO J. *13*, 4798–4806.

Rincon, M., Whitmarsh, A., Yang, D. D., Weiss, L., Derijard, B., Jayaraj, P., and Davis, R. J. (1998). The JNK pathway regulates the *in vivo* deletion of immature $CD4^+CD8^+$ thymocytes. J. Exp. Med. *188*, 1817–1830.

Rivera, R. R., Johns, C. P., Quan, J., Johnson, R. S., and Murre, C. (2000). Thymocyte selection is regulated by the helix-loop-helix inhbitor protein, Id3. Immunity, *12*, 17–26.

Robey, E., Chang, D., Itano, A., Cado, D., Alexander, H., Lans, D., Weinmaster, G., and Salmon, P. (1996). An activated form of Notch influences the choice between CD4 and CD8 T cell lineages. Cell *87*, 483–492.

Rodewald, H. R., Kretzschmar, K., Takeda, S., Hohl, C., and Dessing, M. (1994). Identification of pro-thymocytes in murine fetal blood: T lineage commitment can precede thymus colonization. EMBO J. *13*, 4229–4240.

Rodewald, H. R., Ogawa, M., Haller, C., Waskow, C., and DiSanto, J. P. (1997). Pro-thymocyte expansion by c-kit and the common cytokine receptor γ chain is essential for repertoire formation. Immunity *6*, 265–272.

Saito, H., Kanamori, Y., Takemori, T., Nariuchi, H., Kubota, E., Takahashi-Iwanaga, H., Iwanaga, T., and Ishikawa, H (1998). Generation of intestinal T cells from progenitors residing in gut cryptopatches. Science *280*, 275–279.

Sawada, S., and Littman, D. R. (1991). Identification and characterization of a T-cell specific enhancer adjacent to the murine CD4 gene. Mol. Cell. Biol. *11*, 5506–5515.

Sawada, S., Scarborough, J. D., Killeen, N., and Littman, D. R. (1994). a lineage specific transcriptional silencer regulated CD4 gene expression during T lymphocyte development. Cell *77*, 917–929.

Schilham, M. W., Moerer, P., Cumano, A., and Clevers, H. C. (1997). Sox-4 facilitates thymocyte differentiation. Eur. J. Immunol. *27*, 1292–1295.

Schmedt, C., Saijo, K., Niidome, T., Kuhn, R., Aizawa, S., and Tarakhovsky, A. (1998). Csk controls receptor-mediated development and selection of T-lineage cells. Nature *394*, 901–904.

Schweighoffer, E., and Fowlkes, B. J. (1996). Positive selection is not required for thymic maturation of transgenic gamma delta T cells. J. Exp. Med. *183*, 2033–2041.

Sharp, L. L., Schwarz, D. A., Bott, C. M., Marshall, C. J., and Hedrick, S. M. (1997). The influence of the MAPK pathway on T cell lineage commitment. Immunity *7*, 609–618.

Shinkai, Y., Rathbun, G., Lam, K. P., Oltz, E. M., Stewart, V., Mendelsohn, M., Charron, J., Datta, M., Young, F., Stall, A. M., and Alt, F. W. (1992). RAG-2 deficient mice lack mature lymphocytes owing to inability to initiate V(D)J rearrangement. Cell *68*, 855–867.

Shores, E. W., Ono, M., Kawabe, T., Sommers, C. L., Tran, T., Lui, K., Udey, M. C., Ravetch, J., and Love, P. E. (1998). T cell development in mice lacking all T cell receptor ζ family members (ζ, η, and FcεRIγ). J. Exp. Med. *187*, 1093–1101.

Siu, G., Wurster, A. L., Duncan, D. D., Soliman, T. M., and Hedrick, S. M. (1994). A transcriptional silencer controls the developmental expression of the CD4 gene. EMBO J. *13*, 3570–3579.

Sleckman, B. P., Bardon, C. G., Ferrini, R., Davidson, L., and Alt, F. W. (1997). Function of the TCR α enhancer in αβ and γδ T cells. Immunity *7*, 505–515.

Smiley, S. T., Kaplan, M. H., and Grusby, M. J. (1997). Immunoglobulin E production in the absence of interleukin-4 secreting CD1 dependent cells. Science *275*, 977–979.

Sugawara, T., Moriguchi, T., Nishida, E., and Takahama, Y. (1998). Differential roles of ERK and p38 MAP kinase pathways in positive and negative selection of T lymphocytes. Immunity *9*, 565–574.

Surh, C. D., and Sprent, J. (1994). T-cell apoptosis detected in situ during positive and negative selection in the thymus. Nature *372*, 100–103.

Swan, K. A., Alberola-Ila, J., Gross, J. A., Appleby, M. W., Forbush, K. A., Thomas, J. F., and Perlmutter, R. M. (1995). Involvement of p21ras distinguishes positive and negative selection in thymocytes. EMBO J. *14*, 276–285.

Swat, W., Fujikawa, K., Ganaiatsas, S., Yang, D., Xavier, R. J., Harris, N. L., Davidson, N. L., Ferrini, R., Davis, R. J., Labow, M. A., Flavell, R. A., Zon, L. I., and Alt, F. W. (1998). SEK1/MKK4 is required for maintenance of a normal peripheral lymphoid compartment but not for lymphocyte development. Immunity *8*, 625–634.

Swat, W., Shinkai, Y., Cheng, H. L., Davidson, L., and Alt, F. W. (1996). Activated Ras signals differentiation and expansion of $CD4^+8^+$ thymocytes. Proc. Natl. Acad. Sci. USA *93*, 4683–4687.

Takeda, S., Rodewald, H. R., Arakawa, H., Bluethmann, H., and Shimizu, T. (1996). MHC class II molecules are not required for survival of newly generated $CD4^+$ T cells but affect their long-term life span. Immunity *5*, 217–228.

Tanchot, C., Lemonnier, F. A., Perarnau, B., Freitas, A. A., and Rocha, B. (1997). Differential requirements for survival and proliferation of CD8 naive or memory T cells. Science *276*, 2057–2062.

Ting, C. N., Olson, M. C., Barton, K. P., and Leiden, J. M. (1996). Transcription factor GATA-3 is required for development of the T cell lineage. Nature *384*, 474–478.

Tomita, K., Hattori, M., Nakamura, E., Nakanishi, S., Minato, N., and Kageyama, R. (1999). The bHLH gene hes1 is essential for expansion or early T cell precursors. Genes Dev. *13*, 1203–1210.

Turner, M., Mee, P. J., Costello, P. S., Williams, O., Price, A. A., Duddy, L. P., Furlong, M. T., Geahlen, R. L., and Tybulewicz, V. (1995). Perinatal lethality and blocked B cell development in mice lacking the tyrosine kinase Syk. Nature *378*, 298–303.

Turner, M., Mee, P. J., Walters, A. E., Quinn, M. E., Mellor, A. L., Zamoyska, R., and Tybulewicz, V. L. (1997). A requirement for the Rho-family GTP exchange factor Vav in positive and negative slection of thymocytes. Immunity *7*, 451–460.

Uematsu, Y., Ryser, S., Dembic, S. Z., Borgulya, P., Krimpenfort, P., Berns, A., Von Boehmer, H., and Steinmetz, M. (1988). In transgenic mice the introduced functional T cell receptor β gene prevents expression of endogenous β genes. Cell *52*, 831–841.

Van Kaer, L., Ashton-Rickardt, P. G., Ploegh, H. L., and Tonegawa, S. (1992). TAP1 mutant mice are deficient in antigen presentation, surface class I molecules, and $CD4^-8^+$T cells. Cell *71*, 1205–1214.

Van Kaer, L., Ashton-Rickardt, P. G., Eichelberger, M., Gaczynska, M., Nagashima, K., Rock, K. L., Goldberg, A. L., Doherty, P. C., and Tonegawa, S. (1994). Altered peptidase and viral specific T cell response in LMP2 mutant mice. Immunity *1*, 533–541.

van Meerwijk, J. P., Marguerat, S., Lees, R. K., Germain, R. N., Fowlkes, B. J., and MacDonald, H. R. (1997). Quantitative impact of thymic clonal deletion on the T cell repertoire. J. Exp. Med. *185*, 377–383.

Van Oers, N. S. C., Lowin-Kropf, B., Finlay, D., Connolly, K., and Weiss, A. (1996). αβ T cell development is abolished in mice lacking both Lck and Fyn protein tyrosine kinases. Immunity *5*, 429–436.

Verbeek, S., Izon, D., Hofhuis, F., Robanus-Maandag, E., te Riele, H., van de Wetering, M., Oosterwegel, M., Wilson, A., MacDonald, H.R., and Clevers, H. (1995). An HMG box containing T cell factor required for thymocyte differentiation. Nature *374*, 70–74.

Viville, S., Neefjes, J., Lotteau, V., Dierich, A., Lemeur, M., Ploegh, H., Benoist, C., and Mathis, D. (1993). Mice lacking the MHC class II-associated invariant chain. Cell *72*, 635–648.

von Freeden-Jeffry, U., Solvason, N., Howard, M., and Murray, R. (1997). The earliest T-lineage committed cells depend on IL-7 for Bcl-2 expression and normal cell cycle progression. Immunity *7*, 147–154.

Von Freeden-Jeffry, U., Vieira, P., Lucian, L. A., McNeil, T., Burdach, S. E. G., and Murray, R. (1995). Lymphopenia in IL-7 gene-deleted miceidentifies IL-7 as a non-redundant cytokine. J. Exp. Med. *181*, 1519–1526.

Wallin, J., Eibel, H., Neubuser, A., Wilting, J., Koseki, H., and Balling, R. (1996). Pax 1 is expressed during development of the thymus epithelium and is required for normal T cell maturation. Development *122*, 23–30.

Washburn, T., Schweighoffer, E., Gridley, T., Chang, D., Fowlkes, B. J., Cado, D., and Robey, E. (1997). Notch activity influences the αβ versus γδ T cell lineage decision. Cell *88*, 833–843.

Wiest, D. L., Ashe, J. M., Howcroft, T. K., Lee, H. M., Kemper, D. M., Negishi, I., Singer, D. S., Singer, A., and Abe, R. (1997). A spontaneously arising mutation in the DLAARN motif of murine ZAP-70 abrogates kinase activity and arrests thymocyte development. Immunity *6*, 663–671.

Wilson, A., deVillartay, J. P., and MacDonald, H. R. (1996). T cell receptor δ gene rearrangement and T early α (TEA) expression in immature αβ lineage thymocytes: implications for αβ/γδ lineage commitment. Immunity *4*, 37–45.

Winandy, S., Wu, P., and Georgopoulos, K. (1995). A dominant mutation in the Ikaros gene leads to rapid development of leukemia and lymphoma. Cell *83*, 289–299.

Wu, L., D'Amico, A., Winkel, K. D., Suter, M., Lo, D., and Shortman, K. (1998). RelB is essential for the development of myeloid-related $CD8\alpha^-$ dendritic cells but not of lymphoid related $CD8\alpha^+$ dendritic cells. Immunity *9*, 839–847.

Wu, L., Scollay, R., Egerton, M., Pearse, M., and Spangrude, G. J. (1991). CD4 expressed on earliest T-lineage precursor cells in the adult murine thymus. Nature *349*, 71–74.

Xu, Y., Davidson, L., Alt, F. W., and Baltimore, D. (1996). Function of the pre-T cell receptor alpha chain in T cell development and allelic exclusion at the T cell receptor deta locus. Proc. Natl. Acad. Sci. USA *93*, 2169–2173.

Zerrahn, J., Held, W., and Raulet, D. H. (1997). The MHC reactivity of the T cell repertoire prior to positive and negative selection. Cell *88*, 627–636.

Zhang, Y., Cado, D., Asarnow, D. M., Komori, T., Alt, F. W., Raulet, D. H., and Allison, J. P. (1995). The role of short homology repeats and TdT in generation of the invariant γδ antigen receptor repertoire in the fetal thymus. Immunity *3*, 439–447.

Zhang, W., Sommers, C. L., Burshtyn, D. N., Stebbins, C. C., DeJarnette, J. B., Trible, R. P., Grinberg, A., Tsay, H. C., Jacobs, H. M., Kessler, C. M., Long, E. O., Love, P. E., and Samelson, L. E. (1999). Essential role of LAT in T cell development. Immunity *10*, 323–332.

Zhong, X. P., and Krangel, M. S. (1997). An enhancer blocking element between α and δ gene segments within the human T cell receptor α/δ locus. Proc. Natl. Acad. Sci. USA *94*, 5219–5224.

Zhu, C., Bogue, M. A., Lim, D. S., Hasty, P., and Roth, D. B. (1996). Ku86-deficient mice exhibit severe combined immunodeficiency and defective processing of V(D)J recombination intermediates. Cell *86*, 379–389.

Zijlstra, M., Bix, M., Simister, N. E., Loring, J. M., Raulet, D. H., and Jaenisch, R. (1990). β2-microglobulin deficient mice lack $CD4^-8^+$ cytolytic T cells. Nature *344*, 742–746.

Zuniga-Pflucker, J. C., Jiang, D., and Lenardo, M. J. (1995). Requirement for TNFα and IL-1α in fetal thymocyte commitment and differentiation. Science *268*, 1906–1909.

10

The Activation and Differentiation of Mature T Lymphocytes

In this chapter, we will examine the fate of mature T lymphocytes after they have encountered antigen. Pathways involved in signal transduction in T cells, and the participation of coreceptors and costimulators in T cell signaling have been discussed in some detail in Chapter 4. Some aspects of signal transduction and costimulation will be revisited in this chapter in the context of T cell activation and anergy.

T cells assume an activated phenotype when they are optimally stimulated by cognate major histocompatability complex (MHC)–peptide complexes in conjunction with costimulatory ligands. As discussed in Chapter 4, costimulatory ligands are generally induced on antigen-presenting cells (APCs) by microbes or microbial products, which inform the immune system that "danger" lurks and that it needs to respond in a serious fashion. The concomitant stimulation of mature T lymphocytes by peptide–MHC complexes and costimulatory receptors leads to the proliferation of T cells that are also reprogrammed by these activation events to function as effectors. The activation of naive $CD4^+$ T cells initially leads to the induction of interleukin-2 (IL-2) secretion and the expression at high levels of the IL-2 receptor α chain gene. This leads to the proliferation of helper T cells, which may secrete a range of cytokines capable of promoting both humoral and cell-mediated immunity. Prolonged exposure to certain antigens may facilitate the creation in the host of potentially polarizing milieus dominated by specific inductive cytokines. These milieus may influence the differentiation of newly generated naive helper T cells into Th1 or Th2 type cells. As discussed below, the typical activation of a naive $CD4^+$ cell does not necessarily lead to a polarized Th1 or Th2 type of response.

The activation of a naive $CD8^+$ T cell leads it to assume the phenotype of an effector CTL. Pathways involved in T cell activation have not been as extensively studied in $CD8^+$ cells as in $CD4^+$ cells, but there are clearly many common molecular targets of the activation processes in both these cell types. In general, the activation of $CD8^+$ cells occurs optimally when help is also available from activated $CD4^+$ cells in the vicinity. The participation of APCs and the involvement of the CD40/CD40L pathway in CTL activation will be discussed below.

T cells that are triggered in a suboptimal fashion may either not respond in any discernible way or may be rendered anergic. Triggering may be suboptimal because T cell-receptor (TCR) activation was not accompanied by the ligation of an important costimulatory receptor such as CD28, or because the peptide used to stimulate the T cell represented an "altered ligand," which in combination with the appropriate MHC molecule could only get the TCR to engage selected downstream pathways or to signal in a quantitatively muted fashion.

Once a T cell is activated, it may become an effector cell that contributes to the elimination of antigen. A proportion of these effector cells may become memory cells. The available evidence for $CD8^+$ cells supports such a sequence of events, although it is possible that memory cells may be derived from other activated cells that do not pass through the effector stage. Once antigen is cleared, effector cells might fail to receive further survival signals and may wither away and die. The clearance of antigen is probably an important mechanism by which T cell activation is attenuated. In circumstances where antigen persists, one might predict that the host runs the risk of accumulating huge numbers of activated T cells that would in many ways mimic the onset of a leukemic state. However, this generally does not occur, probably because T cells that are restimulated by persistent antigen are induced to commit suicide by a process known as activation-induced cell death (AICD).

Some of the outcomes that might result from the encounter of $CD4^+$ T cells with antigen are summarized in Figure 10.1.

In this chapter, we will initially consider how a T cell responds to suboptimal and optimal stimuli and how specific signals or combinations of signals determine whether the outcome will reflect activation or anergy. We will then discuss the molecular pathways involved in Th1 versus Th2 cell generation and describe what is known about $CD8^+$ T cell activation. Finally, we will discuss the process of AICD and speculate on how a T cell might choose between suicide and potential immortality as a memory cell.

Activation Versus Anergy

Ligation of the TCR by peptide–MHC complexes alone is generally insufficient for "full" activation and may either lead to the absence of T cell activation or to the generation of nonresponsive or anergic T cells. Of course, circumstances do exist where an MHC–peptide complex might have an extremely high affinity for a particular TCR and may therefore be capable of initiating all the required downstream pathways for "full" activation in the absence of costimulation. In general, however, costimulation is essential for the generation of a proliferative response.

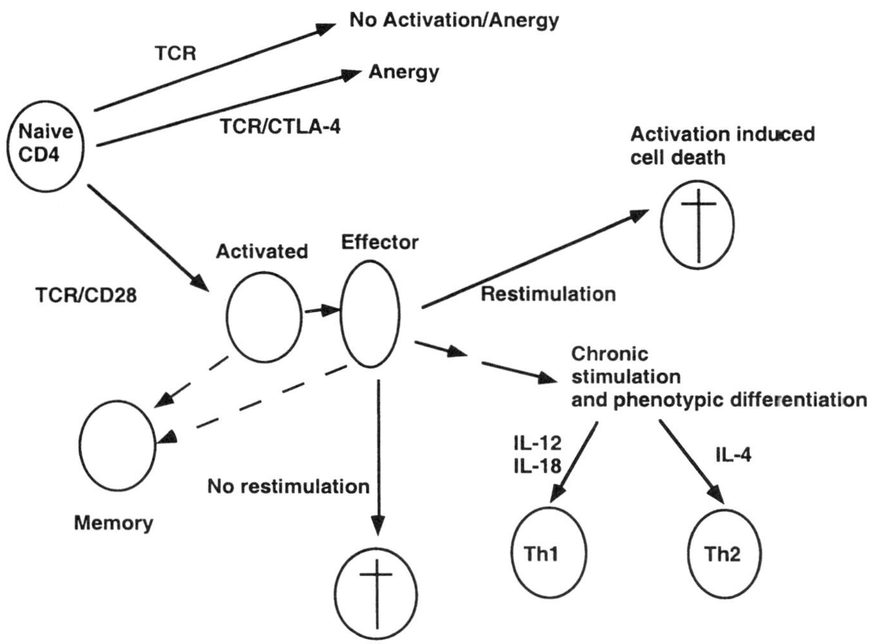

FIGURE 10.1. An overview of some of the decisions that must be made during CD4$^+$ T cell activation and differentiation. Optimal triggering of a T cell with concomitant costimulation primarily via CD28 leads to the generation of activated cells, most of which will become short-lived effector cells. Less than optimal stimulation, usually when costimulation is inadequate, may lead to a failure of activation or to the generation of anergy. The development of anergy may require signaling via CTLA-4. Stimulation of some activated or effector cells may lead to the induction of memory. A large number of effector cells are eliminated by a process known as AICD. Prolonged exposure to antigen such as that seen in a chronic disease may lead to the differentiation of helper T cells into Th1 or Th2 cells. This differentiation event is triggered by the generation of an appropriate inductive cytokine milieu.

It is relatively easy to understand why "incomplete" signaling may fail to activate a T cell. It is, however, unclear why this kind of signaling sometimes leads to the generation of anergy. Indeed, as noted below, it is unclear whether anergy actually is of any significance in "real life."

There are three potential circumstances in which activation might not occur and in which anergy might develop:

1. Signal 1 may be delivered in the absence of Signal 2 (TCR but no CD28).
2. Signal 1 might be "wimpy" (TCR interacts with a low affinity or "altered peptide" ligand).
3. Signal 1 is delivered in conjunction with the wrong type of second signal (TCR in conjunction with cytotoxic T lymphocyte antigen-4, CTLA-4).

In earlier defining studies on anergy that were performed on cloned T cell lines, TCR occupancy in the absence of costimulation not only failed to activate the production of IL-2, but the intracellular signals produced in these circumstances also biochemically altered these T cells in an intriguing way. In these original studies, anergized T cells were unable to produce IL-2 even when restimulated by a combination of Signal 1 and Signal 2. IL-2 production was reduced about 20- to 50-fold and IL-3, granulocyte macrophage colony-stimulating factor (GM-CSF), and interferon-γ (IFN-γ) production were also significantly decreased. The inability to produce IL-2 is a distinguishable characteristic of anergic T cells, and the investigation of T cell activation as well as of anergy has frequently focused on transcriptional regulators that target the IL-2 promoter and enhancer.

As we discussed in Chapter 4, Signal 1 (delivered via the TCR) may induce a number of pathways including the activation of Phospholipase C (PLC)γ and calcium signaling, the induction of the PI3 kinase pathway, the activation of the Ras/Raf/ERK (extracellular signal-regulated kinase) pathway and of the Jun N-terminal kinase (JNK) pathway (albeit with limited effectiveness), among others. Optimal JNK activation however requires CD28 signaling. In the absence of CD28 ligation, PI3 kinase may not be properly activated, because it cannot be recruited to the YXXM site in the C-terminus of CD28. Lipid products generated by PI3 kinase activity may recruit and activate exchange factors for Rac and Cdc42 (these exchange factors contain PH domains that bind to inositol phospholipids phosphorylated in the 3 position of the inositol ring). A critical exchange factor for Rac is Vav. Tyrosine phosphorylation of Vav also contributes to its activation. CD28 might contribute to the activation of Vav via the activation of PI3 kinase. Tyrosine phosphorylation of Vav as well as its (PH domain and PI3 kinase-dependent) recruitment to the membrane may both be required for the activation of this guanine nucleotide exchange factor for Rac. The activation of Vav may represent a crucial step in the induction of the JNK pathway. In terms of the induction of IL-2 transcription, some of the transcriptional activators that are induced downstream of TCR and CD28 signaling include nuclear factor of activated T cells (NFAT), c-Rel, Egr-1, and AP-1. As we have discussed earlier in Chapter 4, calcium signaling downstream of the TCR may be linked to the nuclear translocation of NFAT proteins. Optimal AP-1 activation requires the phosphorylation of critical serine residues of TCF/Elk-1 by ERK and the phosphorylation of critical serine residues on the N-terminus of c-Jun by JNK. Phosphorylated TCF/Elk-1 and cJun are required for the transcriptional induction of c-Fos, and phosphorylated c-Jun is also required for inducing Jun expression. AP-1 (Activation Protein-1) is a heterodimer of Jun and Fos, and AP-1 activity may require that both the ERK and JNK pathways be activated in a T cell. JNK activation also contributes significantly to an increase in the stability of IL-2 mRNA in activated T cells. In

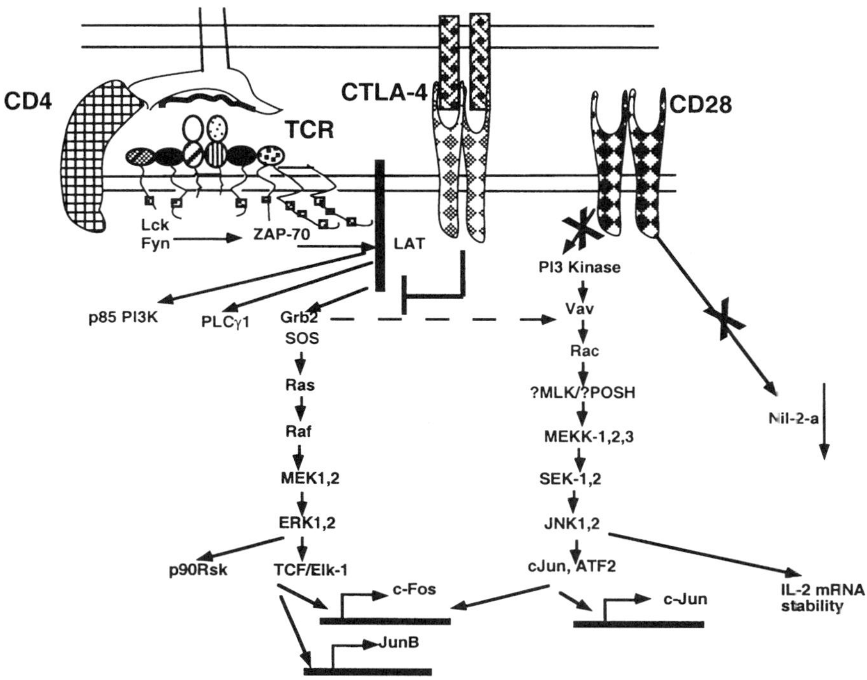

FIGURE 10.2. Selected signaling events in T cells that may be of relevance in considerations of anergy. TCR signaling leads to the activation of nonreceptor tyrosine kinases and the subsequent induction of the Ras/Raf/ERK pathway. ERK enters the nucleus and phosphorylates the TCF/Elk-1 transcription factor, which can induce c-Fos and JunB. Tyrosine phosphorylation activates Vav, an exchange factor for Rac. Vav activation may also require the activation of PI3 kinase activity downstream of CD28. Rac may trigger JNK activation, which contributes to the phosphorylation of c-Jun at serines 63 and 73. JNK also contributes to the stability of IL-2 mRNA. Phosphorylated c-Jun is required for the transcriptional activation of the c-Jun promoter as well as for AP-1 activity in the regulatory elements of genes such as the *IL-2* gene. AP-1 consists of Jun and Fos and requires both ERK and JNK pathways. CD28 signaling may also contribute to shutting off the production of Nil-2-a, which represses IL-2 transcription in anergic cells.

the absence of Signal 2, AP-1 induction is compromised, and this might contribute to the reduced induction of IL-2 transcription. In addition, IL-2 mRNA is rapidly degraded (Figure 10.2).

It is relatively easy to explain how "incomplete signaling" leads to the failure of T cell activation. How and why anergy develops are less well understood.

The Induction of the Anergic State

It is unclear whether "incomplete" or "negative" signaling (involving Signal 1 and CTLA-4) actually induces anergy *in vivo* or merely results in the absence of activation. As pointed out earlier, most studies on T cell anergy have been performed on T cell clones *in vitro*. Although anergy may be generated *in vivo*, it is unclear as to when and why a T cell chooses anergy over slothful slumber.

It is possible that in the absence of "danger" the low levels of B7-1 and B7-2 on APCs can trigger CTLA-4, which binds with high affinity to B7 ligands and may be induced on T cells by Signal 1 alone. Signaling via CTLA-4 may be critical for the induction of the anergic state. *Ctla-4*$^{-/-}$ mice present with a lymphoprolferative disorder, suggesting that negative signaling via CTLA-4 may be important in attenuating an immune response, but direct evidence that CTLA-4 is essential for the generation of anergy has not been obtained. Exactly how CTLA-4 provides inhibitory signals is unclear; it has been reported that CTLA-4 might recruit and activate a tyrosine phosphatase, SHP-2 (SH_2 containing protein tyrosine phosphatase-2), which could potentially dephosphorylate critical activators of the ERK and JNK pathways (such as phosphorylated ITAMs (immunoreceptor tyrosine based activation motifs) Zap-70, LAT (linker for activation of T cells) and Vav).

Stimulation of T cells by altered peptide ligands that induce anergy leads to characteristic alterations in some of the early signaling events downstream of the TCR. Partial tyrosine phosphorylation of the CD3ζ chain (which contains 3 ITAM motifs) has been observed, and this may be linked to the defective recruitment and activation of Zap-70 and a decrease in phosphatidylinositol turnover. Other downstream alterations are discussed below. These early defects in CD3ζ phosphorylation might reflect either the defective activation of Src-family kinases following TCR triggering or dephosphorylation of target proteins by a tyrosine phosphatase, perhaps as a consequence of CTLA-4 ligation. How does incomplete signaling during the anergization process contribute to the biochemical alterations that lead to the anergic state? Why does a failure of Zap-70 activation not lead merely to the absence of T cell activation as opposed to the generation of anergy? Why is this unresponsive state maintained subsequently even when both Signal 1 and Signal 2 are provided to an anergic cell? Answers to these issues have been slow in coming. We are left today with some understanding of the biochemical changes that persist in anergic cells but fewer insights as to how this state is brought about.

Biochemical Characteristics of Anergic Cells

An examination of signaling in anergic T cell clones has revealed the following biochemical alterations when these cells are restimulated using approaches that would normally trigger both Signal 1 and Signal 2 in nonanergic T cells:

1. Ras activation (as evidenced by the ratio of Ras-GTP (guanosine triphosphate) to Ras-GDP (guanosine diphosphate)) is compromised and ERK activation is also defective.
2. JNK activation is defective.
3. AP-1 activity is compromised as a result of 1 and 2, and this might contribute to the inablility to induce IL-2 transcription in these cells.
4. A transcriptional repressor of the IL-2 gene, Nil-2-a (discussed below), is induced in anergic T cells.

It has been suggested that most of the biochemical alterations in anergic T cells might follow from the inability of these cells to activate Ras. While this is clearly an oversimplification, it is possible that the defect in Ras activation reflects the defect in Zap-70 activation discussed above. However, it is unclear why the failure to activate Ras would lead to anergy rather than just quiescence.

The failure to activate the Ras/Raf/ERK pathway and the JNK pathway can readily explain why AP-1 might not be induced in anergic cells as well as in quiescent but nonanergic T cells. Although the *IL-2* enhancer/promoter region contains a number of canonical and potential AP-1 sites, two sites at –150 and –180 have been noted to be involved in transcriptional repression of the *IL-2* gene in anergic T cell clones. In Figure 10.3, the c-Jun component is depicted as being phosphorylated (on serines 63 and 73) in activated cells. In anergic cells, it is unclear whether these critical AP-1/AP-1 like sites actually bind nonphophorylated Jun/Fos heterodimers or whether they associate with Jun/Jun homodimers, both of which are lacking in transcriptional activity.

JNK activity is not only required for the transcriptional activation of the IL-2 promoter, but, as discussed in Chapter 4, this kinase is also required for the stabilization of IL-2 transcripts and may thus contribute in a significant way to the accumulation of IL-2 mRNA.

A major component of the phenotype of anergic cells appears to be the binding of a specific repressor in these cells to a motif at position –105 in the IL-2 enhancer/promoter region. This repressor is known as Nil-2-a, a protein that belongs to the ZEB (Zinc finger E box binding) family of repressors, and the site it binds to has been described at the negative regulatory element-A (NRE-A). Levels of Nil-2-a are quite dramatically increased in anergic T cells. CD28 signaling, in some poorly understood way, may prevent this increase in Nil-2-a.

The induction of anergy probably involves the induction of a critical tyrosine phosphatase that might be linked to CTLA-4 signaling. While defective Ras/ERK activation has been linked to the anergic state, the failure to inactivate Nil-2-a, a repressor of IL-2 transcription, provides one of the better molecular explanations for the induction of anergy. However, the

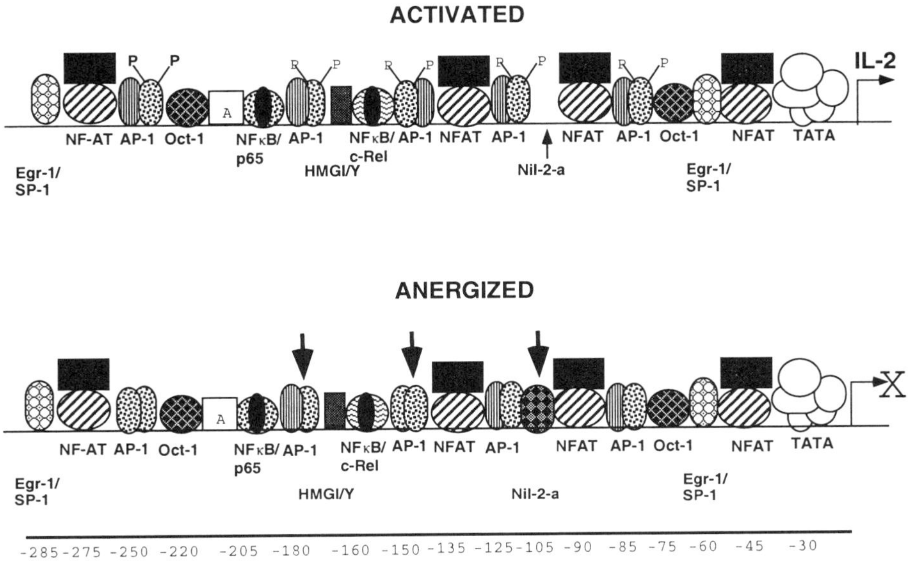

FIGURE 10.3. Factors that bind to the IL-2 enhancer/promoter region in activated and anergic T cell clones. This figure is based on information from both the human and mouse IL-2 promoter/enhancer regions (after Schwartz, 1997). In activated T cells, c-Jun has been depicted as a phosphorylated protein (the sites representing serine 63 and serine 73). In anergized cells, AP-1 or AP-1-like sites at positions –150 and –180 may participate in negatively regulating the activity of this promoter/enhancer region. AP-1 sites are shown ocupied either by Jun-Jun homodimers or nonphosphorylated Jun-Fos heterodimers. A ZEB family repressor, Nil-2-a, binds to a site at position –105 in anergic T cells.

molecular mechanisms involved in rendering T cells nonresponsive remain rather hazy.

Receptor Revision and T Cell Anergy

The reactivation of V(D) J recombination both at the immature B stage and in germinal center B cells has been discussed in Chapters 7 and 8. In the T lineage as well, reactivation of V(D)J recombination might occur in peripheral T cells in response to "incomplete" T cell signaling by self-antigen. In one study, examining mice expressing a Vβ5 transgene that is specific for a weak mouse mammary tumor virus superantigen, peripheral T cells re-expressed RAG proteins and revised their receptors. Interestingly, these peripheral T cells did not express the pTα gene, unlike their controversial B cell counterparts in germinal centers, which express surrogate light chain genes. The importance of receptor revision in T cell biology is unclear.

The Induction of Th1 or Th2 Helper T Cells

During an immune response, $CD4^+$ T cells can be molded to differentiate into helper T cells with fairly well-defined phenotypes. Differentiated effector T helper cells are generally categorized by their ability to secrete distinct patterns of cytokines. Polar helper T cell phenotypes have been described after chronic antigenic stimulation *in vivo* as well as in certain disease states.

Th1 cells secrete IFN-γ (the signature cytokine), IL-2, TNF (tumor neurosis factor), and LT (lymphotoxin). They do not synthesize Th2 cytokines but do express the CCR5 and CXCR3 chemokine receptors. Th1 cells primarily participate in cell-mediated immunity.

Th2 cells secrete IL-4 (the signature cytokine for this subset), IL-5, IL-6, IL-9, IL-10, and IL-13. These cells do not secrete Th1 cytokines. Th2 cells express the CCR3, chemokine receptor that is not expressed on Th1 cells. Th2 cells primarily provide help for humoral immune responses.

Both Th1 and Th2 cells develop from the same category of naive $CD4^+$ precursor cell that is described as a Thp cell. This has been established by studies examining the differentiation process in TCR transgenic mice in which all T cells are specific for a known MHC class II–peptide combination. A multiplicity of factors can influence development toward a Th1 or a Th2 direction APCs and the way they are programmed by exposure to antigen may be the key to subsequent events in T cell activation and differentiation. Human dendritic cells of myeloid origin (called DC1 cells) were found to drive helper T cells to a Th1 fate, whereas plasmacytoid dendritic cells (DC2 cells), which correspond to lymphoid dendritic or LD cells, facilitated Th2 differentiation. These events occurred in a cytokine-independent manner.

The major determining factor in the Th1 versus Th2 decision is the inductive cytokine environment that is generated during a specific immune response. The dose and type of antigen, and the triggering of APCs will obviously influence the nature of the cytokine environment generated, as discussed below. It has been suggested that the type of costimulatory ligand (B7-1 versus B7-2) might determine whether an immune response is driven in a Th1 or a Th2 direction, but in general, these ligands probably influence T helper cell activation in a more complex fashion. The SLAM (signaling lymphocyte activation molecule) costimulatory receptor found on both T and B cells can influence Th1 differentiation in a poorly understood fashion, as mentioned in Chapter 4. The "strength" and "quality" of the signals downstream of the TCR and costimulatory receptors that are induced by specific antigens or pathogens, by specific doses of antigens, and by specific routes of immunization obviously influence cytokine gene expression. Exactly how the appropriate combinations of signals are generated in Thp cells to drive differentiation along the Th1 or Th2 routes remains somewhat hazy.

The critical cytokines that drive Th1 cell differentiation are IL-12 and IL-18, which are induced, usually, by intracellular pathogens or particulate antigens, in macrophages and dendritic cells. IL-12 and IL-18 signaling contribute to the expression of IFN-γ in T helper cells and in natural killer (NK) cells. IFN-γ also subsequently participates in the shaping of the Th1 differentiation program. Until recently, IL-18 was called IGIF (for interferon γ-inducing factor). IL-12 and IL-18 double-knockout mice are markedly impaired in their ability to mount a Th1 response in the face of a challenge by intracellular pathogens such as *Mycobacteriae*. There is a certain symmetry to the induction of a Th1 response by intracellular pathogens that reside in macrophages; the generation of a Th1 response will lead to the T cell-mediated activation of macrophages, which may contribute to the subsequent elimination of the inciting intracellular pathogen.

The critical inducing cytokine for Th2 cells is IL-4. The initial source of IL-4 that induces the differentiation of Thp cells has not been identified with certainty. Natural T (NT) cells were viewed as a potential source, but it is now known that they are not required for the generation of Th2 responses. It appears most likely that the source of initial IL-4 production is a population of $CD4^+$ T cells that might be induced to secrete IL-4 by antigen, or possibly as a consequence of the local generation of IL-6 by APCs. The importance of IL-6 for Th2 differentiation has not been convincingly established. It may be that the type of APCs and the manner in which they are stimulated early in an immune response help determine whether $CD4^+$ T cells will subsequently differentiate into Th1 cells or Th2 cells. If APCs predominantly secrete IL-12, differentiation will proceed along the Th1 pathway. It is possible that an initial burst of IL-6 from APCs may contribute to the release of IL-4 from mature T cells and that antigen-triggered T cells in this milieu subsequently develop along the Th2 pathway.

The genetic background of the host may also critically influence whether the response to a particular immunogen will primarily be of the Th1 or Th2 variety. C57BL/6 mice are resistant to infection with *Leishmania major*. They make a Th1 response and rapidly clear the infection. Balb/C mice are susceptible to this protozoon parasite and develop a Th2 type response that does not protect them from visceral leishmaniasis and death. Balb/C mice are unable to respond to IL-12. Six distinct chromosomal regions have been demonstrated to be involved in the regulation of the immune response to *L. major*. A locus has been described on mouse chromosome 11, which may control the maintenance of IL-12 responsiveness (which is defective in Balb/C mice). Interestingly, the IRF-1 gene (discussed below), which encodes a transcription factor that may participate in the activation of APCs, has been shown to map to this locus.

In molecular terms, signals generated by the TCR and costimulators cooperate with specific cytokine signals to induce a developmental program in a Thp cell, which may drive it along the Th1 or Th2 pathway. While we tend to focus on Th1 and Th2 programs, keep in mind that the Th1 and Th2 phenotypes represent extreme polar forms at the ends of what may well be a continuum of responses. Many T cells may be programmed by some specific combinations of the signals considered above to differentiate into other distinct and functionally important phenotypes. Th0 cells, for instance, may secrete both Th1- and Th2-type cytokines. While they might represent an intermediate stage of differentiation on the way to a Th1 or a Th2 phenotype, it is possible that they represent a distinct differentiation program elicited to deal with pathogens that require both cell-mediated and humoral responses. It has also been suggested that Th0 responses reflect the generation simultaneously of Th1 and Th2 cells. Other smaller subpopulations of $CD4^+$ T cells are also probably generated *in vivo* and *in vitro*. $CD4^+$ cells that are induced by IL-10 and that may secrete IL-5 and/or TGF-β may represent a regulatory/suppressor T cell population that can potentially regulate inflammatory states. T cells that may be involved in the induction of oral tolerance and that secrete TGF-β as well as IL-4 and IL-10 have been sometimes called Th3 cells, though very little is known about these cells. An overview of some of the factors involved in Th1 and Th2 differentiation is provided in Figures 10.4 and 10.5.

Induction of a Th1 Response

Infectious agents or particulate antigens may induce APCs such as macrophages and dendritic cells to secrete IL-12 and IL-18. The activation of APCs might involve the triggering of Fc receptors or signaling via CD40 when an APC interacts with an activated T cell. It is also possible that signals are delivered to the APC by molecules that are usually considered to be ligands during T cell activation but that might generate signals in APCs when they connect with their partners on T cells. These molecules include MHC class II heterodimers, and the costimulatory ligands B7-1 and B7-2. A critical transcription factor that is expressed in APCs and that is required for Th1 cell development is interferon-regulated factor-1 (IRF-1). IRF-1 was originally identified as a DNA-binding protein that could bind to the promoters of the IFN-α and IFN-β genes and that was also involved in the transcriptional induction of a number of genes downstream of IFN-γ. IRF-1-deficient mice have defects in NK cell generation, in $CD8^+$ cell formation, and also in Th1 type responses. These mice are unable to efficiently induce IFN-γ in response to *L. major* infection. They are, however, perfectly capable of mounting Th2-type responses. The reason why IRF-1 deficient mice are apparently unable to mount a Th1 response appears to be the inability of APCs in these mice to induce IL-12. It is known that the expression of one of the subunits of IL-12, IL-12p40, is transcriptionally regulated in macrophages by a cooperative interaction between IRF-1 and NFκB.

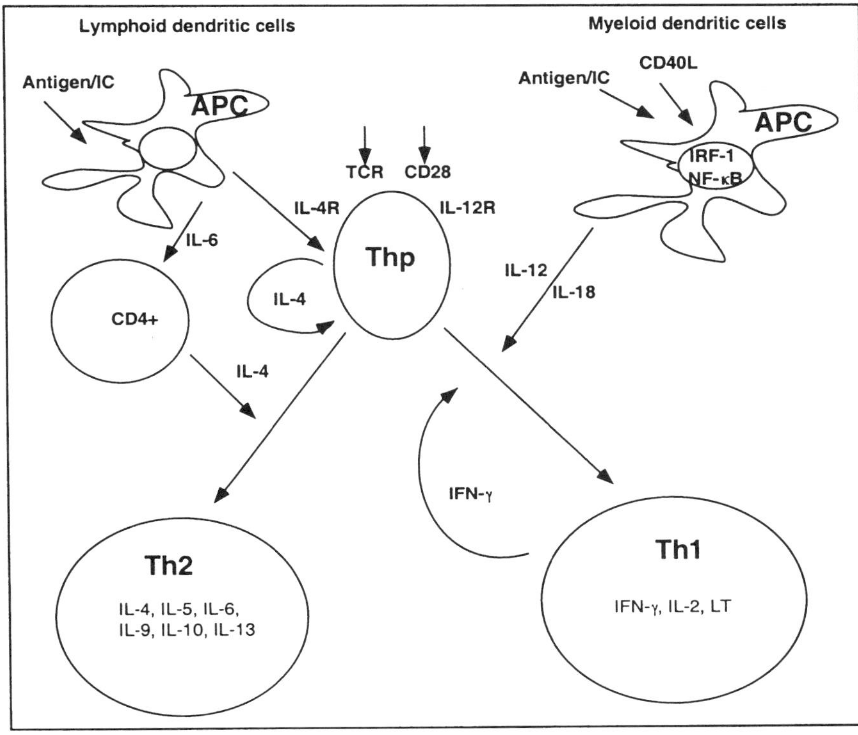

FIGURE 10.4. An overview of some of the triggering events involved in Th1 and Th2 differentiation. Myeloid dendritic cells may favor Th1 differentiation, and lymphoid dendritic cells may facilitate Th2 differentiation. The decision to drive differentiation in the Th1 or Th2 direction might depend on the type of cytokines produced by APCs. IL-12 and IL-18 produced by macrophages and dendritic cells may induce differentiation along the Th1 pathway. Although the initial source of IL-4 that induces Th2 differentiation remains unclear, this cytokine is probably produced by a subset of specific mature $CD4^+$ T cells. Induction of IL-4 synthesis and secretion might depend on the generation of IL-6 by APCs.

Once APCs have been induced to secrete relatively large amounts of IL-12, this cytokine triggers differentiation along the Th1 pathway. The IL-12 receptor (IL-12R) contains two chains, IL-12Rβ1 and IL-12Rβ2, which are structurally similar to gp130 (the signaling chain for the IL-6/IL-11 family) of cytokines (see Chapter 4 for a broad outline of cytokine signaling). The IL-12Rβ2 subunit is not expressed on naive T cells, but it is induced following signaling via the TCR. This expression of IL-12Rβ2 can be influenced by IFN-γ and IL-4. IL-4 can inhibit the expression of IL-12Rβ2, whereas IFN-γ can override the inhibitory effects of IL-4. Early regulation of IL-12R synthesis can thereby influence the Th1 versus Th2 decision at an early "formative" stage.

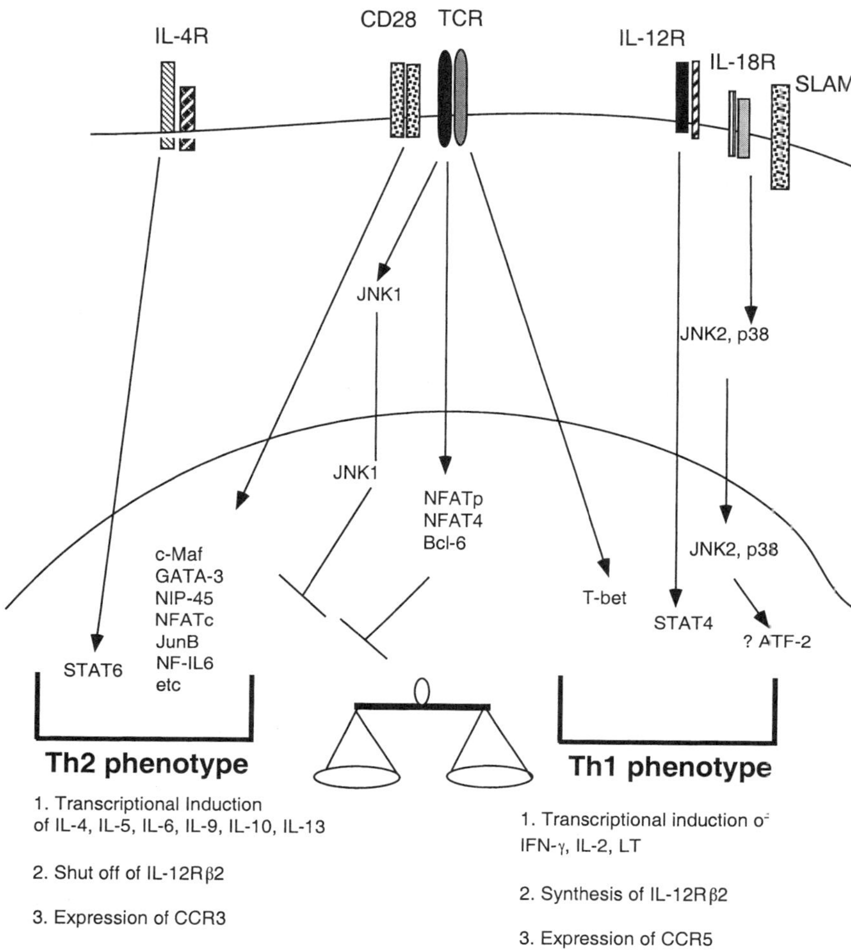

FIGURE 10.5. The appropriate balance of cytoplasmic signals and nuclear factors is required for Th1 and Th2 differentiation. IL-4 induces the dimerization and nuclear translocation of STAT6. JNK1 activity represses Th2 induction. NFATp, NFAT4, and Bcl-6 might also negatively influence Th2 cell differentiation. Other transcriptional regulators that may be involved in the induction or maintenance of the Th2 phenotype include c-Maf, NFATc, and GATA-3. IL-12 signaling contributes to the dimerization and activation of STAT4. T-bet may induce the Th1 phenotype. The SLAM coreceptor may contribute to the generation of Th1 cells. JNK2 and p38 drive Th1 differentiation.

The membrane-proximal portion of the cytoplasmic tail of the IL-12R associates with the Jak2 and Tyk2 tyrosine kinases. These kinases are activated upon IL-12R ligation and activate the STAT-4 transcriptional regulator (by tyrosine phosphorylation and dimerization, as described in Chapter 4). IL-12 and STAT-4 knockout mice are severely compromised in their ability to make Th1-type T cells. Signaling via IL-12 induces the transcription of IFN-γ. However, STAT-4 sites have not been identified in the IFN-γ promoter, and it is unclear as to exactly which relevant target genes are directly induced by this transcription factor in Th1 cells. Transcriptional regulators that may be involved in the induction of IFN-γ during Th1 differentiation include ATF-2, c-Jun, and AP-1. However, the most critical transcriptional regulator so far identified that drives Th1 differentiation is a protein called T-bet (for T box expressed in T cells). T-bet is related to products of the mouse t locus and is induced upon TCR ligation. T-bet can induce progenitor T cells to differentiate into Th1 cells. Exactly how Th1 signals induce T-bet expression is unclear.

Another cytokine that participates in the induction of the Th1 phenotype is IL-18/IGIF. IL-18 is a member of the IL-1 cytokine family that is released by APCs. It is initially synthesized, like IL-1, as an inactive precursor, pro-IL-18, which is proteolytically cleaved by caspase-1/ICE. IL-12 signaling may induce IL-18 receptor synthesis in Th1 cells. Analysis of the IL-18 knockout mouse, the IL-12 knockout mouse, and mice in which both IL-12 and IL-18 have been knocked out has revealed that IL-18 cooperates with IL-12 to mediate the induction of Th1 cells in response to mycobacterial antigens. The IL-18 receptor (also known as the IL-1R related protein) resembles the type I IL-1 receptor (IL-1R) and like IL-1R activates the IL-1R-associated kinase (IRAK) and the NFκB pathway. It is likely therefore that the coordinate activation of transcriptional regulators, such as ATF-2 (activating transcription factor-2) and STAT4 on the one hand and NFκB on the other, leads to the optimal induction of IFN-γ expression and to the assumption by helper T cells of the Th1 phenotype.

Induction of the Th2 Phenotype

While the participation of APCs in the induction of a Th1 response is fairly well understood, it is still unclear as to what kind of antigenic provocation drives a Th2 response and which host cells are involved in this process. As we have discussed above, the initial burst of IL-4 that is required for Th2 differentiation may be produced by a small subpopulation of $CD4^+$ T cells, and IL-6 production by APCs may induce IL-4 expression and secretion in these T cells. NT cells can produce fairly large amounts of IL-4 when triggered, but this specialized population of cells is not absolutely required for Th2 responses.

The current paradigm for Th2 differentiation is that IL-4 itself triggers antigen-activated T cells via the IL-4 receptor and that cooperative signals delivered via the TCR, CD28, and IL-4R work together in Thp cells to induce high-level IL-4 expression. It is possible that continued production of IL-6 by APCs as well as by Th2 cells might generate signals via the IL-6 receptor in Thp cells and thereby contribute to the activation of the Th2 program in uncommitted precursors. The evidence for such a role for IL-6 remains shaky. The regulation of IL-4 gene expression has been extensively studied, and we have some understanding of the transcription factors that are eventually targeted by cell surface signals to drive a naive T cell to assume a Th2 phenotype. Exactly how these various cell surface signals cooperate to induce these nuclear events, however, is less well understood.

The IL-4 receptor (IL-4R) is a member of the γc family (see Chapter 4). Ligand binding leads to the activation of Jak1 and Jak3, and specificity is provided by the association of STAT6 with the α chain of the IL-4R. In the absence of IL-4, IL-4R, or STAT6, induction of Th2 cell differentiation is severely compromised. Apart from the activation of STAT6, the IL-4R can recruit two large docking proteins known as IRS-1 (insulin-receptor substrate-1) and IRS-2, which in turn can contribute to mitogenesis and potentially to signaling events that are relevant during Th2 differentiation. The IL-4 promoter/enhancer region does contain a STAT6 site, although the functional relevance of this site for IL-4 expression has not been conclusively established. In addition, this regulatory region does contain functionally relevant binding sites for c-Maf, NF-IL-6, NFAT, AP-1, and GATA-3. Some of the regulatory motifs in the IL-4 promoter/enhancer region are summarized in Figure 10.6.

c-Maf is encoded by the cellular homologue of the v-maf viral oncogene (the transforming gene of the avian retrovirus AS42). c-Maf is a member of the basic region/leucine zipper (bZip) family of transcription factors, which includes c-Fos and c-Jun. It binds to a c-Maf response element (MARE) in the IL-4 promoter. C-Maf is expressed at very low levels in Thp cells and at high levels in Th2 but not Th1 clones. The expression of c-Maf can be induced by TCR signaling in Th2 but not Th1 cells. Ectopic expression of c-Maf in B cell lines or in Th1 clones leads to the activation of IL-

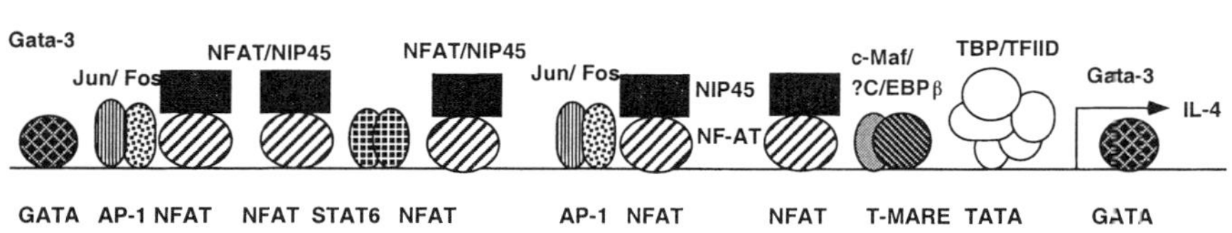

FIGURE 10.6. Some of the cis-acting sites in the IL-4 promoter/enhancer region and proteins that may regulate gene expression by binding to these sites. c-Maf is a critical regulator. It might form heterodimers with C/EBPβ (NF-IL-6) when it binds to the T-MARE site. NFAT may contribute to IL-4 activation, and NIP-45 cooperates with NFAT in a poorly understood fashion. GATA-3 also regulates transcription via this promoter/enhancer region. The functional role of the STAT-6 site in this promoter remains to be established.

4 transcription, and ectopically expressed c-Maf cooperates with NFAT factors in the induction of IL-4 expression in B cells. A transcriptional regulator that modulates IL-6 expression in Th2 cells, NF-IL-6 (also known as C/EBPβ, for CCAAT enhancer-binding protein) may cooperate with c-Maf in inducing Th2-specific transcription. NF-IL-6 is also a bZip protein. In the IL-4 promoter, its binding site overlaps the MARE. NF-IL-6 is expressed in both Th1 and Th2 clones but may form heterodimers with c-Maf and bind to the MARE site to induce transcription of IL-4 specifically in Th2 cells.

NFAT and AP-1 are important nuclear effectors of signaling downstream of the TCR (and of the B cell receptor, BCR) and of costimulatory pathways. We have discussed how a number of signaling pathways contribute to the activation of these nuclear effectors. The existence of multiple NFAT and AP-1 sites in the IL-4 promoter suggests that these transcriptional regulators might be critical for the activation of IL-4 transcription in Th2 cells. Perhaps because there are multiple members of both the Jun and Fos families, it is not too surprising that specific biological roles for AP-1 during T cell activation and Th2 differentiation have been more difficult to establish in a formal sense (by specific knockout phenotypes). AP-1 is induced in both Th1 and Th2 cells, but Th2 cells express particularly high levels of JunB, suggesting that this particular member of the c-Jun family might participate in the Th2 differentiation pathway. While the nuclear translocation of NFAT may be induced in both Th1 and Th2 cells, specific NFAT family members may also be important for driving differentiation toward the Th2 pathway. Three distinct NFAT proteins are expressed in lymphocytes. Mice engineered to lack NFATc1/NFATc/NFAT2 (see Table 10.1 for nomenclature and details of knockout phenotypes) die *in utero*. In Rag-2 chimeras, rescued NFATc1$^{-/-}$ lymphocytes are defective in T and B cell activation and in IL-4 secretion, and there is a reduction in IL-4-driven antibody isotypes (such as IgG1 and IgE). NFATc1 may participate in the differentiation of Th2 cells.

Mice lacking NFATc2/NFATp presented with a surprising phenotype. These mice exhibited enhanced lymphocyte proliferation and increased Th2 cells. NFATc3/NFAT4 null mice had defects in T cell positive selection but normal amounts of Th1 and Th2 cytokines. A double knockout for NFATc2 and NFATc3 presented with a fairly dramatic phenotype involving enhanced lymphocyte activation and proliferation. These mice produced massive amounts of Th2 cytokines, and vast increases in immunoglobulin G1 (IgG1) and IgE were observed. It appears, therefore, that NFATc1 might be a positive inducer of the Th2 phenotype, whereas NFATc2/NFATp and, to a lesser degree, NFATc3/NFAT4 might participate in the repression of lymphocyte activation and possibly in the negative regulation of the Th2 phenotype.

NIP-45 is an NFAT interacting protein, which associates with the Rel homology domain of NFAT proteins and can potentiate NFAT–driven tran-

TABLE 10.1. Functions of NFAT family members

NFAT family member	Phenotype of knockout	References
NFATc1 (NFATc; NFAT2)	Death dE13.5–17.5; congestive cardiac failure because of defects in valve formation; normal lymphoid development;? decreased reconstitution in *Rag-2*$^{-/-}$ chimeras; defective T cell activation and IL-4 secretion; reduced B cell proliferation; reduction in IL-4 driven isotypes.	Ranger et al., 1998a Ranger et al., 1998b Ranger et al., 1998c Yoshida et al., 1998
NFATc2 (NFATp; NFAT1)	Imbalance in IL-4 transcription; Shift to Th2 cytokines; enhanced lymphocyte proliferation; splenomegaly; defective induction of CD40L and FasL; increased chondrogenesis.	Hodge et al., 1996b Xanthoudakis et al., 1996 Ranger et al., 2000
NFATc3 (NFAT4; NFATx)	Normal amounts of Th1 and Th2 cytokines; defect in T cell positive selection	Oukka et al., 1998
NFATc2 and NFATc3	Massive increases in Th2 cytokines; increased secretion of GM-CSF and TNF; lymphoproli feration; marked hyperactivation of T and B cells; nuclear localization of NFATc1; vast increase in IgG1 and IgE; interstitial pneumonitis; blepharitis	Ranger et al., 1998b
NFATc4 (NFAT3)	Not expressed in lymphocytes	

scription of IL-4 in studies involving the transfection of these genes into cell lines that do not express IL-4. In a transfected B cell line, transient overexpression of NFATc1, NIP-45, and c-Maf led to the induction of transcription of the endogenous IL-4 gene.

Another transcriptional regulator that appears to be an important participant in the induction of the Th2 phenotype is GATA-3. In Chapter 9, we considered the important role played by GATA-3 in the early stages of commitment to the T lineage. GATA-3 is expressed at high levels in naive CD4$^+$ T cells and in Th2 cells but at low levels in Th1 cells. Forced transgenic expression of GATA-3 leads to the transcriptional activation of IL-4, IL-6, and IL-10 in Th1 cells. In antisense experiments in which GATA-3 synthesis was markedly reduced, IL-4, IL-6, IL-10, IL-13, and to a lesser extent, IL-5 expression was suppressed. GATA-3 and c-Maf, STAT-6, and possibly NFATc1 are likely to be important for the maintenance of the Th2 phenotype and possibly for its induction as well.

Activation of Cytotoxic T Cells

When a naive CD8$^+$ T cell (CTLp, for cytotoxic T lymphocyte precursor) encounters a cognate MHC class I-peptide complex on a professional APC, it can be activated into a killer cell that expresses high levels of perforin, granzymes and often a range of cytokines including IFN-γ, IL-2, and IL-4 (Figure 10.7). As is true for lymphocyte activation in general, CTLp activation occurs primarily in secondary lymphoid organs (and probably not in infected tissues as is sometimes assumed). For most CTLps to proliferate and to be optimally activated, they must be triggered not only via the TCR but also by costimulatory receptors such as CD28. A professional APC such

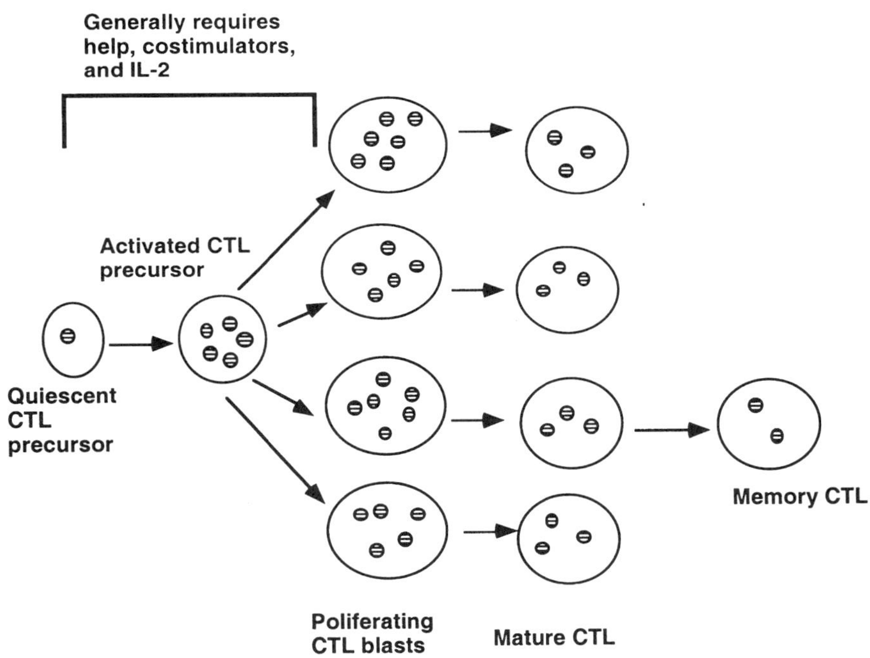

FIGURE 10.7. A naive CD8$^+$ T cell (a quiescent CTLp) may be activated, generally by a "licensed" APC in a secondary lymphoid organ. There is a marked increase in the number of granzyme- and perforin-containing granules in activated CTLs. The APC is "licensed" by a helper T cell to express costimulatory ligands and can activate the CTL precursor via the TCR and CD28 as well. Proliferation probably depends on the induction of IL-2 and the IL-2R. A mature CTL may be triggered to kill by TCR ligation alone in the absence of costimulation or the generation of IL-2. Most activated CTLs die during the course of an immune response probably by a form of AICD (discussed below). A small number may survive to become memory CTLs; alternatively, memory cells may be generated directly from activated cells that do not necessarily pass through an interemediate effector stage.

as a dendritic cell is best suited to provide costimulation. The requirement for costimulation and therefore for "help" may occasionally be bypassed because certain peptide–MHC class I combinations can trigger the TCR very effectively, presumably recruiting signaling pathways (such as the JNK pathway) that normally are optimally recruited only by costimulatory receptors. In general, the activation of CTLps requires helper T cells and the local generation of cytokines such as IL-2 and/or IFN-γ. IL-2 is critical for proliferation, while IFN-γ might participate in the induction of "killer" activity. These cytokines may be generated by the activated CTL itself, and, as we have discussed earlier, the induction of IL-2 by T cells generally requires both Signal 1 and Signal 2. However, the killing of targets by an activated or mature CTL does not require help (or costimulation) and does not require the local presence or generation of IL-2. Two interesting issues come up when considering the biology of CTL activation:

1. If CTLps are primarily activated in lymph nodes and not in infected tissue itself, viral antigens must be expressed appropriately in professional APCs to be presented by the MHC class I pathway. Do all viruses including presumed tissue-tropic pathogens infect dendritic cells?
2. Since CTL activation generally requires helper T cell participation, how is specificity achieved in the process? Must the same APC present peptides in an MHC class I-dependent manner to the CTLp while presenting other pathogen-derived peptides in an MHC class II dependent manner to a $CD4^+$ helper T cell?

It is likely that many viruses are not tissue tropic in an absolute sense and that dendritic cells that are present in tissue sites themselves, or in draining lymph nodes, do often get infected by viruses that may be predominantly targeted to other cells in the host. Under these circumstances, presentation in an MHC class I context would not be an issue, since viral proteins would be synthesized on cytosolic ribosomes and would be in the right location for entry into the MHC class I pathway. However, whether or not a dendritic cell actually gets infected by a pathogenic virus, it will internalize viral antigens either directly or as constituents of immune complexes into endosomes and thereby present peptides in an MHC class II context to cognate helper T cells. An unusual antigen-presentation pathway has been described in some professional APCs, whereby protein antigens that are internalized into endosomes are released into the cytosol. Once they are in the cytosol, these antigens may be processed in proteasomes, transported via TAP transporters into the ER, and then presented in an MHC class I context. Professional APCs, particularly dendritic cells, therefore appear to be capable of presenting peptides from any virus (whether or not the virus actually infects the APC), not just in an MHC class II context but also via the MHC class I pathway.

Although a particularly potent (presumably very high-affinity) interaction between an MHC class I–peptide complex and a CTLp may lead to its activation without an apparent need for CD4$^+$ help, in general, CTLp activation requires helper T cells to be around and to be activated. It is considered extremely unlikely that a given dendritic cell will actually simultaneously find two rare antigen-specific cells, one of the CD4$^+$ type, and the other a CTLp, and thus get them to come together for help to be provided directly (Figure 10.8). An alternative and more logical scenario has emerged. When a viral antigen is presented in an MHC class II context to a CD4$^+$ helper T cell, the activated CD4$^+$ cell not only secretes cytokines but also is induced to express the CD40L. This and other dendritic cells may then be "licensed" by the activated CD4$^+$ T cell, which can trigger dendritic cells via the CD40 pathway. Activation via CD40 leads dendritic cells to be "revved up" and to express high levels of costimulatory ligands such as B7-1 and B7-2. When one of the many activated and infected/antigen-loaded dendritic cells encounters a specific CTLp, it is poised to readily activate the latter because of its prior activation via the CD40/CD40L pathway (Figure 10.9).

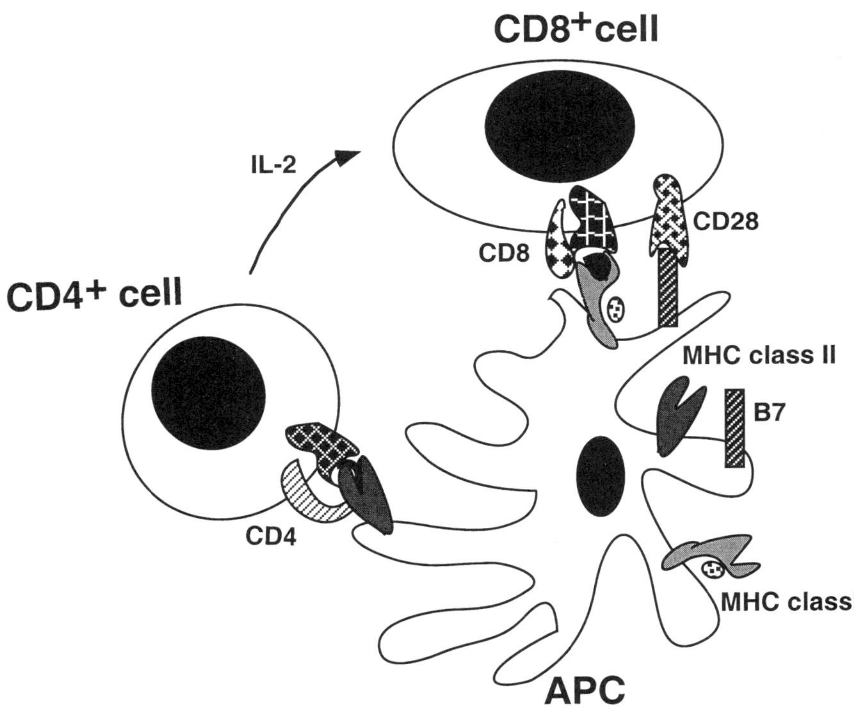

FIGURE 10.8. One model for how CD4$^+$ cells could provide help for CTL activation. In this model, two rare antigen-specific cells, one a CD4$^+$ helper cell and the other a CTL precursor, recognize petide–MHC class II and peptide–MHC class I complexes, respectively, on the same APC. No evidence supports such a model.

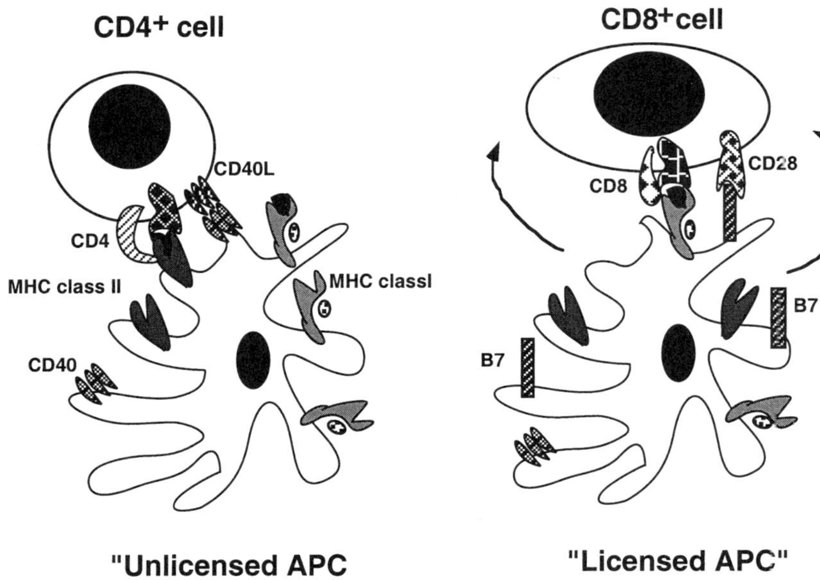

FIGURE 10.9. Specific CD4⁺ cells "license" APCs by triggering CD40 on the latter, thus inducing costimulatory ligand expression and the secretion of cytokines. One of the many licensed APCs is likely to meet a specific CTL precursor and to activate it via the TCR and CD28 and probably by other less well-defined means as well.

There are many similarities between the activation of $CD8^+$ T cells and $CD4^+$ T cells. Like $CD4^+$ T cells, activated $CD8^+$ T cells may induce the expression of cytokines and FasL. However, very little is still known about the specific programs of gene activation that are unique to cells of the $CD8^+$ lineage. The genetic programs involved in the induction of perforins, granzymes, and other molecules that are more specific to $CD8^+$ cells are currently being actively investigated.

Ending T Cell Responses and AICD

By using artificially generated reagents such as labeled MHC class I-peptide tetramers, it has been possible to follow the fate of antigen-specific $CD8^+$ T cells in infected mice and in human patients. It is now recognized that during an active immune response a very large number of the T cells in the host represent the clonal outgrowth of a few antigen-specific $CD4^+$ or $CD8^+$ T cells. Examination of these clones has revealed that a large proportion of activated T cells are in the process of undergoing apoptosis. While large numbers of T cells do get activated by antigen, most of these cells probably die because the amount of antigen is limiting and these cells no longer receive survival signals. Another major way in which activated T cells may be rendered quiescent is by the induction of inhibitory signaling via CTLA-4. Some cells that are restimulated by antigen also receive signals to die.

AICD is probably as important for the elimination of activated $CD8^+$ cells as it is for the elimination of $CD4^+$ cells. However, it has been examined mainly in the context of $CD4^+$ cells. Evidence from *lpr* and *gld* mice, from human lymphoproliferative syndromes involving Fas mutations, and from IL-2 and IL-2Rα chain knockout mice has all come together to suggest the following scenario. When both Signal 1 and Signal 2 combine to generate maximal T cell activation, transcriptional induction of both the IL-2 gene and of the IL-2R is achieved. This form of complete activation leads to signal transduction via the IL-2R and the maximal induction of both CD40L and FasL. During the course of these events, the T cell makes cytokines and may activate specific B cells and professional APCs via CD40L–CD40 interactions as well as by triggering specific cytokine receptors.

The majority of properly activated T cells commit suicide. They do so because the FasL induced on individual T cells triggers Fas receptors expressed by the same cell and thus induces death. We have considered the FasL–Fas pathway in Chapter 5, A potentially important aspect of this pathway is the destruction by caspases of Bcl-2 and other Bcl-2 family survival factors. Signaling via the IL-2 R is critical for the induction of AICD, probably because signals delivered via this cytokine receptor are required for the maximal induction of FasL. In addition, IL-2 signaling may contribute to an inhibition of the cellular levels of FLIP (Flice Inhibitory protein). FLIP is a protein that is structurally similar to caspase-8 but lacks proteolytic activity. It can prevent the recruitment of caspase-8 to Fas-associated death domain (FADD) (see Chapter 5) and thus protect a cell from Fas-induced apoptosis. The reduction of FLIP levels by IL-2 may be key in allowing the suicide process to go through successfully.

In $CD8^+$ T cells, AICD might require signals to be delivered both by IL-2 as well as by TNF. Given that "proper" activation of a T cell leads to death, it is pertinent to ask how memory T cells are generated.

Making the Choice Between Activation and Memory: Memory T Cells

The existence of "true" memory lymphocytes has often been questioned, and while their existence remains controversial, there is a growing acceptance of the view that such cells do exist. The persistence of antigen—either preserved by a low-grade viral infection or as part of long-lived immune complexes sequestered by follicular dendritic cells—could potentially contribute to the identification of recently restimulated cells as "memory" cells. The best evidence for the existence of "true" memory cells has come from studies on the transfer of $CD8^+$ cells in the apparent absence of antigen, which can result in the transfer of immune responsiveness for extended periods. Although naive T cells are maintained by being "tickled" via their TCRs it appears that memory T cells do not require MHC molecules for survival. There is no consensus in the field as to how a memory cell should be defined, and many aspects of the biology of memory T cells are enmeshed in controversy. However, memory T cells may be perhaps best described as long-lived antigen-specific T cells that emerge after activation by specific antigen but which do not require antigen for their extended *in vivo* survival.

The following general features characterize memory T lymphocytes (whether or not they pass the litmus test of antigen-independent survival):

1. Memory cells apparently survive for a long time *in vivo* in the apparent absence of specific antigen. They may survive as long as the life span of some small vertebrates.
2. Memory lymphocytes generally present with an activated phenotype but can be distinguished from effector cells on the basis of size and function. They are smaller than effector cells. Effector $CD4^+$ cells may secrete large amounts of cytokines, whereas memory $CD4^+$ cells may need to be triggered to do so. Effector $CD8^+$ cells may be able to kill *ex vivo* targets directly, while memory cells need to be triggered to be able to kill.
3. Memory T cells can be activated by signals that may be below the threshold for the activation of naive T cells. This may in part be due to the high levels of adhesion molecules expressed on memory T cells.
4. Most memory cells probably live for a long time because they have been programmed by antigen exposure to express high levels of survival factors that include, but may not be restricted to, members of the Bcl-2 family. Survival of these cells *in vivo* may require the triggering of specific cytokine receptors by low levels of circulating cytokines.

In humans, memory T cells may be categorized, as discussed in Chapter 6, into "central" memory cells that express CCR7 and traffic like naive T cells to lymph nodes, and into "effector" memory cells that do not express CCR7 and return to tissue sites. Memory cells as defined in the mouse may be generally considered to be of the "central" variety," whereas cells generally referred to as effector T cells in the mouse may be synonymous with human "effector" memory cells.

The vast majority of effector T cells that are generated during an immune response are eliminated by AICD. A number of models have been proposed to explain how memory cells are generated. The most likely scenario is as follows: Following T cell activation, a very large number of effector T cells are generated. As antigen is cleared, most of the vigorously activated T cells either die because they are no longer being stimulated, or they undergo AICD. A small number of these cells—perhaps those that were not activate as well because they arrived late on the scene or those that were activated by distinct APCs in a slightly less vehement manner—fail to induce the suicide pathway and may go on to become memory cells. For $CD8^+$ cells, it has been established that effector cells go on to become memory cells. It is also possible that certain effectors are induced to "rest" and become memory cells or that distinct activated cells give rise to effector cells and memory cells (Figure 10.10).

The exact molecular pathways that are responsible for memory cell generation remain to be identified. It is very likely that antiapoptotic Bcl-2 family members such as Bcl-2 and Bcl-XL are induced by antigen receptor

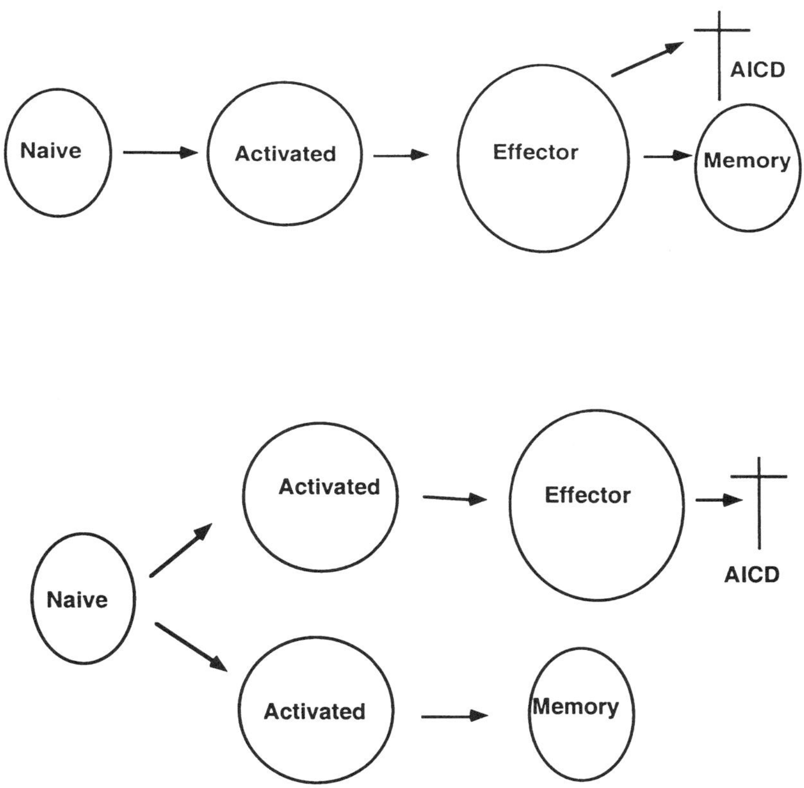

FIGURE 10.10. Models for memory T cell generation. Memory cells share many features of effector cells. However, effector cells tend to be larger than memory cells. In a linear model for memory cell generation (which has experimental support in $CD8^+$ cells), most effectors are lost due to AICD, but a few are "chosen" to survive or come to "rest" as memory cells. In an alternative model, separate activated cells may either become effectors or memory cells.

and costimulatory signals. Costimulation through CD28 may be required for the induction of Bcl-X_L. The induction of cytosolic kinases such as Akt might contribute to the phosphorylation and inactivation of pro-apoptotic Bcl-2 family members such as Bad. It has been recently suggested that calcium/cabnodulin kinase II (CaMKII), a known player in cerebral memory, may also participate in memory T cell generation.

In Figure 10.11, a totally speculative scenario is presented in an attempt to explain how an activated T cell may receive distinct sets of signals that might favor memory versus death or vice versa. If one assumes that memory cells may be presented the same antigen by different APCs, one set of which express high levels of costimulatory ligands and another that do not, it could be predicted that excessive activation might lead to AICD; more circumscribed activation either later in an infection or in a distinct site could potentially lead to the generation of memory. There is absolutely no evidence for any such model, but it is presented as one of many potential and distinct models that could be conceived of to explain how activated T cells might make these decisions.

Much of the memory phenotype might be explained by two critical sets of biochemical changes:

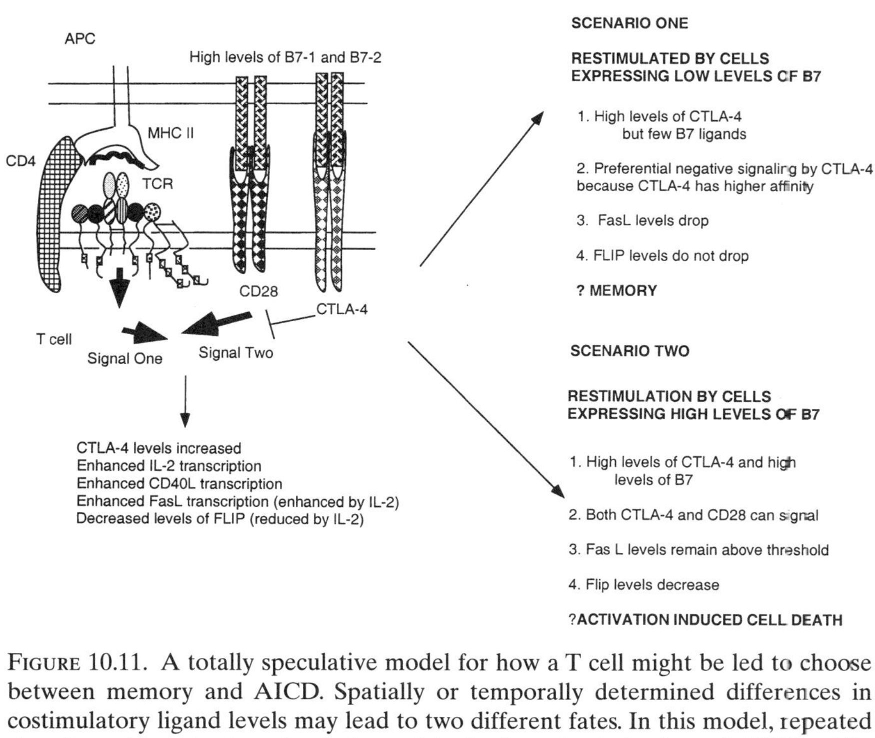

FIGURE 10.11. A totally speculative model for how a T cell might be led to choose between memory and AICD. Spatially or temporally determined differences in costimulatory ligand levels may lead to two different fates. In this model, repeated signaling with potent costimulation might lead to the induction of AICD, while a second triggering with little costimulation might contribute to memory T cell generation.

1. Memory cells can be much more easily triggered through the TCR. This might be because memory cells express higher levels of adhesion molecules than naive cells or because of alterations in intracellular signaling pathways. Genes downsteam of TCR signaling may be in an "open chromatin" configuration in memory cells, thus allowing low-intensity signals to trigger these cells or to dramatically shorten the G1 period in cell cycle progression following a mitogenic stimulus.
2. Possibly as a result of being constantly "tickled by cytokines," memory T cells express higher levels of antiapoptotic proteins or appropriately inactivate proapoptotic proteins.

Distinguishing Naive, Effector, and Memory T Cells

As mentioned above, memory T cells share many features of effector T lymphocytes but are smaller cells that appear to have unwound a bit from the activated state and are generally not cycling. While they are exquisitely sensitive, they still need to be triggered to display a functional and fully activated phenotype. In humans, effector T cells are sometimes referred to as "effector" memory cells, whereas memory cells that home to lymph nodes may be called "central" memory cells.

Naive T cells can be readily distinguished from effector and memory cells. They typically express high levels of L-selectin (CD62L), which facilitates interaction with lymph node high-endothelial venules (HEVs). They also express substantial levels of the high molecular weight CD45 isoforms (RA, RB, and RC). They express low levels of CD44, ICAM-1, LFA-1, CD43, the $\alpha4\beta1$ and $\alpha4\beta7$ integrins.

When T cells are activated, they no longer express high levels of L-selectin and the CD45 RA/RB/RC isoforms. However, they do express high levels of CD44, ICAM-1, LFA-1, CD43, $\alpha4\beta1$, and $\alpha4\beta7$. "Centrol" memory T cells, like marginal zone B cells, do not recirculate efficiently, since they cannot enter lymph nodes via HEVs and tend to accumulate in the blood and the spleen. Some memory T cells do reach lymph nodes, perhaps via afferent lymphatics drawing lymph in from tissue sites, or perhaps because they express CCR7.

Although activated and memory T cells express broadly similar surface markers (with the exception of CCR7, as referred to earlier and in Chapter 6), memory cells may move a little closer to the quiescent state. Memory cells express higher levels of antiapoptotic Bcl-2 family proteins than effector cells. Memory cells also express higher levels of Bcl-X_L than naive T cells. Memory $CD4^+$ cells generally retain a $CD45RB^{lo}CD62L^{lo}CD44^{hi}$ phenotype, although they may also revert to a more naive-type surface phenotype. $CD8^+$ memory cells appear to retain many more features of activated effectors even after transfer into recipients that are free of specific antigen.

Summary and Conclusions

Both helper and cytotoxic T cells are triggered to become effectors when optimally stimulated via the TCR and costimulatory receptors. Incomplete activation can lead to the generation of anergy. The anergic state depends on cytosolic signals and alterations in nuclear regulators. Transcriptional repressors of the IL-2 gene may participate in the induction of the anergic state.

Chronic activation of helper T cells may lead to the generation of specific inducer cytokine milieus that facilitate the generation of polar Th1 or Th2 cells. IL-12 and IL-18 released by APCs contribute to the differentiation of naive precursors into IFN-γ, IL-2, and TNF-β-secreting Th1 cells. IL-4 contributes to differentiation along the Th2 pathway, which culminates in the generation of T cells secreting IL-4, IL-5, IL-6, and IL-13. Induction of the Th1 and Th2 phenotypes is best understood in terms of the transcriptional regulation of the IL-4, IFN-γ, and IL-12Rβ chain genes. Some of the transcription factors involved in the maintenance of the Th1 or Th2 phenotypes have been characterized. Most helper T cells are not triggered to differentiate into these polar forms. Activation of helper and cytotoxic cells leads to proliferation and cytokine production. Most effectors die either as a result of antigenic clearance and the failure of stimulation or because of repeated restimulation and AICD. A small proportion of activated cells may go on to become memory T cells. These cells can be triggered more easily via their antigen receptors and may express higher levels or more active forms of survival factors, possibly as a result of constant low-level triggering of cytokine receptors.

Selected Reviews

Abbas, A. K. (1996). Die and let live: eliminating dangerous lymphocytes. Cell *84*, 655–658.

Abbas, A. K., Murphy, K. M., and Sher, A. (1996). Functional diversity of helper T lymphocytes. Nature *383*, 789–793.

Agarwal, S., and Rao, A. (1998). Modulation of chromatin structure regulates cytokine gene expression during T cell differentiation. Immunity *9*, 765–775.

Ahmed, R., and Gray, D. (1996). Immunological memory and protective immunity: understanding their relation. Science *272*, 54–60.

Glimcher, L. H., and Singh, H. (1999). Transcription factors in lymphocyte development—T and B cell get together. Cell *96*, 13–23.

Kuo, C. T., and Leiden, J. M. (1999). Transcriptional regulation of T lymphocyte development and function. Annu. Rev. Immunol. *17*, 149–187.

Murphy, K. M. (1998). T lymphocyte differentiation in the periphery. Curr. Opin. Immunol. *10*, 226–232.

O'Garra, A. (1998). Cytokines induce the development of functionally heterogeneous T helper cell subsets. Immunity *8*, 275–284.

Rao, A., Luo, C., and Hogan, P. G. (1997). Transcription factors of the NF-AT family: regulation and function. Annu. Rev. Immunol. *15*, 707–747.

Rincon, M., and Flavell, R. A. (1997). T cell subsets: transcriptional control in the Th1/Th2 decision. Curr. Biol. *7*, R729–732.

Sprent, J. (1997). Immunological memory. Curr. Opin. Immunol. *9*, 371–379.

Szabo, S. J., Glimcher, L. H., and Ho, I.-C. (1997). Genes that regulate interleukin-4 expression in T cells. Curr. Opin. Immunol. *9*, 776–781.

Thompson, C. B. (1995). Distinct roles for the co-stimulatory ligands B7-1 and B7-2 in T helper cell differentiation. Cell *81*, 979–982.

Van Parijs, L, and Abbas, A. K. (1998). Homeostasis and self-tolerance in the immune system: turning lymphocytes off. Science *280*, 243–248.

Some Selected References

Bach, E. A., Szabo, S. J., Dighe, A. S., Ashkenazi, A., Aguet, M., Murphy, K. M., and Schreiber, R. D. (1995). Ligand induced autoregulation of IFN-γ receptor β chain expression in T helper cell subsets. Science *270*, 1215–1218.

Barton, K., Muthusamy, N., Chanyangam, M., Fischer, C., Clendenin, C., and Leiden, J. M. (1996). Defective thymocyte proliferation and IL-2 production in transgenic mice expressing a dominant negative form of CREB. Nature *379*, 81–84.

Beals, C. R., Sheridan, C. M., Turck, C. W., Gardner, P., and Crabtree, G. R. (1997). Nuclear export of NF-ATc enhanced by glycogen synthase kinase-3. Science *275*, 1930–1934.

Bennett, S. R., Carbone, F. R., Karamalis, F., Flavell, R. A., Miller, J. F., and Heath, W. R. (1998). Help for cytotoxic-T cell responses is mediated by CD40 signalling. Nature *393*, 478–480.

Bonecchi, R., Bianchi, G., Bordignon, P. P., D'Ambrosio, D., Lang, R., Borsatti, A., Sozzani, S., Allavena, P., Gray, P. A., Mantovani, A., and Sinigaglia, F. (1998). Differential expression of chemokine receptors and chemotactic responsiveness of type I T helper cells Th1s and Th2s. J. Exp. Med. *187*, 129–134.

Brunner, T., Mogil, R. J., LaFace D., Yoo, N. J., Mahboubi, A., Echeverri, F., Martin, S. J., Force, W. R., Lynch, D. H., Ware, C. F., and Green, D. R. (1995). Cell autonomous Fas (CD95)/Fas-ligand interaction mediates activation-induced apoptosis in T cell hybridomas. Nature *373*, 441–444.

Bruno, L., Kirberg, J., and Von Boehmer, H. (1995). On the cellular basis of immunological memory. Immunity *2*, 37–43.

Butz, E. A., and Bevan, M. J. (1998). Massive expansion of antigen-specific $CD8^+$ T cells during an acute virus infection. Immunity *8*, 167–175.

Chen, Y., Inobe, J.-I., Marks, R., Gonnella, P., Kuchroo, V. K., and Weiner, H. L. (1995). Peripheral deletion of antigen reactive T cells in oral tolerance. Nature *376*, 177–180.

Chow, C. W., Rincon, M., Cavanagh, J., Dickens, M., and Davis, R. J. (1997). Nuclear accumulation of NFAT4 opposed by the JNK signal transduction pathway. Science *278*, 1638–1641.

Cocks, B. G., Chang, C. C. J., Carballido, C. M., Yssel, H., deVries, J. E., and Aversa, G. (1995). A novel receptor involved in T cell activation. Nature *376*, 260–263.

Dent, A. L., Hu-Li, J., W. E., P., and Staudt, L. M. (1998). T helper type 2 inflammatory disease in the absence of interleukin 4 and transcription factor STAT6. Proc. Natl. Acad. Sci. USA *95*, 13823–13828.

Dent, A. L., Shaffer, A. L., Yu, X., Allman, D., and Staudt, L. M. (1997). Control of inflammations, cytokine expression and germinal center formation by Bcl-6. Science *276*, 589–592.

Dhein, J., Walczak, H., Baumler, C., Debatin, K.-M., and Krammer, P. H. (1995). Autocrine T cell suicide mediated by APO-1 (fas/CD95). Nature *373*, 438– 441.

Dong, C., Yang D. D., Wysk, M., Whitmarsh, A. J., Davis, R. J., and Flavell, R. A. (1998). Defective T cell differentiation in the absence of Jnk1. Science *282*, 2092–2095.

Guler, M. L., Gorham, J. D. Hsieh, C.-S., Mackey, A. J., Steen, R. G., Dietrich, W. F., and Munphy, K. M. (1996). Genetic susceptibility to Leishmania: IL-12 responsiveness in TH1 cell development. Science *271*, 984–987.

Hamann, D., Baars, P. A., Rep. M. H. G., Hooibrink, B., Kerkhof-Garde, S. R., Klein, M. R., and van Lier, R. A. W. (1997). Phenotypic and functional separation of memory and effector Human CD8⁺ T cells. J. Exp. Med. *186*, 1407–1418.

Heemskerk, M. H. M., Blom, B., Nolan, G., Stegmann, A. P. A., Bakker, A. Q., Weijer, K., Res, P. C. M., and Spits, H. (1997). Inhibition of T cell and promotion of natural killer cell development by the dominant negative helix loop helix factor Id3. J. Exp. Med. *186*, 1597–1602.

Ho, I.-C., Hodge, M. R., Rooney, J. W., and Glimcher, L. H. (1996). The protooncogene c-maf is responsible for tissue specific expression of interleukin-4. Cell *85*, 973–983.

Ho, I.-C., Lo, D., and Glimcher, L. (1998). c-maf Promotes T helper cell type 2(Th2) and attenuates Th1 differentaition by both interleukin-4 dependent and independent mechanisms. J. Exp. Med. *188*, 1859–1866.

Hodge, M. R., Chun, H. J., Rengarajan, J., Alt, A., Lieberson, R., and Glimcher, L. H. (1996a). NFAT driven interleukin-4 transcription potentiated by NIP45. Science *274*, 1903–1905.

Hodge, M. R., Ranger, A. M., de la Brousse, C. F., Hoey, T., Grusby, M. J., and Glimcher, L. H. (1996b). Hyperproliferation and dysregulation of IL-4 expression in NF-ATp deficient mice. Immunity *4*, 397–405.

Ju, S.-T., Panka, D. J., Cui, H., Ettinger, R., El-Khatib, M., Sherr, D. H., Stanger B. Z., Marshak-Rothstein, A. (1995). Fas(CD95)/FasL interactions required for programmed cell death after T cell activation. Nature *373*, 444–448.

Kaplan, M. H., Schindler, U., Smiley, S. T., and Grusby, M. J. (1995a). Stat6 is required for mediating responses to IL-4 and for the development of Th2 cells. Immunity *4*, 313–319.

Kaplan, M., Sun, Y., Hoey, T., and Grusby, M. (1996b). Impaired IL-12 responses and impaired development of Th2 cells in Stat 4 deficient mice. Nature *382*, 174–177.

Kuchroo, V. K., Das, M. P., Brown, J. A., Ranger, A. M., Zamvil, S. S., Sobel, R. A., Weiner, H. L., Nabavi, N., and Glimcher, L. H. (1996). B7-1 and B7-2 costimulatory molecules activate differentially the Th1/Th2 developmental pathways: application to autoimmune disease therapy. Cell *80*, 707–718.

Li, B., Tournier, C., Davis, R. J., and Flavell, R. A. (1999). Regulation of IL-4 expression by the transcription factor JunB during T helper cell differentiation EMBO J. *18*, 420–432.

Lohoff, M., Ferrick, D., Mittrucker, H. W., Duncan, G. S., Bischof, S., Rollinghoff, M., and Mak, T. W. (1997). Interferon regulatory factor-1 is required for a T helper 1 immune response *in vivo*. Immunity *6*, 681–689.

McHeyzer-Williams, M. G., and Davis, M. M. (1995). Antigen-specific development of primary and memory T cells *in vivo*. Science *268*, 106–111.
McMahan, C. J., and Fink, P. J. (1998). RAG reexpression and DNA recombination at T cell receptor loci in peripheral CD4+ T cells. Immunity *9*, 637–647.
Moran, M., and Miceli, M. C. (1998). Engagement of GPI-linked CD48 contributes to TCR signals and cytoskeletal reorganization: a role for lipid rafts in T cell activation. Immunity *9*, 787–796.
Murali-Krishna, K., Altman, J. D., Suresh, M., Sourdive, D. J. D., Zajac, A. J., Miller, J. D., Slansky, J., and Ahmed, R. (1998). Counting antigen-specific CD8+ T cells. A reevaluation of bystander activation during viral infection. Immunity *8*, 167–175.
Murali-Krishna, K., Lau, L. L., Sambhara, S., Lemonnier, F., Altman, J., and Ahmed, R. (1999). Persistence of memory CD8 T cells in MHC class I-deficient mice. Science *286*, 13777–1381.
Noben-Trauth, M., Kropf, P., and Muller, I. (1996). Susceptibility to Leishmania major infection in interleukin 4 deficient mice. Science *271*, 987–990.
Opferman, J. T., Ober, B. T., and Ashton-Rickardt, P. G. (1999). Linear differentiation of cytotoxic effectors into memory T lymphocytes. Science *283*, 1745–1748.
Oukka, M., Ho, I.-C., de la Brousse, C. F., Hoey, T., Grusby, M. J., and Glimcher, L. H. (1998). The transcription factor NFAT4 is involved in the generation and survival of T cells. Immunity *9*, 295–304.
Ranger, A. M., Hodge, M. R., Gravallese, M. R., Oukka, M., Davidson, L., Alt, F. W., de la Brousse, F. C., Hoey, T., Grusby, M., and Glimcher, L. H. (1998a). Delayed lymphoid repopulation with defects in I1-4 driven responses produced by inactivation of NFATc. Immunity *8*, 125–134.
Ranger, A. M., Oukka, M., Rengarajan, J., and Glimcher, L. H. (1998b). Inhibitory function of two NFAT family members in lymphoid homeostasis and Th2 development. Immunity *9*, 627–635.
Ranger, A. M., Grusby, M. J., Hodge, M. R., Gravallese, E. M., de la Brousse, F. C., Hoey, T., Mickanin, C., Baldwin, H. S., and Glimcher, L. H. (1998c). The transcription factor NFATc is essential for cardiac valve formation. Nature *392*, 186–190.
Ranger, A. M., Gerstenfeld, L. C., Wang, J., Kon, T., Bae, H., Gravallese, E. M., Glimcher, M. J., and Glimcher, L. H. (2000). The nuclear factor of activated T cells (NFAT) transcription factor NFATp (NFATc2) is a repressor of chondrogenesis. J. Exp. Med. *191*, 9–22.
Ridge, J. P., Di Rosa, F., and Matzinger P. (1998). A conditioned dendritic cell can be a temporal bridge between a CD4+ T-helper and a T-killer cell. Nature *393*, 474–478.
Rincon, M., Enslen, H., Raingeaud, J., Recht, M., Zapton, T., Su, M. S.-S., Penix, L. A., Davis, R., J., and Flavell, R. A. (1998). Interferon γ expression by Th1 effector T cells mediated by the p38 MAP kinase signaling pathway. EMBO J. *17*, 2817–2829.
Rissoan, M. C., Soumelis, V., Kadowski, N., Grouard, G., Briere, F., de Waal Malefyt, R., and Liu, Y. J. (1999). Reciprocal control of T helper cell and dendritic cell differentiation. Science *283*, 1183–1186.
Sallusto, F., Lenig, D., Forster, R., Lipp, M., and Lanzavecchia, A. (1999). Two subsets of memory T lymphocytes with distinct homing potentials and effector functions. Nature *401*, 708–712.

Schaeffer, E. M., Debnath, J., Yap, G., McVicar, D., Liao, X. C., Littman, D. R., Sher, A., Varmus, H. E., Lenardo, M. J., and Schwartzberg, P. L. (1999). Requirement for Tec kinases Rlk and Itk in T cell receptor signaling and immunity. Science *284*, 638–641.

Schoenberger, S. P., Toes, R. E., van de Voort, E. I., Offringa, R., and Melief, C. J. (1998). T cell help for cytotoxic T lymphocytes is mediated by CD40-CD40L interactions. Nature *393*, 480–483.

Schuh, K., Twardzik, T., Kneitz, B., Heyer, J., Schimpl, A., and Serfling E. (1998). The interleukin 2 receptor α chain/CD25 promoter is a target for nuclear factor of activated T cells. J. Exp. Med. *188*, 1369–1373.

Schwartz, R. H. (1997). T cell clonal anergy. Curr. Opin. Immunol. *9*, 351–357.

Swain, S. L., Hu, H., and Huston, G. (1999). Class II independent generation of CD4 memory T cells from effectors. Science *286*, 1381–1383.

Takeda, K., Tsutsui, H., Yoshimoto, T., Adachi, O., Yoshida, N., Kishimoto, T., Okamura, H., Nakanishi, K., and Akira, S. (1998). Defective NK cell activity and Th1 response in IL-18 deficient mice. Immunity *8*, 383–390.

Taki, S., Sato, T., Ogasawara, K., Fukuda, T., Sato, M., Hida, S., Suzuki, G., Mitsuyama, M., Shin, E. H., Kojima, S., Taniguchi, T., and Asano, Y. (1997). Multistage regulation of Th1-type immune responses by the transcription factor IRF-1. Immunity *6*, 673–679.

Thierfelder, W. E., van Deursen, J. M., Yamamoto, K., Tripp, R. A., Sarawar, S. R., Carson, R. T., Sangster, M. Y., Vignali, D. A., Doherty, P. C., Grosveld, G. C., and Ihle, J. N. (1996). Requirement for Stat 4 in interleukin-12 mediated responses of natural killer and T cells. Nature *382*, 171–174.

Xanthoudakis, S., Viola, J. P. B., Shaw, K. T. Y., Luo, C., Wallace, J. D., Bozza, P. T., Curran, T., and Rao, A. (1996). An enhanced immune response in mice lacking the transcription factor NFAT1. Science *272*, 892–895.

Yang, D. D., Conze, D., Whitmarsh, A. J., Barrett, T., Davis, R. J., Rincon, M., and Flavell, R. A. (1998). Differentiation of CD4$^+$ T cells to Th1 cells requires MAP kinase JNK2. Immunity *9*, 575–585.

Yoshida, H., Nishina, H., Takimoto, H., Marengere, L. E. M., Wakeham, A. C., Bouchard, D., Kong, Y.-Y., Ohteki, T., Shahinian, A., Nachmann, M., Ohashi, P. S., Penninger, J. F., Crabtree, G. R., and Mak, T. W. (1998). The transcription factor NF-ATc1 regulates lymphocyte proliferation and Th2 cytokine production. Immunity *8*, 115–124.

Zheng, W.-P., and Flavell, R. A. (1997). The transcription factor GATA-3 is necessary and sufficient for Th2 cytokine gene expression in CD4 T cells. Cell *89*, 589–596.

Zhu, J., Shibasaki, F., Price, R., Guillemot, J.-C., Yano, T., Dotsch, V., Wagner, G., Ferrara, P., and McKeon, F. (1998). Intramolecular masking of nuclear import signal on NF-AT4 by casein kinase I and MEKK1. Cell *93*, 851–862.

11

The Development of Natural Killer Cells

Natural killer (NK) cells represent an important but evolutionarily more recent component of the innate immune system in vertebrates. These cells may have co-evolved with other lymphoid cells in vertebrates. These $CD3^-Ig^-TCR^-Fc\gamma RIII^+$ lymphocytes are a major source of interferon γ (IFN-γ) early in the course of viral infections. They can be triggered by a number of cytokines including tumor necrosis factor (TNF), interleukin-12 (IL-12), IL-15, IL-18, and the interferons. Apart from their cytokine driven role in innate immunity, NK cells have a more specific function in identifying and killing cells which have lost major histocompatibility complex (MHC) class I expression (as a consequence of viral infection or oncogenic transformation) and which are therefore not readily targeted by cytotoxic T lymphocytes (CTLs). NK cells can identify cells that express reduced levels of MHC class I molecules in general or are missing a specific self-MHC class I allele. In this chapter we will briefly discuss the development of NK cells. From a developmental standpoint there appear to be two major issues to consider. What are the signals that commit a progenitor to the NK lineage? What is the basis for the emergence of subsets of NK cells, and how are they educated during development?

A bipotential progenitor has been described in the thymus that can give rise to the T and NK lineages (but might still have the ability to differentiate into lymphoid dendritic or LD cells). A distinct type of fetal progenitor cell has also been described that can give rise to NK cells, expresses lymphotoxin β, and is required for the induction of lymph nodes and Peyer's patches. This latter $CD4^+CD3^-IL\text{-}7R\alpha^+$ cell does not migrate to the thymus and probably accounts for most NK cell development. This progenitor fails to develop in $Id2^{-/-}$ mice (see Chapter 6), and in its absence, NK cells are markedly diminished. NK cells are therefore derived in part from a thymic lymphoid progenitor but primarily from a hematopoietic progenitor that does not give rise to either T or B lymphocytes. These killer cells are considered to be lymphocytes, even though they do not rearrange antigen-receptor genes and develop normally in *Rag* knockout or *scid* mice. Like T cells, NK cells have receptors that recognize MHC molecules, and they kill targets in much the same way that CTLs do.

Both inhibitory and activating receptors are critical for NK cell function. The possible existence of inhibitory receptors that recognize self-MHC class I and prevent NK cells from being activated was predicted by the "missing-self" hypothesis. These inhibitory receptors define the NK repertoire. Activating receptors may direct NK cells to specific targets and trigger them to kill using machinery similar to that used by CTLs (perforin and granzymes). We will discuss inhibitory receptors in some detail because they form the basis for all considerations of subsets and of NK cell education.

Inhibitory NK Cell Receptors

Inhibitory receptors on NK cells fall into two broad structural categories: C-type (Ca^{++} dependent) lectin-like receptors and members of the Ig superfamily known as killer inhibitory receptors (KIRs). Each family of "inhibitory" receptors, however, contains a few members with subtle differences in structure that function as activating receptors. All the inhibitory receptors contain cytosolic immunoreceptor tyrosine inhibitory motifs (ITIM), which, when phosphorylated on tyrosine residues, can recruit SH2 domain-containing tyrosine phosphatases such as SH2 containing PTP-1 (SHP-1) and thus negatively regulate signaling. Activating receptors lack ITIM motifs but can associate with DAP12, a CD3ζ-like disulfide linked homodimer that contains ITAM motifs and facilitates "positive" signaling.

C-type lectin-like receptors (some of which are inhibitory receptors, while others are not) include members of the murine Ly49 family, the NKR-P1 family, and the CD94/NKG2 family (originally described in humans). These receptors are disulfide-linked dimers of type II membrane proteins in which the extracellular C-terminal portions contain the C-type lectin domain. The lectin domains (and the adjoining stalk region of these proteins) bind to MHC class I molecules in a carbohydrate-independent manner. These lectin domains can also bind to polysaccharides; this latter recognition ability might be more relevant for family members that function as activating receptors. Ly49 proteins have only been found in rodents and exist as homodimers. The cytoplasmic tails of most Ly49 proteins contain ITIM motifs that participate in inhibitory signaling (Figure 11.1).

The NKR-P1 family of genes is only expressed in certain mouse strains, and the proteins they encode are therefore not considered essential for NK cell development or function. The prototypic NK cell antigen, NK 1.1, is encoded by an *NKR-P1* gene. These C-type lectins are also disulfide-linked homodimers. Most NKR-P1 proteins are activators of NK cells. NKR-P1B

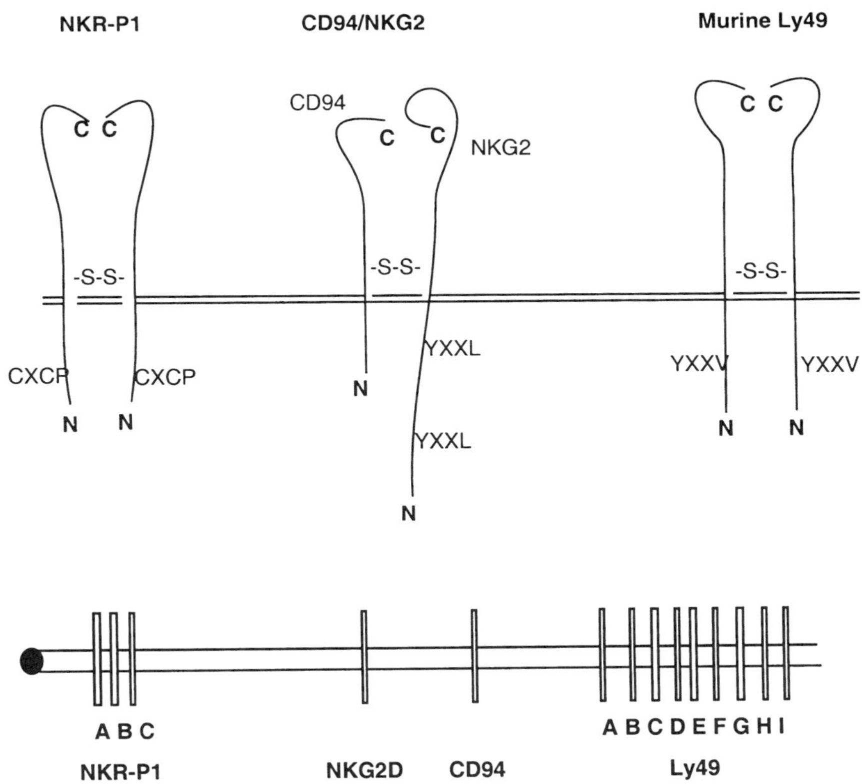

FIGURE 11.1. The NK gene complex on mouse chromosome 6 (and human chromosome 12p12–p13) encodes C-type lectins of the NKR-P1, CD94/NKG2, and Ly49 families. The mouse NK complex is shown in this figure. In humans, a single *NKR-P1* gene has been identified, four *NKG2* genes are known, but no *Ly49* family members have been identified. All the above C-type lectins are type II membrane proteins (with cytosolically disposed N-termini) and exist as disulfide-linked dimers. NKR-P1 and Ly49 proteins are homodimers. The cytoplasmic tail of human NKR-P1A does not contain the CXCP motif found in all mouse NKR-P1 family members. Mouse NKR-P1B contains an ITIM motif in its cytoplasmic tail, suggesting that it might function as an inhibitory receptor. The NKG2 family member schematically depicted in the upper portion of the figure, disulfide linked to CD94, is human NKG2A, which contains two cytosolically disposed ITIM motifs. Ly49 chains each contain an ITIM motif.

contains an ITIM motif in its cytoplasmic tail and may potentially serve an inhibitory function. All murine NKR-P1 molecules (A, B, and C) contain a CXCP motif in their cytoplasmic tails that may be involved in binding Lck (this motif is important for Lck binding by CD4 and CD8). Rat NKR-P1 has been found to associate with Lck, but human NKR-P1A lacks a CXCP motif.

The CD94 and NKG2 proteins were originally found in humans but have now been identified in the mouse and rat as well. These receptors are heterodimers consisting of an "invariant" CD94 chain disulfide linked to an NKG2 polypeptide. CD94 contains only seven amino acids in its cytoplasmic tail, and signal transduction depends on the associated NKG2 chain. There are four distinct NKG2 proteins in humans (NKG2A, NKG2C, NKG2E, and NKG2D/F), each of which has distinct extracellular and cytoplasmic tails. NKG2A contains two ITIM motifs in its cytoplasmic tail. A single murine NKG2 gene has been identified.

A number of distinct Ly49 genes encode NK receptors, most of which are inhibitory, while others such as Ly49D can activate this category of lymphocytes. Ly49, NKRP1, and NKG2 genes are clustered on chromosome 6 in mice and chromosome 12p12–p13 in humans (Figure 11.1). This cluster of genes is known as the NK gene complex. There are at present nine known members of the Ly49 family, and allelic variants of individual Ly49 family members have been described.

KIRs have only been described in humans. These Type I membrane proteins contain immunoglobulin (Ig)-like domains in their extracellular portions (Figure 11.2). The KIRs are broadly divided into two families, depending on whether they contain two or three Ig-like domains. Most KIRs are monomeric, while some are homodimers. Some KIR family members (in spite of their names) are not inhibitory receptors and may potentially serve an activating function. These receptors are noncovalently associated with disulfide-linked DAP12 homodimers, each chain of which contains an ITAM motif. Inhibitory KIRs contain two ITIM motifs in their intracellular portions. KIRs in humans appear to mediate the functions sybserved by Ly49 family receptors in the mouse, in that they provide specificity and sensitivity to NK cell recognition of MHC class I ligands. The KIRs are present on human chromosome 19q13.4. Although KIRs are not found in mice, they are part of a huge superfamily of inhibitory receptors that are widely expressed in mice and humans. These receptors are all located on human chromosome 19q13.4 or a syntenic region of mouse chromosome 7. These Ig superfamily members include CD22 (which inhibits B cell receptor signaling), leukocyte associated Ig like receptor-1 (LAIR-1), the paired inhibitory receptor PIR-1, and many other genes whose functions extend to almost all known cell types.

All inhibitory receptors on NK cells recognize MHC class I ligands. Receptors of the CD94/NKG2 family appear to be more "general" in their recognition capabilities. They may have evolved to recognize cells in which all MHC class I molecules are expressed at low or undetectable levels. In human NK cells, CD94/NKG2A heterodimers recognize an MHC class Ib protein, HLA-E, when its peptide binding groove is occupied by signal peptides from HLA-A, HLA-B, HLA-C, or HLA-G proteins. More specific or restricted recognition of MHC class I molecules might be provided by the

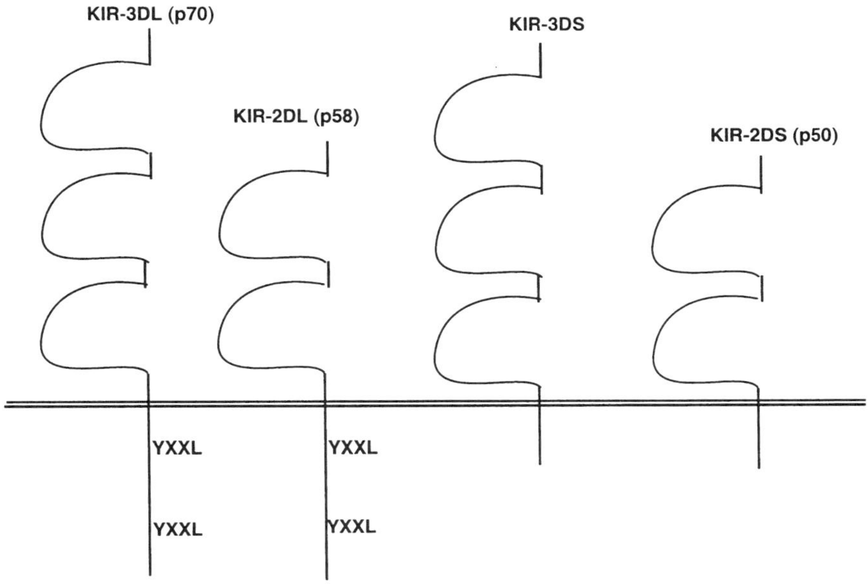

FIGURE 11.2. NK cell receptors of the KIR family include proteins with two or three Ig domains. Most proteins exist as monomers, while one disulfide-linked homodimer (of a KIR-3DL type protein monomer) has also been described. KIRs containing cytosolic ITIM motifs are inhibitory receptors. KIRs with short cytoplasmic tails may serve an activating function. Over a dozen KIR genes have been identified in humans, and these genes are apparently found only in primates. However, KIRs are members of a very large superfamily of inhibitory genes that are found in many species.

Ly49 family in the mouse and by KIRs in humans. Ly49A, for instance, recognizes H-2D^d and H-2D^k. Ly49G2 recognizes H-2D^d and H-2L^d. Both Ly49 and KIR genes have apparently evolved as diverse gene families with many allelic forms, to permit the development of distinct NK cell subsets specific for individual self-MHC class I alleles. It is important that NK cells not be self-reactive and that they should be activated when confronted with targets that have reduced self-MHC class I expression or are missing a specific self-MHC class I allele.

Early Steps in NK Cell Development

Most NK cell development probably occurs in the bone marrow, although a significant number of NK cells are made in the thymus. Some NK cell development may occur in lymph nodes as well. As mentioned above, two distinct types of progenitors have been described that can give rise to NK cells in the mouse. A relatively minor proportion of these cells probably develop from a common T/NK progenitor cell in the thymus that is derived from common lymphoid progenitors (CLPs), and that can give rise to all T cells (including natural T or NT cells), NK cells, and LD cells (discussed in Chapter 9). Studies on thymic progenitors revealed the existence of bipotential cells that express NKR-P1 proteins such as NK1.1. These cells express high levels of c-Kit when they commit to the T lineage and lose NK1.1 expression. Cells that commit to the NK lineage retain NK 1.1 expression, express low levels of c-Kit, and differentiate in an IL-15–dependent manner into mature NK cells. IL-15 probably influences both the differentiation process as well as the proliferation of NK 1.1^+ "pre-NK" cells. In studies on bipotential progenitors from human fetal thymus that can develop *in vitro* into NK cells or T cells, blocking the activity of bHLH transcription factors by the constitutive expression of Id3 (which contains a bHLH DNA binding domain but no activation domain) blocks T cell development and promotes NK cell development.

Bone marrow stromal cells secrete IL-15, and transcription of IL-15 in these cells requires the IRF-1 transcription factor. Development of NK cells does not occur in the absence of IL-15 or in mice in which *IL-2Rβ*, *γc*, or *IRF-1* have been targeted by homologous recombination. IL-2Rβ and γc are components of both the IL-2 and IL-15 receptors. NK cell development, however, is not compromised in IL-2 or IL-2Rα knockout mice, indicating that IL-2 is not required during NK ontogeny. Not surprisingly, NK cell development is also compromised in Jak3-deficient mice. It is unclear whether most NK cells that develop in the bone marrow do so from a T/NK type progenitor in the bone marrow or from a distinct progenitor.

A distinct $CD4^+CD3^-LT\beta^+\alpha4\beta7^+$ fetal progenitor has also been described that gives rise to dendritic cells, poorly defined follicular cells that are believed to be FDCs, and possibly the majority of NK cells. This progenitor migrates into developing fetal lymph nodes, which transiently express the Peyer's patch addressin, MadCAM-1, and is believed to be critical for the generation of secondary lymphoid organs. This progenitor fails to develop in the absence of Id2, and in *Id2*$^{-/-}$ mice, NK cell generation is markedly reduced. This suggests that this $CD4^+$ precursor described in the fetus probably also emerges in the bone marrow in adult life and may be the major source of NK cells. In keeping with this possibility, membrane lymphotoxin (LT) has been shown to be required for NK cell development. Further development of NK cells requires exposure to IL-15. Initially, these cells do not express Ly49 family proteins but begin to do so as they mature. IL-12 and IL-18 contribute to the activation and maturation of NK cells, perhaps activating similar target genes to those activated in Th1 cells (Figure 11.3).

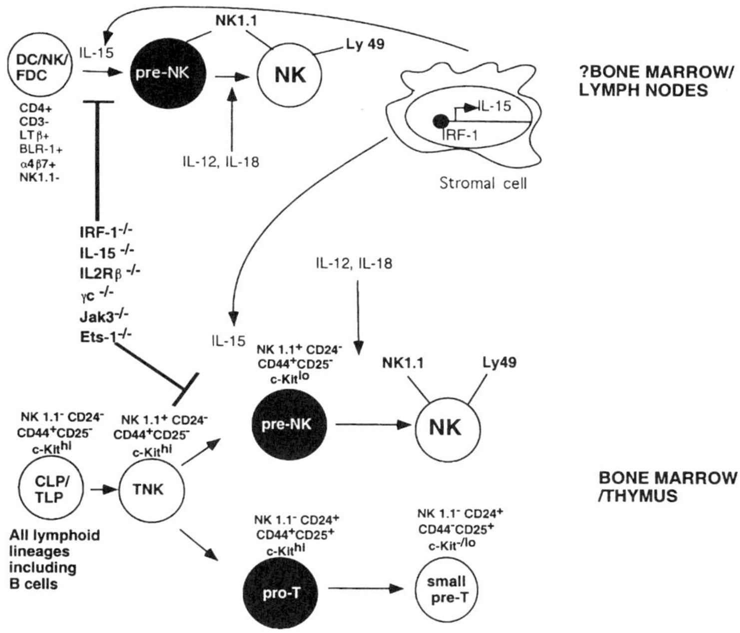

FIGURE 11.3. Most NK cells develop in the bone marrow (though some may develop in the thymus and possibly in lymph nodes as well) from a CLP-derived progenitor or from a distinct fetal $CD4^+CD3^-$ progenitor. The latter cell expresses LTβ and can enter primordial lymph nodes in the fetus and give rise to dendritic cells, NK cells, and less well-defined follicular cells that are probably follicular dendritic cells (FDCs). Development of NK cells requires the local presence of IL-15. In bone marrow stromal cells, interferon-regulated factor-1 (IRF-1) contributes to the transcriptional induction of IL-15. "Pre-NK" cells probably expand in response to IL-15 and express NKR-P1 gene products. They do not express Ly49 family receptors at this stage of development. The acquisition of Ly49 receptors will be discussed below. IL-12 and IL-18 contribute to maturation and acquisition of NK activity.

NK Cell Education and Subsets

It is important that every NK cell express at least one inhibitory receptor that can be recognized by host MHC class I molecules. The repertoire of NK inhibitory receptors must therefore be sufficiently broad to ensure that there are enough receptors to cover all the possible MHC class I molecules in any given host. Genes for inhibitory receptors and MHC class I molecules are not coinherited in a linked fashion. The potential NK repertoire in any individual is quite broad; some educational process must ensure that each NK cell possesses a receptor specific for a self-MHC class I allele.

The following "syllabus" must be kept in mind during NK cell education:

1. Every NK cell should express at least one receptor that recognizes self-MHC class I molecules with moderate avidity.

2. Cells with no, or low, avidity to self-MHC class I molecules are potentially dangerous, since they may be autoaggresive. They may need to be negatively selected or educated so that they will subsequently recognize self-MHC class I molecules with moderate avidity.

3. Cells with inhibitory receptors that have a very high avidity for self-MHC class I are not likely to be useful (but pose no danger). They may not be useful, because they might be inhibited by host cells expressing very low levels of MHC class I molecules. They might also require to be educated to recognize self in a less avid manner.

4. NK cells with too many different anti-self MHC class I inhibitory receptors are also not likely to be useful. When these cells express receptors against two (or more) different self-MHC class I molecules, they might not be able to attack host cells that have shut off only one of those class I molecules.

One of the limitations of most studies performed to date on NK cell education is that reliable reagents are not available to identify each of the Ly49 proteins in the mouse. Some of the few available reagents have limited specificity. It appears that each Ly49 receptor is expressed on a subset of NK cells. Most NK cells do express multiple Ly49 receptors. At least three have been discovered on a single NK cell with the available reagents. It is unclear if two different receptors specific for different self-MHC class I molecules are ever expressed on the same NK cell. It is believed that this should not happen, but formal evidence for such specificity remains to be obtained. Studies on human NK cells indicate that every cell expresses at least one inhibitory receptor (either a KIR or CD94/NKG2A) that is directed against a self-MHC class I ligand.

It appears that in the mouse different *Ly49* genes are transcriptionally induced in different NK cells in a stochastc fashion. Both alleles of a given *Ly49* gene are generally not turned on in the same developing NK cell, and allelic exclusion of Ly49 receptors, like that Ig κ light chains, is probably achieved at the level of transcription. A *sequential gene activation model* has been proposed to explain how the Ly49 repertoire might be generated (Figure 11.4). A given NK cell might sequentially induce the expression of Ly49 family members in a random manner, and this sequential induction might continue until the cell expresses a self-reactive Ly49 receptor. It has been suggested that signals delivered via the first self-MHC class I-recognizing inhibitory receptor expressed on a given cell lead to the termination of the maturation process and shut off the random induction of Ly49 gene expression in this cell. In effect, MHC class I molecules may deliver signals via specific inhibitory receptors to terminate the differentiation process. Such a scheme of events would ensure that every NK cell

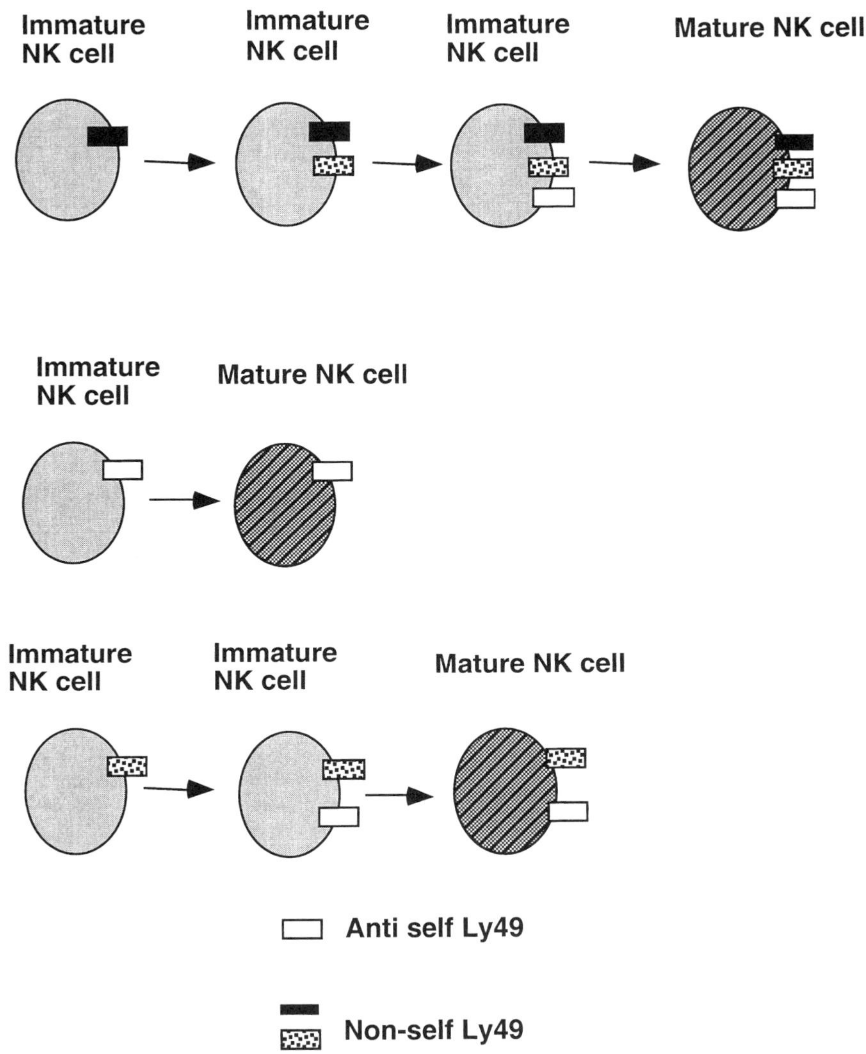

FIGURE 11.4. A sequential gene activation model for NK cell repertoire formation. Ly49 receptors are randomly turned on in a sequential manner in individual cells. When a given cell expresses a Ly49 gene whose product recognizes a self-MHC class I molecule, the cell matures, and no new Ly49 genes are induced. As a result, a given NK cell may express only a single type of specific self-reactive inhibitory receptor. In human NK cells, a similar process might regulate the generation of the KIR repertoire.

expresses a self-MHC class I specific Ly49 receptor, and that the expression of more than one type of antiself-MHC class I-recognizing Ly49-type receptor would not be likely to occur on any individual NK cell.

An alternative model for NK cell education is a *cell selection model* (Figure 11.5). In this model, different Ly49 genes, or sets of genes, are turned on stochastically at around the same stage of development in different cells. A positive selection event is postulated to drive the maturation of cells that express receptors capable of recognizing self-MHC class I molecules with moderate avidity. In addition, in this model, cells expressing receptors that recognize self-MHC class I molecules with very high avidity are selected against. These high-avidity cells may be reeducated rather than eliminated. In such a scheme, it could also be argued that if a single type of self-reactive Ly49 receptor is expressed on a given cell it may be positively selected. If two different self-reactive Ly49 molecules are expressed, their combined levels might generate an excessive amount of inhibitory signaling, which leads to negative selection of such a cell. Both the sequential gene activation model and the cell selection models are consistent with *the receptor calibration model*, which suggests that NK cells expressing lower levels of self-MHC class I-reactive Ly49 receptors are selected for during

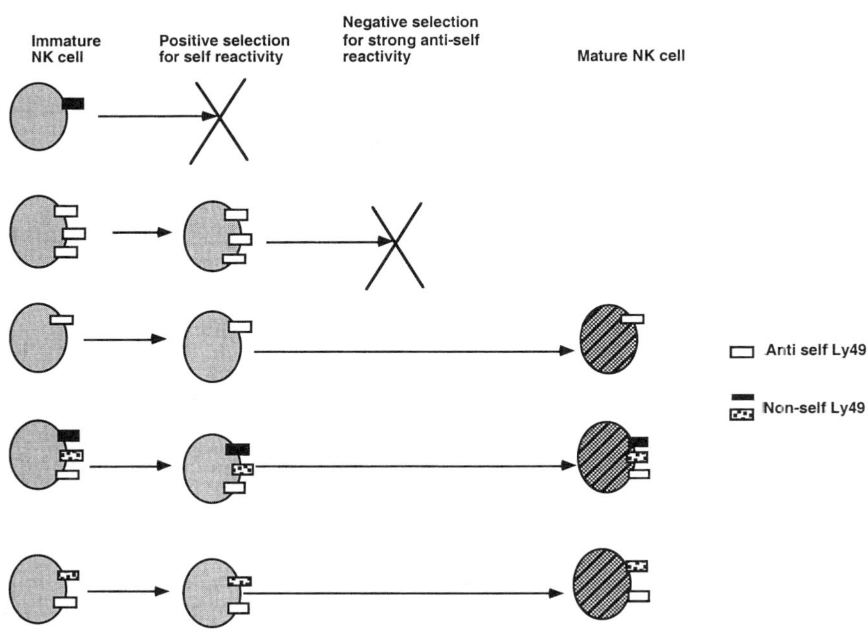

FIGURE 11.5. A cell selection model for NK cell education (also described as a two-step selection model, Raulet et al., 1997). Cells that express moderate levels of self-reactive inhibitory receptors are positively selected, and cells that express very high levels of self-reactive receptors are selected against.

development. It is unclear why allelic exclusion of Ly49 genes is useful during NK cell development. It might be that allelic exclusion ensures that only a limiting "haploid" level of any given Ly49 gene would be expressed in a given cell and that this low expression level provides a threshold signal for positive selection.

While models such as those discussed above have formed the basis for ongoing research in the field, virtually nothing is known as to how NK cells are actually educated and how signaling via inhibitory receptors influences the maturation process in developing cells. However, considerable data do indicate that host MHC class I proteins can regulate both the levels of a specific NK-inhibitory receptor on a given cell as well as the total number of cells expressing inhibitory receptors. Studies using bone marrow chimeras as well as the examination of mice expressing *Ly49A* transgenes suggest that specific MHC class I molecules can reduce the level of expression of their cognate NK-inhibitory receptors. This reduction of expression is a post-transcriptional event that remains to be studied in molecular detail. In addition, the overall proportion of cells expressing a specific inhibitory receptor may be reduced as a consequence of education by a specific MHC class I molecule. At the present time, clearcut molecular explanations cannot be provided for these phenomena, and mechanistic issues in NK cell education remain very poorly understood.

About 20% of NK cells express Ly49A. It is known that the *Ly49A* promoter is regulated by T cell factor-1 (TCF-1) a transcription factor of known importance in T cell development (see Chapter 9). TCF-1 is required for the development of NK clones that express Ly49A but not for the expression of other genes of the *Ly49* family expressed by other clonal populations. How exactly TCF-1 is turned on in selected clones of NK cells and whether Wnt signaling and β-catenin are required for TCF-1 activity in these cells remains unclear.

An F1 hybrid animal expresses class I molecules from both parents. Any parentally derived MHC class I molecule might induce the termination of differentiation of an NK cell that expresses a cognate inhibitory receptor. As a result, in any F1 animal, a subset of NK cells develops that may be inhibited by MHC class I molecules of one parent but not of the other. NK cells from the F1 animal thus can reject parental bone marrow grafts in apparent violation of the laws of transplantation. This phenomenon is known as *hybrid resistance* or the hybrid histocompatibility (Hh) phenomenon.

The Activation of NK Cells

NK cells may be activated when they recognize target cells that lack ligands for inhibitory receptors, and they may also be activated by cytokines. While inhibitory receptors play a defining role in categorizing NK cells, there are many gaps in our understanding of activating NK cell receptors and what they see. The best understood activating receptor is FcγRIII, or CD16, which is the low-affinity receptor for IgG. This receptor is involved in antibody-dependent cellular cytotoxicity (ADCC)-type responses mediated by NK cells and activated macrophages. CD16 is an Ig superfamily member that associates noncovalently with an immunoreceptor tyrosine-based activation motif (ITAM)-containing signaling chain. In mouse NK cells, the associated chain is FcεRIγ, (the γ subunit of the high-affinity IgE receptor that is involved in activating mast cells, basophils, and eosinophils), whereas in human NK cells, either FcεRIγ or CD3ζ may be involved in signaling. The signal transduction machinery downstream of CD16 is very similar to that downstream of the TCR. Signaling involves the tyrosine phosphorylation of ζ or FcεRI-γ by Src-family kinases followed by the recruitment and activation of Zap-70. Activation of the Ras/Raf/ERK (extracellular signal-regulated kinase) pathway, of the PI3 kinase pathway, of phospholipase C (PLC) γ1 and γ2, the induction of calcium signaling and the nuclear translocation of NFAT proteins have all been described following CD16 ligation. NKG2D-DAP10 complexes also serve as activating receptors on NK cells, which respond to stress-inducible MICA proteins (distantly related to MHC class I molecules) and which have been shown to be ligands for certain γδ TCRs as well.

While we do understand in some detail how NK cells mediate ADCC responses and why NK cells do not attack normal host cells that express self-MHC class I molecules, we do not know how exactly NK cells are activated when they encounter targets that are "missing" self-MHC class I expression. The activating members of the Ly49, CD94/NKG2, and KIR families also recognize MHC class I molecules. As mentioned above, we now know that some of these activating family members associate with the ITAM-containing DAP12 signal-generating protein. Certain KIR isoforms, namely KIR2DS and KIR3DS, lack ITIM motifs but contain a basic lysine residue in their transmembrane domains. These basic residues may be important in allowing these activating KIRs to associate noncovalently with DAP12 homodimers. Similarly, CD94/NKG2C heterodimers are activating receptors that recognize HLA-E (human leukocyte antigen-E) molecules. NKG2C also contains a basic transmembrane residue, and this receptor is also noncovalently associated with DAP12.

Why do activating receptors bind to self-MHC class I molecules? It has been suggested that when an infected cell has lost certain self-MHC class I molecules, it might still express HLA-E, which could activate NK cells via the CD94/NKG2C receptor. It is even possible that virally derived signal peptides might on occasion contribute to HLA-E expression even as MHC class I expression is generally downregulated. The biological reasons for the recognition of MHC class I molecules by activating NK receptors remains a matter of conjecture at this point.

Most members of the NKR-P1 family of C-type lectins are also activating receptors, as discussed earlier. While these receptors might recognize carbohydrates on target cells, very little is known about how they recognize targets and are activated *in vivo*.

A receptor that is found both on T and NK cells and that contributes to NK cell activation is CD2. CD2 is a member of the Ig superfamily and is also noncovalently linked to CD3ζ. The major ligand protein (which would have to be expressed on target cells) for CD2 is a transmembrane protein, CD58/LFA-3. CD2 probably functions as a signal augmentor or costimulator on NK cells. CD2 ligation leads to the activation of all the pathways downstream of CD3ζ described for T cell-receptor (TCR) or CD16 signaling. Inactivation of the CD2 gene, however, does not compromise NK cell development or function. This might reflect some redundancy in the function of CD2 both in T cells and NK cells. Given the postulated role of CD2 in inducing cytoskeletal rearrangements and the formation of an immunological synapse between T cells and antigen-presenting cells (APCs), it is possible that CD2 also participates in similar processes in NK cell–target cell interactions.

2B4, another member of the Ig superfamily that is structurally related to CD2 and SLAM (signaling lymphocyte activation molecule) is expressed on all NK cells and some T cells. This molecule is a counter-receptor for CD48 (also a member of the CD2 family that includes SLAM, CD48, CD84, LFA-3/leukocyte function antigen-3 and CD2). 2B4 is an activating receptor on NK cells and some T cells. The cytoplasmic tail of 2B4, like the cytoplasmic tail of SLAM, has been shown to associate with the SH2 domain containing SLAM associated protein (SAP) which is mutated in patients with the X-linked lymphoproliferation syndrome.

Interestingly, a number of receptors and signaling proteins have been found to be clustered some distance from the NK gene complex on mouse chromosome 6 and human chromosome 12. These linked genes include *Hcph* which encodes the SHP-1 protein tyrosine phosphatase, *Cd4*, and *Lag3* (*lymphocyte activation gene 3*), which encodes a CD4-like protein. A point mutation in the Hcph is responsible for the *motheaten* mouse mutation in which SHP-1 is defective. SHP-1 is required for inhibitory receptor function, and NK cells are defective in *motheaten* mice. LAG3 is expressed on activated NK and T cells, and is an Ig family protein with four extracellualr Ig-like domains, which structurally resembles CD4. Mice lacking LAG3 have defective NK cells that are unable to lyse tumor targets. These NK cells, however, are perfectly capable of lysing cells with MHC class I disparities. LAG3 is therefore likely to represent either an NK-activating receptor or coreceptor that is selectively required during NK cell-mediated lysis. Table 11.1 lists some of the genes shown to be relevant for NK cell development by gene-targeting approaches.

TABLE 11.1. Factors required for NK cell development

Factors	References
Transcriptional regulators in progenitors	
Ikaros	Georgopoulos et al., 1994
PU.1	Scott et al., 1994
? GATA-3	Ting et al., 1996
Ets-1	Barton et al., 1998
Id2	Yokota et al., 1999
TCF-1 (for Ly49A clones)	Held et al., 1999
Transcriptional regulators in stromal cells	
IRF-1	Ohteki et al., 1998
Cytokines	
IL-15	Puzanov et al., 1996 Carson et al., 1997 Ogasawara et al., 1998
IL-12	Magram et al., 1996
IL-18	Takeda et al., 1998
LT-β	Iizuka et al., 1999
Factors involved in IL-15 signaling	
IL-15Rα	Lodolce et al., 1998
IL-2Rβ	Suzuki et al., 1997
γc	Cao et al., 1995 Di Santo et al., 1995
Jak3	Thomis et al., 1995 Nosaka et al., 1995 Park et al., 1995

Summary and Perspective

The study of NK cell development is in its infancy. These lymphocytes are involved in innate immunity and can recognize cells with reductions in self-MHC class I expression. Most NK cells probably develop from $CD4^+CD3^-$ IL-7$R\alpha^+$ stem cells in the bone marrow, lymph nodes, and gut. In the thymus (which is not the major site of NK cell development) as well as in the bone marrow, these cells may develop from a common T/NK progenitor. Bone marrow stromal cells require the IRF-1 transcriptional regulator to produce IL-15. Signaling via the IL-15 receptor on poorly identified bone marrow progenitors is required for the development of NK1.1-expressing immature NK cells. These cells then turn on the expression of inhibitory receptor genes in a stochastic fashion. NK cell education ensures that each NK clone expresses a self-MHC class I recognizing inhibitory receptor. In mice, the Ly49 inhibitory receptors of the C-type lectin family contribute to the recognition of specific MHC class I alleles. In humans, Ly49 family receptors have not been identified, but the KIRs serve a similar function. NK cell education ensures that each developing NK cell expresses an inhibitory receptor with a moderate avidity for self-MHC class I molecules.

Selected Reviews

Hoglund, P., Sundback, J., Olsson-Alheim, M. Y., Johansson, M., Salcedo, M., Ohlen, C., Ljunggren, H.-G., Sentman, C. L., and Karre, K. (1997). Host MHC class I control of NK cell specificity in the mouse. Immunol. Rev. *155*, 11–28.

Lanier, L. L. (1998). NK cell receptors. Annu. Rev. Immunol. *16*, 359–393.

Leibson, P. J. (1997). Signal transduction during natural killer cell development: inside the mind of a killer. Immunity *6*, 655–661.

Long, E. O., and Wagtmann, N. (1997). Natural killer cell receptors. Curr. Opin. Immunol. *9*, 344–350.

Moore, T. A., Bennett, M., and Kumar, V. (1996). Murine natural killer cell differentiation: past, present, and future. Immunol. Res. *15*, 151–162.

Raulet, D., Held, W., Correa, I., Dorfman, J., Wu, M.-F., and Corral, L. (1997). Specificity, tolerance, and developmental recognition of natural killer cells defined by expression of class I specific Ly49 receptors. Immunol. Rev. *155*, 41–52.

Raulet, D. H. (1999). Development and tolerance of natural killer cells. Curr. Opin. Immunol. *11*, 129–134.

Trinchieri, G. (1989). Biology of natural killer cells. Adv. Immunol. *47*, 187–376.

Valiante, N. M., Lienert, K., Shiling, H. G., Smits, B. J., and Parham, P. (1997). Killer cell receptors: keeping pace with MHC class I evolution. Immunol. Rev. *155*, 155–164.

Some Relevant References

Barton, K., Muthusamy, N., Fischer, C., Ting, C.-N., Walunas, T. L., Lanier, L. L., and Leiden, J. M. (1998). The Ets-1 transcription factor is required for the development of natural killer cells in mice. Immunity *9*, 555–564.

Bauer, S., Groh, V., Wu, J., Steinle, A., Phillips, J. H., Lanier, L. L., and Spies, T. (1999). Activation of NK cells and T cells by NKG2D, a receptor for stress inducible MICA. Science *285*, 727–729.

Cao, X., Shores, E. W., Hu-Li, J., Anver, M. R., Kelsall, B. L., Russell, S. M., Drago, J., Noguchi, M., Grinberg, A., and Bloom, E. T. (1995). Defective lymphoid development in mice lacking expression of the common cytokine receptor γ chain. Immunity *2*, 223–238.

Carlyle, J. R., Michie, A. M., Furlonger, C., Nakano, T., Lenardo, M. J., Paige, C. J., and Zuniga-Pflucker, J. C. (1997). Identification of a novel developmental stage marking lineage commitment of progenitor thymocytes. J. Exp. Med. *186*, 173–182.

Carlyle, J. R., and Zuniga-Pflucker, J. C. (1998). Requirement for the thymus in αβ T lymphocyte development. Immunity *9*, 187–198.

Carson, W. E., Fehniger, T. A., Haldar, S., Eckhert, K., Lindemann, M. J., Lai, C. F., Croce, C. M., Baumann, H., and Caliguri, M. A. (1997). A potential role for interleukin-15 in the regulation of human natural killer cell survival. J. Clin. Invest. *99*, 937–943.

DiSanto, J. P., Muller, W., Guy-Grand, D., Fischer, A., and Rajewsky, K. (1995). Lymphoid development in mice with a targeted deletion of the interleukin-2 receptor γ chain. Proc. Natl. Acad. Sci. USA *92*, 377–381.

Dorfman, J. R., and Raulet, D. H. (1998). Acquisition of Ly 49 receptor expression by developing natural killer cells. J. Exp. Med. *187*, 609–618.

George, T., Yu, Y. Y. L., Liu, J., Davenport, C., Lemieux, S., Stoneman, E., Mathew, P. A., Kumar, V., an Bennett., M. (1997). Allorecognition of murine natural killer cells: lysis of T lymphoblasts and rejection of bone marrow grafts. Immunol. Rev. *155*, 29–40.

Georgopoulos, K., Bigby, M., Wang, J.-H., Molnar, A., Wu, P., Winandy, S., and Sharpe, A. (1994). The Ikaros gene is required for the development of all lymphoid lineages. Cell *79*, 143–156.

Heemskerk, M. H. M., Blom, B., Nolan, G., Stegmann, A. P. A., Bakker, A. Q., Weijer, K., Res, P. C. M., and Spits, H. (1997). Inhibition of T cell and promotion of natural killer cell development by the dominant negative helix loop helix factor Id3. J. Exp. Med. *186*, 1597–1602.

Held, W., Kunz, B., Lowin-Kropf, B., van de Wetering, M., and Clevers, H. (1999). Clonal acquisition of the Ly49A NK cell receptor is depenent on the trans-acting factor TCF-1. Immunity *11*, 433–442.

Iizuka, K., Chaplin, D. D., Wang, Y., Wu, Q., Pegg, L. E., Yokoyama, W. M., and Fu, Y. X. (1999). Requirement for membrane lymphotoxin in natural killer cell development. Proc. Natl. Acad. Sci. USA *96*, 6336–6340.

Kapasi, Z. F., Qin, D., Kerr, W. G., Kosco-Vilbois, M. H., Shultz, L. D., Tew, J. G., and Szakal, A. K. (1998). Follicular dendritic cell precursors in primary lymphoid tissues. J. Immunol. *160*, 1078–1084.

Karre, K., Ljunggren, H. G., Piontek, G., and Kiessling, R. (1986). Selective recognition of H-2 deficient lymphoma variants suggests alternative immune defence strategy. Nature *319*, 675–678.

Lanier, L. I., Corliss, B. C., Wu, J., Leong, C., and Philips, J. H. (1998). Immunoreceptor DAP12 bearing a tyrosine based activation motif is involved in activating NK cells. Nature *391*, 703–707.

Lodolce, J. P., Boone, D. L., Chai, S., Swain, R. E., Dassopoulos, T., Trettin, S., and Ma, A. (1998). IL-15 receptor maintains lymphoid homeostasis by supporting lymphocyte homing and proliferation. Immunity *9*, 669–676.

Magram, J., Connaughton, S. E., Warrier, R. R., D.M., C., Wu, C. Y., Ferrante, J., Stewart, C., Sarmiento, U., Faherty, D. A., and Gately, M. K. (1996). IL-12 deficient mice are defective in IFN gamma production and type 1 cytokine responses. Immunity *4*, 471–481.

Mebius, R. E., Rennert, P., and Weissman, I. L. (1997). Developing lymph nodes collect $CD4^+CD3^-$ $LT\beta^+$ cells that can differentiate to APC, NK cells, and follicular cells but not T or B cells. Immunity *7*, 493–504.

Nosaka, T., van Deursen, J. M. A., Tripp, R. A., Thierfelder, W. E., Wittuhn, B. A., McMickle, A. P., Doherty, P. C., Grosveld, G. C., and Ihle, J. N. (1995). Defective lymphoid development in mice lacking Jak3. Science *270*, 800–802.

Ogasawara, K., Hida, S., Azimi, N., Tagaya, Y., Sato, T., Yokochi-Fuda, T., Walkmann, T. A., Taniguchi, T., and Taki, S. (1998). Requirement for IRF-1 in the microenvironment supporting development of natural killer cells. Nature *391*, 700–703.

Ohteki, T., Yoshida, H., Matsuyama, T., Duncan, G. S., Mak, T. W., and Ohashi, P. S. (1998). The transcription factor interferon regulatory factor-1 (IRF-1) is important during the maturation of natural killer 1.1^+ T cell receptor $\alpha\beta^+$ ($NK1^+T$) cells, natural killer cells and intestinal intraepithelial cells. J. Exp. Med. *187*, 967–972.

Park, S. Y., Saijo, K., Takahashi, T., Osawa, M., Arase, H., Hirayama, N., Miyakie, K., Nakauchi, H., Shirasawa, T., and Saito, T. (1995). Developmental defects of lymphoid cells in JAK3 kinase deficient mice. Immunity *3*, 771–782.

Puzanov, I. J., Bennett, M., and Kumar, V. (1996). IL-15 can substitute for the marrow microenvironment in the differentiation of natural killer cells. J. Immunol. *157*, 4282–4285.

Raulet, D., Held, W., Correa, I., Dorfman, J., Wu, M.-F., and Corral, L. (1997). Specificity, tolerance, and developmental recognition of natural killer cells defined by expression of class I specific Ly49 receptors. Immunol. Rev. *155*, 41–52.

Sanchez, M. J., Muench, M. O., Roncarolo, M. G., Lanier, L. L., and Phillips, J. H. (1994). Identification of a commonT/natural killer cell progenitor in human fetal thymus. J. Exp. Med. *180*, 569–576.

Scott, E. W., Simon, M. C., Anastasi, J., and Singh, H. (1994). Requirement of transcription factor PU.1 in the development of multiplehematopoietic lineages. Science *265*, 1573–1577.

Suzuki, H., Duncan, G. S., Takimoto, H., and Mak, T. W. (1997). Abnormal development of intestinal intraepithelial lymphocytes and peripheral natural killer cells in mice lacking the IL-2 receptor β chain. J. Exp. Med. *185*, 499–505.

Takeda, K., Tsutsui, H., Yoshimoto, T., Adachi, O., Yoshida, N., Kishimoto, T., Okamura, H., Nakanishi, K., and Akira, S. (1998). Defective NK cell activity and Th1 response in IL-18 deficient mice. Immunity *8*, 383–390.

Thomis, D. C., Gurniak, C. B., Tivol, E., Sharpe, A. H., and Berg, L. J. (1995). Defects in B lymphocyte maturation and T lymphocyte activation in mice lacking Jak3. Science *270*, 794–797.

Ting, C. N., Olson, M. C., Barton, K. P., and Leiden, J. M. (1996). Transcription factor GATA-3 is required for development of the T cell lineage. Nature *384*, 474–478.

Williams, N. S., Moore, T. A., Schatzle, J. D., Puzanov, I. J., Sivakumar, P. V., Zlotnik, A., Bennett, M., and Kumar, V. (1997). Generation of lytic Natural Killer 1.1^+, Ly^-49^- cells from multipotenitial murine bone marrow progenitors in a stroma free culture: definition of cytokine requirements and developmental intermediates. J. Exp. Med. *186*, 1609–1614.

Wu, J., Song, Y., Bakker, A. B. H., Bauer, S., Spies, T., Lanier, L. L., and Phillips, J. H. (1999). An activating immunoreceptor complex formed by NKG2D and DAP10. Science *285*, 730–732.

Wu, L., Nichogiannopoulou, A., Shortman, K., and Georgopoulos, K. (1997). Cell-autonomous defects in dendritic cell populations of Ikaros mutant mice point to a developmental relationship with the lymphoid lineage. Immunity *7*, 483–492.

Yokota, Y., Mansouri, A., Mori, S., Sugawara, S., Adachi, S., Nishikawa, S.-I., and Gruss, P. (1999). Development of peripheral lymphoid organs and natural killer cells depends on the helix-loop-helix inhibitor Id2. Nature *397*, 702–706.

12

Naturally Occurring Disorders of Lymphocyte Development

Although a very large number of engineered defects in murine lymphocyte development have been generated in the last decade, naturally occurring disorders of lymphocyte development have (not too surprisingly) been most thoroughly investigated in humans. In this chapter, the focus will primarily be on naturally occurring human immunodeficiencies that arise because of mutations in genes critical for lymphocyte development. We will examine the molecular mechanisms involved in inherited disorders of lymphocyte development in the light of information covered in previous chapters. The genetic disorders that we will discuss here have somewhat arbitrarily been divided into diseases of lymphocyte development and immune dyscrasias involving aberrant lymphocyte activation. As will be obvious to most readers, there is likely to be considerable overlap between the processes of development and activation. Mutations in *Rag-1* or *Rag-2*, for instance, that drastically compromise antigen-receptor rearrangement may lead to a severe combined immunodeficiency phenotype, while other mutations in these very same genes permit partial V(D)J recombinase activity and result in the very distinct clinical presentation of Omenn syndrome.

Differences Between Lymphocyte Development in Humans and Mice

Human and murine lymphocytes are not identical; however, lymphocyte development in humans and mice are remarkably similar. We have stressed some of the subtle and not-so-subtle differences in murine and human lymphocyte development from time to time in the course of this book. These differences have been most apparent when a gene that is mutated in a naturally occuring human immunodeficiency disease has been knocked out in the mouse and the disease and engineered phenotypes have been non-identical. By and large, however, most murine knockouts mimic their human disease counterparts. Generally, knockouts tend to represent null mutations, whereas a human disease involving a single gene may present with a wide spectrum of phenotypes that may reflect the range of functional defects in a single protein that can be generated by different mutations. Often, phenotypes may vary because of the effects of modifier genes. Some of the differences in human and mouse lymphocyte development will be revisited in this chapter when discussing X-linked agammaglobulinemia (which involves mutations in *Btk*), X-linked severe combined immunodeficiency (SCID)—with defects in interleukin-7 (IL-7) and IL-15 signaling—and T cell immunodeficiency linked to mutations in *Zap-70*.

Disorders of Lymphocyte Development

The last decade has seen the mechanistic unraveling of many members of a group of specific immunodeficiency disorders that were often lumped together before the causative genes for some of them were identified. Clinically, children with immunodeficiency disorders may present with recurrent infections. Recurrent pyogenic infections with bacteria such as *Streptococcus pneumoniae* and *Hemophilus influenzae* are commonly seen in disorders of B cell development or function. In disorders in which T cell development or function is compromised, opportunistic pathogens such as *Pneumocystis carinii* may be responsible for symptoms, and the likelihood of succumbing to intracellular pathogens such as *Mycobacteria* or viruses is greatly increased.

A number of inherited immunodeficiency disorders that affect more than one lineage are commonly divided into two subsets, the "severe combined" and "combined" immunodeficiencies (Table 12.1). While for simplicity's sake we maintain this classification, the distinction between "severe combined" and "combined" immunodeficiencies is made on the basis of clinical rather than mechanistic considerations.

Severe Combined Immunodeficiencies

The SCIDs in general represent diseases that are fatal early in life. They also represent a set of disorders that are ideal candidates for gene therapy and that may therefore prove completely curable in the next few decades. As a group, these deficiencies are noted in about 1 in 75,000 live births, are likely to prove lethal during the first year of life, and all forms involve a profound block in T cell differentiation. Infants present with opportunistic infections, often caused by live vaccines, and a failure to thrive. There at least eight variant forms of SCID.

Reticular Dysgenesis

As seen in Figure 12.1, some of these disorders represent defects in the development of early lymphoid progenitors prior to commitment to the T

TABLE 12.1. Primary immunodeficiencies involving defects in lymphocyte development

I. Defects in lymphocyte development
  A. Severe combined immunodeficiencies
    1. Absence of all myeloid and lymphoid cells (reticular dysgenesis)
    2. Absence of all lymphoid populations (adenosine deaminase deficiency)
    3. Absence of T and NK cells (X-linked SCID; γc deficiency)
    4. Absence of T and NK cells (autosomal SCID; Jack-3 deficiency)
    5. Absence of T and NK cells (autosomal SCID, unknown mechanism)
    6. Absence of T and B cells (alymphocytosis; *Rag* gene defects)
    7. Absence of T and B cells (alymphocytosis; non-RAG recombination defect)
    8. Absence of T cells ($T^-B^+NK^+$ SCID; IL-7R deficiency)
  B. Combined Immunodeficiencies
    1. DiGeorge syndrome
    2. Purine nucleotide phosphorylase deficiency
    3. Zap-70 mutations
    4. DNA ligase I deficiency
    5. CD3 ε deficiency
    6. CD3 γ deficiency
    7. HLA class II deficiency (CIITA, RFX5)
    8. Class I deficiency (TAP2)
    9. Ataxia telangiectasia
    10. Cartilage hair hypoplasia
  C. Humoral Immunodeficiencies
    1. X-linked agammaglobulinemia (Bruton's disease)
    2. Human surrogate light chain, μ heavy chain, and Blnk mutants
    3. IgG subclass deficiencies
    4. κ chain deficiency
    5. X-linked hyper-IgM syndrome
    6. Hyper-IgM syndrome (autosomal)
    7. Selective IgA deficiency
    8. Common variable immunodeficiency
II. Disorders involving aberrant lymphocyte activation
  A. Wiskott-Aldrich syndrome
  B. Autoimmune lymphoproliferative syndrome
  C. "Accelerated phase" disorders
    1. Chediak-Higashi syndrome
    2. Familial hemophagocytic lymphohistiocytosis
    3. X-linked lymphoproliferation
    4. Immunodeficiency with partial albinism
  D. Omenn's syndrome
  E. Hyper-IgE syndrome
  F. Idiopathic disseminated mycobaterial infections

and B lineages. In patients with a form of SCID known as reticular dysgenesis, no T, B, natural killer (NK), or myeloid cells are seen. The defect resembles that observed in the PU.1 knockout mouse (see Chapter 2), in which, as in reticular dysgenesis, erythroid and platelet differentiation remain unaffected.

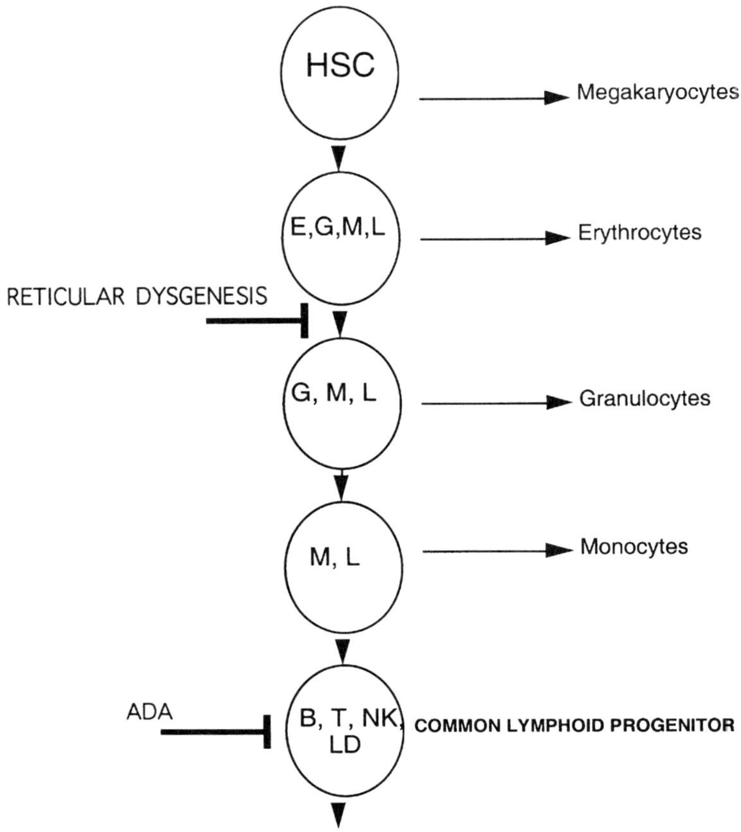

FIGURE 12.1. Some forms of SCID involve defects in progenitor cells. The presence of platelets (P) and erythrocytes (E) in reticular dysgenesis is reminiscent of the mouse PU.1 knockout. G refers to granulocytes, M to macrophages, and L to lymphocytes. In adenosine deaminase (ADA) deficiency, the susceptible cell may be the common lymphoid progenitor, although the most affected cell is probably a T cell progenitor. LD refers to lymphoid dendritic cells.

Combined Immunodeficiencies

Adenosine Deaminase Deficiency

A fairly common form of SCID occurs in infants with an inherited deficiency in adenosine deaminase (ADA). During the catabolism of purines (Figure 12.2), phosphate residues are removed from nucleotide monophosphates by nucleotidase. Inosine, xanthosine, and guanosine are further degraded by the action of purine nucleoside phosphorylase (PNP) and subsequently converted to uric acid. Intermediates in the degradation process may be reused to form nucleotides by salvage reactions. Adenosine and

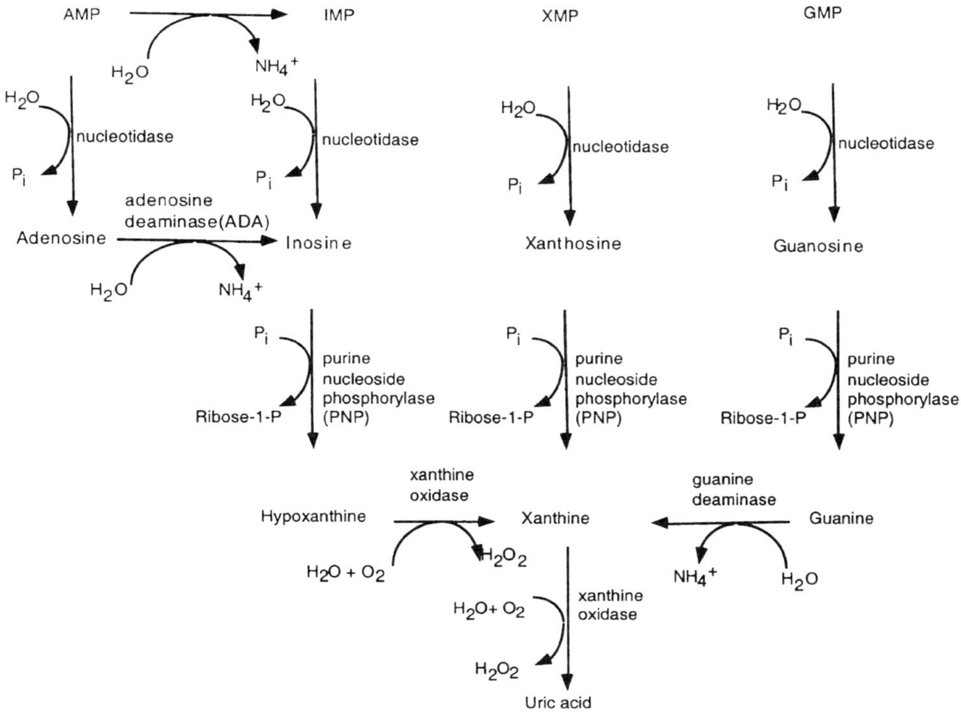

FIGURE 12.2. An overview of purine catabolism in mammalian cells. Mutations in adenosine deaminase result in the accumulation of phosphorylated dATP in lymphoid progenitors that inactivate ribonucleotide reductase and thus arrest cell cycle progression. While ADA deficiency affects B, T, and NK cell development, mutations in purine nucleotide phosphorylase primarily compromise T cell function.

deoxyadenosine are not degraded in mammalian cells by PNP, but instead are deaminated by ADA to yield inosine and deoxyinosine. The latter are degraded by PNP and xanthine oxidase, the final product being uric acid. ADA has been crystallised and forms an eight-stranded α/β barrel with the active site at the C-terminal end of the β barrel. Eight different mutant forms of ADA have been described in SCID patients, all of which compromise the active site of the enzyme.

In the absence of functional ADA, deoxyadenosine is phosphorylated to produce very high levels of dATP. These high levels of dATP (deoxy adenosine triphosphate) inhibit ribonucleotide reductase, which is essential for deoxy nucleoside triphosphate (dNTP) synthesis. DNA synthesis is thereby choked off, and cells cannot proliferate. Patients with ADA deficiency have defects in the the generation of T, B, and NK cells, suggesting a defect at the level of the putative common lymphoid progenitor or in the early developmental stages of each of these three lineages. Indeed, NK cell function is often unaffected in patients with ADA, and B cell development may be minimally compromised, suggesting that the defect may be largely confined to T cell precursors. Why are only these hematopoietic cells affected? It appears that the phosphorylation of deoxyadenosine occurs particularly efficiently in lymphoid progenitors, and it is likely that the earliest cell in which toxic levels of dATP accumulate might be the common lymphoid progenitor cell but that the most susceptible cell may be a T cell progenitor. The manifestations of ADA deficiency may not be restricted to lymphoid cells; patients also exhibit rib cage and other skeletal abnormalities, presumably in response to toxic effects of dATP and the inhibition of proliferation of other selected cell types.

When ADA is injected intravenously, it is cleared very rapidly by the liver. However, when polyethylene glycol (PEG) is covalently coupled to ADA, the enzyme survives in the bloodstream for 1 or 2 weeks, and these adducts can be used to treat ADA deficiency fairly effectively. Preliminary attempts at gene therapy have yielded promising results, and it is very likely that this will prove to be the therapeutic intervention of choice in the near future.

X-Linked SCID

The common γ chain (γc) of the interleukin-2 (IL-2), IL-4, IL-7, IL-9, and IL-15 receptors is encoded by an X-linked gene (located at Xq12–13.1). Defects in this gene lead to an X-linked form of SCID in which T and NK cell development are compromised, but there is no defect in B cell generation. This is the most common form of SCID, and a large variety of mutations have been discovered in γc. This is a lethal disease, and there is clearly a very high rate of new mutations being generated. In humans, IL-7 is required for T cell development but is not critical for B cell development, whereas in the mouse IL-7 is required both during T and B cell ontogeny. IL-15 is required for NK cell development and the combined loss of IL-7– and IL-15–derived signals leads to the failure of development of these lineages (Figure 12.3).

Autosomal Recessive SCID With Jak3 Deficiency

Mutations in an autosomal gene encoding the Jak3 kinase result in a very similar phenotype to that seen in X-linked SCID. This is not at all surprising, since Jak3 and γc are functionally linked (Figure 12.3) and identical phenotypes are seen in mice in which these genes are separately targeted. Jak1 is also a critical player in this pathway in the mouse, but mutations in Jak1 have yet to be described in patients with autosomal recessive forms of SCID. IL-7 plays a demonstrable role in B cell development in mice but not in humans, and this accounts for the differences in phenotype seen between the human disorders involving γc and Jak3 and the corresponding knockout mice lines.

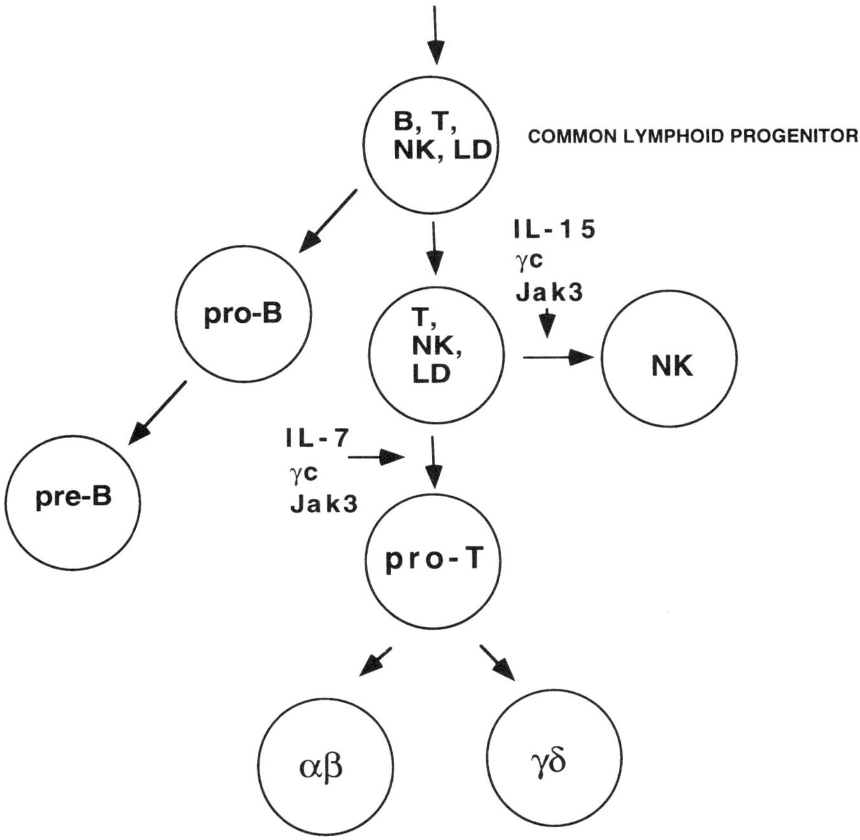

FIGURE 12.3. IL-7 and IL-15 influence the development of T cells and NK cells, respectively, in humans. Signals downstream of the receptors for IL-2, IL-4, IL-7, IL-9, and IL-15 are transduced via γc, Jak3, and probably Jak1 as well. LD refers to lymphoid dendritic cells.

Autosomal Recessive SCID With IL-7R Deficiency

In a form of SCID in which there is a defect in T cell development with normal numbers of B and NK cells ($T^-B^+NK^+$SCID), inherited mutations in the gene encoding the IL-7Rα chain have been described. This disease underscores the fact that in humans IL-7Rα-derived signals are not required for B cell development and that IL-7 is not required for NK cell development in mice or humans (although IL-15 is crucial).

Other Autosomal Recessive Forms of SCID

The four autosomal recessive forms of SCID considered above are reticular dysgenesis, Jak3 deficiency, ADA deficiency, and an IL-7Rα chain defect.

Two autosomal recessive forms of SCID have been described in which B and T cells are not seen. In one of these forms, mutations in *Rag-1* or *Rag-2* severely compromise V(D)J recombination. In an additional form, V(D)J recombination is compromised, but no mutations have as yet been described either in the *Rag* genes or DNA repair-related components such as DNA-PK, Ku-70, or Ku-80. DNA PK is defective in SCID mice and in an equine form of SCID seen in Arabian foals. Certain mutations in the *Rag-1* and *Rag-2* genes result in partially compromised V(D)J recombination and an unusual clinical phenotype described as Omenn syndrome, which will be discussed later in this chapter.

A form of autosomal recessive SCID similar in phenotype to X-linked SCID and Jak3 deficiency (with no T or NK cells) has been described. The defective gene in this form has not yet been identified, but it is likely to encode a protein in the γc/Jak1/Jak3 pathway (other than Jak3 or γc).

"Combined" Immunodeficiencies

The combined immunodeficiencies are characterized by a defect usually in T cell development that does not lead to as severe a phenotype as is seen in SCID. Defects either in B cell development or function also may be seen (often because of the absence of T cell help). As discussed earlier, the distinction between SCID and "CID" is operational, and there is considerable overlap between these groups of diseases.

DiGeorge's Syndrome

DiGeorge's syndrome is characterized by thymic and parathyroid hypoplasia. In its most severe form, it can resemble SCID, but the defect does not involve the B lineage. This syndrome results from dysmorphogenesis of the third and fourth pharyngeal pouches. The defect in thymic development results from a presumed failure of the neural crest mesenchyme to induce morphogenetic changes in the third pharyngeal pouch-derived endoderm. The gene that is defective in the DiGeorge syndrome has been located to 22q11 but remains to be cloned.

Purine Nucleoside Phosphorylase Deficiency

Nezelof's syndrome is a form of combined immunodeficiency in which B cell development is normal but T cell development is impaired. A large number of patients with this syndrome have a defect in an enzyme that is critical for purine degradation, purine nucleoside phosphorylase (PNP) (see Figure 12.2). Exactly why PNP deficiency targets T cell development is unclear. Inhibition of PNP by pharmacological means can prevent the hyperactivation of T cells seen in some autoimmune diseases, but the mechansism by which PNP activity is linked to T cell development or T cell activation remains to be elucidated.

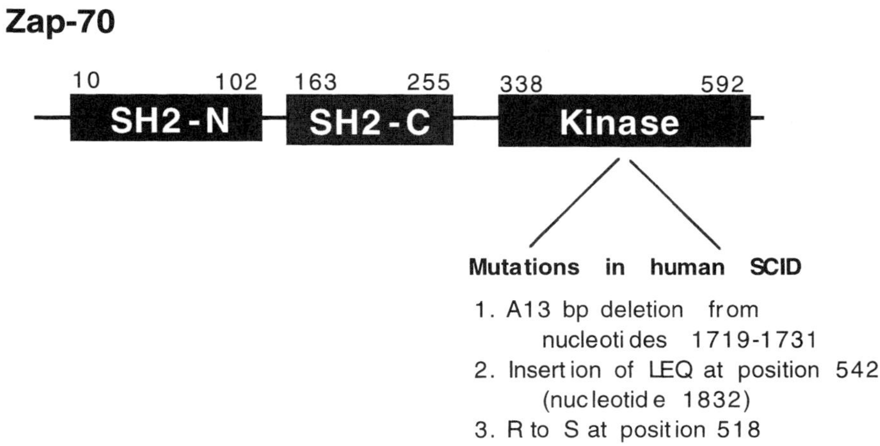

FIGURE 12.4. Mutations in *Zap-70* compromise human CD8 T cell development.

Mutations in *Zap-70* and CD3

A form of T cell deficiency seen in humans results from mutations in the *Zap-70* tyrosine kinase, a member of the Syk family. These mutations occur in the kinase domain of this enzyme (Figure 12.4)

Unlike *Zap-70* null mice, which lack both $CD4^+$ and $CD8^+$ T cells because of a failure of TCR signaling during positive selection II, in humans, *Zap-70* mutations result only in a failure of $CD8^+$ T cell development. However, while $CD4^+$ cells do emerge in patients with inherited defects in *Zap-70*, these cells cannot be readily activated in response to T cell-receptor (TCR) triggering. These children have recurrent opportunistic infections in the first year of life. They may sometimes be categorized as SCID patients, and these distinctions are somewhat arbitrary. Why do patients with *Zap-70* defects not have a problem in positively selecting $CD4^+$ T cells? It has been suggested that CD4 more readily associates with Lck than CD8, and that, as a result, *Zap-70* is more readily activated during positive selection II in $CD4^+$ cells than in $CD8^+$ cells. Alternatively, Syk may be more readily activated in $CD4^+$ T cells than in $CD8^+$ cells and may thus compensate for the absence of *Zap-70* in the former.

Defects in CD3ε and CD3γ also result in defects in T cell development, presumably because they compromise signaling via the pre-TCR and TCR.

T Cell Immunodeficiencies Associated With DNA Ligase I Mutations

It is likely, as discussed above, that some SCID or CID phenotypes will be associated with mutations in genes that are involved in double-strand break repair such as Ku70 and Ku80. Defects in T cell development have been noted in isolated patients who have inherited mutations in DNA ligase I, which may be required during V(D)J recombination (although DNA ligase IV appears to be the major DNA ligase involved in this process).

The "Bare Lymphocyte Syndrome": MHC Class II Deficiency

As expected inherited disorders in which major histocompatability complex (MHC) class II expression is compromised result in a defect in $CD4^+$ T cell development, and defects in MHC class I expression result in abnormalities in $CD8^+$ T cell positive selection. MHC class II expression defects have been noted in patients with the bare lymphocyte syndrome who have inherited defects in transcriptional regulators of MHC class II (HLA DP, DQ, and DR) expression. These patients also fail to express the invariant chain (Ii) and HLA-DM, a class II-like heterodimer that facilitates the exchange of the invariant chain-derived CLIP (Class II associated invariant chain peptides) peptides that remain bound to MHC class II proteins, with peptides derived from the proteolysis of endocytosed proteins. The failure to express HLA class II molecules in the thymic cortex during positive selection results in an absence of $CD4^+$ T cells. Patients may belong to any one of four complementation groups. Three of the four regulatory genes involved in this syndrome, CIITA, RFX5, and RFXAP, have been isolated. As seen in Figure 12.5, RFX5 and RFXAP are components of the multimeric RFX complex, which binds to the X box in MHC class II, Ii, and HLA-DM promoters. Although RFX is ubiquitously expressed, its activity depends on a co-activator CIITA, which may be expressed constitutively in professional antigen-presenting cells (APCs) but can also be induced by IFN-γ in certain cell types. CIITA is regulated by multiple promoters that result in distinct CIITA transcripts with different first exons. The promoter PI is responsible for constitutive expression in dendritic cells, and the promoter PIII is required for constitutive B cell expression. The PIV promoter regulates IFN-γ mediated induction of CIITA. Regulation of this promoter depends on the binding of phosphorylated STAT1 to a GAS site, which cooperates with the ubiquitously expressed bHLH protein USF1, which binds to an adajcent E-box. STAT1 is phosphorylated by Jak1/2 as a consequence of IFN-γ signaling. One of the other regulatory elements is an IRF-1 binding site. During B cell

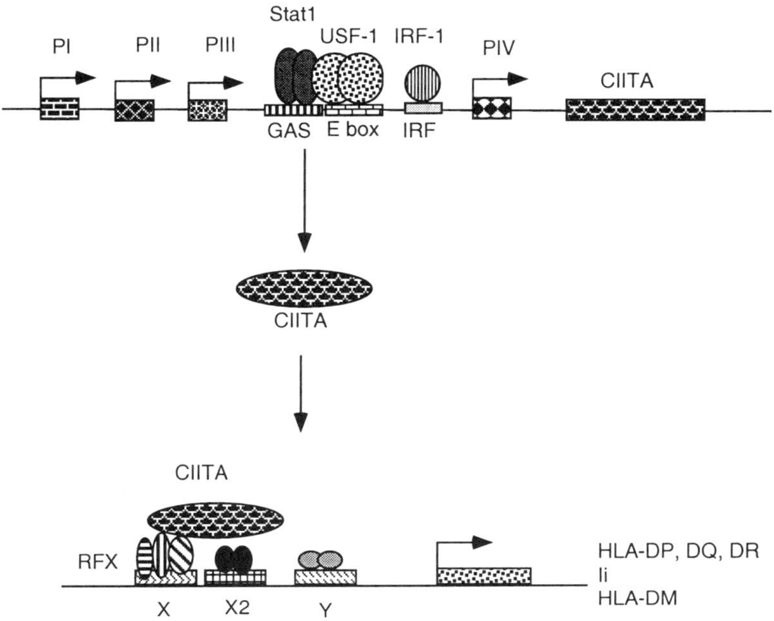

FIGURE 12.5. An overview of the regulation of HLA class II gene expression in humans.

development, the extinction of MHC class II expression observed in plasma cells is also achieved by repression of CIITA.

An inherited deficiency in MHC class-I (HLA-A, -B, and -C) gene expression is noted in patients who have inherited defects in the expression of the TAP2 peptide transporter, an ABC cassette-containing protein that, as part of the TAP1/TAP2 heterodimer, is responsible for the translocation of proteasomally cleaved peptides from the cytosol into the endoplasmic reticulum.

Immunodeficiency Associated With Ataxia Telangiectasia

One fairly complex and poorly understood form of combined immunodeficiency is observed in patients with ataxia telangiectasia. Patients present with a host of neurologic, immunologic, endocrine, cutaneous, and hepatic abnormalities. These include progressive cerebellar ataxia, oculocutaneous telangiectasias, chronic sinopulmonary disease, a high incidence of malignancy, and immunodeficiency. Between 50% and 80% of patients may present with a selective absence of IgA. IgE and IgG levels are frequently decreased. Cell-mediated immunity is impaired, and the thymus has been noted to be hypoplastic. The gene that is defective in ataxia telangiectasia is located on 11q22–23 and encodes a protein known as ataxia

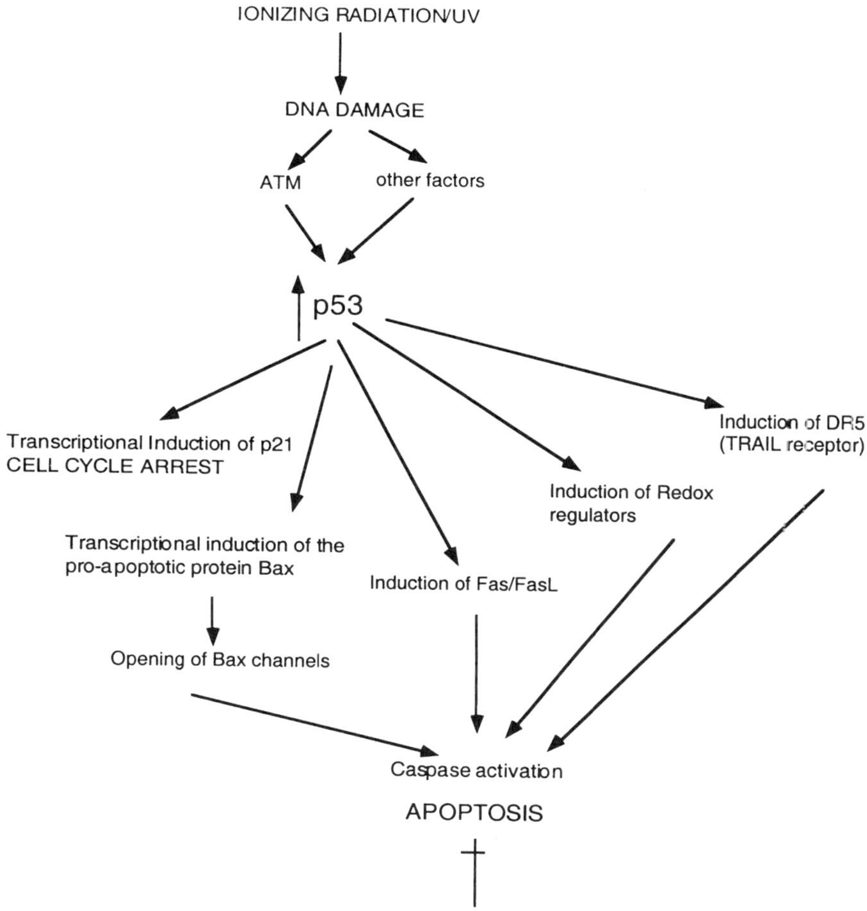

FIGURE 12.6. ATM may be part of a DNA damage checkpoint involving p53. ATM binds to p53 and phosphorylates it on serine-15, thus stabilizing the p53 protein.

telangiectasia mutated (ATM). This protein is a member of the PI3 kinase family and is a dual serine/threonine protein kinase/lipid kinase. ATM is induced by DNA damage and may contribute to the induction of p53 and the induction of apoptosis (Figure 12.6), thus functioning as part of a cell cycle checkpoint. ATM can bind to p53 and phosphorylate it on serine-15, thus activating and stabilizing p53.

In a poorly understood way, this kinase is believed to be critical for maintaining the integrity of recombination. While V(D)J recombination itself is not impaired, the integrity of the joining process may be compromised, leading to an extremely high frequency of translocations in B and T lymphocytes. While this presents a rather unsatisfying explanation at present for the immunodeficiency and cancer seen in this disease, this disease and the ATM gene are being very aggressively studied at present, and better answers are likely to be at hand in the near future.

There is a potential link between V(D)J recombination and the maintenance of chromosomal integrity, and this is reflected in a few genetic disorders that result in immunodeficiency as well as in chromosomal translocations and the malignant transformation of lymphocytes.

The *nude* mouse presents with a naturally occurring defect in thymus development and has been discussed in detail in Chapter 9.

Humoral Immunodeficiencies

In disorders in which B cell development is primarily impaired (see Table 12.1), antibodies are not made, and patients may be particularly susceptible to bacterial infections.

X-linked Agammaglobulinemia

The earliest developmental stage at which B cell ontogeny is blocked in a human disease is at the pro-B to pre-B transition, presumably because of compromised pre-B receptor signaling. In X-linked agammaglobulinemia (XLA) or Bruton's disease, numerous different mutations have been described in Btk, a protein tyrosine kinase discussed in Chapters 4 and 7. These mutations include point substitutions and deletions and can result in a range of altered proteins or no protein at all. In different XLA patients, point substitutions have been described in the PH domain, SH2 domain, SH3 domain, and kinase domain, suggesting important functional roles for each of these domains. The severity of the disease presumably depends on the extent to which catalytic activity is compromised and in some patients may be influenced by the epistatic role played by unidentified "modifier" genes. In its most severe form, no peripheral B cells are seen, and after the first year of life, boys carrying this defect have recurrent pyogenic infections usually with *H. influenzae* and *S. pneumoniae*. Very low levels of serum immunoglobulin (Ig) are observed, and the treatment at present primarily involves frequent injections of polyclonal human serum Ig. Gene therapy is also being considered. The primarily defect in this disease is almost certainly linked to the failure to generate pre-B receptor-derived checkpoint signals during development. It is particularly interesting that a phenotype similar to that observed in XLA has been described in children carrying mutations in the human *14.1* gene, which encodes the human homologue of the $\gamma 5$ gene, an important component of the pre-B receptor. A similar autosomal recessive agammaglobulinemia is also observed in females as well as males with mutations that compromise the structure and function of the μ Ig heavy chain gene, and also in patients with mutations in the *Blnk* gene which encodes the BLNK (B cell linker protein) adapter.

In the mouse, Btk is not absolutely required for positive selection I at the pre-B receptor-mediated checkpoint. However, Btk does participate in peripheral B cell maturation, as described in Chapter 7. A naturally occurring Btk mutation occurrs in the Xid mouse, which has been discussed in some detail in Chapter 7. In some patients with XLA, milder phenotypes that mimic those seen in the Xid mouse might reflect defects in peripheral B cell maturation and survival.

The X-linked Hyper-IgM Syndrome

Another X-linked disease in which B cell function is compromised is the X-linked hyper-IgM syndrome. In this disease, young boys present with recurrent pyogenic infections generally with *H. influenzae* and *S. pneumoniae*, just as in patients with XLA, but they may also have opportunistic infections, particularly involving *Pneumocystis carinii*. These latter opportunistic infections suggest a defect in cell-mediated immunity. In this syndrome, mutations have been described in the X-linked gene that encodes the CD40 ligand. CD40L on activated T cells is critical for T-dependent activation of B cells and for the initiation of B cell proliferation, class switching, germinal center formation, and somatic hypermutation, as discussed in Chapter 8. As a result, boys with this syndrome tend to make largely IgM antibodies and do not expand their repertoires via affinity maturation. CD40L also participates in T cell activation (Figure 12.7), as discussed in Chapter 10, resulting in the cell-mediated immune defects noted in patients with this syndrome. The activation of dendritic cells and other APCs via CD40 signaling can induce costimulatory ligands such as the B-7 proteins. These activated cells may facilitate the release of IL-12 by macrophages and the activation of a TH1 type of response. In patients with the X-linked hyper-IgM syndrome, the activation of T cells, probably as a consequence of defective APC activation, is compromised.

Disorders Involving Aberrant Lymphocyte Activation

In a number of inherited immune dyscrasias, the major defect appears to be inappropriate activation of the immune system. In a sense, all allergic and autoimmune disorders might be lumped in this category, but they generally involve more subtle abnormalities of the immune system and represent multigene disorders that are not inherited in a Mendelian fashion. In the diseases that we will consider here, mutations in genes involved in some aspect of lymphocyte development result in the generation of large numbers of activated lymphocytes that contribute to pathology.

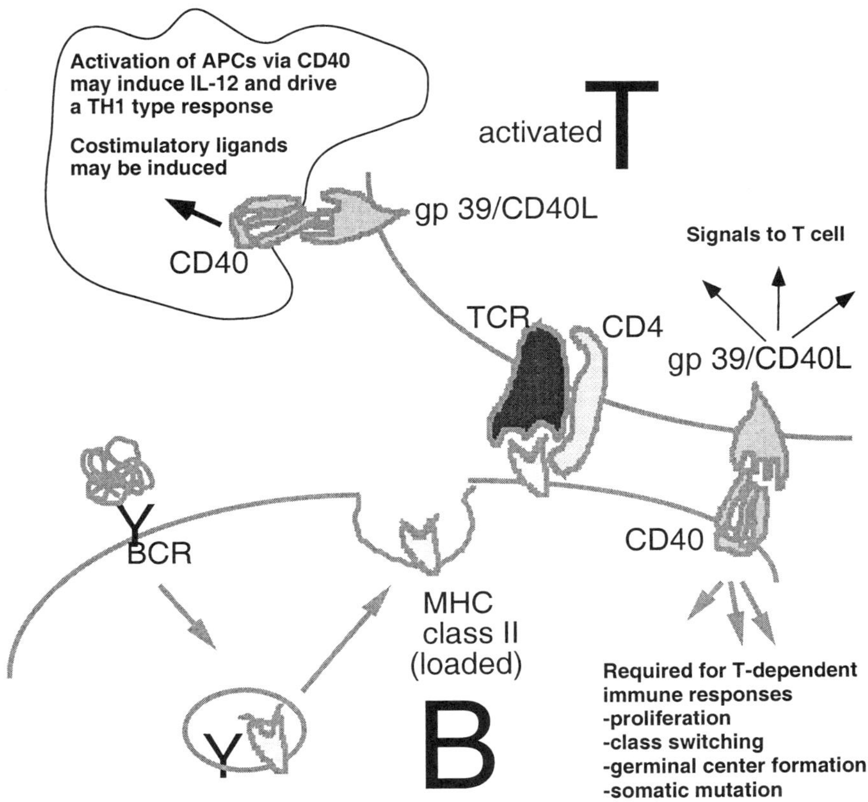

FIGURE 12.7. CD40L is required both for the activation of B cells and T cells. Triggering of CD40 on dendritic cells permits the induction of B7 molecules, which facilitate the generation of Signal 2 for proper T cell activation (see discussion on "licensing" in Chapter 10).

The Wiskott-Aldrich Syndrome

The Wiskott-Aldrich syndrome is an X-linked disease characterized by thrombocytopenia, defective platelet activation, pyogenic and opportunistic infections, and eczema. IgM levels tend to be low, IgA and IgE levels are generally elevated, and IgG levels are usually in the normal range. There is a failure to respond to polysaccharide antigens, and T cells have signaling and cytoskeletal defects. The X-linked gene that is defective is this disease encodes a protein known as WASP (Wiskott Aldrich syndrome protein). WASP is a proline-rich protein with multiple sites capable of interacting with SH3 domains. The WASP protein interacts with the Rho family small GTPase, Cdc42, with Src-family kinases, with the Nck adaptor protein, with proline serine threonine phosphatase interacting protein (PSTPIP), and with a protein known as WIP (WASP-interacting protein), among others.

Cdc42 is involved in actin rearrangements, as is WIP and PSTPIP. WASP is a homolog of the *S. cerevesiae* cortical actin rearrangement inducing protein Bee1. In activated lymphocytes, platelets, and other cells, the function of WASP appears to be to help orchestrate the rearrangement of cortical actin. A critical function of Cdc42 and of WASP in T cells is the formation of a polarized synapse-like interface with antigen-presenting cells (see Chapter 4). Almost all of the defects seen in the Wiskott Aldrich syndrome can potentially be tied to the failure of WASP to help reorganize cortical actin.

It is possible that, in B cells triggered by T-independent polysaccharide antigens, Btk signaling involves the tyrosine phosphorylation of WASP. A site on WASP that appears to be tyrosine phosphorylated *in vivo* closely resembles the autophosphorylation site on Btk itself. WASP patients do have defective responses to polysaccharide antigens, and this B cell defect closely resembles that seen in the Xid mouse.

Autoimmune Lymphoproliferative Syndrome

Another interesting but very rare disorder involving aberrant lymphocyte activation is the autoimmune lymphoproliferative syndrome in which Fas mutations have been found. Infants and young children with this disease present with lymphadenopathy, splenomegaly, and hypergammaglobulinemia. These lymphoid organs are overpopulated with $CD3^{+}CD4^{-}CD8^{-}$ T cells. In addition, common features of this syndrome include episodic autoimmune anemias, thrombocytopenia, autoantibody-mediated neutropenias, and urticarial rashes. This syndrome closely mimics the phenotype of naturally occurring mutations in the mouse of of Fas (in the *lpr* mouse) and the Fas ligand (in the *gld* mouse) briefly discussed in Chapters 5 and 10. Activated T cells are not eliminated by activation-induced cell death, and as explained in Chapter 8, bystander and potentially self-reactive B cells, which should normally be eliminated via Fas signals, proliferate and may contribute to autoimmune manifestations.

There are two interesting features to be noted about the inheritance of this extremely rare disease. The mutations in Fas are generally observed in heterozygotes, and homozygous mutations are rare. Given that Fas is a trimeric protein (discussed in Chapter 5), the likelihood that all three subunits in a given trimer will be "normal" in a heterozygote (assuming equivalent abundance of both normal and mutant proteins) is only 1/8. The second interesting feature related to the inheritance of this rare disease is that the heterozygous mutant chromosome is usually transmitted from a normal parent. This indicates that modifier genes play an important role in contributing to this syndrome. Interestingly, the genetic background is also a critical contributor to the autoimmune phenotype of *lpr* and *gld* mice. In a variant of this disease, (autoimmune lymphoproliferative syndrome, Type 1b) mutations in *FasL* are inherited. Another interesting variant of Autoimmune lymphoproliferative syndrome (Type II) has been described in which *caspase-10* mutations are inherited.

Disorders Involving a Postinfectious "Accelerated Phase"

A small number of genetic disorders result in what appears to be an accelerated and uncontrolled end-stage of TH1 cell and macrophage activation probably following a viral infection. In these diseases, which include the Chediak-Higashi syndrome (CHS), familial hemophagocytic lymphohistiocytosis, partial albinism with immunodeficiency, and possibly the X-linked lymphoproliferative syndrome, a defect in NK cell activity is also frequently seen. The common feature in this otherwise diverse subset of diseases is probably a basic inability to counter viral infections, and the consequent triggering of an accelerated phase of TH1 type T cell activation.

Chediak-Higashi Syndrome

The CHS in humans has a murine counterpart, the *beige* mouse. This syndrome is characterized by partial cutaneous and ocular albinism, defective cytotoxic T lymphocyte (CTL) and NK cell activity, increased susceptibility to infections, and the accumulation in many cells of large intracellular granules. There is a defect in granule discharge (accounting for the defects in CTLs and NK cells) and in the fusion of endosomes with primary lysosomes. The defective gene on human chromosome 1 has been identified and is believed to encode a cytosolic protein that is involved in a vesicular trafficking function—probably regulating the guided movement of vesicles along microtubules during endocytosis and exocytosis. Most patients with CHS develop an accelerated phase, which generally leads to death. During the accelerated phase, T cells and macrophages are activated, and cytokine production spins out of control.

Familial Hemophagocytic Lymphohistiocytosis

A similar accelerated phase response is observed in an autosomal recessive condition known as *familial hemophagocytic lymphohistiocytosis* (FHL). FHL is characterized by polyclonal T lymphocyte and macrophage activation that begins in early childhood and invariably has fatal consequences. Macrophage activation presumably occurs because T cells secrete large amounts of interferon-γ. The serum levels of IFN-γ, tumour neorosis factor (TNF), and IL-1 are greatly elevated. These activated macrophages ingest red cells and, along with the activated T cells, infiltrate the liver, spleen, bone marrow, and the central nervous system. The activated lymphocytes that acumulate in FHL patients tend to be $CD8^+$ Fas^+ human leukocyte antigen-DR^+ (HLA-DR^+) cells. However, defective T and NK cell cytotoxicity is a consistent feature of this disease. FHL is a heterogeneous disease and has been linked in some patients to 9q21.3–22 and in others to

10q21–22. Patients in whom FHL has been linked to 10q21 have mutations in the *perforin* gene. Perforin is critical for CTL and NK cell mediated cytolysis but also plays a homeostatic role in the immune system.

X-Linked Lymphoproliferative Syndrome

In the X-linked lymphoproliferative (XLP) disorder, also known as Duncan's syndrome, young boys are prone to succumb to Epstein-Barr virus infections after entry into an accelerated phase. The causative gene on Xq25 encodes a small protein called SAP (for SLAM-associated protein). SAP contains an SH2 domain, but not much else, and is expressed primarily in T cells. SLAM, as mentioned in Chapters 4 and 10, is a costimulatory molecule found on activated ($CD45RO^+$) T cells, thymocytes, and some B cells. SLAM is activated by homotypic interactions, can induce lymphocyte proliferation, and can facilitate the generation of Th1 responses. This member of the Ig superfamily is structurally related to CD2, and its cytoplasmic tail contains four potential tyrosine phosphorylation sites. Some of these sites, even in an unphosphorylated state, can bind to SAP with a very high affinity. SAP presumably prevents other SH2-containing signaling molecules from being recruited to one or more of these sites in T cells and may be an attenuator of T cell activation or may impair "inhibitory" signals. The cytoplasmic tail of SLAM can potentially recruit a range of SH2 domain-containing proteins. The primary defect in the disease probably involves T cells, since SAP is expressed only in the T lineage. It is likely that the underlying defect is the uncontrolled activation of T cells by viral antigens. Most of the immunological abnormalities noted in this disease may be manifest after infection with Epstein Barr Virus. This disorder may therefore fit into the broader "accelerated phase" category discussed above. SAP knockout mice are unable to suppress T cell activation after infection with the lymphocytic choriomeningitis virus (LCMV) and may resemble *scurfy* mice, which carry a naturally occurring X-linked mutation.

Apart from its association with the cytoplasmic tail of SLAM, SAP also associates with the cytoplasmic tail of another CD2 family member, the 2B4 activation receptor which is found on all NK cells and some T cells. SAP could potentially be an inhibitor of signaling mediated by the SLAM costimulator or the 2B4 receptor. 2B4 is an inducer of NK activation and possibly contributes to the activation of some T cells. The absence of SAP in XLP patients could potentially deregulate 2B4 signaling either positively or negatively. The exact function of SAP remains to be elucidated.

Omenn Syndrome

Omenn syndrome is a rare autosomal recessive disease. Affected infants present with severe erythrodermia, alopecia, diarrhea, and a failure to thrive. Large numbers of activated T cells infiltrate the skin, the gut epithelium, and the lamina propria. While there is a profound depletion of T and B lymphocytes in the blood and lymph nodes, the massive tissue infiltrations of T cells in this disease appear to be generated from a few T cell clones that have taken over the whole immune system. These few T cells may be γδ T cells or αβ T cells that are $CD4^+$, or $CD8^+$ or DN. Functionally, some of the oligoclonal T cell expansions may be of the TH2 type, since eosinophilia and high blood levels of IL-4 and IL-5 are characteristically noted.

The underlying defect in a number of families with Omenn syndrome is an inherited missense mutation in the *Rag-1* or *Rag-2* gene that contributes to an incomplete abrogation of V(D)J recombination. This partial defect results in an almost complete block in B and T cell generation with the occasional developmental escape of a few T cell clones. Some of these T cells are apparently driven to unremitted expansion in these immunodeficient infants, and this results in a "TH2"-type accelerated phase disease with fatal consequences.

Disorders Involving Cytokines and Cytokine Receptors (Other Than IL-7 and IL-15)

It is likely that the coming years will see the identification of an increasing number of cytokine and cytokine receptor mutations in a range of disorders involving lymphocyte development or function. Defects have been discovered in the genes encoding the IFN-γRI and the IL-12Rβ2 chain in patients with "idiopathic" disseminated mycobacterial and *Salmonella* infections. Both these receptors are required for driving the Th1 versus Th2 decision (discussed in Chapter 10). The IFN-γR is also required for macrophage activation in the course of Th1 type responses. Defects in the Il-7R and in IL-15R signaling have been discussed above.

Summary and Perspective

Although there are some differences in the details of lymphocyte development as studied in mice and in humans, the genes and mechanisms involved in lymphoid ontogeny are remarkably similar in these two species. Clearly, the enormous amount of information on lymphocyte development obtained from knockout and knock-in strategies in the mouse will continue to grow. The identification of molecular defects that lead to disorders of human lymphocyte development has only just begun—but what a beginning it has been!

Selected Reviews

Conley, M. E. (1995). Primary immunodeficiencies: a flurry of new genes. Immunol. Today *16*, 313–315.

Fischer, A., Cavazzana-Calvo, M., De Saint Basile, G., DeVillartay, J. P., DiSanto, J. P., Hivroz, C., Rieux-Laucat, F., and LeDeist, F. (1997). Naturally occurring primary deficiencies of the immune system. Annu. Rev. Immunol. *15*, 93–124.

Fischer, A., and Malissen, B. (1998). Natural and engineered disorders of lymphocyte development. Science *280*, 237–243.

Glimcher, L. H., and Kara, C. J. (1992). Sequences and factors: a guide to MHC class II transcription. Annu. Rev. Immunol. *10*, 13–49.

Mach, B., Steimle, V., Martinez-Soria, E., and Reith, W. (1995). Regulation of MHC class II genes: lessons from a disease. Annu. Rev. Immunol. *14*, 301–331.

Rosen, F. S., Cooper, M. D., and Wedgwood, R. J. P. (1995). The primary immunodeficiencies. N. Engl. J. Med. *333*, 431–440.

Some Selected References

Allen, R. C., Armitage, R. J., Conley, M. E., Rosenblatt, H., Jenkins, N. A., Copeland, N. G., Bedell, M. A., Edelhoff, S., Disteche, C. M., Simoneaux, D. K., Fanslow, W. C., Belmont, J., and Spriggs, M. K. (1993). CD40 ligand gene defects responsible for X-linked hyperIgM syndrome. Science *259*, 990–993.

Altare, F., Durandy, A., Lammas, D., Emile, J.-F., and Lamhamedi, S., et al. (1998). Impairment of mycobacterial immunity in human IL-12 receptor deficiency. Science *280*, 1432–1435.

Arnaiz-Villena, A., Timon, M., Corell, A., Perez-Aciego, P., Martin-Villa, J. M., and Regueiro, J. R. (1992). Primary immunodeficiency caused by mutations in the gene encoding the CD3-γ subunit of the T lymphocyte receptor. N. Engl. J. Med. *327*, 529–533.

Arpaia, E., Shahar, M., Dadi, H., Cohen, A., and Roifman, C. M. (1994). Defective T cell receptor signaling and $CD8^+$ thymic selection in humans lacking Zap-70 kinase. Cell *76*, 1–12.

Aruffo, A., Farrington, M., Hollenbaugh, D., Li, X., Milatovich, A., Nonoyama, S., Bajorath, J., Grosmaire, L. S., Stenkamp, R., Neubauer, M., Roberts, R. L., Noelle, R. J., Ledbetter, A., Francke, U., and Ochs, H. D. (1993). The CD40 ligand, gp39, is defective in activated T cells from patients with X-linked hyper-IgM syndrome. Cell *72*, 291–297.

Barbosa, M. D. F. S., Nguyen, K. A., Tcherney, V. T., Ashley, J. A., Detter, J. C., Blaydes, S. M., Brandy, S. M., Chotai, D., Hodgman, C., Solari, R. C. E., Lovett, M., and Kingsmore, S. F. (1996). Identification of the homologous beige and Chediak-Higashi syndrome genes. Nature *382*, 262–265.

Blunt, T., Finnie, N. J., Taccioli, G. E., Smith, C. M., Demengeot, J., Gottlieb, T. M., Mizuta, R., Varghese, A. J., Alt, F. W., Jeggo, P. A., and Jackson, S. P. (1995). Defective DNA dependent protein kinase is linked to V(D)J recombination and DNA repair defects associated with the murine scis mutation. Cell *80*, 813–823.

Bosma, G. C., Custer, R. P., and Rosman, M. J. (1983). A severe combined immunodeficiency mutation in the mouse. Nature *301*, 527–529.

Buckley, R. H., Schiff, R. I., Schiff, S. E., Markert, M. L., Williams, L. W., Harville, T. O., Roberts, J. L., and Puck, J. M. (1997). Human severe combined immunodeficiency: genetic, phenotypic, and functional diversity in one hundred and eight infants. J. Pediatr. *130*, 378–387.

Chan, A. C., Kadlecek, T. A., Elder, M. E., Filipovich, A. H., Kuo, W. L., Iwashima, M., Parslow, T. W., and Weiss, A. (1994). ZAP 70 deficiency in an autosomal recessive form of SCID. Science *264*, 4599–4602.

Coffey, A. J., Brooksbank, R. A., and Brandau, O. (1998). Host response to EBV infection in X-linked lymphoproliferative disease results from mutations in an SH2 domain encoding gene. Nat. Genet. *20*, 129–135.

de Jong, R., Altare, F., Haagen, I.-A., Elfereink, D. G., and de Boer, T. et al. (1998). Severe mycobacterial and salmonella infections in interleukin-12 receptor deficient patients. Science *280*, 1435–1438.

Derry, J. M. J., Ochs, H. D., and Francke, U. (1994). Isolation of a novel gene mutated in Wiskott-Aldrich syndrome. Cell *78*, 635–644.

DiSanto, J. P., Bonnefoy, J. Y., Gauchat, A., Fischer, A., and de Saint Basile, G. (1993). CD40 ligand mutations in X-linked immunodeficiency with hyper IgM. Nature *361*, 541–543.

Elder, M. E., Lin, D., Clever, J., Chan, A. C., Hope, T. J., Weiss, A., and Parslow, T. G. (1994). Human SCID due to a defect in ZAP 70, a T cell tyrosine kinase. Science *264*, 4596–4599.

Fisher, G. H., Rosenberg, F. J., Straus, S. E., Dale, J. K., Middleton, L. A., Lin, A. Y., Strober, W., I enardo, M. J., and Puck, J. M. (1995). Dominant interfering Fas gene mutations impair apoptosis in a human autoimmune lymphoproliferative syndrome. Cell *81*, 935–946.

Khanna, K. K., Keating, K. E., Kozlov, S., Scott, S., Gatei, M., Hobson, K., Taya, Y., Gabrielli, B., Chan, D., Less-Miller, S. P., and Lavin, M. F. (1998). ATM associates with and phosphorylates p53: mappingthe region of interaction. Nat. Genet. *20*, 398–400.

Kirchgessner, C. U., Patil, C. K., Evans, J. W., Cuomo, C. A., Fried, L. M., Carter, T., Oettinger, M. A., and Brown, J. M. (1995). DNA dependent kinase (p350) as a candidate gene for the murine scid defect. Science *267*, 1178–1182.

Korthauer, U., Graf, D., Mages, H. W., Briere, F., Padayachee, M., Malcolm, S., Ugazio, A. G., Notarangelo, L. D., Levinsky, R. J., and Kroczek, R. A. (1993). Defective expression of T cell CD40 ligand causes X-linked immunodeficiency with hyper IgM. Nature *361*, 539–541.

Macchi, P., Villa, A., Giliani, S., Sacco, M. G., Frattini, A., Porta, F., Ugazio, A. G., Johnston, J. A., Candotti, F., O'Shea, J. J., Vezzoni, P., and Notarangelo, L. D. (1995). Mutations of JAK-3 genes in patients with autosomal severe combined immunodeficiency. Nature *377*, 65–68.

Minegishi, Y., Coustan-Smith, E., Wang, Y. H., Cooper, M. D., Campana, D., and Conley, M. E. (1998). Mutations in the human λ-5/14.1 gene result in B cell deficiency and agammaglobulinemia. J. Exp. Med. *187*, 71–77.

Minegishi, Y., Rohrer, J., Coustan-Smith, E., Lederman, H. M., Pappu, R., Campana, D., Chan, A. C., and Conley, M. E. (1999). An essential role for BLNK in human B cell development. Science *286*, 1954–1957.

Nichols, K. E., Harkin, P., Levitz, S., et al. (1998). Inactivating mutations in an SH2 domain containing gene in X-linked lymphoproliferative syndrome. Proc. Natl. Acad. Sci. USA *95*, 13765–13770.

Noguchi, M., Yi, H., Rosenblatt, H. M., Filipovitch, A. W., Adelstein, S., Nodi, W. S., McBride, O. W., and Leonard, W. J. (1993). Interleukin-2 receptor γ chain mutation results in X-linked severe combined immunodeficiency in humans. Cell *3*, 147–156.

Puel, A., Ziegler, S. F., Buckley, R. H., and Leonard, W. J. (1998). Defective IL-7R expression in $T^-B^+NK^+$ severe combined immunodeficiency. Nat. Genet. *20*, 394–397.

Rieux-Laucat, F., Le Diest, F., Hivroz, C., Roberts, I. A. G., Debatin, K. M., Fischer, A., and De Villartay, J. P. (1995). Mutation in fas associated with human lymphoproliferative syndrome and autoimmunity. Science *268*, 1347–1349.

Russell, S. M., Tayebi, N., Nakajima, H., Riedy, M. C., Roberts, J. L., Aman, M. J., Migone, T. S., Noguchi, M., Markert, M. L., Buckley, R. H., O'Shea, J. J., and Leonard, W. J. (1995). Mutation of Jak3 in a patient with SCID: essential role of Jak3 in lymphoid development. Science *270*, 797–800.

Sayos, J., Wu, C., Morra, M., Wang, N., Zhang, X., Allen, D., van Schalk, S., Notarangelo, L., Geha, R., Roncaraolo, M. G., Orttgen, H., DeVries, J. E., Aversa, G., and Terhorst, C. (1998). The X-linked lymphoproliferative disease gene product SAP regulates signals induced through the co-receptor SLAM. Nature *395*, 462–469.

Schultz, L. D., Schweitzer, P. A., Rajan, T. V., Yi, T., Ihle, J. N., Matthews, J., Thomas, M. L., and Beier, D. R. (1993). Mutations at the murine motheaten locus are within the hematopoietic cell protein tyrosine phosphatase (Hcph) gene Cell *73*, 1445–1454.

Schwarz, K., Gauss, G. H., Ludwig, L., Pannicke, U., Li, Z., Lindner, D., Friedrich, W., Seger, R. A., Hansen-Hagge, T. E., Desiderio, S., Lieber, M. R., and Bartram, M. R. (1996). RAG mutations in human B cell negative SCID. Science *274*, 97–99.

Soudais, C., De Villartay, J. P., Le Deist, F., Fischer, A., and Lisowska-Grospierre, B. (1993). Independent mutations of the human CD3-ε gene resulting in a T cell receptor/CD3 complex immunodeficiency. Nat. Genet. *3*, 77–81.

Stepp, S. E., Dufourcq-Lagelouse, R., Le Deist, F., Bhawan, S., Certain, S., Mathew, P. A., Henter, J.-I., Bennett, M., Fischer, A., de Saint Basile, G., and Kumar, V. (1999). Perforin gene defects in familial hemophagocytic lymphohistiocytosis. Science *286*, 1957–1959.

Takahashi, T., Tanaka, M., Brannan, C. I., Jenkins, N. A., Copeland, N. G., Suda, T., and Nagata, S. (1994). Generalized lymphoproliferative disease in mice caused by a point mutation in the Fas ligand. Cell *76*, 969–976.

Tsukada, S., Saffran, D. C., Rawlings, D. J., Parolini, O., Allen, R. C., Klisak, I., Sparkes, R. S., Kubagawa, H., Mohandas, T., Quan, S., Belmont, J. W., Cooper, M. D., Conley, M. E., and Witte, O. N. (1993). Deficient expression of a B cell cytoplasmic tyrosine kinase in human X-linked agammaglobulinemia. Cell *72*, 279–290.

Vetrie, D., Vorechowsky, I., Sideras, P., Holland, J., Davies, A., Flinter, F., Hammarstrom, L., Kinnon, C., Levinsky, R., Bobrow, M., Smith, C. I. E., and Bentley, D. R. (1993). The gene involved in X-linked agammaglobulinemia is a member of the src-family of protein tyrosine kinases. Nature *361*, 226–233.

Villa, A., Santagata, S., Bozzi, F., Giliani, S., Frattini, A., Imberti, L., Gatta, L. B., Ochs, H. D., Schwarz, K., Notarangelo, L. D., Vezzoni, P., and Spanopoulou, E. (1998). Partial V(D)J recombination activity leads to Omenn syndrome. Cell *93*, 885–896.

Wang, J., Zheng, L., Lobito, A., Chan, F. K.-M., Dale, J., Sneller, M., Yao, X., Puck, J. M., Straus, S. E., and Lenardo, M. J. (1999). Inherited human caspase 10 mutations underlie defective lymphocyte and dendritic cell apoptosis in autoimmune lymphoproliferative syndrome type II. Cell *98*, 47–58.

Watanabe-Fukunaga, R., Brannan, C. I., Copeland, N. G., Jenkins, N. A., and Nagata, S. (1992). Lymphoproliferation disorder in mice explained by defects in Fas antigen that mediates apoptosis. Nature *356*, 314–317.

Yel, L., Minegishi, Y., Coustan-Smith, E., Buckley, R. H., Trubel, H., Pachman, L. M., Kitchingman, G. R., Campana, D., Rohrer, J., and Conley, M. E. (1996). Mutations in the μ heavy chain in patients with agammaglobulinemia. N. Engl. J. Med. *335*, 1486–1493.

Index